Biomedical Science Practice

TRANSFUSION &
TRANSPLANTATION SCIENCE

EDITED BY Robin Knight

BIOMEDICAL SCIENCE
PRACTICE

EXPERIMENTAL & PROFESSIONAL SKILLS

EDITED BY Nadine Glencross, Nessar Ahmed,
Chris Smith & Qiuyin Wong

CYTOPATHOLOGY

EDITED BY Behdad Shambayati

CLINICAL BIOCHEMISTRY

EDITED BY Nessar Ahmed

IMMUNOLOGY

EDITED BY Angela Hall & Christine Yates

HAEMATOLOGY

Andrew Blann, Gavin Knight,
& Gary Moore

MEDICAL
MICROBIOLOGY

EDITED BY Michael Ford

BIOLOGY OF DISEASE

EDITED BY Roz Gibbs and Sheelagh Heugh

HISTOPATHOLOGY

EDITED BY Phil Warren & Tony Sims

fundamentals OF
biomedical science

Biomedical Science Practice

Experimental and professional skills

DATE DUE

GAYLORD   #3523PI   Printed in USA

...ited by

...cross
...Science

...hmed
...niversity

...Wang
Manchester Metropolitan University

OXFORD
UNIVERSITY PRESS

UNIVERSITY PRESS

Great Clarendon Street, Oxford OX2 6DP

Oxford University Press is a department of the University of Oxford.
It furthers the University's objective of excellence in research, scholarship,
and education by publishing worldwide in

Oxford New York

Auckland Cape Town Dar es Salaam Hong Kong Karachi
Kuala Lumpur Madrid Melbourne Mexico City Nairobi
New Delhi Shanghai Taipei Toronto

With offices in

Argentina Austria Brazil Chile Czech Republic France Greece
Guatemala Hungary Italy Japan Poland Portugal Singapore
South Korea Switzerland Thailand Turkey Ukraine Vietnam

Oxford is a registered trade mark of Oxford University Press
in the UK and in certain other countries

Published in the United States
by Oxford University Press Inc., New York

British Library Cataloguing in Publication Data
Data available

Library of Congress Cataloging in Publication Data
Data available

Typeset by MPS Limited, a Macmillan Company
Printed in Great Britain by Bell & Bain Ltd, Glasgow

ISBN 978-0-19-953329-9

5 7 9 10 8 6 4

Contents

An introduction to the Fundamentals of Biomedical Science series

Biomedical Scientists form the foundation of modern healthcare, from cancer screening to diagnosing HIV, from blood transfusion for surgery to food poisoning and infection control. Without Biomedical Scientists, the diagnosis of disease, the evaluation of the effectiveness of treatment, and research into the causes and cures of disease would not be possible.

However, the path to becoming a Biomedical Scientist is a challenging one: trainees must not only assimilate knowledge from a range of disciplines, but must understand—and demonstrate—how to apply this knowledge in a practical, hands-on environment.

The *Fundamentals of Biomedical Science* series is written to reflect the challenges of biomedical science education and training today. It blends essential basic science with insights into laboratory practice to show how an understanding of the biology of disease is coupled to the analytical approaches that lead to diagnosis.

The series provides coverage of the full range of disciplines to which a Biomedical Scientist may be exposed – from microbiology to cytopathology to transfusion science. Alongside volumes exploring specific biomedical themes and related laboratory diagnosis, an overarching Biomedical Science Practice volume provides a grounding in the general professional and experimental skills with which every Biomedical Scientist should be equipped.

Produced in collaboration with the Institute of Biomedical Science, the series

- Understands the complex roles of Biomedical Scientists in the modern practice of medicine.

- Understands the development needs of employers and the Profession.

- Places the theoretical aspects of biomedical science in their practical context.

Learning from this series

The *Fundamentals of Biomedical Science* series draws on a range of learning features to help readers master both biomedical science theory, and biomedical science practice.

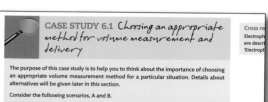

CASE STUDY 6.1 Choosing an appropriate method for volume measurement and delivery

The purpose of this case study is to help you to think about the importance of choosing an appropriate volume measurement method for a particular situation. Details about alternatives will be given later in this section.

Consider the following scenarios, A and B.

Case studies illustrate how the biomedical science theory and practice presented throughout the series relates to situations and experiences that are likely to be encountered routinely in the biomedical science laboratory. Answers to questions posed in some Case Studies are available in the book's Online Resource Centre.

Additional information to augment the main text appears in **boxes**

> ┌─────┐
> │   │   **BOX 9.1** *Ohm's Law*
> └─────┘
>
> Ohm's Law can be rearranged to describe the current, voltage, or resistance. The easiest
> way to remember these relationships is to use a **VIR triangle** as shown in Figure 9.2. To
> use this, simply use your finger to hide the value that you want to calculate. The remain-
> ing two values then show you how to do the calculation.

Method boxes walk through the key protocols that the reader is likely
to encounter in the laboratory on a regular basis.

> **METHOD** *How to set up and use a pipettor*
>
> ■ Choose a pipettor appropriate to the task in hand. For a given model, it is essential to use the pipettor only within the volume range indicated. Use beyond this range can damage the pipettor mechanism.
>
> ■ Organize the work area so that all materials and items of equipment to be used are conveniently placed. The work surface needs to be at a comfortable height. If nec-essary this can be achieved by using a height adjustable
>
> P1000, 085 represents 850 μL. The m[...] digits on the display varies with the m[...] must be checked otherwise an incor[...] delivered.
>
> ■ To measure and deliver a volume of s[...] is first depressed as far as the first st[...] be felt) with the pipettor held awa[...] Next the tip should be placed belo[...]

**Further features are used to help consolidate and extend
students' understanding of the subject**

Key points reinforce the key concepts that the reader should master
from having read the material presented, while **Summary** points act
as an end-of-chapter checklists for readers to verify that they have remembered correctly the principal themes and ideas presented within
each chapter.

> The *concentration* of a solution is determined by the *quantity* of solute and the *volume* of solvent as described in Box 6.6.
>
> *Key Points*
>
> Solutions consist of a solute dissolved in a solvent; the concentration of the solution is determined by the relative proportions of solute and solvent.
>
> The different ways in which concentration may be expressed will now be considered. The commonest form is **molarity** and this along with the concept of the **mole** will be considered

Key terms provide on-the-page explanations of terms with which the
reader may not be familiar; in addition, each title in the series features
a **glossary**, in which the key terms featured in that title are collated.

> [...]l lines represent a phase boundary (electrode/electrolyte interface)[...]ines the salt bridge. Note that according to this notation the negative[...]ed to the left, the positive electrode (**cathode**) is placed to the right,[...]refore flow from left-to-right in the connecting wire.
>
> [...]of electrochemical techniques
>
> [...]n be categorized as **galvanic** or **electrolytic.** A galvanic cell is[...]s occur spontaneously at the electrodes when they are connected[...]tor. In an electrolytic cell, however, the electrochemical reactions
>
> **A cation** (pronounced cat-ion) is an ion that has fewer electrons in its electron shells than protons in its nucleus. It thus carries a net positive charge and will move towards the negative electrode during electrolysis.
>
> **An anion** (pronounced an-ion) is an ion that has more electrons than protons. Thus it has a net negative charge and will move towards the positive electrode

Self-check questions throughout each chapter and extended questions at the end of each chapter provide the reader with a ready means
of checking that they have understood the material they have just encountered. Answers to these questions are provided in the book's Online Resource Centre; visit www.oxfordtextbooks.co.uk/orc/glencross

> A potentiometric electrode is said to have Nernstian characteristics if it obeys Nernst's law. If the changes in potential relative to concentration are less than the theoretical values, it may indicate an electrode malfunction or interference from other ions.
>
> **SELF-CHECK 9.3**
>
> Given that the universal gas constant is 8.314 J mol^{-1} per K, T is the temperature in Kelvin (=273 + T°C) and that the Faraday constant is 96,485 coulomb mol^{-1}, calculate the change in potential in mV for a ten-fold change in the concentration of H$^+$ when the temperature of the system is 35°C.

Cross references help the reader to see biomedical science as a uni-
fied discipline, making connections between topics presented within
each volume, and across all volumes in the series.

> **Cross reference**
> Measuring HCO$_3^-$, PCO$_2$, and pH in clinical samples are described in Chapter 6 of *Clinical Biochemistry.*
>
> The system is allowed to achieve equilibrium, at which time the pH is compared with a calibration curve of pH against known PCO$_2$ values.
>
> Severinghaus's modifications doubled the sensitivity of the original devi[...]ity, and improved the response time. In 1958, Severinghaus described [...] both his PCO$_2$ electrode and a Clark O$_2$ electrode, which you can rea[...] to form a combined blood gas analyser. Commercial manufacture of t[...] electrode began in 1959 and such electrodes are still found in PCO$_2$ ana[...] ratories today.
>
> The Severinghaus electrode is a *gas-sensing electrode* and as it is based[...] the electrical output is a *logarithmic function* of PCO$_2$. At 37°C an outpu[...] for a 10-fold change in PCO$_2$. The response is also temperature depend[...]

Online learning materials

online resource centre

The *Fundamentals of Biomedical Science* series doesn't end with the printed books. Each title in the series is supported by an Online Resource Centre, which features additional materials for students, trainees, and lecturers.

www.oxfordtextbooks.co.uk/orc/fbs

Guides to key experimental skills and methods

Multimedia walk-throughs of key experimental skills—including both animations and video—to help you master the essential skills that are the foundation of Biomedical Science practice.

Biomedical science in practice

Interviews with practising Biomedical Scientists working in a range of disciplines, to give you valuable insights into the reality of work in a Biomedical Science laboratory.

Digital Microscope

A library of microscopic images for you to investigate using this powerful online microscope, to help you gain a deeper appreciation of cell and tissue morphology.

The Digital Microscope is used under licence from the Open University.

'Check your understanding' learning modules

A mix of interactive tasks and questions, which address a variety of topics explored throughout the series. Complete these modules to help you check that you have fully mastered all the key concepts and key ideas that are central to becoming a proficient Biomedical Scientist.

We extend our grateful thanks to colleagues in the School of Health Science at London Metropolitan University for their invaluable help in developing these online learning materials.

Answers to self-check, case study, and end of chapter questions

Answers to questions posed in the book are provided to aid self-assessment.

Lecturer support materials

The Online Resource Centre for each title in the series also features figures from the book in electronic format, for registered adopters to download for use in lecture presentations, and other educational resources.

To register as an adopter visit **www.oxfordtextbooks.co.uk/orc/ glencross** and follow the on-screen instructions.

FIGURE 7.1
Diagram of the skin.

Any comments?

We welcome comments and feedback about any aspect of this series.
Just visit **www.oxfortextbooks.co.uk/orc/feedback/** and share your views.

Acknowledgements

I would like to thank the following individuals and organizations for their help, without whom it would have been difficult to complete this project.

Katy Linton and Sidonie Kamtchouang for the case studies reproduced in Chapter 1. The Health Professions Council for their advice on regulatory matters and their degree approval process, for permission to use the HPC logo and to reproduce their documents. I would like to thank Chris Smith for his comments and support during the various stages of planning and writing, and would especially like to thank my fellow editors Qiuyu Wang and Nessar Ahmed whose help and advice have proved invaluable to an editorial novice. Particular thanks must also go to my assistant Sandra Shevlin without whom it would have been far more difficult to complete this project.

Finally, I would like to thank my wife Jill, for her support and encouragement.

Hedley Glencross
London, April 2010

I would like to thank Chris Smith for his friendship over the years. His support and guidance whilst working on this and other books, has been invaluable. He has been an excellent mentor and I have learnt a lot from him for which I shall be eternally grateful. I would also like to take this opportunity to acknowledge the friendship, companionship, and help of the late Professor Ed Wood. Finally, I would like to acknowledge my family for their support, patience, and encouragement.

Dr Nessar Ahmed
Manchester, April 2010

I would like to thank my co-editors, Nessar and Hedley for their invaluable help and advice and for giving me such an enjoyable co-editing experience. I would like to express my appreciation to Jonathan Crowe of OUP for his organizational skills and constant support and help. My thanks are also due to Mr Mick Hoult for the beautiful illustrations he has provided. Lastly, but not least, is my gratitude to my husband Chris Smith for his morale boosting and academic support.

Qiuyu Wang
Manchester, April 2010

Contributors

Nessar Ahmed
School of Healthcare Science, Manchester Metropolitan University

Andrew Blann
Haemostasis, Thrombosis and Vascular Biology Unit, University Department of Medicine, City Hospital, Birmingham

John Gaffney
School of Healthcare Science, Manchester Metropolitan University

Hedley Glencross
Institute of Biomedical Science, London

Ian Graham
Formerly of School of Biology, Chemistry and Health Science, Manchester Metropolitan University

Tim James
Department of Clinical Biochemistry, Oxford Radcliffe Hospitals NHS Trust

Georgina Lavender
Laboratory Training Officer and freelance tutor

Helen Montgomery
Kratos Analytical, Manchester

Elaine Moore
Pathology Department, St Peter's Hospital, Surrey

Joyce Overfield
School of Healthcare Science, Manchester Metropolitan University

Lynda Petley
Partnership Pathology Services, Frimley Park Hospital Foundation Trust

Peter Robinson
Institute of Postgraduate Dental Education, University of Central Lancashire, Preston

Tony Sims
Formerly of the Histopathology Department, University Hospitals of Leicester NHS Trust

Mark Slevin
School of Healthcare Science, Manchester Metropolitan University

Chris Smith
Formerly of School of Biology, Chemistry and Health Science, Manchester Metropolitan University

Jan Still
Department of Chemical Pathology, Watford General Hospital

Alison Taylor
Histology Department, Basingstoke and North Hampshire Hospital

Qiuyu Wang
School of Healthcare Science, Manchester Metropolitan University

Christine Yates
Immunology Department, Hull and East Yorkshire Hospitals NHS Trust

Online materials developed by

Sheelagh Heugh, Principal Lecturer in Biomedical Science, Faculty of Human Sciences, London Metropolitan University

Dr Ken Hudson, Lecturer in Biomedical Science, Faculty of Human Sciences, London Metropolitan University

William Armour, Lecturer in Biomedical Science, Faculty of Human Sciences, London Metropolitan University

Abbreviations

% v/v	volume–volume percentage		ccc	supercoiled covalently closed circular
% w/v	mass–volume or weight–volume		CCD	cooled charge-coupled device
% w/w	mass percentage		CD	cluster of differentiation
μ	electrophoretic mobility		cDNA	complementary DNA
A&E	Accident and Emergency department		CDC	Centers for Disease Control
a	absorptivity		CE	capillary electrophoresis
A	absorbance		CEP	Centre for Evidence-based Purchasing
ABTS	azino-*bis*-ethylbenzthiazoline sulfonic acid		CF	cystic fibrosis
ACDP	Advisory Committee on Dangerous Pathogens		CH100	total haemolytic complement test
ACOP	Approved Codes of Practice		CHART	continuous hyperfractionated radiotherapy
ADR	accord européen relative au transport international des marchandises dangereuses parroute		CHIP	Chemicals (Hazard, Information and Packaging for Supply) Regulations
AGE	agarose gel electrophoresis		CI	chemical ionization
AHG	anti-human globulin		CIC	Connectivity Industry Consortium
AIDS	acquired immune deficiency syndrome		CID	collisional dissociation cell
ALP	alkaline phosphatase		CJD	Creutzfeldt–Jakob disease
AMP	adenosine 5'-monophosphate		CNS	central nervous system
Amp^R	ampicillin resistance gene		CNST	Clinical Negligent Scheme for Trusts
ANCA	anti-neutrophil cytoplasmic antibodies		COSHH	Control of Substances Hazardous to Health
ANOVA	analysis of variance		CPA	Clinical Pathology Accreditation
AP	alkaline phosphatase		CPD	continuing professional development
ATP	adenosine 5'-triphosphate		CPSM	Council for Professions Supplementary to Medicine
BAC	bacterial artificial chromosome		CSF	cerebrospinal fluid
BIVDA	British *In-Vitro* Diagnostics Agency		CSU	catheter sample of urine
BJBS	*British Journal of Biomedical Science*		CT	computed tomography
bp	base pairs		CTP	cytidine 5'-monophosphate
BRI	Bristol Royal Infirmary		CV	coefficient of variation
BSA	bovine serum albumin		d	distance
BSE	bovine spongiform encephalopathy		dAMP	deoxyadenosine 5'-monophosphate
BSHI	British Society of Histocompatibility and Immunogenetics		dATP	deoxythymidine 5'-monophosphate
			dCTP	deoxycytidine 5'-monophosphate
BSI	British Standards Institute		ddNTP	dideoxynucleoside triphosphate
BSQR	Blood Safety and Quality Regulations		DDT	dichloro-diphenyl-trichloroethane
c	concentration		DGA	DiGeorge anomaly
c	velocity of light		dGMP	deoxyguanosine 5'-monophosphate

DGSA	Dangerous Goods Safety Advisor
DIG	digoxygenin
DNA	deoxyribonucleic acid
DPA	Data Protection Act
dps	disintegrations per second
dRib	deoxyribose
DSE	display screen equipment
DSEAR	Dangerous Substances and Explosive Atmosphere Regulations
DSS	3-trimethylsilyl 1-propane suphonate
dTDP	deoxythymidine 5′-diphosphate
dTMP	deoxythymidine 5′-monophosphate
E	electric field or potential gradient; energy of electromagnetic radiation
E	electronic component of electromagnetic radiation
ECD	electron capture detector
EDTA	ethylenediamintetraacetic acid
EI	electron ionization
EIA	enzyme immunoassay
EIm	electron impact
ELISA	enzyme-linked immunosorbent assay
ELISPOT	enzyme-linked immunospot
EQA	external quality assessment
EQF	European Qualification Framework
ESI	electrospray ionization
ESR	erythrocyte sedimentation rate
EU	European Union
eV	electron volt $(1.6 \times 10^{-19} \, \text{J})$
F	Faraday constant
f	frictional coefficient
FAB	fast atom bombardment
FACS	fluorescence activated cell sorter
FBC	full blood count
FID	flame ionization detector
FISH	fluorescent *in situ* hybridization
FOI	Freedom of Information
FTP	fitness to practise
FVU	first voided urine
g	earth's gravitational field

G	centrifugal field or acceleration
GALT	gut associated lymphoid tissue
GHS	globally harmonized system
GLC	gas–liquid chromatography
GLP	good laboratory practice
GMP	guanosine 5′-monophosphate
H&E	haematoxylin and eosin
h	Planck's constant
H	height of a theoretical plate in a chromatography column; magnetic component of electromagnetic radiation
H_0	applied magnetic field
HASAWA	The Health and Safety at Work Act
Hb	haemoglobin
Hb^S	haemoglobin S
HB^A	adult haemoglobin
HETP	height equivalent to a theoretical plate
HIS	Hospital Information System
HIV	human immunodeficiency virus
HPC	Health Professions Council
HPCE	high performance capillary electrophoresis
HPLC	high performance liquid chromatography
HRP	horseradish peroxidase
HSA	human serum albumin
HSC	Health and Safety Commission
HSD	Higher Specialist Diploma
HSE	Health and Safety Executive
HSG	Health and Safety Guidance publications
HTA	Human Tissue Act (also used for Human Tissue Authority)
I	intensity of transmitted light
I_0	intensity of incident light
IBMS	Institute of Biomedical Science
ICR	ion cyclotron resonance
IEF	isoelectric focusing
Ig	immunoglobulin
IIF	indirect immunofluorescence
IMS	industrial methylated spirits
INR	international normalized ratio

IOSH	Institute of Occupational Safety and Health		MM	multiple myeloma
IPTG	isopropylthio-β-galactoside		mol	mole – the SI unit denoting the amount of a substance
IQC	internal quality control		MOSFET	metal oxide semiconductor field effect transistors
IQR	inter-quartile range			
IR	infrared		M_r	relative molecular mass
IRMA	immunoradiometric assay		MRI	nuclear magnetic resonance imaging
IRT	immunoreactive trypsin		mRNA	messenger RNA
IS	internal standard		MRS	magnetic resonance spectroscopy
ISE	ion selective electrode		MRSA	methicillin resistant *Staphylococcus aureus*
ISFET	ion selective field effect transistor		MS	mass spectrometry
ISO	International Standards Organization		MSDS	Material Safety Data Sheet
IT	information technology		MSU	midstream urine
IUPAC	International Union of Pure and Applied Chemistry		N	Avogadro's number
IVD	*in vitro* diagnostic device		N	the number of theoretical plates in a chromatography column
JBL	journal-based learning		NAS	sodium (Na), aluminium (Al), silicon (Si) glass used in ion selective electrodes
Kbp	kilobase pairs			
K_d	distribution coefficient		NEQAS	National External Quality Assessment Service
K_{ow}	octanol–water partition coefficient		NHS	National Health Service
KSF	Knowledge and Skills Framework		NHSLA	NHS Litigation Authority
L or l	path length		NMR	nuclear magnetic resonance
L	length		NPSA	National Patient Safety Agency
LAS	lithium (Li), aluminium (Al), silicon (Si) glass used in ion selective electrodes		nt	number of nucleotides
			oc	open circular or nicked conformation
LDH	lactate dehydrogenase		OH	occupational health
LET	linear energy transfer		OJEU	*Official Journal of the European Union*
LEV	local exhaust ventilation		p	probability
LIMS	laboratory information management system		PAC	phage artificial chromosomes
LIS	laboratory information system		PAGE	polyacrylamide gel electrophoresis
LOLER	Lifting Operations and Lifting Equipment Regulations		PAH	phenylalanine hydroxylase
			PAS	periodic acid-Schiff
m	mass		PASA	NHS Purchasing and Supply Agency
mAB	monoclonal antibody		PAT	portable appliance testing
MALDI	matrix-assisted laser desorption ionization		PBC	primary biliary cirrhosis
MALDI-MS	matrix-assisted laser desorption ionization-mass spectrometry		PCR	polymerase chain reaction
			PCV	packed cell volume
MALT	mucosa associated lymphoid tissue		PDP	personal development plan
MBL	mannose-binding lectin		PET	positron emission tomography
MHO	manual handling operations		PFGE	pulsed field gel electrophoresis
MHRA	Medicines and Healthcare products Regulatory Agency		Phage	bacteriophage

pI	isoelectric points
PKU	phenylketonuria
PMF	peptide mass fingerprinting
POCT	point of care testing
PPE	personal protective equipment
PQQ	pre-qualification questionnaire
PUWER	Provision and Use of Work Equipment Regulations
PVA	polyvinyl alcohol
PVC	polyvinyl chloride
PVDF	polyvinylidene difluoride
PX	partial pressure (concentration) of the gas X
Q	quadrupole
QA	quality assurance
QAA	Quality Assurance Agency for Higher Education
QC	quality control
QCDA	Qualifications and Curriculum Development Agency
QIT	quadupole ion-trap
QMS	quality management system
QP	quality policy
qPCR	quantitative real time polymerase chain reaction (PCR). Also called real-time (PCR)
R	electrical resistance
R	universal gas constant
Rad	radiation absorbed dose
RBE	relative biological effectiveness
RCA	root cause analysis
RCF	relative centrifugal field
RE	restriction endonuclease
Real-time PCR	real-time polymerase chain reaction. Also called quantitative real time or qPCR
Rem	roentgen equivalent man
rev min^{-1}	revolutions per minute
RF	radio frequency
R_f	relative to front
RFLP	restriction fragment length polymorphisms
Rh	Rhesus blood group
RIA	radioimmunoassay
Rib	ribose
RIDDOR	The Reporting of Injuries, Diseases, and Dangerous Occurrences Regulations
RNA	ribonucleic acid
rRNA	ribosomal RNA
RSI	repetitive strain injury
RT-PCR	reverse transcriptase-PCR
s	sedimentation coefficient
S	Nernst factor
SABRE	serious adverse blood reactions and events
SABS	Safety Alert Broadcast System
SCOT	support coated open tubular
SD	standard deviation
SDS	sodium dodecyl sulphate
SDS-PAGE	sodium dodecyl sulphate polyacrylamide gel electrophoresis
SE	standard error
SETS	Standards of Education and Training
SHOT	serious hazard of transfusion
SI	Le Système International d'Unités
SLE	systemic lupus erythematosus
SOP	standard operating procedure
SRID	single radial immunodiffusion
t	retention time
T	transmitted light
t_o	dead time
$t_{1/2}$	half-life
TAG	triacylglycerol
TAT	turnaround time
Tb	turbidity
TB	tubercle bacilli/tuberculosis
TCA	trichloroacetic acid
TEM	transmission electron microscopy
TEMED	N,N,N',N'-tetramethylethylenediamine
TLC	thin layer chromatography
TMB	tetramethylbenzidine
TMS	tetramethylsilane (Si(CH$_3$)$_4$)
TOF	time of flight
TREM	transport emergence card
Tris	2-amino-2-hydroxymethylpropane-1,3-diol
tRNA	transfer RNA

TAT	turnaround time	α	selectivity
UKAS	United Kingdom Accreditation Service	β^-	positron
UMP	uridine 5'-monophosphate	ε	molar extinction coefficient
UPS	uninterruptible power supply	λ	wavelength of electromagnetic radiation
UV	ultraviolet	ρ	density of the solvent
v	frequency of electromagnetic radiation		
v	(sedimentation) velocity	**Units**	
v'	wave number		
$\bar{v}$	partial specific volume	Ω	Ohm
V_0	void volume	A	Ampere – the SI unit of electrical current
VDU	visual display unit	Å	angstrom (10^{-10} m)
V_e	elution volume	Bq	becquerel
Vis	visible	C	coulomb
VLCFA	very long chain fatty acids	Ci	curie
VOSA	Vehicle and Operators Service Agency	Da	dalton
V_t	total volume	Gy	gray
W	width at the base of a peak	Hz	hertz
WCOT	wall coated open tubular	nm	nanometre
WEL	workplace exposure limit	ppm	parts per million
W_h	width at half the height of a peak	s	second
WRULD	work-related upper limb disorders	Sv	Sievert
X-gal	artificial substrate, 5-bromo-4-chloro-3-indolyl-β-galactoside, of β-galactosidase	S	Svedberg unit
		T	temperature in kelvin
z	charge	V	volt

1

Biomedical science and biomedical scientists

Hedley Glencross

Learning objectives

After studying this chapter, you should be able to:

- Define biomedical science as a scientific discipline
- Outline the academic and scientific contents of biomedical science degree programmes
- Discuss the training required to become a biomedical scientist
- Discuss how the education and training of a biomedical scientist are interrelated

Introduction

Biomedical science is often described as the application of the basic sciences, but especially the biological sciences, to the study of medicine – in particular, the causes, consequences, diagnosis, and treatment of human diseases. Biomedical scientists are registered practitioners of the subject, who normally work in departments of clinical pathology. Thus, they are vital healthcare, scientifically qualified professionals who, for example, perform diagnostic tests on clinical samples of blood and urine. Given their role(s), biomedical scientists are typically active in more laboratory-based work and tend to have a relatively limited contact with patients compared with other healthcare professionals.

Approximately 70% of medical decisions or interventions – from diagnoses to monitoring medical treatments to screening populations at risk of a particular clinical condition – are based on the activities of biomedical scientists or require the knowledge and skills of biomedical science. This may be as simple as a preoperative blood test or something more complex, such as using an image-guided needle biopsy to remove tissue samples to help diagnose breast or prostate cancers. Unfortunately, while everyone knows their doctor and, indeed, doctors and nurses have a high degree of recognition in society, in many ways the same is not true of biomedical scientists. Perhaps considering the following scenario may reinforce the point. A 52-year-old, overweight male smoker presents to his local Accident & Emergency department with chest pains, shortage of breath, and complaining of a generalized weakness and tiredness. It is apparent that he may be experiencing any one of a number of problems. These include, for example, a panic attack, a chest infection, an allergic reaction, or he may be at the early stage of a heart attack. How can an appropriate clinical decision be reached? Biomedical scientists earn their living by helping in these decision-making processes!

Routine tests, such as measuring the patient's height, weight, temperature, and blood pressure, as well as taking X-rays and performing detailed laboratory investigations on samples from a patient may all help provide answers. The laboratory investigations may include blood tests to look for increased activities of enzymes or substances that act as markers for a heart attack, microbiology tests to identify infection-causing microorganisms, while immunological tests may help eliminate the possibility of an allergic reaction. Hence, clinical laboratories are often busy places (Figure 1.1).

The education and training needed to become a biomedical scientist are detailed and lengthy. However, if this is your career choice, then the knowledge you learn and the skills you acquire have the potential to impact greatly upon many individuals and their lives. They will also give you an interesting and worthwhile career.

1.1 **What is biomedical science?**

Biomedical science is more than a broad collection of scientific topics.

Instead, it refers to a defined area of study related to the clinical investigation of disease within a degree programme. Degree programmes of these types are designed to allow the student to meet the educational standards and, in some cases, the professional skills required of a registered **biomedical scientist** by the **Health Professions Council** (HPC). Only those individuals

Cross reference

The HPC will be discussed further in Chapter 2 'Fitness to practise'.

FIGURE 1.1
A typical, modern clinical laboratory.

who have met the appropriate standards of the HPC will be legally allowed to call themselves a biomedical scientist in the UK. The HPC is a statutory regulatory body for a variety of therapeutic and scientific health professions.

The definition of biomedical science that this book and the accompanying volumes in the series will be using is that of the **Quality Assurance Agency for Higher Education** (QAA) and is described in Section 1.2 below.

1.2 Biomedical science degree programmes

Biomedical science degrees are concerned with the study of a variety of subjects including:

- Biology
- Chemistry
- Pathobiology of human disease
- Analytical methods

Although biology and chemistry are key elements of the academic programme, **pathobiology** is a fundamental component of a biomedical science degree. Pathobiology involves studying anatomy, physiology, biochemistry, genetics, immunology, microbiology, cell biology, and molecular biology. It is also concerned with knowledge of disease processes, which are investigated in the clinical laboratory specialities of histology, cytology, biochemistry, immunology, haematology, transfusion science, and microbiology. Any student successfully completing such a course of study will have a graduate level proficiency in the science of the causes, consequences, diagnosis, treatment, and management of human diseases.

Biomedical science requires multidisciplinary approaches to the study of human diseases and, as such, you must know why and how the development of diseases affects the normal structures and functions of the human body. You will also need to know the current methods used for investigating, diagnosing, and monitoring diseases, as well as research methodologies, which will include an appreciation of how new methods to combat diseases are developed, evaluated, and integrated into clinical practice. Biomedical science also requires knowledge of several scientific and clinical subject areas, which have been summarized by the QAA. You can see a list of these subjects in Table 1.1.

TABLE 1.1 Quality Assurance Agency for Higher Education (QAA) subject areas in biomedical science

Scientific subject areas	Clinical subject areas
Human anatomy and physiology	Histology
Cell biology	Clinical biochemistry
Biochemistry	Cytology
Genetics	Clinical immunology
Molecular biology	Haematology
Immunology	Transfusion science
Microbiology	Medical microbiology

Practical and transferrable skills

Biomedical science is very much a *hands-on* science and acquiring some of the discipline or subject-specific skills associated with its practice is an essential part of a biomedical science degree programme. These skills include:

- Learning how to handle patient samples in a safe and responsible manner when preparing these samples for clinical investigation
- Using aseptic procedures
- Using laboratory instruments safely and appropriately to ensure accuracy, precision, and reproducibility
- Interpreting patient-related data in an appropriate manner

Biomedical science, as described by the QAA, is an *honours* degree programme of study and so this will incorporate a **project** or **dissertation**, which are the definitive components of honours degree programmes. They are individually undertaken by a student although, of course, they are under the guidance of and studied with the support of professional staff. An honours degree will help in the development of transferrable skills, such as literature searching, appraisal of literature, experimental design, scientific communication, research methodology related to ethics, governance, audit, and statistical analysis.

As part of the assessment of your dissertation or project you may be required to present this work to a group of peers (fellow students).

You have now seen the extensive and complex nature of biomedical science degrees in terms of their scientific and professional contents, but these degree programmes are not only just about science. They will enable you to develop a wide range of generic skills related to team-working, communication, negotiation, numeracy, data analysis, and information technology. Developing these generic skills is as essential as the gaining of underlying scientific knowledge upon which biomedical science is based. Although the knowledge and skills acquired from biomedical science degrees are essential for work in National Health Service (NHS) laboratories of the UK, they also provide a sound academic and professional base and the potential to work in any of the organizations and areas you can see in Table 1.2.

SELF-CHECK 1.1

What generic skills may be acquired during the study for a biomedical science degree?

1.3 **What is a biomedical scientist?**

This section title may seem a slightly odd question to ask as you may naturally infer that graduates with a biomedical science degree course are all biomedical scientists. However, as we have already explained in Section 1.1, the title *biomedical scientist* is legally protected and only those persons who are registered as such with the HPC are permitted to use it within the UK. Thus, the title describes a person who has met the appropriate minimum or threshold standards of proficiency required by the HPC (and have demonstrated that their character and health are satisfactory, too). These standards are met by the successful completion of an HPC approved degree programme.

One of the more important HPC proficiency standards for biomedical scientists to consider is *standard 3.a1*, which states: 'a registrant biomedical scientist must know the key concepts of

TABLE 1.2 Summary of organizations and work areas for biomedical science graduates

Organizations
The National Blood Service
The Health Protection Agency
Independent pathology laboratories
Veterinary laboratories
Agricultural laboratories
The World Health Organization
Universities

Areas of practice
Clinical genetics
Forensic science
Research
Pharmaceutical diagnostics
Medical devices
Clinical trials
Commerce
Education
Food and food safety
Biotechnology

the biological, physical, social, psychological and clinical sciences which are relevant to their profession-specific practice'. To meet this standard you will need to demonstrate an understanding of the knowledge and practice of *all* of the major areas of biomedical science. The underlying scientific knowledge was outlined in Section 1.2 and the associated professional skills will be discussed in Section 1.4.

Amongst the HPC approved programmes are a number of honours degrees that are also accredited by the **Institute of Biomedical Science** (IBMS), which is the professional body that biomedical scientists may join; its offices are shown in Figure 1.2.

The IBMS offers professional qualifications and sets standards of professional behaviour for its members. As part of its work, the IBMS has also developed a training programme using a portfolio of professional evidence (the *registration portfolio*) that leads to the award of a **Certificate of Competence**, which you can see in Figure 1.3. The certificate may be used as suitable evidence when applying to the HPC for registration. The concept of a portfolio should not be new to most readers. However, a portfolio worthy of the award of a Certificate of Competence needs to focus on the evidence necessary to an external verifier that you have acquired the knowledge and skills to meet the threshold standards of proficiency necessary to join the HPC.

Cross reference

The HPC, their Standards of Proficiency, and the degree approval process will be discussed further in Chapter 2 'Fitness to practise'.

Cross reference

The Institute of Biomedical Science, their professional guidelines, and the degree approval process will be discussed further in Chapter 2 'Fitness to practise'.

FIGURE 1.2
Institute of Biomedical Science, 12 Coldbath Square, London.

1.4 **Professional skills**

The professional training of a biomedical scientist can only be delivered and the relevant skills acquired by physically working in a *clinical laboratory* to gain the necessary hands-on experience.

Key Points

Throughout this chapter and the rest of the accompanying volumes in the series, *clinical laboratory* will be used to describe a *collection* of departments. *Department* refers to an individual discipline-specific area of work.

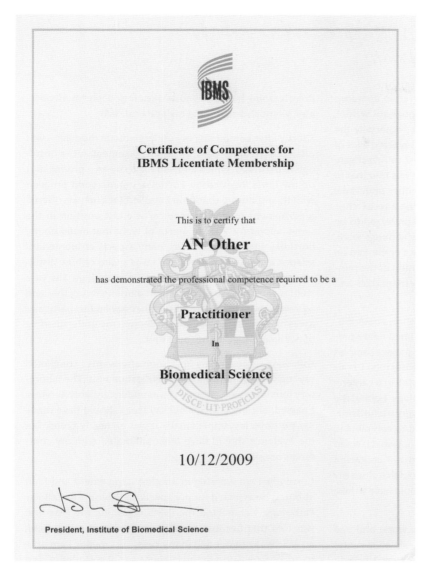

FIGURE 1.3
Institute of Biomedical Science Certificate of Competence. Copyright © Institute of Biomedical Science.

Professional skills may be acquired in a variety of ways. One way may be through the completion of a full-time HPC approved degree. However, these skills may also be gained during employment, while simultaneously completing an IBMS accredited part-time degree programme and the portfolio of professional evidence described in Section 1.3. Professional skills may also be acquired after graduation, during employment in a relevant department by completing the portfolio of professional evidence, whilst also undertaking some discipline-specific training. This is the route followed if you have gained an unaccredited degree whose content needs to be supplemented with further study. Case study 1.1 illustrates how this was achieved by a now successful biomedical scientist.

CASE STUDY 1.1 Supplementary study and professional skills

I graduated with a first class honours degree in anatomy. Biology had always been my strength throughout school, and I enjoyed every minute of my degree, especially the pathology elements. My final year project involved a lot of practical histology. Consequently, I decided to investigate histology as a potential career option and I was told that to do histology for a living, I would have to either undertake a medical degree, or 'do' a PhD and become a researcher. I chose the latter option, as I have always preferred to be 'behind the scenes', finding out how and why things work, rather than dealing with the public directly. I embarked on a PhD studying the morphology and formation of the mineral phase of bone, and how it alters in osteoporosis and osteoarthritis.

Part-way through my PhD, I found information about biomedical science and biomedical scientists. I had never previously given much thought to who tested patient samples in hospitals, assuming those people to be doctors. I realized that this was what I had been looking for in a career – investigating anatomy, histology, and pathology 'behind the scenes'. Deciding that the PhD would still be very useful to me, and given that I was enjoying it and wanted to complete it, I continued with my research, while finding out what I would need to do to become a biomedical scientist. I contacted the IBMS, who assessed my BSc, and I was told that since my degree was not accredited by the IBMS, I would have to take a course of supplementary education.

Once I had completed my PhD laboratory work, and had begun the daunting task of writing up my thesis while studying as an undergraduate again! I undertook six 10-credit undergraduate modules in clinical microbiology, haematology and immunohaematology, transfusion science, immunology, clinical biochemistry, and cytopathology. While studying and writing up, I contacted a local laboratory to ask if they would be willing to have me visit for work experience. The laboratory agreed, and I spent one day each week there learning how a diagnostic histology department operates, whilst also refreshing my practical skills.

Shortly after completing my undergraduate modules, I was offered a full-time, paid, 14-week placement in the histology department, which allowed me to make a proper start on my IBMS Registration Portfolio. I made good progress with it, and during that time I applied for and was offered a permanent trainee biomedical scientist position in the same laboratory. For the pieces of work that made up my portfolio, I tried to pull in as many aspects of biomedical science as I could, to show that I was aware of how histology related to the other disciplines in pathology. The case studies were particularly useful in demonstrating this, and by researching how other disciplines applied to particular areas of pathology, I was able to deepen my understanding of histopathology at the same time.

Following success in my PhD, soon afterwards I completed my registration portfolio and verification visit. The laboratory tour went well, although I was understandably nervous. I passed, but my assessor did say that I should have made my portfolio less impersonal – it was all neatly typed, but the evidence should have been annotated with my comments or reflections.

Currently, I am working in another department and I am about to be assessed for my Specialist Diploma in Cellular Pathology. I contacted the IBMS regarding my MSc and was told that because my PhD was relevant to my field, an MSc was not needed in my case. This means I am able to move directly on to the Higher Specialist Diploma, and I plan to spend the next 3 years gaining experience and gathering evidence for this, before sitting the examinations. Eventually, I hope to train in histological dissection, as to me this is the most exciting and appropriate use of the varied skills I have developed over the years of my training.

Finally, these skills can be gained through full-time study and completion of an accredited biomedical science degree programme that includes both academic study and professional training through a series of laboratory placements. Case Study 1.2 shows a typical reflective essay written at the end of a laboratory placement. A Certificate of Competence is then awarded upon graduation.

CASE STUDY 1.2 *Example of a placement reflective essay*

My placement took place at a District General Hospital in Scotland, in a multidisciplinary laboratory consisting of the histopathology, biochemistry, haematology & blood transfusion, and microbiology departments. The placement was scheduled for the second semester of the third year of the course and the outcome was to gather a training portfolio over a 12-week period. Being trained in a multidisciplinary laboratory was particularly advantageous for me in that I was able to experience work in different fields, so this gave me an insight on what were my disciplines of preference.

It all started with an induction week, which involved tutorials on important laboratory requirements such as laboratory accreditation, the roles of regulatory and professional organizations (HPC and IBMS), quality management and quality assurance, health and safety, and data protection, and this list is not exhaustive. My view of that first week was a little bit of apprehension because not only was the '9 to 5' routine difficult to cope with, most of the concepts were new, very prescriptive, and appeared to have nothing to do with what was being taught at university. However, it didn't take long for me to realize that the life of patients is what is at the end of the line, explaining why the biomedical profession is so tightly regulated.

Two weeks discipline-specific training were then received in each of the departments and it was interesting to see how the knowledge and skills gained at university were applied to work in the laboratory, this in terms of test principles, test procedure and techniques, interpretation, and statistical analysis of results. My first 2 weeks of training were in the biochemistry department where I was introduced to the modular analysers. Despite the fact that the system was highly automated, standard operating procedures (SOPs) indicated that different test procedures and detection principles were exactly the same as those performed manually in practical classes at university. For instance, most analytes (for example, glucose, lipids) were measured using colour reactions and spectrophotometer readings; hormones, enzymes, and tumour markers were measured using immuno-assay systems, and depending on their molecular size, either sandwich methods (measurement of Troponin T) or competitive assays (as for free thyroxin) were used. Although these methods were also automated and the detection principles varied (fluorescence,

chemiluminescence), the basic principles behind the test procedures were the same as those taught in the immunology module. I made similar observations when I moved on to other departments.

It was a relief not to have to perform manual cell counting in haematology or in microbiology as at university; meanwhile, it was good to see that the university was in line with new advances in biomedical science, such as molecular methods, which are increasingly used in clinical laboratories. It was also interesting to see how most of the techniques learned at university could be used interchangeably across disciplines, namely molecular techniques, microscopic techniques, and immunological methods. My only reservation was about the content of a particular university practical class where no practical work was actually done (not even the haematoxylin and eosin staining routinely done in histopathology laboratories) except for microscopic identification on ready-made slides.

The importance of maintenance, calibration of equipment, and quality control (QC) was much more emphasized than at university. Other important notions, such as prioritization, effective communication, time management, and cost effectiveness are only some of the other topics with which students are not really familiarized; it was a privilege to be introduced to these at an early stage and this gave me a feel of what is expected in the working place.

To sum it up, I would say that going on to a placement was an excellent opportunity for me to appreciate the practical relevance of every single module taught in the biomedical science course, from the first year of study. It is evident that without the basic knowledge and skills gained from clinical modules it would be very difficult to understand work in the laboratory. Working in a NHS laboratory is not just about applying a SOP to patient samples to obtain some results. I've learned that a biomedical scientist should be able to know why a test is being requested, when and how the sample should be properly collected, what sort of additives (if indicated) are needed and what will be their impact on the tests, what test methods are more appropriate for a particular specimen, and what precautions should be taken during processing. He/she must also be able to analyse the results obtained and make informed comments about their significance.

The rewards of the placement were, in fact, greater than I initially thought, as it wasn't until I started working on my honours project that I realized how much I had learned in just a few weeks. I became conscious as soon as I started the project that my planning, organizational, and practical skills were much more enhanced. I also had a totally independent and more confident approach to it compared with the execution of my previous laboratory assignments.

In addition, the placement turned out to be an integral work experience on my CV, which helped earn me a job in a NHS laboratory so I'm happy today to say that it was worth doing.

SELF-CHECK 1.2

How are professional skills acquired?

1.5 Initial employment as a biomedical scientist

Following registration with the HPC, it is likely that newly qualified biomedical scientists will be employed in a discipline-specific department, such as histology, biochemistry, cytology, haematology, immunology, microbiology, blood transfusion, or virology. However, it is becoming apparent that the divisions between these traditional disciplines are becoming blurred and new departments are beginning to emerge through advances in science and technology or through changes in clinical practice. For example, some laboratories combine the routine diagnostic investigations performed in haematology or biochemistry into a department often called *blood science* or *blood sciences*. Further, the use of **point of care testing** (POCT) by bench-top or hand-held devices in settings outside of the laboratory, including a patient's own home, is increasing (Box 1.1). Often POCT devices allow simple microbiological, biochemical, or haematological tests to be performed in combination.

Cross reference

Point of care testing is explored in Chapter 18.

It is only after registration and employment that you will acquire the full range of more detailed discipline-specific skills. This may involve completing an IBMS Specialist Portfolio, which is explained in Chapter 2 *Fitness to practise*.

Training

Properly structured and delivered training programmes are essential to the development of a biomedical scientist. Whether the training is part of a work placement or as a new employee

BOX 1.1 Point of care testing

Point of care testing involves using hand-held or bench-top analysers to test patient samples in settings other than a clinical laboratory. These relatively simple analysers often use capillary blood sampling and may be encountered in hospital areas such as intensive care or within a primary care setting, for example, an anticoagulation clinic. Increasingly, POCT devices and tests are being offered directly to patients for part of self-testing or as part of the self-management of chronic diseases such as diabetes.

in a laboratory, all new biomedical scientists will be assigned a training lead who will facilitate their training. All of the training will be guided by the contents of an individual training programme and the HPC requirements for minimum standards of proficiency that are illustrated in Figure 1.4.

Training programmes are, by their nature, individual to the person concerned because of a variety of issues, such as prior knowledge and expertise, and the individual's own learning style. However, any one programme is likely to contain the basic underlying knowledge needed, together with an indication of the level of skill, a definition of competency and the assessment processes. As training is a continuous process, any training programme will need to be negotiated and agreed by a trainee and the training lead. Training is also an interactive

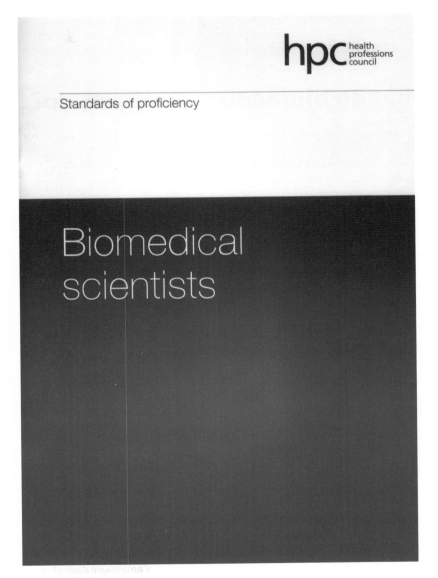

FIGURE 1.4
Front cover of the Health Professions Council Standards of Proficiency for Biomedical Scientists. Copyright © Health Professions Council.

process that requires equal commitment from the trainee and the employer and must take place without affecting the service provided by the laboratory. So its delivery must be in a manner that minimizes its effects on the normal working of a department, but maximizes the benefits to the trainee.

During training it may be necessary to work alongside support staff or undertake some of the basic work of the department, such as loading machines, making solutions, and filing. Hence, professional training may have to fit around the needs of an individual department. Such tasks should not affect the delivery of training. In fact, these types of tasks are likely to benefit a trainee, as they will increase his or her understanding of all the functions of a laboratory and the associated benefits of collaborative work. All individuals are members of a team, and need to help colleagues and support all the different areas and activities of a department. This will also provide time for more experienced colleagues to deliver professional training to others. After successful completion of such a training programme, the trainee should have become a competent, confident, and independent practitioner known as a *registered biomedical scientist*.

1.6 **What do biomedical scientists do?**

Until now we have considered the knowledge and skills that are necessary to be a biomedical scientist, but what will you do as a professional biomedical scientist in your daily working life? Much of the work undertaken by a laboratory is determined by a combination of the population served by the hospital, the associated range of illnesses that this population may develop, and the clinical specialities that are practised within the hospital. For example, in a *teaching* hospital, you would probably undertake more specialized work or research-related activities, along with the routine work associated with investigating, diagnosing, and treating disease. In a *general* hospital, the work may be less specialized, but there may also be opportunities for involvement in other activities. Wherever you work, it is likely that you will be busy, and it will be expected that you and the other biomedical scientists perform your duties and fulfil your roles in a helpful and professional manner, remembering that all samples originated from a *patient* and, in the end, all test results affect him or her.

Laboratory tests

It is estimated that over 70% of clinical or medical decisions or interventions are based wholly or partly on the results of laboratory tests. The significance of these tests cannot be underestimated.

Laboratory tests can be divided into three general categories, diagnostic, screening, and monitoring, which you can see outlined in Table 1.3.

Diagnostic tests are used to investigate a patient who has signs or symptoms of illness. For example, someone who experiences pain on urinating may have an infection of the renal tract. Samples of urine from the patient must be investigated for evidence of disease-causing micro-organisms. Screening tests are used to identify an 'at risk' individual from a defined population, who does not demonstrate symptoms, but may have an undiagnosed illness or condition. Thus, for example, in the UK all females of reproductive age (and slightly beyond) are invited to have a sample taken from their cervix to investigate for any potential cellular changes that may indicate the development of cancer. Monitoring tests are used to check the status of an individual before a medical procedure, following diagnosis and treatment, or to monitor his or her progress during a long-term condition. For example, someone who goes into hospital

TABLE 1.3 Types and role of laboratory tests

Type of test	Role of test
Diagnostic	Investigating an individual showing signs or symptoms of an illness
Screening	Identification of 'at risk' individuals within a defined population who do not show any symptoms, but may have an underlying undiagnosed clinical condition
Monitoring	Checking the health status of an individual before undertaking a medical procedure Assessing an individual following diagnosis and treatment Monitoring the progress of long-term or chronic conditions

for an operation will be subjected to blood and urine tests designed to give information about their overall health and physiological functions. Increasingly, microbiological investigations form part of this test profile to ascertain if an individual has one of the so-called 'super bugs', such as methicillin-resistant *Staphylococcus aureus* (MRSA). It may be that the blood and urine tests show the patient's liver, kidneys, and heart are healthy, and are not a bar to the planned operation. However, the microbiology tests may show he or she is suffering an infection that needs to be treated before an operation can take place. This type of situation shows why it is essential for biomedical scientists to understand the work of all the main departments and why training programmes (discussed above) emphasize these aspects of biomedical science.

We will now, briefly, consider the work carried out in each of the main discipline-specific departments.

Histopathology

Histology involves the microscopic study of tissues to identify disease that may be present. Tissues may be removed from living individuals following an operation or from deceased individuals at post-mortem. The tissue samples range from small biopsies to whole organs. Slides are prepared from these samples, which are stained and examined using a microscope. Diseases such as inflammation, infections, benign and malignant tumours can all be diagnosed using histology, as you can see in Figure 1.5. Biomedical scientists are involved at all stages of the preparation processes and work closely with medical histopathologists in this respect.

Cross reference
You can read more about the roles of biomedical scientists in histology in the companion text, *Histopathology*.

Clinical biochemistry

Clinical biochemistry involves the study of biochemicals and biochemical mechanisms, and their imbalances within the body. Clinical biochemistry tests are usually performed on blood or urine samples, but other fluids may also be tested, for example, cerebrospinal fluid or fluids that abnormally collect around the heart, lungs, or abdominal organs. Many of the tests are performed automatically on machines (Figure 1.6) that use robotics to manage many samples simultaneously. Malfunctions of organs or diseases, such as diabetes, heart attacks, and kidney failure, may be diagnosed using biochemical tests. Other aspects of biochemistry are related to screening for drug use and abuse, which may involve investigating living patients or deceased individuals. Again, biomedical scientists are involved at all stages of the preparation process, and work closely with clinical scientists or chemical pathologists in this respect.

Cross reference
You can read more about the work of clinical biochemists in *Clinical Biochemistry*.

FIGURE 1.5
Thin section of breast stained
with haematoxylin and eosin
showing abnormal cells
infiltrating connective tissue.

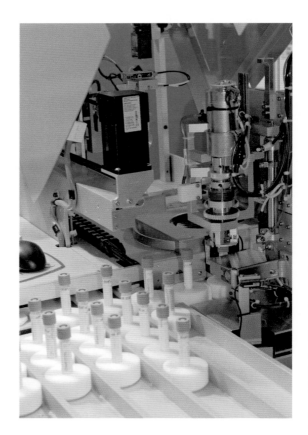

FIGURE 1.6
Samples waiting to be loaded
onto a chemistry analyser.

Cytopathology

Cytology (like histology) involves microscopic studies, but in this case cells are studied, rather than tissues, to identify disease that may be present. The cells are obtained through natural shedding, aspiration, or mechanical abrasion, and often involve the analysis of body fluids. Slides are prepared from these samples and examined using a microscope. Diseases such as inflammation, infections, benign and malignant tumours are diagnosed using cytology. Probably the best-known use of cytology is the examination of samples obtained from the uterine cervix as part of the UK cervical screening programme as can be seen in Figure 1.7. Once again, biomedical scientists are involved at all stages of the preparation and interpretation processes, and work very closely with medical cytopathologists and clinical medical staff in this respect.

Cross reference
You can learn more about clinical cytology in *Cytopathology*.

Immunology

Immunology involves the study of the immune functions of the body, as a system or its responses to a stimulus. Immunological specimens include blood (Figure 1.8) and other fluid or tissue samples that are subjected to a variety of manual or automated clinical tests. Diseases of the immune system include immunodeficiency where it fails to respond to appropriate stimuli, autoimmune disease when the immune system 'attacks' the tissues it is meant to protect, or hypersensitivity when it responds inappropriately to harmless stimuli. Immunology is also involved in tissue and organ transplantation through the matching of donor and recipient tissues, or the determination of rejection. Biomedical scientists, clinical scientists, and medical immunologists are all concerned with such studies.

Cross reference
Read more about clinical aspects of immunology in the companion volume, *Immunology*.

FIGURE 1.7
Normal and abnormal cells in a cervical smear stained by Papnicolaou's method.

FIGURE 1.8
White blood cells showing green fluorescence after being labelled with fluorescein.

Haematology

Cross reference
You can read more about
haematological investigations in
Haematology.

Haematology is the study of the morphology and physiology of blood. Haematological tests are performed, as you might expect, on blood samples; such tests are increasingly carried out by automated instruments. Diseases such as anaemia or leukaemia can be diagnosed following analyses of the numbers of erythrocytes or the haemoglobin content, or the numbers of leukocytes present in the sample haematology investigations. Other aspects of haematology are related to abnormalities of blood coagulation or the monitoring of individuals on long-term medication using anticoagulants. Biomedical scientists and medical haematologists collaborate during these studies. If you examine Figure 1.9 you can see the presence of normal blood cells on a microscope slide of a stained blood film.

Transfusion science

Transfusion science involves the provision of blood and blood products to patients who have anaemia, blood clotting problems, and those who are suffering from acute blood loss or have been involved in an accident. Donated blood is the primary source of blood for these purposes. Acquiring donations of blood for transfusion requires its appropriate collection, storage, preparation, and the issue of suitable blood and blood products. Other work in transfusion science is related to the collection, storage, and preparation of donated cells and tissues. Work in blood transfusion is highly regulated and is subject to EU Directives that are now part of UK law. Biomedical scientists tend to work very closely with clinical staff due to the nature of the supply aspects of these products. Figure 1.10 shows blood that has been cross-matched in order to establish its ABO and Rhesus blood group.

Cross reference
The companion volume,
*Transfusion and Transplantation
Science* will give you more
detailed information on
this topic.

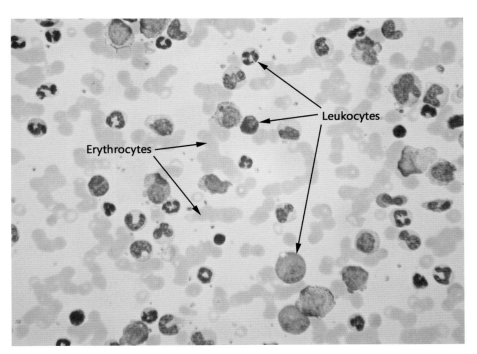

FIGURE 1.9
Normal blood cells in a blood film stained by May–Grünwald and Giemsa's method.

Microbiology and virology

Microbiology and virology are involved in the detection, isolation, and identification of micro-organisms, including bacteria, fungi, parasites, and viruses that cause disease. Samples may be taken from any part of a body, which may be of a living or deceased individual. Traditionally, microorganisms were grown on nutrient agar plates as shown in Figure 1.11, in broths or cell cultures, but increasingly the use of molecular detection methods (see Chapter 15) is becoming more common. Tests that used to take weeks to confirm the identity of a microorganism

FIGURE 1.10
Cross-matched blood showing blood group types.

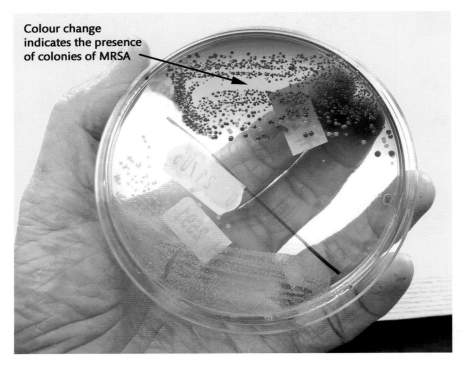

Colour change
indicates the presence
of colonies of MRSA

FIGURE 1.11
Chromogenic (colour-identifying) agar plate showing the presence of colonies of
methicillin-resistant *Staphylococcus aureus* (MRSA).

Cross reference
You can read more about the
roles of biomedical scientists in
medical microbiology in *Medical
Microbiology*.

can now be done in days. Often when examining a bacterium, its sensitivity to antibiotics is
also determined, to ensure the optimal treatment for the patient is chosen and to minimize
the misuse of these drugs. Medical microbiologists and virologists work closely with biomedical scientists in microbiology and virology investigations.

SELF-CHECK 1.3

Using histopathology as the discipline, briefly explain the uses of diagnostic tests.

 SUMMARY

■ Biomedical science is a term used to describe the study of the causes, consequences,
  diagnosis, and treatment of human diseases

■ Biomedical science is also an academic subject and can be studied as part of an honours
  degree programme, the content of which has been defined by the QAA

■ To become a registered biomedical scientist, an individual must meet defined minimum
  standards of proficiency by successfully completing an approved degree programme that
  includes a period of professional training

- The study of biomedical science will include acquiring knowledge and professional skills in all of the major discipline-specific areas

- Once registered, biomedical scientists normally pursue their career by professional practice and the extended study of one of the major discipline-specific areas

FURTHER READING

- Halushka PV, Krug EL. **Is the training of biomedical scientists at a crossroads?** *Academic Medicine*, 2009; **84**: 459–463.

- Burnett D, Poyser KH, Sherwood RA. *A Practical Guide to Accreditation in Laboratory Medicine*. ACB Venture Publications, London, 2002.

Useful Website

- Clinical Pathology Accreditation. **www.cpa-uk.co.uk**

QUESTIONS

1. Which of the following statements is/are **TRUE**?

    (a) Screening tests are used to investigate signs and symptoms

    (b) Populations are subjected to screening tests

    (c) Screening tests can identify illnesses

    (d) Screening tests are used to identify cancers only

    (e) Screening tests are 100% accurate

2. Discuss why it is necessary for prospective HPC registrants to study all of the major areas of biomedical science.

3. If the outcomes are similar, why do training programmes differ between organizations and individuals?

4. Which of the following are role(s) of a test to monitor health?

    (a) Checking the general health status of an individual prior to performing a medical procedure.

    (b) Identification of at-risk individuals who have no symptoms

    (c) Investigating an individual who is showing symptoms

    (d) Assessing the progress of long-term clinical conditions

    (e) Follow-up assessments of treated individuals

Answers to self-check and end of chapter questions are available in the Online Resource Centre accompanying this book.

 Go to www.oxfordtextbooks.co.uk/orc/glencross

2

Fitness to practise

Hedley Glencross

Learning objectives

After studying this chapter, you should be able to:

- Define the functions of a regulatory authority
- Discuss the roles of a professional body
- Explain how a regulatory body determines and maintains the fitness to practise of its members
- Demonstrate how a professional body may help you meet the requirements imposed by regulatory authorities
- Explain how a professional body may help you develop your career

Introduction

Fitness to practise is the ability of an appropriately trained and registered individual to perform his or her professional tasks in a safe manner, and to make appropriate judgements with a degree of autonomy. In Chapter 1 you were introduced to the Health Professions Council (HPC) and the Institute of Biomedical Science (IBMS). In this chapter we describe these organizations (Figure 2.1a,b), discuss how their respective, but complementary approaches to fitness to practise differ, and explore their contributions to the training and career development of biomedical scientists.

Cross reference
Chapter 1 'Biomedical science and biomedical scientists'.

2.1 Healthcare regulation

Biomedical scientists have been regulated by legal statutes since 1960 under the Professions Supplementary to Medicine Act, when biomedical scientists, or rather medical laboratory technicians as they were then known, were registered and directed by the **Council for Professions Supplementary to Medicine** (CPSM). The CPSM regulated nine professions and maintained a list of individuals who had met their threshold standards. However, the CPSM did not set other professional standards, so that once a biomedical scientist was admitted to the then CPSM *state register* there was no requirement to demonstrate any continuing competence or

bib\GLE

FIGURE 2.1
Logos of (a) Health Professions Council (HPC) and (b) Institute of Biomedical Science (IBMS). Both logos reproduced with permission from the respective organizations.

further education to remain on the register. As a consequence, it was difficult for the CPSM to remove someone from the register, other than for a criminal offence. However, following a number of then well-publicized incidents, such as the murder of four children by the nurse Beverley Allitt in 1991 and the murders of many elderly patients by the general practitioner Harold Shipman between 1995 and 1998, the issue of regulation of all healthcare professionals (including medical staff) was brought into focus during the 1990s. Consequently, in 2001 a Parliamentary Order for the Health Professions, which included provision for the formation of the HPC, was approved. The HPC was given (and maintains) a UK-wide remit for the registration of competent healthcare professionals.

> **Cross reference**
> Chapter 4 'Health and safety' discusses some of the legal framework relevant to healthcare professionals.

2.2 **Health Professions Council**

The HPC is principally concerned with patients and their safety. Indeed, it was established as a regulator of healthcare standards with a duty to protect the public. Thus, the work of the HPC is designed to safeguard the public from being harmed by the actions or inactions of healthcare professionals. Parliament has granted legal powers for the HPC to carry out its statutory roles by establishing and maintaining a **register** of **competent** individuals who meet its standards of training, professional skills, behaviour, and health. Competence is defined as the ability of a registered individual to repeatedly perform any action related to his or her profession to a defined standard, and in a safe and effective manner. Any individual who wishes to be admitted to the HPC register must demonstrate minimum or threshold *standards of proficiency* and provide evidence that they are of excellent character and in good health. Appropriate evidence that these professional standards have been satisfied would be by successfully completing an approved degree course, as we discussed in Chapter 1. Students who have not studied in the UK may apply to the HPC for registration by its international route, as they will need to have their qualifications, experience, and competence individually assessed.

> **Cross reference**
> Chapter 1 'Biomedical science and biomedical scientists'.

The HPC currently registers 15 professions, including therapeutic professionals, such as physiotherapists and radiographers, as well as scientific professionals, for example clinical scientists and biomedical scientists (Table 2.1). The HPC fulfils its principal role of regulating the activities of these professionals by restricting the use of protected titles, such as physiotherapist or biomedical scientist to those individuals it deems competent. The HPC also has the legal rights to investigate and remove from its register any individuals who it finds to be incompetent. Individuals may be removed from the register because of unprofessional conduct, inappropriate personal actions, or because of criminal activities.

TABLE 2.1 Professionals regulated by the HPC

Therapeutic professionals	Scientific professionals
Arts therapists	Biomedical scientists
Chiropodists/podiatrists	Clinical scientists
Dieticians	
Hearing aid dispensers	
Occupational therapists	
Operating department practitioners	
Orthoptists	
Paramedics	
Physiotherapists	
Practitioner psychologists	
Prosthetists and orthotists	
Radiographers	
Speech and language therapists	

The HPC sets both generic and profession-specific standards in the following areas:

- Proficiency
- Education and training
- Conduct, performance, and ethics
- Continuing professional development

Standards of proficiency are the knowledge and skills that a registrant must demonstrate before he or she can be admitted to the register. To remain registered, all registrants must maintain their knowledge and skills as appropriate to their developing role during the progress of their careers. The standards of education and training (SETS) are those that providers of education and training must meet to ensure all students who complete an HPC approved educational programme meet their standards of proficiency. Successful students are then able to apply to HPC for registration. Standards of conduct, performance, and ethics apply to registrants. These are the codes of behaviour, attitude, and aptitude to which registrants must adhere if they are to remain a registered professional. It is the continuing enforcement of these standards that allows the public, patients, and the registrants' colleagues to be confident of their suitability to practise and remain on the register. Standards of continuing professional development (CPD) also apply to registrants. Individual registrants are required to demonstrate that they are keeping up to date in their developing role. They must also show, through a range of CPD activities, how these activities benefit themselves professionally, the wider health service, and patients alike. The Institute of Biomedical Science also contributes to CPD as outlined in Box 2.1.

Cross reference
Chapter 20 'Personal development'.

The SETS and CPD will be discussed in more detail in Chapter 20. The standards of proficiency and the standards of conduct, performance, and ethics are explored in more detail in Section 2.3.

BOX 2.1 Continuing professional development

Upon joining the IBMS, new members are automatically enrolled on the IBMS Continuing Professional Development Scheme (although there is an option to opt out). The variety of activities associated with the scheme and their independent, annual verification with its systematic and structured approach means it has been recognized by the HPC as appropriate for their purposes. Thus, registrants can meet many of the HPC requirements for CPD without needing to provide evidence of each individual activity undertaken.

HPC approval of biomedical science degrees

The HPC has developed standards of education and training that education providers must satisfy if their biomedical science degree courses are to achieve HPC approval. These generic standards ensure that any individual who successfully completes an approved programme will also reach a minimum standard of proficiency. As a consequence, he or she will then be eligible to apply to be admitted to the HPC register.

Key Points

The major advantage of successfully graduating from an HPC approved degree is that you will be able to apply immediately for admittance to the HPC Register.

The key areas considered when approval for a biomedical science degree is sought are:

- Level of entrance qualifications
- Programme admissions
- Programme management and resources
- Curriculum
- Practice placements
- Assessments

The HPC approval process is designed to be robust, but not too onerous for the education providers. The approval of a new degree programme by the HPC will require that they make an initial visit to the education provider. The HPC delegation may visit as part of a wider assessment group or in isolation. Once approval has been granted by the HPC, it is 'open ended' and subject to satisfactory monitoring. For example, there is an annual review of the degree programme to ensure its structure and content have not been subject to major revisions and that it continues to meet the HPC Standards of Education and Training. If there have been major changes in an approved programme, the HPC will consider their effects on the standards. Thus, an annual review may be the trigger for further visits by the HPC for the re-approval of significantly altered biomedical science degree programmes.

Although degree approval is normally done on behalf of individuals who wish to gain HPC registration, a number of post-registration degree programmes are also approved by the HPC.

These include, for example, prescribing or local anaesthetic programmes, and do not apply to biomedical scientists.

2.3 HPC Standards of Conduct, Performance and Ethics

The HPC has developed 14 'duties' that define their Standards of Conduct, Performance and Ethics (Figure 2.2) and to which all registrants from the 15 regulated professions must conform. These duties are listed in Table 2.2. The duties describe in general terms what is expected of registrants as an autonomous and accountable professional. A rather simpler summary is that the standards describe the levels of honesty, integrity, and probity required of a registered individual at all times in his or her personal and professional life. These standards have necessarily been written in a rather stilted style as they are used (in conjunction with the standards of proficiency) to decide on a course of action if any concerns are raised about the practice of a specific individual.

Of the duties listed in Table 2.2, only 9 and 14 are outside the current normal practice of biomedical scientists, given that they are unlikely to treat patients or advertise their services.

These standards may be met in a number of ways. For example, a registrant may seek professional advice and support from professional bodies as part of their roles in representing and promoting the interests of their members. This may be as simple as joining and maintaining membership of a professional body or it may be by seeking expert advice from that professional body. As will be discussed later (Section 2.6), the IBMS also defines good practice for its members and other individuals by providing guidance to help clarify professional or scientific matters.

Your duties as a registrant

Standards of
conduct,
performance
and ethics

FIGURE 2.2
Front cover of *HPC Standards of Conduct, Performance and Ethics*. Copyright © Health Professions Council.

TABLE 2.2 Health Professions Council standards of conduct, performance and ethics

Duty	You must:
1	Act in the best interests of service users
2	Respect the confidentiality of service users
3	Keep high standards of personal conduct
4	Provide any important information about your conduct and competence
5	Keep your professional knowledge and skills up to date
6	Act within the limits of your knowledge, skills, and experience and, if necessary, refer the matter to another practitioner
7	Communicate properly and effectively with service users and other practitioners
8	Effectively supervise tasks that you have asked other people to carry out
9	Get informed consent to give treatment (except in an emergency)
10	Keep accurate records
11	Deal fairly and safely with the risks of infection
12	Limit your work or stop practising if your performance or judgement is affected by your health
13	Behave with honesty and integrity, and make sure your behaviour does not damage the public's confidence in you or your profession
14	Make sure any advertising you do is accurate

To meet duties 3 and 12 (Table 2.2), individuals are required at registration to provide a signed reference from an appropriate person. Duty 3 is often satisfied by a reference from an HPC-registered individual as they are familiar with the level of professional conduct required. However, with regard to duty 12, it is most likely to be provided by a GP or other medical practitioner who is able to comment on the applicant's current state of health.

Once an individual has joined the register, he or she is only required to sign (every 2 years at renewal of registration) a declaration that their ability to practise has not been affected by any changes in his or her health status. This declaration covers all aspects of health and abilities, in particular the decision-making skills and awareness required by registered individuals. It is expected that the HPC will be informed immediately if the health status of a registrant changes in a way that affects their role(s) as a registered individual. From 2007 this declaration has also included a statement regarding CPD, as discussed in Section 2.4.

HPC Standards of Proficiency

The HPC has also developed standards of proficiency for each of the 15 professions (Figure 1.4). Some of the standards are generic in that they apply to all of the regulated professions, while others are related to individual professions. However, both groups of standards cover three areas:

1. Expectations of a health professional
2. Skills required for the application of practice
3. Knowledge and understanding of these skills

Each of these areas is subdivided into a number of individual statements that describe the details regarding each standard. The top-level generic standards are illustrated in Table 2.3.

TABLE 2.3 Health Professions Council Standards of Proficiency

1. Expectations of a health professional	2. Skills required for the application of practice	3. Knowledge, understanding, and skills
Registrant biomedical scientists must:	Registrant biomedical scientists must:	Registrant biomedical scientists must:
1a.1 Be able to practise within the legal and ethical boundaries of their profession	2a.1 Be able to gather appropriate information	3a.1 Know and understand the key concepts of the bodies of knowledge that are relevant to their profession-specific practice
1a.2 Be able to practise in a non-discriminatory manner	2a.2 Be able to select and use appropriate assessment techniques	3a.2 Know how professional principles are expressed and translated into action through a number of different approaches to practice, and how to select or modify approaches to meet the needs of an individual, groups, or communities
1a.3 Understand the importance of and be able to maintain confidentiality	2a.3 Be able to undertake or arrange investigations as appropriate	3a.3 Understand the need to establish and maintain a safe practice environment
1a.4 Understand the importance of and be able to obtain informed consent	2a.4 Be able to analyse and critically evaluate the information collected	
1a.5 Be able to exercise a professional duty of care		
1a.6 Be able to practise as an autonomous professional, exercising their own professional judgement		
1a.7 Recognize the need for effective self-management of workload and resources and be able to practise accordingly		
1a.8 Understand the obligation to maintain fitness to practise		

1. Expectations of a health professional	2. Skills required for the application of practice	3. Knowledge, understanding, and skills
1b.1	**2b.1**	
Be able to work, where appropriate, in partnership with other professionals, support staff, service users, and their relatives and carers	Be able to use research, reasoning, and problem-solving skills to determine appropriate actions	
1b.2	**2b.2**	
Be able to contribute effectively to work undertaken as part of a multidisciplinary team	Be able to draw on appropriate knowledge and skills in order to make professional judgements	
1b.3	**2b.3**	
Be able to demonstrate effective and appropriate skills in communicating information, advice, instruction, and professional opinion to colleagues, service users, their relatives, and carers	Be able to formulate specific and appropriate management plans including the setting of timescales	
1b.4	**2b.4**	
Understand the need for effective communication throughout the care of the service user	Be able to conduct appropriate diagnostic or monitoring procedures, treatment, therapy, or other actions safely and skilfully	
	2b.5	
	Be able to maintain records appropriately	
	2c.1	
	Be able to critically evaluate the impact of or response to the registrant's actions	
	2c.2	
	Be able to monitor and review the ongoing effectiveness of a planned activity and modify it accordingly	
	2c.3	
	Be able to audit, reflect on, and review practice	

In summary, these standards describe the minimum level of knowledge and skills required by a registered biomedical scientist to be able to practise. Indeed, it is essential that these standards are met as they are the threshold standards needed for an individual to be admitted to the HPC register and describe the **scope of practice** for a newly registered biomedical scientist. Scope of practice defines the range of activities in which a registered individual has been trained, assessed, and deemed competent to perform. The initial scope of practice defines the threshold standards for admittance to the HPC register, which are met on successfully completing an HPC approved degree or obtaining an IBMS Certificate of Competence (Figure 1.3). Although the statements (standards) are too numerous to fully explore within this book, the education and training programmes undertaken by a prospective biomedical scientist will provide the knowledge, skills, and evidence to ensure that these minimum standards of proficiency are met.

Cross reference

We described the Certificate of Competence in Chapter 1 'Biomedical science and biomedical scientists'.

SELF-CHECK 2.1

Why is it considered necessary that the HPC apply standards of proficiency for prospective registrants?

Scopes of practice are not static and change during a registrant's career. Thus, as a career progresses the associated scope of practice also evolves as new skills are acquired or new roles undertaken. However, any new scope of practice also provides a different context for the HPC Standards of Proficiency and their Standards of Conduct, Performance and Ethics. For instance, if a registrant no longer practises in the laboratory, but works as, for example, a researcher, lecturer, or manager, and can show that he or she continues to meet the standards of the HPC, they can remain on the register. It is only when these standards cannot be met that a registrant may wish to voluntarily remove themselves from the register or be removed by the HPC itself.

If a registrant has voluntarily come off the register or been removed from it, the HPC does allow a suitable individual to be readmitted and has developed processes, which include remedial training, supervised practice, or a combination of both, for this purpose.

SELF-CHECK 2.2

How may registrants who no longer practise in a laboratory, continue to meet the HPC standards?

HPC sanctions on practice

The HPC has statutory powers to apply sanctions or restrictions that limit a registered individual's ability to practise as a biomedical scientist (or as 1 of the other 15 appropriately registered professionals). Information regarding the competency of a registered individual will result in an investigation by the HPC. If this raises sufficient concern, it will be followed by a *Fitness to Practice* (FTP) hearing on either competence or health grounds. You can see an outline of the FTP hearing in Figure 2.3. Depending on the outcome, the hearing can lead to sanctions against the individual concerned, which include restrictions on the right to practise, supervision of practice, suspension from the register or, ultimately, removal from the register. Fitness to practise hearings are detailed and rigorous, but apply the lesser civil, not criminal, court burden of proof, as these hearings relate only to professional practice and public safety.

Summaries of three FTP hearings are given in Case studies 2.1, 2.2, and 2.3 that indicate the detail in which the HPC investigates all allegations made against registrants from all of its 15 registered professions.

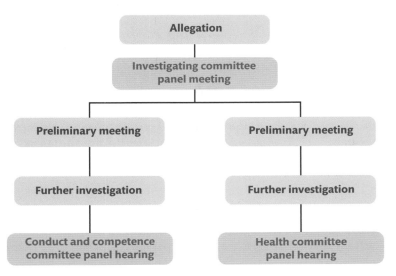

FIGURE 2.3
Schematic illustrating the HPC Fitness to Practise process.

CASE STUDY 2.1 *Example of FTP hearing relating to an occupational therapist*

An occupational therapist was struck off the register following allegations that he or she failed to maintain adequate records, provided inappropriate treatment to patients, wrote up case notes retrospectively, falsely wrote up case notes, and incorrectly closed cases that required further assessment.

The panel determined that the misconduct found to be proven was wide ranging, covered a period of time, and concerned basic competencies. The allegations relating to note keeping, which included the falsification of records, demonstrated a marked lack of honesty and integrity. Furthermore, the panel concluded that the registrant had not shown insight into his or her failings, or the consequences arising from them.

CASE STUDY 2.2 *Example of FTP hearing relating to an operating department practitioner*

An operating department practitioner was struck off the register for self-administration of the drug propofol having accessed the employer's drug store without authorization. The registrant had also received a police caution for this offence.

The panel took into account the fact that for the police caution to have been given, a full admission of the allegation had to have been made. Accordingly, the panel were satisfied that the theft of the drugs had occurred. The panel were also satisfied that the registrant had self-administered the drug.

The panel considered that a caution order would not reflect the severity of the matter and that a 'conditions of practice' order would not be appropriate, given that the registrant was not present at the hearing. It was not known if the registrant was working, with the result that conditions could not be considered. The panel gave careful consideration to imposing a suspension order, but concluded that there had been a serious breach of trust on the registrant's part, which had the effect of putting patients and colleagues at risk.

CASE STUDY 2.3 *Example of FTP hearing relating to a biomedical scientist*

A biomedical scientist was struck off the register following allegations that the registrant had failed to maintain adequate records and not ensured staff were adequately trained, had behaved in an intimidating and aggressive manner towards staff, and discussed confidential personal details of staff with others. These behaviours had continued over a significant period, but at no stage had the registrant shown any insight into their inappropriate nature, or remorse for the consequences of these actions on patients and colleagues. The panel concluded that these behaviours were misconduct and showed a clear breach of several of the Standards of Conduct, Performance and Ethics.

All registered professionals who are found guilty of a criminal offence will have the details of the offence, trial, and any sentence passed to the HPC for investigation, hearing, and possible sanctions as described above.

Full transcripts of all FTP hearings, including those for biomedical scientists, are made public and may be found on the HPC website.

SELF-CHECK 2.3

List five reasons for which a biomedical scientist may be suspended or removed from the HPC register.

2.4 Renewal of HPC registration

To remain on the HPC register, an individual registrant is required to sign a personal declaration every 2 years confirming that he or she continues to meet the HPC Standards of Conduct, Performance and Ethics and their Standards of Proficiency. Since November 2007 this declaration by biomedical scientists also included a statement that they also continue to meet the HPC standards of health and CPD. To check the validity of this latter statement, 2.5% of the total registrants from each of the regulated professions are required to provide evidence of their CPD activities during the preceding 2 years. Failure to provide sufficient evidence within a given time period results in action by the HPC against those individuals, such as suspension or even removal from their register, although further time to submit evidence may also be given. All new registrants are automatically exempt from this requirement for the first 2 years of their registration period. The first audits for biomedical scientists occurred in 2009.

The HPC register currently contains over 200,000 professionals. Whilst it is mandatory for individuals providing services to the NHS to be registered (or in training to be registered or appropriately supervised by a registered individual if unregistered), membership of a professional body is a matter of personal choice. The professional body for biomedical scientists is the Institute of Biomedical Science.

2.5 Institute of Biomedical Science

The Institute of Biomedical Science (IBMS) is the **professional body** for biomedical scientists in the United Kingdom and, as such, it:

- Consists of a number of individuals who have voluntarily joined the Institute to maintain the knowledge and skills that defines biomedical science
- Sets standards of professional conduct appropriate to individuals who practise biomedical science
- Administers the award of qualifications appropriate to this knowledge and skills

Within the UK, the IBMS has a highly developed network of regions and branches that provide a forum for the discussion of scientific and/or professional matters. Commonly, there is also a CPD officer and a network of local hospital representatives, who may also be approached for information or help. Some regions and branches also support discipline-specific discussion groups that are able to provide another arena where scientific or professional matters may be discussed. Upon joining the IBMS, an individual is assigned to a specific branch, which is normally determined by the work address of the individual. It is possible to change branch if he or she wishes to join a branch closer to their home address.

The vast majority of IBMS members reside and practise in the UK; however, it also has members in many countries and supports branches in Cyprus, Gibraltar, and Hong Kong.

The IBMS promotes and develops biomedical science and those who practise biomedical science, and is therefore principally concerned with professional matters and issues related to its members; you should not, however, infer that it is not concerned about patients or with patient safety. The activities of the IBMS are supportive and allow appropriate members to meet the HPC standards described in Section 2.2.

Classes of IBMS membership

The IBMS has four classes of membership: *Associate, Licentiate, Member, and Fellow*. You can see a summary of the classes of membership and how to progress through them in Table 2.4. As an undergraduate student, an individual is admitted to the class of Associate upon joining the IBMS. Following graduation and the award of a Certificate of Competence (Section 2.2 and Chapter 1) he or she will be able to upgrade to Licentiate membership. Once admitted as a Licentiate, further upgrade of membership class to Member and ultimately to Fellow is encouraged. Equivalent routes for entry and membership class upgrade for all corporate classes of IBMS membership exist for those individuals whose education and training are different to what might be described as 'standard'.

Whatever their membership class, members of the IBMS are expected to apply the same principles of good professional conduct and practice in biomedical science.

SELF-CHECK 2.4

How may the structure of the IBMS membership help biomedical scientists develop their careers?

2.6 **IBMS Code of Conduct and Good Professional Practice**

The *IBMS Code of Conduct* is embedded within the IBMS publication, *Code of Good Professional Practice for Biomedical Scientists*. Like the HPC Standards of Conduct, Performance and Ethics (Section 2.3), this publication states in general terms what is expected of a member of the

TABLE 2.4 Classes of the Institute of Biomedical Science membership effective from 1 January 2009

Membership class	Comments and progression
Licentiate	Licentiate is the initial class of corporate membership for someone who has been awarded the IBMS Certificate of Competence.
	Licentiates are entitled to use the letters LIBMS after their name.
	Licentiates of the IBMS have the opportunity to work towards an IBMS Specialist Diploma.
	This qualification will allow individuals to apply for the next class of IBMS membership: *Member*
Member	Member is the next class of corporate membership for someone who is already a licentiate and holds an IBMS Specialist Diploma.
	Members are entitled to use the letters MIBMS after their name.
	Members of the IBMS have the opportunity to work towards an IBMS Higher Specialist Diploma.
	This qualification will allow individuals to apply for the highest class of IBMS membership: Fellow.
Fellow	Fellow is the highest corporate class of IBMS membership for someone who is already a Member and holds an IBMS Higher Specialist Diploma.
	Fellows are entitled to use the letters FIBMS after their name.
Associate	Associate is the non-corporate class of IBMS membership for someone who is not eligible for corporate membership.
	Categories of Associate include: students on accredited biomedical science degree courses and graduate trainees employed in an IBMS approved training laboratory.

IBMS in both their personal and professional lives. The IBMS Code of Conduct is reproduced in Table 2.5.

Compare the seven major points of the code with the 12 of the 14 statements in the Standards of Conduct, Performance and Ethics of the HPC (Table 2.2) pertinent to biomedical scientists; you can see that both are essentially similar. However, the ways in which they are written are

TABLE 2.5 Institute of Biomedical Science Code of Conduct

All members of the Institute of Biomedical Science will at all times:
1. Exercise their professional judgement, skill, and care to the best of their ability.
2. Fulfil their professional role with integrity, refraining from its misuse to the detriment of patients, employers, or professional colleagues.
3. Seek to safeguard patients and others, particularly in relation to health and safety.
4. Treat with discretion all confidential and other information requiring protection, and avoid disclosing to any unauthorised person the result of any investigation or other information of a personal or confidential nature gained in the practise of their profession.
5. Act in good faith towards those with whom they stand in professional relationship and conduct themselves so as to uphold the reputation of their profession.
6. Strive to maintain, improve, and update their professional knowledge and skill.
7. Promote the study and development of biomedical science, and the education and training of practitioners of biomedical science.

rather different. For example, duty 6 of the HPC requires biomedical scientists to act within the limits of their knowledge and skills, and to refer problems to another (senior) practitioner as necessary. This duty is incorporated into statements 1, 2, 5, and 6 of the IBMS code.

SELF-CHECK 2.5

Which statements in the IBMS Code of Conduct would allow you to meet the HPC standard 1 that requires a biomedical scientist to act in the best interests of service users?

The IBMS *Code of Good Professional Practice for Biomedical Scientists* is only one of a number of its publications (Box 2.2).

BOX 2.2 IBMS publications and website

The IBMS produces a variety of publications and leaflets, however, its two principal journals, which have been published for over 50 years and are sent to its members are:

1   *The Biomedical Scientist*

2   *British Journal of Biomedical Science*

The Biomedical Scientist is published monthly and contains news, articles of scientific interest, CPD activities, IBMS activities, and a classified advertising section. The *British Journal of Biomedical Science (BJBS)* is published quarterly and is the largest worldwide circulation peer-reviewed journal dedicated to biomedical science (Figure 2.4). It includes wide-ranging papers on basic science, scientific, and clinical practice of biomedical science.

The website of the IBMS is of increasing interest to members and visitors alike. You can see the welcome page in Figure 2.5

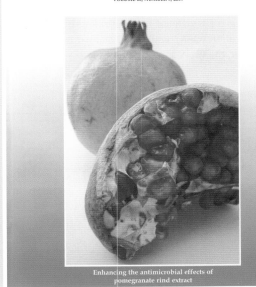

FIGURE 2.4

Front cover of an issue of the *British Journal of Biomedical Science*.

FIGURE 2.5

Welcome page of the IBMS website (www.ibms.org).

Good professional practice

The IBMS has defined what it regards as good professional practice in the context of bio-medical science and the behaviour of biomedical scientists. Its code of conduct describes the attributes and the professional behaviour expected of IBMS members. This code comple-ments that of the HPC Standards of Conduct, Performance and Ethics, which we described in Section 2.3. The IBMS code is relatively comprehensive as you can see from the areas listed in Table 2.6, and is explained in more detail in the IBMS publication, *Code of Good Professional Practice for Biomedical Scientists* (Figure 2.6). However, in many ways the IBMS code does not describe the full range of good professional practice. This would be difficult anyway, since observations as to what is good or bad practice are subjective, and it can often be difficult to give an objective viewpoint. In this volume, we will discuss in detail health and safety, quality assurance, and professional development in the relevant chapters. However, the education and training programme required of a biomedical scientist in the UK will ensure an individual works within the context of the IBMS Code of Good Professional Practice and meet the various standards of the HPC.

Cross reference

Health and safety, quality assurance and management, and personal development are discussed in Chapters 4, 19, and 20, respectively.

2.7 **IBMS and education**

The IBMS has been involved in formal educational activities since its founding in 1912; indeed, the IBMS motto is '*Learn that you may improve*'. Its major educational activities in higher educa-tion institutions are the accreditation of appropriate degree programmes and its involvement in the provision of professional qualifications.

Degree accreditation

The programme of biomedical science degrees is defined for registration purposes by the Quality Assurance Agency for Higher Education (QAA), while their standards are subject to approval by the HPC (Section 2.2). However, the IBMS, in its role as a professional body also

TABLE 2.6 Areas to which the IBMS code of practice applies

Functions of a biomedical scientist
Professional relationships
Management and leadership
Staffing and skill mix
Staffing and workload
Records and archives
Training
Professional development
Quality assurance
Performance

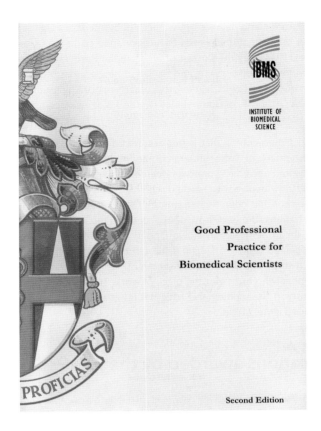

FIGURE 2.6
Front cover of *IBMS Code of Good Professional Practice.* Copyright © Institute of Biomedical Science.

accredits the contents of degrees in biomedical science. **Accreditation** of a professional degree is a process of peer review set against a number of defined learning outcomes. The IBMS accredits both undergraduate and postgraduate degree programmes for membership purposes. Accreditation of any particular degree in biomedical science by the IBMS is not designed to replace or replicate the HPC approval process or, indeed, the internal validation exercises undertaken by the universities concerned. The IBMS considers the content of the units or modules constituting the curriculum of the degree and uses the definitions and subject areas already discussed in Chapter 1, as the basis for their undergraduate accreditation process.

Accreditation by the IBMS requires the formation of a local liaison group consisting of academic staff of the university, NHS employees (largely hospital laboratory staff), and any independent part-time teachers who may be involved. This group is required to meet on a regular basis to oversee matters relevant to the course. Written evidence regarding the formation of the group and its meetings must be provided for scrutiny by the IBMS. It is generally expected that senior and experienced laboratory staff from local hospitals will contribute to the delivery of both undergraduate and postgraduate courses through part-time lecturing as 'experts' in clinical subjects, particularly where expertise in those areas cannot be provided by university staff.

Successfully completing an IBMS degree at an accredited university (Figure 2.7) offers a number of advantages to the graduate. Thus, it eases the entry of a graduate into the IBMS and contributes towards registration with the HPC.

The accreditation of a specific degree in biomedical science provided by a particular university may only apply to the academic component. This allows successful students to work in a variety of occupations or areas, which we listed in Chapter 1. It also allows the IBMS to tailor

FIGURE 2.7
Logo for use on, for example, stationary by an IBMS accredited university.

its accreditation of postgraduate education and degrees to the needs of biomedical scientists during their professional careers. A number of universities have developed co-terminus or applied biomedical science degrees. These degrees consist of an accredited programme of academic study together with laboratory-based teaching and training during a series of work placements. The advantage of completing this type of degree is that professional training is integrated into the degree programme. A Certificate of Competence (Figure 1.3) is awarded upon graduation allowing an individual to immediately apply for inclusion on the HPC register.

Graduates who have completed a degree course that is not accredited by the IBMS may be required to undertake supplementary or extra undergraduate studies to make their non-accredited degree equivalent to an accredited honours degree in biomedical science. Normally, the major areas of extra study involve the clinical disciplines, which are often not fully covered in non-accredited degrees. If a degree programme of study does not include any laboratory-based professional training, the Certificate of Competence must be obtained by a different route. For example, it can be obtained by working in a laboratory following graduation. Alternatively, it can be gained while working in a laboratory and studying for an appropriate degree part-time. Such study pathways require the registration portfolio to be verified by IBMS appointed examiners, who must visit the place of work. The verifiers must also assess the role of the laboratory in the training and have the remit to award training approval to appropriate laboratories (Figure 2.8).

FIGURE 2.8
Logo for use on, for example, stationery by an IBMS approved training laboratory.

Professional qualifications awarded by the IBMS

The IBMS has developed the following professional qualifications:

- Specialist Diploma
- Certificate of Expert Practice
- Higher Specialist Diploma
- Diploma of Expert Practice
- Advanced Specialist Diploma

These qualifications are linked to classes of Institute membership (Section 2.5) and also to a career in the National Health Service (NHS). The IBMS qualification structure is outlined in Figure 2.9.

The *Specialist Diploma* (Figure 2.10) is likely to be the first IBMS qualification encountered following registration at the beginning of a career in biomedical science. Qualification for a diploma involves the professional study of a single discipline. Diplomas are awarded in the principal, discipline-specific areas of biomedical science:

- Clinical biochemistry
- Cytology
- Histology
- Haematology with hospital transfusion practice
- Histocompatibility and immunogenetics
- Immunology
- Microbiology
- Transfusion science
- Virology

HCS Career framework stage			Additional optional qualifications	Membership class
9	Professional Doctorate			
8	**Advanced Specialist Diploma**			
	⬆		Diploma of Expert Practice	*Fellow*
7	**Higher Specialist Diploma**			
6	⬆ MSc		Certificate of Expert Practice	*Member*
	Specialist Diploma			
5	⬆			*Licentiate*
	Certificate of Competence			
	⬆ BSc (Hons) Biomedical Science			

FIGURE 2.9

Hierarchy of IBMS membership classes related to academic and professional qualifications.

Some of these programmes have been produced in partnership with other organizations, such as the British Society of Histocompatibility and Immunogenetics (BSHI). The content of all specialist diplomas have a common element related to professional roles and responsibilities, health and safety, receipt and processing of samples, laboratory computers, laboratory equipment, and quality assurance. However, they also include discipline-specific elements that naturally differ from subject to subject.

The IBMS recommends that 12–18 months of study be undertaken to collect the evidence required for the award of a specialist diploma. Its award requires an assessment by an IBMS appointed assessor. The assessment process consists of three parts:

1. Candidate presentation
2. Assessment of the portfolio
3. Candidate-led tour of the relevant hospital department

Each process will assess knowledge, practice, and skills, but is also designed to test an individual's confidence in demonstrating these points to the assessor.

The presentation by the candidates to the assessor will be expected to last 20 minutes and demonstrate that a transition from a newly qualified and generically trained individual to a specialist biomedical scientist has occurred. The candidate will also need to explain how their training relates to the daily practice within a single discipline of biomedical science.

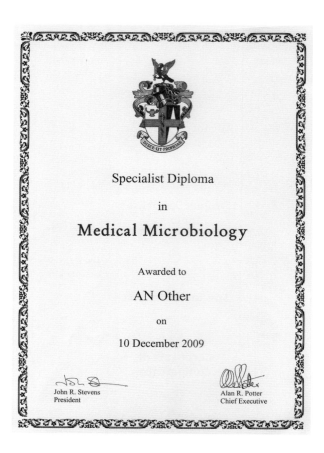

Specialist Diploma

in

Medical Microbiology

Awarded to

AN Other

on

10 December 2009

John R. Stevens
President

Alan R. Potter
Chief Executive

FIGURE 2.10
IBMS Specialist Diploma.

The assessor will examine the portfolio of evidence to ensure all sections have been completed. The candidate will then take the assessor on a tour of their department during which they may be asked questions about their activities or the functions of the department. Finally, there will be feedback to the candidate, their training co-ordinator, and department manager.

If successful, the candidate will be notified of this on the day.

Certificates of Expert Practice are intended to recognize scientific practice in specific areas that would only be performed by a few individuals who have chosen to specialize, and for whom alternative qualifications do not exist. They are at a level commensurate with a specialist diploma, but in areas that do not warrant the development of a full portfolio. In this way, they complement the broader discipline-specific post-registration awards already undertaken.

Examples of such qualifications include the *Certificate of Expert Practice in Electron Microscopy*, *Certificate of Expert Practice in Quality Management* and *Certificate of Expert Practice in Training*.

The *Higher Specialist Diploma* (HSD) is also a discipline-specific qualification, although an HSD in *Management and Leadership* has also been developed. The HSD is designed to test and demonstrate a deeper professional knowledge of a discipline than that at Diploma, and may not be entered without first holding an appropriate postgraduate degree either an IBMS accredited MSc or an IBMS approved masters or doctorate-level qualification.

The HSD is at the centre of the IBMS qualification process in that although the HSD is the normal requirement for the upgrade from Member to Fellow, it represents only an intermediate step in terms of professional practice. The HSD assessment runs annually over two days and includes case studies or scenarios as a major component of the examination. In this way, the ability to draw strands of information or practice together and demonstrate a deeper under-standing is assessed. Ironically, although related to practice, this qualification has no practical component in its assessment!

Diplomas of Expert Practice and *Advanced Specialist Diplomas* of the IBMS recognize knowl-edge and skills at the highest level of scientific or clinical practice within areas of biomedical science or in the application of biomedical science. They are designed for those who wish to practise at the level of a consultant biomedical scientist.

Examples of Diplomas of Expert Practice qualifications include the *Diploma of Expert Practice in Clinical Transfusion, Advanced Specialist Diploma in Cervical Cytology,* and *Advanced Specialist Diploma in Ophthalmic Pathology.* The cervical cytology advanced diploma was introduced in 2001 and holders have seen their role expand in the intervening period from a starting point of reporting of abnormal cervical cytology results to a number of other areas including case presentations at multidisciplinary team meetings, audit of invasive cancers, and commissioning.

 SUMMARY

- The principal duty of the Health Professions Council is to protect the public.

- Biomedical scientists are regulated by the Health Professions Council, which has the legal powers to admit individuals to its register.

- The Health Professions Council posseses the legal powers to restrict practice and remove from its register those registrants who it considers not competent to practise.

- Biomedical scientists may join their professional body, the Institute of Biomedical Science.

- The Institute of Biomedical Science acts for its members by helping them meet the require-ments of the Health Professions Council but also provides other support and benefits to members.

- The functions and activities of the Health Professions Council and Institute of Biomedical Science are complementary, and they work closely together.

 FURTHER READING

- Wood J. **The roles, duties and responsibilities of technologists in the clinical labora-tory.** *Clinica Chimica Acta* 2002; **319**: 127–132.

- Burnett D, Poyser KH, Sherwood RA. *A Practical Guide to Accreditation in Laboratory Medicine.* ACB Venture Publications, London, 2002.

- Institute of Biomedical Science. *Good Professional Practice for Biomedical Scientists*, 2nd edn. Institute of Biomedical Science, London, 2005.

Useful Websites

- Health Professions Council. **http://www.hpc-uk.org/index.asp**
- HPC international route. **http://www.hpc-uk.org/apply/international/**
- HPC degree approval. **http://www.hpc-uk.org/education/programmes/**
- HPC readmittance process. **http://www.hpc-uk.org/apply/readmission/**
- HPC Fitness to Practise. **http://www.hpc-uk.org/complaints/making/**
- Institute of Biomedical Science. **http://www.ibms.org/**
- IBMS Code of Good Professional Practice. **http://www.ibms.org/go/professional:good-professional-practice**
- IBMS degree accreditation process. **http://www.ibms.org/go/education-development:ibms-courses:accreditation**
- IBMS portfolio verification process. **http://www.ibms.org/go/registration:reg-qualification-assess**
- IBMS laboratory training approval. **http://www.ibms.org/go/registration:lab-standards**
- IBMS regions and branches. **http://www.ibms.org/go/ibms-networks**
- IBMS CPD Scheme. **http://www.ibms.org/go/education-development:cpd**
- Department of Health England. **http://www.dh.gov.uk/en/index.htm**
- Health of Wales Information Service. **http://www.wales.nhs.uk/**
- Scotland's Health on the Web. **http://www.show.scot.nhs.uk/**
- Department of Health, Social Services and Public Safety Northern Ireland. **http://www.dhsspsni.gov.uk/**

 QUESTIONS

1. Which of the following statements are applicable to biomedical scientists?

    (a) Must be able to practise in a non-discriminatory manner

    (b) Are not required to use problem-solving skills

    (c) Be able to critically evaluate collected information

    (d) Be able to review their individual knowledge and skills

    (e) Be able to maintain appropriate records

    (f) Have to always practise under direct supervision

    (g) Always work in isolation

2. Which of the following statements from the HPC are applicable to biomedical scientists?

    (a) You must keep your professional knowledge and skills up to date

    (b) You must get informed consent to give treatment (except in an emergency)

    (c) You must keep accurate records

    (d) You must limit your work or stop practising if your performance or judgement is affected by your health

(e)   You must make sure any advertising you do is accurate

(f)   You must provide any important information about your conduct and competence

3.   Why is it considered necessary for an individual to be registered?

4.   Why should it be appropriate for registrants to join a professional body, if this is not mandatory?

5.   Discuss the value of continuing professional development.

Answers to self-check and end of chapter questions are available in the Online Resource Centre accompanying this book.

 Go to www.oxfordtextbooks.co.uk/orc/glencross

3

Communications in laboratory medicine

Georgina Lavender

Learning objectives

After studying this chapter, you should be able to:

- Explain the need for clear policy for communication
- Explain why there is a need for a generic telephone policy
- Discuss the attributes required of a biomedical scientist when communicating results
- Discuss the regulations that apply to the storage and transmission of electronic data with reference to the current guidelines from the Institute of Biomedical Science
- Explain the roles of regulatory bodies that have interests in clinical laboratory record keeping

Introduction

Communication skills, including written, verbal, and electronic communications, are an essential part of being a biomedical scientist. Biomedical scientists are required to communicate clearly and precisely on all aspects of their work. They must communicate within their local workplace with their colleagues, within the organization that employs them, and they must be able to communicate with professionals and any of the public with whom they may have indirect contact.

Communication from the clinical laboratory includes the results of laboratory tests, specimen requirements, changes to methodology, and their effects on reference ranges, changes to service provision, significant changes in laboratory personnel, the introduction of new policies and procedures, or any other information that may affect the relationships between the laboratory and its service users.

One of the principal roles of biomedical scientists is to provide information to appropriate healthcare professionals that may affect the diagnosis and treatment of patients. They provide information regarding prescription drugs, what drugs are suitable and in what amounts for a variety of diseases. They also contribute to screening programmes for large numbers of 'well' individuals to detect disease in the early stages before symptoms are noticed. The whole process of receiving specimens, performing analytical work, and reporting results from start to finish relies on adequate communications between all the groups involved. This communication may not immediately be obvious to less experienced members of a laboratory team; hence, the purpose of this chapter is to highlight areas of both verbal and **nonverbal communication** that may impact on the relationships between laboratory staff, and other professionals and users. As biomedical scientists, you must be aware of your limitations and the consequences of your actions.

This chapter highlights the importance of adequate communication skills within the clinical laboratory and how communications by laboratory personnel can affect the perceptions of users of the service they provide.

3.1 **Communication and the clinical laboratory**

Laboratory medicine has a reputation for being one of the more under-rated services within a hospital environment. However, studies and audits have shown how essential the clinical laboratory is in diagnosis and in the treatment expected in the demanding healthcare organizations of today. It has been estimated that the clinical laboratory will have an input into 70% of patients receiving healthcare. Laboratory medicine plays a vital role in the overall function of the healthcare system, but its role may be overlooked, particularly by the general public. This has come about because, traditionally, clinical laboratory personnel appear to have little contact with the other hospital staff and with patients. Little has been known about the work carried out behind the laboratory doors, with patients largely accepting that the 'path lab' is a somewhat unpleasant entity where doctors dig around with bits of dead people. However, the reality could not be further from these preconceived ideas. The *World Wide Web* has raised public awareness of healthcare to new levels, and evidence-based medicine and audit have increased the importance of clinical laboratory investigations with the other healthcare professional groups within the patient care pathways. Laboratory medicine is now recognized by a wider audience within the healthcare community as an essential discipline and given due appreciation. Staff in clinical laboratories have a much higher profile in the provision of patient care. The corollary is that clinical laboratory staff must be aware of the impression that they give to patients, health professionals, and laboratory users (those who either request tests or receive results as part of their duties) during the course of their work. Just because the laboratory staff are not necessarily seen in person does not detract from the importance of being aware of the often latent signals that are picked up by other groups with which they have dealings.

Laboratory staff in public areas of hospitals

Clinical laboratory personnel rarely wear a uniform, and are often not immediately recognizable as hospital staff when in public areas, such as shops, dining rooms, and hospital corridors. As healthcare professionals, individuals have a responsibility to dress and act in a manner that will not cause offence to others, breach their hospital trust policy, or bring the profession into question by any aspect of their behaviour.

Some organizations have a **dress code** to which all staff are expected to adhere. Dress codes are often viewed as draconian by laboratory staff who (a) are rarely seen outside the laboratory and (b) are usually largely covered with a white laboratory coat. However, there are attempts by employers to bring standardization and professionalism to their workforces. Clothes are perceived, rightly or wrongly, as a form of non-verbal communication. Staff outside the laboratory, patients, and their relatives and visitors are quickly judgemental of the standard of dress worn throughout the hospital. Biomedical scientists must remember that the general public may not share their personal views regarding what is or is not an acceptable standard of dress at work, and that a dress code has been written by an organization with its *total* staff in mind. When considering standards of dress within a hospital, aspects of infection control and health and safety have to be considered, and the reason behind certain aspects of dress codes may not immediately be apparent. While biomedical scientists may not usually be present in clinics and on wards, they are still members of the wider hospital community and cannot be given preferential treatment. Where biomedical scientists are free to dress as they please, they should remember to be clean, tidy, and not wear clothing that may cause offence to others, or that may constitute a breach of the Health Professions Council (HPC) standards of conduct, performance, and ethics. These were introduced in Chapters 1 and 2. Likewise, if they are provided with a full uniform, as opposed to a laboratory coat and other personal protective equipment (PPE), then it would be seen that the dress code of the department is to wear that uniform as and when appropriate.

The HPC has clear standards of behaviour that are expected from its registrants at all times in their working life, and also in their private and personal life. Standards of behaviour are also viewed as non-verbal means of communication, and any unacceptable private behaviour may spill over and be extended to the type of behaviour that should be shown to a member of the public. The biomedical scientist involved in a street brawl may possibly be deemed as having a short temper and be capable of assaulting a patient under certain circumstances. Such behaviour will be investigated by the HPC where a criminal offence has been proven to have taken place and could jeopardize the registrant's fitness to practise.

Communications within the clinical laboratory

Clinical laboratories are increasingly subject to extensive legislation. This has led to descriptive procedures and documents being made available to all personnel within the laboratory. The advantage of having such documentation is that it ensures all staff follow procedures to high professional and scientific standards. The advantage of strict written instructions or **standard operating procedures** (SOPs) is that all staff who are required to carry out a particular task or procedure do so in the same way. Not only are such SOPs necessary for Clinical Pathology Accreditation (CPA) purposes, there is also the guarantee that if a member of staff is run over by a bus in their lunch hour, another member of staff will know exactly what to do to complete the day's work, even though there has been no direct communication between the two individuals!

Communications *within* a clinical laboratory are varied. They include the transfer of instructions and information between staff who work together to provide the end products of the laboratory. These products are the final results of the various measurements and tests performed on clinical specimens. **Teamwork** is the way in which a group of individuals must interact together to achieve a common goal. Thus, maintaining a team structure and working together in an appropriate fashion is crucial. Most working days within a clinical laboratory follow a given pattern in that the work is generated and must be carried out to specific deadlines. It is essential that every member of the team is aware of their responsibilities and the responsibilities of others around them. Each person working in a laboratory, regardless of his or her grade, has their own, sometimes unique place in that team. Team members have a responsibility to meet deadlines.

Cross reference
Chapter 4 'Health and safety'.

Cross reference
Chapter 1 'Biomedical science and biomedical scientists' and Chapter 2 'Fitness to practise'.

Cross reference
You can read more about SOPs in Chapter 19 'Quality assurance and management'.

Communication within the team is necessary so that the team leader is able to maintain appropriate control and the rest of the team recognize his or her leadership. It is essential that any information that may affect the processing of specimens is communicated throughout the team, both upwards to the most senior member with the responsibility of the service, and downwards to the support staff. Figures 3.1 (a and b) illustrate the ends of the spectrum when information needs to be communicated effectively between members of staff in the same department.

For example, the major failure of a large piece of equipment in the clinical chemistry laboratory may only seem important to those involved. Such events, while unlikely, require a number of immediate remedial actions in addition to repairs to the equipment; all staff must know what is required of them. Any major delays in the processing of samples must be reported to staff in specimen reception, so that they are aware of the situation and can inform other departments or service users as necessary that clinical samples may not be processed with the usual speed and efficiency. Staff involved in the issuing of results and dealing with telephone calls asking for results must also be informed so that they are able to explain to service users why the results they require may not be ready. There is also the need for clear written

(a)

(b)

FIGURE 3.1
(a) Appropriate and (b) inappropriate ways of informing staff of departmental or procedural changes.

Cross reference
Chapter 18 'Point of care testing'.

communication in the laboratory to invoke contingency plans that may involve staff who do not directly work in the laboratory, such as those involved in **point of care testing** (POCT) using equipment in remote locations or using another laboratory on a different hospital site, or even in another hospital or organization.

Other examples where effective communication is essential include the following:

- Occasions when staff are temporarily absent and their duties must be covered by others
- The reporting of failures of **quality control procedures**
- Clinical samples that must be processed immediately because test results are required urgently
- Stock control where there is a danger of a particular reagent running out before a new delivery is expected

All these factors may critically affect the routine functioning of the laboratory and will almost certainly require attention from more than one member of staff.

It is also essential to communicate any updated or changed information within the laboratory team, no matter how trivial it may appear. Thus, the introduction of a new method may require a change in the stated reference range of a particular analyte, and this, of course, has implications for the final result appearing to be within normal limits. A change in the batch of a particular reagent on a large analyser may require its recalibration and a re-evaluation of the performance of the method. Even something as apparently simple as a broken door may result in a fire exit becoming unusable and require the selection of an alternative route.

SELF-CHECK 3.1

Why is it necessary to write down any changes in the consignment lot numbers of reagents used in laboratory procedures?

Procedures are often changed by senior members of staff who are not actually involved in the performance of the daily tasks. In practice, often new or inexperienced members of staff are the persons to point out a flaw in a change because they are the ones who can most easily see possible consequences of any changes. However, it is also possible that junior members of staff may feel that particular tasks have no real benefit, and are irrelevant and so stop doing them. This may have adverse effects further along the laboratory chain. It is essential that any changes to SOPs are discussed fully by the whole team and only implemented when there is a written instruction to do so.

Cross reference
Chapter 19 'Quality assurance and management'.

Communication within the laboratory also includes completing critical paperwork, such as **quality control charts**, **reagent books**, and **troubleshooting sheets** in the case of an instrument failure. If a staff member carries out maintenance or runs successful quality control, and does not record it in the correct manner in accordance with recognized laboratory procedure, other members of staff will not know that the tasks have already been carried out.

Communications between clinical laboratory staff and service users

Service users may be described as any person or group of people who interact with the laboratory. This, of course, includes patients and their relatives. It also includes doctors or other healthcare professionals (both within and outside of a hospital trust), and visitors to clinical laboratories. Table 3.1 gives you an indication of possible service users; it is not exhaustive and should only be used as a guide.

Patients and their relatives form a group that are generally described as *direct* service users given the manner in which results and information generated by the laboratory affect them. They will be given information about their own health, or that of a relative or friend that may have a profound impact on their lives. They are rarely medically qualified and communication with this group must be appropriate to their level of understanding, but without being in any way patronizing. The communication must also comply with local policy, professional codes of conduct, and current legislation.

Hospital staff groups are varied and information given to them must also be appropriate to their qualifications. It also depends on their *need-to-know* as part of their own professional duties; not all hospital employees need to have access to confidential information about patients. Different departments will have their own rules regarding who needs to know certain pieces of information, and if there is any doubt regarding the level of information that can be passed between staff groups, it is better to be cautious and give too little in the first instance. More detailed information can always be given at a later date, but it is far more difficult to retract something inappropriate.

Community medicine groups are healthcare professionals who do not necessarily have direct connections to the clinical laboratory, but are *external* users of the laboratory services. They need to communicate with clinical laboratories regarding patient samples and clinical test results as part of their daily activities. Levels of information and communications between laboratories with these groups depend on local policies.

Visitors to the clinical laboratory are wide ranging, from healthcare professionals to patients. The common link in this group is that they are entering an environment with which they are unlikely to be familiar. Their levels of knowledge are equally widespread and may be very specialized. Each visitor must be treated appropriately to the needs of the visit.

TABLE 3.1 List of possible service users

Hospital staff	Community medicine groups	Visitors to the laboratory
Doctors	General practitioners	Company representatives
Nurses	Community nurses	Research scientists and doctors
Radiographers	Midwives	
Physiotherapists	Surgeries	Instrument engineers
Occupational therapists	Administrative staff	Formal inspectors
Cardiographers		Patients
Pharmacists		
Dieticians		
Midwives		
Phlebotomists		
Porters		
Domestic staff		
Maintenance staff		
Ward support staff		
Secretarial staff		

It is often difficult for biomedical scientists to build relationships with user groups that are not based on preconceived ideas. Biomedical scientists are often disadvantaged in a hospital structure because there are largely unseen, and so is the environment in which they work. Often they will be judged by an unseen, audience responding to telephone conversations and **reports**. However, as biomedical scientists, we must also realize that the majority of the work done during the course of a normal day passes in and out of the laboratory without incident, and is processed in a timely manner in accordance with the expectations of the service users. There is no need for a service user to have any direct contact with laboratory staff. It is only when there are deviations from the normal process, such as when a patient's condition becomes critical and a result is needed urgently, a sample is lost and a result not available as expected, or there is a major failure of a piece of laboratory equipment and test results are therefore not available in the usual timeframe. Such occurrences may prompt a user to contact the laboratory staff. It can be that service users who need to communicate with the laboratory are under more pressure or stress than usual, and may be abrupt or rude. However, biomedical scientists should remain professional and act appropriately at all times.

Key Points

It is in everyone's interest that good working relationships are fostered with all service users.

3.2 Communication and confidentiality

In direct face-to-face contact or over the telephone, laboratory staff must quickly establish exactly who they are communicating with. It is fundamental for a biomedical scientist to know to whom they are speaking before releasing information or patient details that could be considered confidential or sensitive. Confidential information must only be given to certain groups of users. For example, it is usually inappropriate for a biomedical scientist to discuss the results of clinical tests or give such results to a patient, a patient's relative, or members of the public unless there are specific circumstances within written guidelines. Such behaviour, even inadvertent, would be in breach of the standards laid down by the HPC, and could lead to both disciplinary action by the employer and investigation of the action by the HPC, and that could then result in the biomedical scientist being removed from the professional register.

Cross reference

You can read more about the HPC in Chapter 2 'Fitness to practise'.

Key Points

In some cases patients may need to know the result of their test, including glucose measurements in diabetes or when monitoring anticoagulation.

Cross reference

You can read more about point of care testing in Chapter 18, and *Haematology* in the series companion book.

These considerations may also apply in some circumstances to commercially sensitive information. For example, discussions about pricing structures or costing regarding laboratory equipment with competitor companies as this may be sensitive information if a major purchase is about to go out to tender.

3.3 Methods of communication

The major form of communication between laboratory staff and most of the user groups listed in Table 3.1 is verbal, but this is not exclusive as written material is still exchanged. Verbal communication may be one-to-one, an example being telephoning a doctor with a set of test results concerning a particular patient. It may also be used to inform a large group of a change in policy or procedure. The introduction of a new method may be presented to a large group of hospital staff in the form of a case study to demonstrate why the new method is preferable over the old. Written communications are used to transfer both clinical test results and messages, and may be in paper or electronic forms. Whether meant for a single recipient or for a wide audience, the message must be clear and unambiguous, and presented at a level suitable for the intended recipient.

The most important interface for the transfer of results through a hospital is the nurses' station (Figure 3.2) on a ward or clinic.

Telephone communications

Telephone communications are often the means by which a laboratory is judged by its users. While few people have face-to-face contact with laboratory staff, a relatively large number use the telephone as the primary means of contacting laboratory staff. Their reasons for telephoning may include:

FIGURE 3.2

A typical nursing station on a children's ward. This normally extremely busy area is the major interface between patients and the clinical laboratory. Photograph courtesy of Central Manchester University Hospitals NHS Foundation Trust, UK.

- Incoming **requests**
- Incoming questions that can be answered in general terms
- Incoming requests that need to be answered by a specific person
- Incoming calls for a named person
- Outgoing results
- Outgoing information of a general nature
- Specific information to a named individual in response to an outstanding request

Individual laboratories will each have their own telephone policies, although all such policies should conform to certain basic principles. All workers must be familiar with the telephone policy and associated administration of their own department and strictly adhere to it. Figures 3.3 (a and b) illustrate the correct and incorrect manner in which to deal appropriately with a telephone enquiry.

Generally the following basic rules apply: biomedical scientists and other laboratory staff must remember that the same rules apply to the telephone as to face-to-face and paper transfers

(a)

"Certainly, I will e-mail those test results to you immediately"

(b)

"Your call is important to us and I'll deal with your emergency tomorrow"

FIGURE 3.3
(a) Appropriate and (b) inappropriate ways of answering telephone enquiries.

of confidential and sensitive information. It is essential that the recipient of any telephone conversation be identified at the start of a call, either incoming or outgoing, so that only appropriate discussions take place. Information of a clinical nature should never be given to patients, their relatives, or members of the public. Nor should messages that relay clinical and confidential details be left on an answer phone.

General incoming requests for specific patients

All incoming requests, such as those relating to urgent work, additional clinical tests, patient demographic changes, and similar instructions must be recorded, together with the time and date of the request, the name of the requestor, and the name of the person recording the message. This is necessary so that all requests can be subsequently tracked and are traceable for audit purposes. You can read about audit in Chapter 19.

Cross reference
Chapter 19 'Quality assurance and management'.

An example of the use of retaining a paper or electronic record may be when the clinical testing of a patient's sample becomes urgent due to the deterioration in his or her condition. If a sample from a patient is already in the laboratory, this will save time. However, should the requesting clinician telephone the laboratory and the individual fails to record the details of this call, it is possible that sample urgency will not be known. This may cause an inappropriate delay in the patient's treatment. An **adverse incident** complaint may then be made by the doctor against the laboratory. If the doctor is able to provide the name of the person to whom he or she spoke, then this may be sufficient evidence that the member of staff is not following laboratory protocol.

Cross reference
See Chapter 1 'Biochemical investigations and quality control' in the accompanying textbook, *Clinical Biochemistry* for more information on laboratory management systems.

Many **laboratory information management systems** (LIMSs) have a facility that automatically records the time, date, and person entering information, as well as auditing any changes to requests. This useful facility allows a LIMS manager to see who has done what to a particular request, and will also bring changes to the attention of other staff without the need for conversation or passing paper notes. However, information held on LIMS is only as accurate as the person who inputs the data.

Incoming questions that can be answered in general terms

General enquiries, such as *What colour tube is used for a particular test?* do not usually require documenting, but it is important for the receiver of the call to realize that there is always a laboratory user at the other end of the telephone. When talking to users, the biomedical scientist must remain professional at all times, no matter how irrelevant the enquiry may seem. However, it is also important that the identity of the user be established and any information given by telephone is appropriate to the level of understanding of the user.

Incoming requests that require answering by a specific or named person

Incoming calls that request specific information are sometimes beyond the professional expertise of the person taking the call. These requests should promptly be passed to a more relevant staff member, with as much detail as possible, so that the requestor does not have to repeat themselves. It can be irritating to a service user to have to keep repeating the same information when handed from person to person on the telephone, and reflects poorly on the communication strategies within the department.

If someone receives a call asking for a named member of staff it is advisable to find out if the named individual is available to receive the call. If so, the requestor should be informed, particularly if the call needs to be transferred to another extension. Again, it is frustrating for a service user to be transferred to an ever-ringing extension with no prospect of an answer. The caller then has no other option, but to put down the telephone and start the process again when there may have been an opportunity to ask for someone else at an early stage.

Outgoing results

The telephone policy of all organizations may cover giving results or this may be guided by a separate policy. At a minimum, when telephoning a result you must identify the person taking the call. This individual should write down the result, the time, and date of the call, and a record of the person giving the result. This information should then be read back to the caller in the laboratory to ensure the correct information has been communicated and understood. This type of information may form a paper record, but is more likely to be held electronically on a LIMS. This forms part of an audit trail and, while at first glance may seem somewhat lengthy, it is necessary in case something goes wrong. It is essential to ensure that any contradictions are corrected. If there are any doubts about the abilities of the receiver to correctly transcribe and communicate the results, then the biomedical scientist should request that someone more suitable be asked to take over the call. This is essential for the safety of the patient; biomedical scientists must be able to demonstrate independent and confident practice, and not feel intimidated in challenging the competence of the receiver of the results. A cautionary example is outlined in Case study 3.1.

CASE STUDY 3.1 Transposing results

A post-operative patient had a short stay on the intensive care unit following major heart surgery before being deemed well enough to be transferred to a surgical ward. Prior to transfer, the ward clerk on intensive care telephoned the laboratory to ask for the most recent test results on the urea and electrolyte concentrations of the patient. Results were given by the support worker in accordance with the laboratory protocol and recorded by the ward clerk in accordance with the intensive care protocol. The electrolytes and creatinine were normal and the urea was 7 mmol dm^{-3} (with a reference range of 3–8). Although this value is near the top of the reference range it is not an unexpected finding after major surgery. In the afternoon, a telephoned request came into the clinical biochemistry laboratory urgently requiring repeat determinations of urea and electrolytes on the patient, following treatment of the **hyperkalaemia** (high concentration of K$^+$ in the plasma) together with measurements of blood insulin and glucose concentrations. The laboratory noticed that there had been no previous hyperkalaemia and quickly assayed the sample, to find the K$^+$ to be 1.2 mmol dm^{-3} (reference range 3.8–5.2). So, what had happened and where had communications broken down?

A check of the record held in the LIMS revealed that the correct procedures had been carried out by the support worker, and a check of the results book on intensive care revealed that the ward clerk had also followed the correct procedures, and the results matched. The nurse transferring the patient had taken the results from the record book on intensive care and written them onto a scrap of paper, which she had taken with the patient to the surgical ward. She then verbally gave the results to the nurse receiving the patient, who also wrote the results on a scrap of paper. Later this nurse transferred the results into the patient's notes. Both nurses agreed that they had written results on paper, but no record of these handwritten notes remained. The patient notes revealed that there had been a transposition of the values of urea and K$^+$ results when compared with the electronic record held in the LIMS. The consequence was that when the surgical junior doctor came to review the patient, he noticed an apparent hyperkalaemia, which he immediately treated, reducing the concentration of K$^+$ in the patient's blood.

This course of events ended with the patient receiving an infusion of K$^+$ and suffering no ill effects. However, this could easily have resulted in the tragic death of the patient and highlights the importance of following an SOP when communicating results verbally, either by telephone or face to face.

Outgoing information of a general nature

There are occasions where the laboratory has to use the telephone as a means of quickly disseminating specific or general information to its users. This use will be covered by a telephone policy but, as always, the persons receiving the information and whether it has been understood must be established. Any questions relating to the information should be answered professionally and courteously, no matter how trivial they may seem. Sometimes it is necessary to tell users that there is a major breakdown in, for example, equipment or information technology, so that the normal service is likely to be disrupted. Again, the recipient of this information may well react to it in a discourteous or unprofessional manner; however, laboratory staff must maintain professional standards of conduct, even when severely provoked to do otherwise.

Returning specific calls

On occasions, laboratory staff will be asked to return a call to a particular individual. It is essential that all such calls are returned, even if the matter seems unimportant or the reason for the call cannot be dealt with. Not responding as promised is the lack of communication that reflects badly on the laboratory service. There is little that is more frustrating to a service user than waiting for an expected piece of information that never arrives. Unfortunately, users are far more likely to remember the one negative experience, rather than the 10 positive times the laboratory was helpful.

SELF-CHECK 3.2

What is the simplest check one can make to ensure data given out over the telephone has been correctly received?

Paper reports

In our increasingly paperless society, the number of hard-copy laboratory reports is in decline. It is felt by many (and there have been many studies to show this) that the majority of paper reports never reach their final destination, usually the patient's written medical record, and is largely a waste of resources and the time needed to produce them in quantity. However, because the majority of reports are available electronically does not mean that the presentation of laboratory data as a written document, as a means of visual communication can be disregarded or dismissed as unimportant. The written report is often the professional standard that the outside world uses to judge the status of the laboratory. Poor reports imply poor mechanisms to produce the data. The production of pathology reports of appropriate quality may not be considered as a means of communication in the first instance but, for many users, the request form and the final report are the only direct interface they have with the clinical laboratory service. The report must be clear and unambiguous, and written with the correct use of English grammar and language. This is particularly important for added comments. Many laboratories now only offer coded comments or only allow certain groups of staff to add comments. While an individual may understand the comment they are writing, it is essential to ask how that comment may be viewed by the reader. Is the comment of relevance and does it add value to the report? Is it written in good English? Will the service user understand the comment? Has the correct medical terminology been used?

When information is presented as a paper report, all laboratory data should be accompanied by appropriate units, reference ranges, useful clinical information, details of the issuing laboratory, as well as the obvious information, such as complete patient demographics and specimen collection details. When such information is not complete, time and effort will be wasted by the recipient of the report in trying to find the missing data, as indicated in Box 3.1.

BOX 3.1 Details are required

As examples, a patient with sepsis is likely to have a variety of microbiological investigations. The results of a swab with no details of the site may render the clinical information available useless. Likewise, serial blood samples are often taken (a poorly controlled diabetic may have several blood glucose samples taken over a 24-hour period) and the date on a report with no indication of the time the blood sample was taken may make the serial results of little value.

Electronic data

As the generation of paper communication is decreasing, so the transfer of electronic data is increasing. The transfer of such data is the primary method of communication in many laboratories. This allows requests to enter the laboratory using ward–based requesting programs linked to pathology networks. Results can also be returned automatically to the patient electronic record and downloaded directly to general practitioner surgeries and outreach clinics. The use of websites and email to 'cascade' information is reducing the need for writing letters, telephone conversations, and the need to print information on paper.

Electronic data must be viewed as being the same as written communication. The rules applied to paper reports about, for example, including units, reference ranges, useful clinical information, details of the issuing laboratory, complete patient demographics and specimen collection details, also apply to electronic reports. Again, only standard, recognized abbreviations should be used, and any possibly ambiguous language such as 'text speak' must be avoided. The NHS and the Department of Health have standardized acronyms for medical and other frequently applied terminology. Previous references to use of LIMS in respect to coded comments and standardization apply, as in the same way a paper report may be the only user–laboratory interface, for some groups of users, the electronic copy of the report may be the only interface. The use of coded comments through a LIMS in the laboratory makes for much easier standardization.

Email is a popular method of communication. It is fast, easy to use, and an excellent mechanism for cascading information. However, there is no excuse for not writing emails in correct English, with complete spelling and no shorthand. Emails should be proofread before sending, as the standard of the English used may be the only basis a service user has to make a judgement on the overall educational standards of a member of the laboratory staff. While electronic data is a boon in today's busy society, it also has serious consequences when things go awry and, like all methods of communication, they are not 100% reliable. All systems can fail and data be lost or fall into the wrong hands. Thus, steps must always be taken to ensure that procedures are in place to support the protection and recovery of electronic data.

All electronic data must be held on a back-up system. Anyone who has lost personal work due to the failure of a personal computer has experienced the frustration and extra work that could so easily have been prevented by saving the data on a secondary data storage system! Whereas CD-ROMs and memory sticks may be useful ways for students to protect their personal educational material, this method of storage is unsuitable for patient information.

Indeed, government data was lost in two high-profile cases involving the Ministry of Defence, who admitted losing personal details of service employees, and the Child Benefit offices, which lost bank details of some claimants. Such cases involving the loss of personal information demonstrate the need for the secure storage of sensitive data pertaining to the general public. Imagine the scale of such a system failure if the data, such as that stored on a LIMS, or on a hospital-wide patient information system, was lost due to insufficient back-up mechanisms or stolen due to poor security. All large systems within the NHS have a back-up mechanism that implements some type of housekeeping and back-up of all accessed files at regular time intervals. This frequently goes unnoticed by the majority of staff, but the usual practice is that a named person has overall responsibility for the security and protection of laboratory data and sets protocols that must be followed to prevent loss of data or even to prevent data falling into the wrong hands. The NHS technology systems have carefully controlled access, with all data being password protected and tracked with audit trails. It is essential that all staff follow all local and Trust-wide policies to ensure the protection and security of electronic data.

The storage and transfer of electronic data is covered by the Data Protection and the Freedom of Information (FOI) Acts, and the consequences of both pieces of legislation require their examination in detail.

3.4 **Data protection**

The Data Protection Act (DPA) 1998 governs the use of personal information by businesses and other organizations. It came into law in March 2000, and regulates how a business or organization may store and use personal details about individuals, including their names and addresses, bank details, medical records, and opinions they may have expressed. A specific problem arising from the Act is that it only applies to living individuals. Following their death, any data stored about them need no longer be protected in the same way as when they were living. This is seen as a loophole in the existing legislation that is particularly pertinent to health records.

The DPA 1998 applies eight principles for the purpose of protection of personal information. You can see these listed in Table 3.2.

TABLE 3.2 Summary of principles listed in the Data Protection Act (DPA)

Processed fairly and lawfully
Processed for one or more specified and lawful purpose, and not further processed in any way that is incompatible with the original purpose
Adequate, relevant, and not excessive
Accurate and, where necessary, kept up to date
Kept for no longer than is necessary for the purpose for which it is being used
Processed in line with the rights of individuals
Kept secure with the appropriate technical and organizational measures taken to protect the information
Not transferred outside the European Economic Area (The European Union member states, plus Norway, Iceland, and Liechtenstein), unless there is adequate protection for the personal information being transferred

These principles ensure that personal information may only be held and used for the purpose for which it was originally intended and cannot be used for anything else subsequently. It also gives individuals the right to view the data that is held about themselves. Thus, for example, the DPA gave individuals the following legal rights:

- The right to a copy of the data
- The right to prevent processing likely to cause damage or distress
- The right to prevent processing for the purpose of direct marketing
- The right to have inaccurate data rectified, blocked, erased, or destroyed
- Rights in relation to automated decision making

In addition, the Act classifies some personal information as *sensitive*, about which there are even stricter rules. Sensitive information includes:

- Racial or ethnic origin
- Political opinions
- Religious or similar beliefs
- Trade union membership
- Physical or mental health or conditions
- Sexual orientation
- Offences or alleged offences committed
- Proceedings related to these offences or alleged offences

An organization is only allowed to use sensitive information where they are able to conform with a narrow set of conditions to ensure that such information is only used where the organization has an essential need to use it, or that the individual has given explicit consent for its use.

It is immediately clear that records held in a clinical laboratory, either in paper or electronic forms, and also any paper reports issued by the laboratory, are considered sensitive personal information for the purpose of the DPA and must be treated accordingly. Considering the scale of sensitive information held within a hospital, and indeed the NHS, it is a significant task, and to that end, in 1997, a report was commissioned by the Chief Medical Officer of England, and a committee established under the chair of Dame Fiona Caldicott. A report followed, referred to as the Caldicott Report, highlighting six key principles and making 16 specific recommendations to be applied to all data held by the NHS.

The six key principles as set out in the Caldicott Report are:

1. Justify the purpose
2. Do not use patient-identifiable information unless it is absolutely necessary
3. Use the minimum necessary patient-identifiable information
4. Access to patient-identifiable information should be on a strict need to know basis
5. Everyone with access to patient-identifiable information must be aware of their responsibilities
6. Understand and comply with the law

These principles mean that every single use or transfer of patient-identifiable information that moves within or outside of an organization, such as the NHS, should be necessary and its purpose defined. All transfers of such information must be regularly scrutinized and reviewed by an appropriate person to establish that the continued transfer really is necessary. All NHS trusts have a named Caldicott Guardian, who is usually a senior member of the medical staff with the overall responsibility for ensuring compliance within the trust.

Patient-identifiable information that is not strictly necessary should be excluded unless it would be dangerous to remove that data. Patients should only be identified where it is essential to the successful movement of that data and, even then, the minimum amount of identification should be used. Current legislation often dictates the amount and type of data needed. Once data has been collected and stored, that data must only be accessed by those people to whom the data is absolutely necessary and those with access must be sure of their own responsibilities to protect the data and how they themselves fit into the legal requirements of the Caldicott principles.

So, having described the legislation, we will now consider how the information held in pathology departments is held in relation to the Caldicott principles. First, *justify the purpose*. Clinical requests must be made on the basis of providing information for diagnosis and treatment and not for research purposes unless consent has been specifically sought. It is clearly paramount that the correct reports be identified with the correct patient, so in terms of using little or no identifying information, this must always be balanced against patient safety. It is generally accepted that patients be identified by full name, date of birth, and a unique number (hospital allocated or NHS number, Box 3.2) as minimum information for the purpose of pathology records. Problems do arise when using data transfer mechanisms, such as fax machines or electronic links, where there is a danger that personalized data may not end up in the hands of the person for whom it was intended. There are circumstances where it is necessary to code or encrypt information so that the true identity can only be recognized by the intended recipient. An exception to this general rule, where sets of results need not have a true identity, are data arising from clinical trials or research programmes; in these cases, anonymous reports are acceptable. Access to information on a need to know basis has already been discussed in this chapter in relation to use of the telephone as a method of communication (Section 3.3). All healthcare professionals and laboratory support workers have policies and protocols that outline their specific responsibilities with regard to patient information.

BOX 3.2 Use of the NHS number

The NHS number is now the identifier of choice across different NHS organizations. It is a 10 digit number that is unique to every individual who is registered with the NHS in England and Wales. The NHS is now required to use this number on all formal patient documentation. The NHS number is issued to all babies born in England and Wales, and will be given to anyone not born in those countries that is eligible to register with a general practitioner and use NHS facilities.

SELF-CHECK 3.3

Why is the NHS number now the recognized means of identifying an individual?

Freedom of Information Act

The Freedom of Information Act 2000 (FOI) came into force in January 2005. The Act gives a general right of access to all types of recorded information held by public authorities and places obligations on them to disclose information (subject to a range of exemptions); indeed, the information applied for may be retrospective in nature. The FOI gives applicants two related rights:

1. The right to be told whether the information exists
2. The right to receive the information within 20 working days, where possible in the manner requested

Thus, a public sector organization has a statutory duty to implement the Act and regulates how an organization, such as the NHS, must respond to a request for information that it may have stored. The Act applies to everyone within the organization of a Public Authority (inclusive of all aspects of the NHS) and has a huge impact on the public sector. The purpose of the FOI Act is to contribute to breaking down barriers within the NHS and help organizations learn from each other, as well as improving on their own performance. There are specific exceptions to the data that can be released that need to be discussed.

It does appear at first glance that the DPA and the FOI Act are at odds with one another, but really they are complementary in nature. Under the DPA, individuals have a right to access information about themselves, but third parties do not have rights to personal information. Under the FOI Act, individuals are not able to access to any personal information, either about themselves or others, but are able to access most other types of information. The main strategic objectives of the combined Acts are listed, and are not specific to the NHS or the hospital environment.

- To improve people's knowledge and understanding of their rights and responsibilities
- Seek to encourage an increase in openness in the public sector
- Monitor the Code of Practice on Access to Government Information
- Develop a data protection policy, which properly balances personal information privacy with the need for public and private organizations to process personal information.

Both the DPA and FOI Act are the responsibility of the Lord Chancellor's Department at Government level, namely the Information Commissioner.

Informed consent

The 2000 NHS Plan pledged that proper consent would be sought from all NHS patients and research subjects. The plan promised a review of consent procedures and the establishment of good practice in seeking consent throughout the NHS. The recommendations are that before a doctor or other healthcare professional treats a patient, they need to obtain their consent. The consent may be verbal, but it is better in writing, and outlines the key points and decisions of the procedure. A patient has the right to withdraw his or her consent at any time. The healthcare professional seeking consent has an obligation to ensure that the patient knows enough to be able to make an informed decision, and has had the opportunity to ask questions and consider alternative treatments and procedures. Patients must be told in advance if their treatment is likely to involve the removal of part of their body, and also if invasive sampling, such as taking blood samples is likely to be part of their care plan. Sample taking during operations may routinely be used for teaching, research, or public health monitoring and, if this is the case, the patient must be made aware of the possible use of such samples and their consent obtained. If prior consent has been provided in writing, particularly in the case of the retention and storage of histological samples, there is no longer ambiguity on the outcome of a given series of events. Likewise, regarding consent to taking blood samples. It is readily assumed that if a patient has a request form and has allowed the phlebotomist to take a blood sample, they have given consent for those tests marked on the form, but to no others. It is no longer considered appropriate to use excess samples for research purposes or for additional tests to be added without discussion with the requesting clinician and in some circumstances, with the patient.

Human Tissue Acts

The Human Tissue Act (HTA) 2004 (England, Wales, and Northern Ireland) or HTA and the Human Tissue (Scotland) Act 2006 have replaced a variety of earlier legislation, including The HTA 1961, the Anatomy Act 1984, and the Human Organ Transplant Act 1989. All consent for the use of pathological samples for prolonged storage and/or research activities must be obtained in writing, and the paperwork be sufficiently detailed to establish that the patient is fully aware of the nature of the consent sought. The consent must be given voluntarily, be accompanied by sufficient information, be made by a legally competent individual, and be given with no form of coercion. Research must have the dignity, rights, safety, and well-being of the participant as the primary considerations. All studies must have appropriate written, informed consent, with clear arrangements about how that consent should be obtained. It is for the protection of the researcher that the consent record shows that the process has been two way, with the participant having the opportunity to ask questions and receive answers and, even better, if a friend or relative of the participant was also present at the interview and allowed to participate in the questioning. It needs to be understood that informed consent is a process and not simply a signature on a form, and that the informed consent form indicates that there has been adequate communication between the healthcare professional and the participant or patient. It is increasingly common that, rather than signing consent to a list of inclusive events, patients are asked to sign a general disclaimer; thus, specific consent is not required at a later date.

The HTA was passed by Parliament in the wake of public response to the activities at the Bristol Royal Infirmary and Alder Hey Hospital, which are discussed further in Chapter 19, where specific consent to remove, store, and carry out research on tissues retrieved at post-mortem from children having undergone unsuccessful major surgery had not been requested. It was clear at that time, that many of the public had little grasp of post-mortem procedures, nor the fairly common practices of storing tissues for research and teaching purposes. The HTA has made the processes of post-mortem and retrieval of tissue much more transparent. Although Scotland has its own Act, the content of both Acts are expressions of the same principles, such that the HTA Act:

Cross reference
Chapter 19 'Quality assurance and management'.

- Regulates removal, storage, and use of human tissues, which are defined as material that has come from a human body and consists of, or includes, human cells. Now the HT Act is in force, it is unlawful to carry out these licensable activities without an appropriate licence

- Created a new offence of deoxyribose nucleic acid (DNA) 'theft', which is having human tissue with the intention of analysing its DNA without the consent of the person from whom the tissue came, from 1 September 2006

- Made it lawful to take minimum steps to preserve the organs of a deceased person for use in transplantation, while steps are taken to determine the wishes of the deceased or, in the absence of their known wishes, obtain consent from someone in an appropriate relationship

- Give specified museums in England discretionary power to move human remains out of their collections, if the remains are reasonably believed to be those of a person who died less than 1000 years ago

These acts also have a list of enforceable offences, with penalties ranging from fines to imprisonment or exceptionally both. These offences include:

- Removing, storing, or using human tissue for Scheduled Purposes without appropriate consent
- Storing or using human tissue donated for a *scheduled* purpose for something else
- Trafficking human tissue for transplantation purposes
- Carrying out licensable activities without holding a licence from the HTA (with lower penalties for related lesser offences such as failing to produce records or obstructing the HTA in carrying out its legal responsibilities)
- Storing or using human tissue, including hair, nail, and gametes with the intention of its DNA being analysed without the consent of the person from whom the tissue came or of those close to them if they have died

The scheduled purposes mentioned in the above list is a general term used to cover the description given to the patient or relative of the procedures and purposes for which that tissue was collected. Scheduled purposes differ between tissue samples and also between patients; they include whether the tissue is to be used only for diagnostic purposes or also for research, whether the tissue can be used by a single institution or be shared, and if it can be stored after use. If is it to be stored, then the scheduled purpose will include details of the storage including its site and duration.

Clearly, the HTA and the need for informed consent go hand-in-hand. The written communication between healthcare professional and patient/participant acts as a legal document to provide the necessary proof that informed consent has been properly sought and agreed with regard to blood, fluid, and tissue samples requiring investigation, processing, and storage within laboratory medicine. This is particularly important in histopathology when acquired tissue from surgery or post-mortem (Figure 3.4) may be further used for teaching or

FIGURE 3.4
Interior of a modern mortuary. The removal of any tissues must comply with the Human Tissue Act. Photograph courtesy of the London Borough of Haringey, UK.

research purposes. Tissue must only be used and stored in the manner for which prior consent has been obtained.

3.5 Record keeping

While record keeping has been mentioned briefly as a method of communication within the laboratory (Section 3.1) and is also an HPC standard, it deserves consideration in its own right within the remit of the Data Protection and the FOI Acts. It has already been mentioned that communication within the laboratory also includes completion of documents such as quality control charts, reagent log books, and troubleshooting sheets in the case of an instrument failure. The primary purpose of this type of documentation is so all laboratory staff can assess situations at a glance, whether certain basic procedures have been followed, whether a particular instrument is functioning correctly, or whether the availability of certain stock is below a certain level without having to find and question a specific member of staff. Record keeping in the laboratory is not only limited to practical procedures carried out on benches, but also covers complete patient and staff records.

> ### Key Points
>
> When completing patient records, clearly it is essential that a report correctly identifies the patient but it is also necessary that an historical record be kept of all the laboratory investigations pertaining to a patient.

When accessing a new set of results on a LIMS for authorization, there is frequently a so-called **delta check** performed, where the new result is compared with the previous existing record. If there is a significant change, the result may not be automatically passed for authorization and release directly by the LIMS, but may be held in a queue and require manual validation or authorization by a biomedical scientist who can investigate further (see Case study 3.2).

Another area of laboratory medicine where it is essential that full historical records be maintained is within the discipline of blood transfusion. If a patient develops an antibody to a particular red cell antigen, there is a possibility that the titre of that antibody may change with time. Examples of these are the antibodies of the Kidd system, such as anti-Jka. Not all antibodies to erythrocyte antigens remain detectable in the serum at a consistent titre and some may decrease with time. A patient may have presented several years previously with a positive

CASE STUDY 3.2 Identifying samples is essential

A patient with a blood haemoglobin of 140 g dm^{-3} (reference range 140–180) 24 hours later has a haemoglobin concentration of only 75 g dm^{-3}, and therefore requires a blood transfusion. If the patient has undergone major surgery, such as requiring a prolonged period of time on a cardiac bypass machine, or has just had a massive gastrointestinal bleed, the discrepancy between the two results is both explained and appropriate. If there has been no such clinical event to explain the change, then it is unlikely that the samples have originated from the same patient and there has been an error in correctly identifying the sample.

Cross reference

You can read more about immunology in Chapter 15 'Immunological techniques'.

antibody screen and such an antibody identified. When returning in the present day for major surgery and possible transfusion, the antibody screen may be negative, as titres are so low as to be undetected by routine screening methods. If that patient required a blood transfusion, it would still be necessary that Jka negative blood be used since the introduction of the antigen would promote a massive response in terms of antibody production, and the patient would be likely to suffer an incompatible transfusion reaction that could possibly be fatal.

Staff records

Staff records must also be kept by departmental managers. Records must exist for human resource purposes, such as sickness and absence rates, the availability of staff on different parts of shift systems, names of staff on professional registers, and details of mandatory training, competence assessment, and continuous personal development, that may be needed to fulfil professional responsibilities.

External bodies

Record keeping in the clinical laboratory also has to satisfy the many external regulatory bodies that have a bearing on current laboratory practices. There are many legal requirements to be met, as well as guidelines for good practice. Such bodies include Clinical Pathology Accreditation (UK) Ltd (CPA), Medicines and Healthcare products Regulatory Agency (MHRA), the Health and Safety Executive (HSE), and the NHS Cervical Screening Programme. We will examine briefly the role of each in this respect.

Clinical Pathology Accreditation (UK) Ltd

The CPA is a peer reviewed organization that provides an external review of clinical pathology services and external quality assessment schemes. Currently, CPA accreditation is desirable, rather than mandatory, but this situation may change in the future. CPA assessors visit laboratories and examine services to confirm that defined standards of practice are being met in all areas of laboratory medicine. Thus, all laboratory records must comply with the most up-to-date and relevant of the CPA Standards to meet the requirements of the inspectors.

Medicines and Healthcare products Regulatory Agency

The MHRA is an agency of the Department of Health and thus a government organization, that was established to protect the health and safety of the public in relation to the dispensation of medicines and use of medical devices. The activities of the MHRA are particularly relevant in the field of blood transfusion, as blood products are classed as pharmaceuticals and administered to a patient following suitable prescription by a qualified person. The MHRA assesses against several European Union directives (the majority of which are now part of UK law) that have set standards of practice in blood transfusion, and which apply to blood establishments and hospital blood banks. Inspectors for the MHRA require evidence that these directives are being met for quality systems of working, the use of equipment and consumables, and that there are suitable traceability and record-keeping requirements for blood banks and facilities that may receive blood products for transfusion.

Health and Safety Executive

The HSE protects people against risks to health or safety arising out of work activities. Its role is discussed more fully in the following chapter. The HSE is a government body, which regulates all employers to ensure that all aspects of health and safety law are met in full.

The HSE inspectors have a right to visit workplaces, and inspect practices and records to ensure that all relevant UK and European Union legislation is being adhered to in full. The HSE has the power, in extreme circumstances, to close down and prosecute any employer found to be in breach of these regulations, particularly if previous warnings have been issued. All records concerning health and safety, which constitute UK and European Union law, must be up to date and available for health and safety inspectors to access at any time.

Cross reference
Chapter 4 'Health and safety'.

SUMMARY

- Effective communication both within a clinical laboratory, and between its staff and a variety of external people and organizations is necessary. This is because biomedical scientists work as part of a laboratory team, must follow local policy and procedures and comply with legislation, and are involved in a process that provides essential information contributing to patient care

- Biomedical scientists are healthcare professionals, and must never act in a manner that causes offence to others or to bring the biomedical science profession into disrepute by inappropriate communications of any means

- Biomedical scientists and other laboratory staff are required to work in structured teams, following set procedures and protocols, and must co-operate and communicate appropriately with others within the laboratory environment

- Biomedical scientists must communicate in a professional manner and at an appropriate level when dealing with the public, patients, and the wider healthcare groups of professionals and support workers

- Biomedical scientists may communicate by non-verbal or verbal means, and verbal communication may be written or spoken, and indirect communication may be the only way that a biomedical scientist is judged by the wider healthcare community

- Biomedical scientists must conform to current legislation, such as the Data Protection, Freedom of Information, and the Human Tissue Acts, and be aware of the consequences of their actions if inappropriate records are kept and audit trails are not maintained

- Biomedical scientists must ensure that all work carried out within laboratory medicine has appropriate consent and that confidentiality is always maintained

FURTHER READING

- **Divan A. *Communication Skills for the Biosciences*. Oxford University Press, Oxford, 2009.**
  Useful guide on how to effectively communicate scientific knowledge.

- **Pitt SJ, Cunningham JM. *An Introduction to Biomedical Science in Professional and Clinical Practice*. Wiley-Blackwell, Chichester, 2009.**
  Chapter 3. 'Communication for biomedical scientists' is worth browsing through.

- **Standards of Conduct, Performance and Ethics. HPC, London, 2008.**
- **Standards of Proficiency – Biomedical Scientists. HPC, London, 2007.**

Useful Websites

- **www.ibms.org**
- **www.hpc-uk.org**
- **www.cpa-uk.co.uk**
- **www.mhra.gov.uk**
- **www.hse.gov.uk**
- **www.hta.gov.uk**
- **www.businesslinks.gov.uk**
- **www.nhsprofessionals.nhs.uk**

 QUESTIONS

1. Guidelines issued by the Institute of Biomedical Science concerning the communication of results by telephone:

    (a)   Are a legal requirement
    (b)   Can be adapted for individual situations
    (c)   Are an example of best practice
    (d)   Must be adhered to by all members of the Institute of Biomedical Science
    (e)   Are only of relevance to NHS employees

2. Under the Data Protection Act, sensitive information is defined as:

    (a)   Information that relates to the age of an individual
    (b)   Data that cannot be sold on to another party
    (c)   Information that can be released to a third party under the Freedom of Information Act
    (d)   Information that can only be held with explicit consent
    (e)   Information that must be removed from a database after 5 years

3. The key principles of the Caldicott report do not include:

    (a)   The appointment of a Caldicott Guardian
    (b)   Information held outside of the NHS
    (c)   Instructions stating which personnel should have access to medical records held by the NHS
    (d)   Instructions informing NHS staff of their legal obligations under the Data Protection Act
    (e)   Instructions that all staff must comply with the law

4. Indicate which of the following statements are **TRUE**:

    (a)   The NHS number is given to all babies born in the UK
    (b)   Telephone calls to the laboratory requesting tests to be added to a previous sample must be documented

(c)   Acronyms for laboratory tests may be used freely, providing the laboratory responds with a published list that explains what they mean

(d)   The Data Protection Act applies only to those members of NHS staff who are named on a professional register

(e)   Electronic data on a screen must be viewed as a written method of communication

5.   The text refers to point of care blood glucose concentrations at a diabetic clinic and international normalized ratio (INR) results at a coagulation clinic as examples where it may be appropriate to give results to patients. List other occasions where results may be given directly to the patient.

6.   With reference to stored electronic data, what are the potential consequences for an individual if medical data are lost?

Answers to self-check and end of chapter questions are available in the Online Resource Centre accompanying this book.

 Go to www.oxfordtextbooks.co.uk/orc/glencross

4

Health and safety

Alison Taylor

Learning objectives

After studying this chapter, you should be able to:

- Identify hazards that represent a risk to yourself or to the safety of others
- List your statutory duties under the Health and Safety at Work Act
- Discuss the regulations that drive health and safety practices in the workplace
- Conduct a basic risk assessment and be able to outline more specialized assessments
- Use personal protective equipment safely and effectively, and be aware of their limitations and maintenance requirements

Introduction

Occupational health and safety are any system or procedure established to ensure the health, safety, and welfare of every person coming into a work place. **Health** is preventing illness occurring to the body or mind, and **safety** is the prevention of physical injury. The term **welfare** describes facilities, such as sanitation, heating, lighting, and changing rooms, which are not direct causes of ill health or injury, although deficiencies in these areas could and almost certainly would, contribute to poor health and safety.

The law defines standards of health and safety (Health and Safety at Work Act, Section 4.3) and, once identified any **hazards** or deficiencies must be corrected. In health and safety practice, a key way to identify deficiencies is by **workplace inspection**, which is a specialized audit, and by performing an assessment of the risk (Section 4.4). Risks are the combined likelihood of a hazardous event occurring, and the severity of injury or ill health that it could cause. Any deficiencies or hazards identified by either route must have corrective actions or control measures implemented to ensure they do not cause ill health or injury. The effectiveness of these controls needs to be assessed when the inspections and/or risk assessments are repeated or reviewed. The regulatory body for health and safely legislation is the Health and Safety Executive (HSE), which carries out these types of inspections and publishes guidance to help employers and employees comply with the law (Figure 4.1a). The HSE also produce reference materials and institute advertising campaigns to raise awareness of health and safety risks. Within the UK, the Institute of Occupational Safety and Health (IOSH) is the recognized

FIGURE 4.1
Logos of (a) Health and Safety Executive (HSE) and (b) Institute of Occupational Safety and Health (IOSH). Logos reproduced with kind permission from the respective organizations.

body for health and safety professionals, providing training, guidance, and information for professionals in this field (Figure 4.1b).

This chapter will examine the legal requirements and regulations governing occupational health and safety, including the methods used to assess and control risks. In combination, these measures attempt to ensure the health and safety of everybody in a workplace. Naturally, however, we will focus on the risks and controls particular to biomedical science laboratories.

4.1 **Hazards**

A hazard is any substance, activity, or process that may cause harm. On a daily basis, you may encounter many potential hazards (for example, crossing a busy road). However, by using subconscious risk assessments and applying actions to minimize the dangers of the hazard, you will (we hope!) avoid injury and ill heath. Unfortunately, in working environments, hazards, especially those found in clinical laboratories, may not be easily identifiable. Thus, it is difficult to take actions to minimize their effects. Training, labelling of substances, using standard operating procedures, and acting on all the relevant information, such as risk assessments, will assist in identifying hazards related to each task undertaken.

In this section we will identify common laboratory hazards and discuss the process of conducting formal risk assessments to assess these hazards, the risks they pose, and the controls required to reduce or eliminate them.

Hazards may be classified into six main types:

1. Biological
2. Physical, mechanical, or equipment
3. Chemical
4. Electrical
5. Fire
6. Environmental

Biological hazards

Biological hazards take five main forms:

- Bacteria
- Viruses

- Fungi
- Moulds
- Prions.

Cross reference
You can read more about the roles of biomedical scientists in Chapter 1 'Biomedical science and biomedical scientists'.

Bacteria are single-celled prokaryotic microorganisms that can live independently or dependent on host organisms. Examples of bacteria include the causative organisms of tuberculosis and legionella. Viruses are nucleoprotein complexes, which are about 200 times smaller than bacterial cells. All viruses are obligate parasites in that they can only grow or replicate within a host cell. Examples of viruses include agents that cause influenza and acquired immune deficiency syndrome (AIDS). Fungi are single-celled or multicellular eukaryotic organisms, which have cell walls rich in chitin. They reproduce by producing spores. Fungi cause skin conditions such as athlete's foot, some forms of eczema, or they may cause allergic reactions. Moulds are microscopic fungi, which can have similar effects on health to the above, but can also lead to respiratory sensitization because their small size allows them to evade filtration by the respiratory system. Thus, they can cause or contribute to conditions such as asthma. Prions are protein molecules that adopt toxic conformations and cause infectious diseases such as scrapie, bovine spongiform encephalopathy (BSE), and Creutzfeldt–Jakob disease (CJD).

The Advisory Committee on Dangerous Pathogens (ACDP) classifies all biological hazards into one of four categories or groups (1 to 4). These are published by the **Health and Safety Commission** (HSC). The group assigned to an agent defines the level of containment required to control the risks associated with it, which are defined in **Control of Substances Hazardous to Health** (COSHH, Section 4.5) regulations.

Group 1 hazards are biological agents that are unlikely to cause human disease. In contrast, hazard group 2 are those agents that can cause human diseases. However, effective treatments and/or prophylaxes are available to combat these agents and so this group is unlikely to spread in the community. Biological agents of hazard group 3 can cause severe human disease and present a high risk to health. There is a risk of the disease spreading in the community but there are usually effective prophylaxis or treatments available. Hazard group 4 agents can cause severe human disease and present a serious risk. There is also a likelihood of the disease spreading in the community because usually effective prophylaxis or treatments are not available.

Physical, mechanical, and equipment hazards

Physical hazards or those associated with mechanical devices or other equipment include: entanglement, entrapment, being struck by or striking an object (impact injuries), abrasion, cutting, and shearing. These hazards most often cause injury but in some cases lead to ill health. Slips/trips/falls also come into this category and are common causes of occupational injuries. Health problems associated with extensive computer or display screen equipment (DSE) work or repetitive muscular actions are included in this group.

Chemical hazards

Chemical hazards can be present in a number of physical forms; these include solids, dusts, gases, **vapours**, mists, fumes, and liquids. The severity of ill health effects they cause can vary considerably.

Chemicals agents can enter the body in a variety of ways, dependent on their physical form. For example, an **irritant** chemical in *solid* pellet form is unlikely to enter the body but on contact with the skin, can lead to relatively minor health effects. Hence, its effects are easily controlled by wearing safety gloves (Section 4.7). However, in powder form the same chemical could enter the body by inhalation and have significant effects, such as shortness of breath and asthma. The control of this greater hazard may possibly require the use of a ventilation hood. **Dusts** are categorized separately as they have significant specific risks dependent on the size of the particle. A flammable substance in the form of a fine dust will mix easily with air and can become explosive. Fine dusts, otherwise known as respirable dusts, present a significant hazard because of the ease with which they are inhaled.

Liquids exist in different forms depending on the ambient temperature and their boiling points. Liquids are less predictable and more difficult to handle than solids. Splashes and spills increase the chance of contact with the skin or entry into the body by ingestion or through the eyes. Gases and vapours are produced by heating liquids or when working in higher temperatures. They can enter the body through the skin, gastrointestinal tract, and eyes, or by inhalation. Spraying processes can form mists, even at low temperatures, and this increases the chances of the liquid causing ill health. **Fumes** are mists or vapours containing very small metallic particles. In a laboratory, a fume could be formed when the bulb in a mercury microscope 'blows'.

The different types of chemical hazards and their classification are discussed under the Control of Substances Hazardous to Health (COSHH) in Section 4.5. Some chemicals are also radiation hazards and so are classified differently; their use is governed by the Radiation Protection Act.

> **Cross reference**
> Safety issues concerning radiation are discussed in Chapter 10 'Radioactivity and radiation'.

Electrical hazards

Any electrical equipment, either fixed position or portable, has the potential to cause an electric shock. This can be particularly so in a clinical laboratory when electricity and water are often used in close proximity. They can pose electrical hazards, with consequences ranging from minor to serious injury or even death. Portable appliance testing (PAT) must be carried out regularly to check the safety of the instrument, but there must also be a more frequent inspection of plugs, cables, and casing. Any potential hazards arising from the use of electrical equipment in close proximity to chemicals need assessing and controlling.

Fire hazards

Fires can only start if all parts of the fire triangle, oxygen, fuel, and a heat source, are present (Section 4.6). Therefore, flammable substances, including liquids and gases, or substances or equipment giving off heat can all fall into this group of hazards.

Environmental hazards

Environmental hazards include loud noises, vibrations, light, excessive humidity, and/or temperatures, and a lack of ventilation. Sometimes these are the most difficult hazards to identify, as they are often experienced daily and so become accepted or tolerated. These hazards often cause welfare-related issues. In extreme situations, or when particularly vulnerable people are affected, they can result in ill health.

SELF-CHECK 4.1

List at least two examples of each of the major types of hazard discussed above.

4.2 **Routes of entry into the body by chemical and biological agents**

Chemical and biological hazards can only cause harm if they enter the body. Controlling these hazards therefore requires knowledge of their possible routes of entry.

Airborne biological and chemical agents can gain entry into the lungs following inhalation through the mouth or nose. Here, they can be absorbed and then transported around the body in the bloodstream. Thus, they can potentially affect every organ in the body. The skin prevents the absorption of many agents into the bloodstream. However, some chemicals can enter the body this way through pores on the surface of the skin. More commonly, absorption occurs through wounds, burns, or breaks in the skin in, for example, conditions like eczema. A number of biological and chemical agents can enter the body through the gastrointestinal tract following their ingestion, and potentially gain access to the bloodstream. This is not a common route of entry into the body, but may occur if good hand hygiene and basic laboratory health and safely practices are not followed. Finally, some agents can enter the body by injection due to accidents with hypodermic needles and compressed air lines, which can puncture the skin giving direct access into the bloodstream.

SELF-CHECK 4.2

To what types of hazards would a person collecting a venous blood sample be potentially exposed? What are the possible routes of entry of the hazard(s) into the body?

There are many regulations governing health and safety; however, in this chapter we will focus on the principal ones most likely to affect you while working in a biomedical laboratory within the UK. Health and safety across Europe is governed by the European Union (EU) through directives, which are incorporated into the laws of the member states.

4.3 **Statutory framework for health and safety**

Prior to 1974, health and safety legislation in the UK was only targeted at specific industries, leaving many workers unprotected. Other people, such as contractors and visitors entering a work place were simply not protected in law. The introduction of new regulations was too slow to reflect changing practices leading to a system that was reactive, rather than proactive. However, the legislative framework changed with the Health and Safety at Work Act.

Health and Safety at Work Act 1974

The Health and Safety at Work Act (HASAWA) resulted from a review of occupational health and safety in 1970 by Lord Robens. The HASAWA is an enabling act, which means the Secretary of State can pass new regulations without the lengthy process of passing them into law by Acts of Parliament. Thus, the process of developing health and safety legislation became more proactive.

The HASAWA established the HSC whose duties are to draw up new regulations and to enforce them through the actions of the HSE. In April 2008, the two bodies merged under the name, HSE.

The HSE is responsible for issuing **Approved Codes of Practice** (ACOP) for most regulations. The role of ACOPs is to explain the regulations and the detailed requirements necessary to comply with them. The HSE produces two types of **health and safety guidance** (HSG) **documents**, legal and best practice. The legal guidance series of booklets covers the technical aspects of the regulations and usually includes the regulations and ACOP. Best practice guidance is published in the HSG series for specific areas; an example being the Decontamination of Equipment Prior to Inspection, Service and Repair HSG(93)26. Both ACPOs and HSGs are good sources of information and are readily available on the HSC website, which is listed at the end of the chapter.

The HSE has teams of inspectors to cover all types of workplaces, including hospitals and laboratories, assisted by local authority inspectors covering premises such as hotels, shops, and restaurants. They can inspect at any time and have similar powers to the police in being able to collect evidence, and take and retain samples and photographs. One of two levels of enforcement notification is issued if deficiencies or dangerous practices are identified. An **improvement notice** follows contraventions of a regulation, together with a set date for its resolution. Alternatively, a **prohibition notice** halts the activity until a satisfactory resolution of the problem is established.

Should the HSE prosecute a health and safety offence successfully, the penalties imposed range from fines of up to £5000 for employees and £20,000 for employers, if the case is tried in a magistrates' court. Failure to comply with an enforcement notice can result in imprisonment for up to 6 months. In the crown courts, fines are unlimited with imprisonment for up to 2 years for the same offence.

SELF-CHECK 4.3

List the role(s) of the HSC.

Duties of employers

The HASAWA places a duty on employers 'to ensure, so far as reasonably practicable, the health, safety and welfare of all employees'. It also details what that may entail; such as providing a safe place in which to work and ensuring all procedures associated with work are safe to perform. In addition, the act sets out the need for set policies with associated processes for organizing arrangements such as risk assessments and consultation with employees. The act protects visitors to a workplace, outside contractors, and people living near it or passers-by, by placing a duty of care on the employers to protect all those affected by its activities. Duties are also placed on the employer to ensure all materials, including equipment and substances, are without risk to health. The Act also affects the practice of suppliers by placing on them a duty to design, manufacture, import, and supply equipment or substances that are, again, as far as is reasonably practicable, without risk to health.

Duties of employees

The HASAWA places two main duties on employees:

1. To take reasonable care for the health and safety of yourself and others affected by your acts or omissions

2. As an employee, you must co-operate with your employer to enable them to fulfil their legal obligations

A further section, often known as the 'horseplay section', impresses a duty on everyone not to interfere or misuse safety equipment. The act also states that employers must provide staff with all **personal protective equipment** (PPE, Section 4.7) required to fulfil their work duties free of charge.

The 'six pack'

The UK is part of the EU and European law, passed by the European Commission, is therefore an integral part of our legal system. The resulting uniform approach to health and safety in the workplace across the EU is designed in part to remove barriers to trade between its member states. The first effect of the EU on UK health and safety law occurred in 1992, when six sets of regulations were introduced:

1. Management of Health and Safety at Work Regulations
2. Workplace (Health, Safety and Welfare) Regulations
3. Health and Safety (Display Screen Equipment) Regulations
4. Personal Protective Equipment at Work Regulations
5. Manual Handling Operations
6. Provision and Use of Work Equipment Regulations

These regulations are contained in a set of documents familiarly called the 'six pack', which you can see in Figure 4.2.

FIGURE 4.2
Documents contained in the 'six pack'; the six sets of regulations that originated in the European Union and which became part of United Kingdom health and safety legislation in 1992. © Crown copyright.

Management of Health and Safety at Work Regulations

The Management of Health and Safety at Work Regulations (known colloquially as the 'management regs') were updated in 1999. They provide the framework for the risk assessment process that we will discuss in the following section. None of these regulations contradicts those of the HASAWA but, in general, expands on them and redefines some of the duties of both employer and employee. Thus, the employer must undertake suitable and sufficient risk assessment, and establish health and safety arrangements that include planning, control, monitoring, and review of all work-associated procedures. To achieve effective health and safety management, the employer should use competent specialists to advise them on how to comply with their legal duties.

Employers must ensure that all health and safety information, particularly findings of risk assessments are communicated to all employees. They must also provide specific health and safety training. Employers must put in place emergency procedures and ensure all workers are aware of them. The employer must also co-operate on health and safety matters with other employers sharing a workplace.

With regard to employees, Management of Health and Safety at Work Regulations require that they follow their training and instructions whenever they use equipment or handle substances. They must also report anything that may be dangerous, and any problems with health and safety equipment or arrangements to their manager or health and safety officer.

SELF-CHECK 4.4

Give an appropriate example of something that has to be reported to a manager under the duties imposed on employees by the Management of Health and Safety at Work Regulations.

Workplace (Health, Safety and Welfare) Regulations

The Workplace (Health, Safety and Welfare) Regulations, known as the 'welfare regs', give the minimum requirements for a workplace to ensure the welfare of employees. These are generalized regulations, and cover elements of the working environment such as its ventilation, temperature, lighting, floors, furniture, wastes, and accommodation space (room dimensions). The associated ACOPs give information that is more detailed. The regulations also cover the use of corridors and staircases within premises that give access and egress to and from places of work.

The *safety* aspects covered by the welfare regs include maintenance of equipment, windows, doors, and traffic routes. Maintenance also includes general cleaning and the emphasis should be on planned maintenance, which is preventative, rather than repair after breakdown. The *welfare* section covers the provision of sanitary and washing facilities, drinking water, facilities where workers can rest and eat meals, and secure space to store items such as outdoor coats.

Health and Safety (Display Screen Equipment) Regulations

The Health and Safety (Display Screen Equipment) Regulations, usually known as the 'Display Screen Equipment (DSE) Regs' are designed to establish systems that prevent ill health occurring when using a computer or any instrument that incorporates a display screen or visual display unit (VDU). The regulations define a user as someone who works continuously with DSE in excess of 1 hour each working day. They entitle a user to an eye test to be paid for by the employer. If glasses are required specifically for DSE work, only the employer must pay for them.

The regulations prescribe suitable and sufficient risk assessment of the workstation, its surrounding area, the software being used, and the user.

The regulations also set minimum specifications for the workstation, giving details on the type of chair, dimensions of the desk or workspace, and how adjustable the DSE must be. The DSE regulations have been written to help prevent the occurrence of ill health, such as visual and musculoskeletal problems and stress. Therefore, taking regular short breaks, away from the DSE when possible, is a requirement. Regular short breaks, to answer a phone or to talk to someone in a different work area are good examples of this, but if these opportunities do not occur, *formal* breaks must be taken. There is also a duty on the employer to provide information on these aspects of the work, and to train staff to be aware of the risks and how to avoid them.

The study of how well a person is suited to their work in combination with their workstation is known as **ergonomics**. Ergonomic studies assess the physical and mental capabilities of the individual, including their perceptions, interactions, movements that occur while doing the job, and the surrounding work area. This includes elements of the task such as the positioning of equipment, how easily the individual can reach it without excessive stretching that over time might cause discomfort or injury, and avoiding stress to the individual by allowing them, rather than the process, to control the speed of the work. Workstations must be adjustable, as individuals have different physical and mental capabilities. The study of variations in physical characteristics, such as height, arm, and leg length, is known as **anthropometry** and is a key component to ergonomic assessments. Poor ergonomics can lead to work-related upper limb disorders (WRULDs) such as repetitive strain injury (RSI), tenosynovitis (inflammation of the tendons), carpal tunnel syndrome (pain in the wrist and hand), and frozen shoulder.

SELF-CHECK 4.5

Name the elements that should be considered in a DSE risk assessment?

Personal Protective Equipment at Work Regulations

The Personal Protective Equipment at Work Regulations place a duty on the employer to provide and maintain PPE when its requirement is identified by risk assessment. Personal protective equipment, by definition, only protects an individual and so potentially leaves others in the workplace at risk. Thus, it should only be used as a last resort, although it can be combined with other control measures such as working in a fume cupboard or other local exhaust ventilation (LEV) system. As with the other six-pack regulations, there is a duty on the employer to provide information, instruction, and training to anyone who is required to use PPE. These must include the reasons why PPE should be used, including the risks it is designed to control, and how the equipment should be used and maintained. The regulations place a duty on employees to use PPE as they have been instructed and trained to do so, to maintain and store it after use, and to report any obvious defects.

Any item of PPE that is provided to the employee must be to control the risk associated with the process, and be appropriate for the conditions of use and the characteristics of the wearer. For example, face masks provided to prevent inhalation of a dust may not adequately protect someone with facial hair, which prevents them fitting correctly. The regulations place a duty for these considerations to be formally assessed for risk and, when multiple PPE is required, that all of the items must be compatible and not introduce additional hazards. The risk assessment must be reviewed if any part of the process changes to ensure that PPE is still required and that what is provided is adequate for the purpose.

All PPE must be regularly maintained; the level of maintenance should be proportional to the potential risk and type of PPE. It may also be necessary to keep a written record of all maintenance. For example, a respirator requires regular inspections and planned preventative maintenance that must be recorded, whereas safety gloves may only require a visual inspection by the user. The use of disposable PPE can simplify this process, but all users must be aware of the necessity to discard and replace disposable PPE items. All PPE must be stored appropriately.

SELF-CHECK 4.6

What factors should be considered when choosing PPE for a task?

Manual Handling Operations

Manual handling operations (MHO) cover those actions required to move any load by human effort. These include pushing, pulling, lifting, putting down, and carrying loads. The regulations place a duty on the employer to put in place processes that avoid the need for MHO. When this is not possible, the MHO must have been subject to a suitable and sufficient risk assessment. Risk assessments for MHOs must cover five areas:

1. Environment
2. Task
3. Individual
4. Load
5. Equipment used

These can be remembered using the acronym ETILE. To make an assessment of the possible load, manufacturers are required to indicate its total weight and the heavier side of uneven loads. Once a risk assessment has taken place, there is a duty on the employer to reduce the risk of injury by introducing appropriate controls such as task layout, work rotation, PPE, teamwork, and training. As with all risk assessments (Section 4.4), they must be reviewed if any of the processes change.

These MHO regulations also place duties on employees to follow the safe systems of work and use any equipment provided in a proper manner. In practice, in a laboratory setting, MHOs are not usually a significant part of the work, but MHO should be considered when ordering stock to ensure materials arrive in units that can be moved, stored, and used without risk of causing injury.

SELF-CHECK 4.7

What does the acronym, ETILE stand for?

Provision and Use of Work Equipment Regulations

The Provision and Use of Work Equipment Regulations (PUWER) are designed to ensure that work equipment is safe to use no matter which country it originated in or how old it is. The Lifting Operations and Lifting Equipment Regulations (LOLER) 1998 apply to equipment used such as hoists and trolleys, for lifting or lowering loads, but the duties placed upon employers and employees are similar to those in PUWER.

The PUWER regulations place a duty on the employer to ensure all work equipment is suitable for its purpose and, when deciding what equipment to purchase, the health and safety of its users is taken into consideration. Furthermore, all work equipment must be maintained in an efficient state and working order, and kept in good repair with planned preventative maintenance as required. All maintenance activities and inspections must be recorded. The regulations are prescriptive about numerous features including systems such as emergency stop controls, guarding of dangerous parts, protection against high temperatures, and procedures to isolate equipment from power supplies. As with all the six-pack regulations, the employer has a duty to provide information, instruction, and training to anyone who uses the equipment so that they are fully aware of its operation, risks, and emergency procedures.

4.4 Risk assessment

There is a legal requirement to carry out specific risk assessments for certain types of work such as DSE and MHOs (Section 4.3). The general duty for an employer to conduct risk assessments is imposed by the Management of Health and Safety at Work Regulations. Risk assessments and other arrangements must be fully described in writing ('documented') when there are more than four employees at a workplace.

To clarify the processes involved in risk assessment we should first consider some of the definitions used:

- **Hazards** are any substance, equipment, or process with the potential to cause harm
- **Risk** is the likelihood that the substance, equipment, or process will cause harm

Using the terms in these specific ways makes it possible to clearly define an activity, for example, a process could present a high hazard, but a low risk because the hazard is well controlled. The level of risk is governed by the severity of the harm that could be caused, often referred to as **impact**, and the likelihood that harm will occur.

The phrase used to describe the level of risk assessment required in many of the regulations is *suitable and sufficient*. *All* risk assessments must identify the major risks and ignore trivial ones, identify controls and list them in order of importance. Assessment should remain valid for a reasonable period of time. In practice, this means that the level of detail in a risk assessment must reflect the degree of risk, so that high risk activities have more detailed assessments that focus on the most dangerous hazard. Risk assessments must be regularly reviewed. The timescale for review should reflect the level of risk, with high risk activity assessments being reviewed more frequently than those of low risk. However, if any part of the process changes, which may include a change in the person performing the activity, the current risk assessment must be reviewed immediately. This cycle is shown in Figure 4.3.

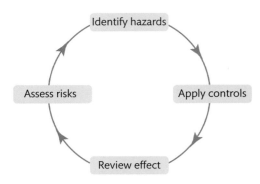

FIGURE 4.3
Outline of the risk assessment cycle.

Types of risk assessment

Risk assessment can either be *qualitative*, using terms such as high, medium, and low, or *quantitative* giving a numerical value. Quantitative risk assessments (Figure 4.4) are usually done for high risk activities, but also in large organizations when a guidance tool of numerical values to be used for different levels of risk is provided to give consistency.

Increasingly, risk assessments are carried out at an organizational level. These assessments look at the same hazards, but also consider the risks to the service or business, and service users or patients. For example, if a fire hazard in a laboratory that stored flammable liquids was poorly controlled, a fire might occur. If a fire does start, it is unlikely that anyone would be seriously hurt provided that the employees followed the correct evacuation procedures. However, a fire that is not rapidly extinguished may cause considerable damage to the laboratory, which may mean it could not operate for some time. The cost of this in terms of repair, effect upon patient care, and the need for a contingency plan would form a part of a business risk assessment.

Risk assessment process

The HSE publishes guidance that outlines the five steps on how to carry out a risk assessment:

1. Identify the hazards
2. Decide who might be harmed and how
3. Evaluate the risks and decide on precautions
4. Record your findings and implement them
5. Review your assessment and update if necessary

Some systems regard step 3 as two separate steps, because once evaluation of the risks has occurred, they should be arranged in order of severity ('prioritized') before precautions are considered.

Generally, at least two people should perform a risk assessment: one who is familiar with the activity and the workplace, and another who is less familiar with both, as it is easy to overlook potential, but well controlled problems if one is used to the activity and environment. A person less familiar with the workplace will see it with 'fresh eyes'. They may be able to identify hazards and suggest alternative controls that those working in the area had not considered. The use of two contrasting assessors therefore increases the likelihood of identifying hazards,

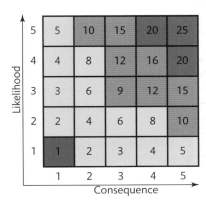

FIGURE 4.4
Example of a 5 × 5 risk matrix used to perform quantitative risk assessments.

which must be recorded together with details of the persons carrying out the assessment, and details of the activity and work area being assessed.

When deciding who might be at risk, it is necessary to consider vulnerable persons, such as young workers or expectant mothers, in addition to anyone else that comes into the work place including visitors and contractors. The HSE provides a sample form in their written guidance, although many organizations have their own format.

Hazards should be evaluated in terms of their potential danger, and then arranged in order of decreasing danger. In most systems, controls already in place are also considered when evaluating risks, but these processes must be checked as part of the assessment. Any actions and controls must have completion dates or priorities set to them. The degree of risk associated with the process (high, medium, or low), and the action completion dates will indicate the timescale for reviewing the assessment. A specific individual must take responsibility for any required action and the individual identified in writing.

Involving all staff in risk assessment is essential and, indeed, a legal requirement. It also identifies and introduces controls that will be workable and have the support of the team needing to implement them.

Hierarchy of risk control

The following options must be considered when deciding what are appropriate controls for an identified hazard. If possible, the risk should be removed, but if this is impossible, something less hazardous should be used as a substitute, for example, by replacing a dangerous chemical with one that is non-hazardous or less harmful. When this is not possible, access to the hazard needs to be prevented, such as by placing hazardous liquids in drip bottles, rather than in open baths. Alternatively, exposure should be reduced, for example, by enclosing the process within a fume cabinet or minimizing the time spent doing the activity. As always, defining and following safe systems of work, using PPE, providing adequate information and training, all help to eliminate risks. In reality, a combination of all these approaches is usually required to control the hazards associated with any one particular activity, and provision of training and information is always essential.

Legal interpretation

The HASAWA impresses a duty on the employer to 'ensure, so far as reasonably practicable, the health, safety and welfare of all employees'. This is the most common level of duty in health and safety law, and was first described by Judge Asquith in 1949 in the case of Edwards versus the National Coal Board. That case decided that 'reasonably practicable' was different to 'physically possible'. This means that the level of risk has to *balance* against the sacrifice involved in implementing its control. If the sacrifice grossly outweighs the risk, the employer would have discharged his duty if he did not implement the control, as it was not reasonably practicable. Put in another way, if the risk is relatively small compared with the cost of the measures required to control it no action is necessary. The level of risk must consider both likelihood and severity and the cost of implementing the control should include time, trouble, and expense.

During any risk assessment process, this balance needs consideration for each of the types of controls set out in the hierarchy of control. At some point, the cost of the control will be at an acceptable balance to the risk; this gives the minimum level of control required. You can see in Figure 4.5 a diagrammatic representation of how time and cost must be considered along with the degree of risk when deciding what controls to put in place.

Risk
(Likelihood
and severity)

Cost
(Time, expense,
and trouble)

FIGURE 4.5
Schematic indicating how costs should be considered against the degree
of risk when deciding which controls should be established.

Specialist risk assessments

Some of the regulations we have discussed, have a requirement for specialized risk assessments that consider specific risks and control hierarchy. In addition to these, the Management of Health and Safety at Work Regulations recognize that some types of people may be more susceptible to risk. Specific risk assessments are necessary for such people because they have different capabilities or may be more susceptible to hazards, therefore increasing the risk of the activity. These categories of people include the young, expectant, or new mothers, and lone workers who do not have the support of a team, and so are subject to additional hazards and higher levels of risk; hence, their need for different systems of work and emergency procedures.

Reporting of incidents

An **incident** is an unplanned event that causes an adverse outcome, such as injury, ill health, and/or damage to property, equipment, or the environment. **Near misses** are events that could have had an adverse outcome. Incident reporting and an appropriate investigation of any incident are essential parts of a health and safety management system. This can turn into a reactive approach, but the system can be proactive, contributing to improving health and safety, when used to report near misses and hazards. Reporting and investigating near misses are essential as generally, 10 near misses occur for every minor injury and the number of minor injuries relates to the occurrences of serious injuries. The Birds Accident Triangle of 1969 shows how many minor incidents occur for every major event (Figure 4.6). Thus, near misses are 'free lessons' that help prevent more serious incidents.

There is a legal requirement, under the Social Security (Claims and Payments) Regulations for employers to keep a written record of any accidents that occur in the workplace. Stationary offices produce an accident book, but the records can be in any form as long as sufficient detail about the accident and persons affected is recorded. The employer is also required to investigate the cause of the accident and enter this into the record.

All organizations have their own systems for reporting incidents and near misses. In general, information about the event, the people involved, and contact details are collated and recorded. In some cases, damaged equipment is quarantined because it may be necessary to examine it as part of the investigation. Often immediate actions are necessary to make the area and/or activity safe, and these must be noted. If a medical practitioner is consulted, he or she must record details of any injuries.

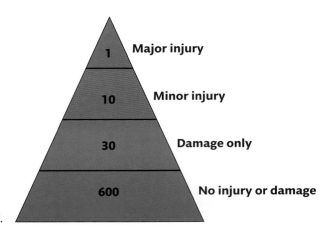

FIGURE 4.6
Bird's accident triangle (1969) showing how many minor occurrences underpin every major accident. See also Figure 19.7.

The severity of an incident is often quantified by scoring its impact against the likelihood of re-occurrence in the same way that risk levels are assessed. Near misses are assessed in the same way, by scoring the likely impact against the likelihood of occurrence. If the resulting score is relatively high, a more formal investigation, including specialist methods, such as root cause analysis, can determine the cause and suitable controls. Root cause analysis is a structured approach to investigating an incident, with the aim of identifying and examining its base cause, as implementing corrective actions at this level will produce the most effective control. If a specimen has been mislabelled, the person responsible may have simply made a mistake; however, checking training and competency records may identify deficiency in the training or induction process. Corrective actions aimed at improving training and induction will affect all staff in the area and potentially prevent them making a similar error. Hopefully, this will lead to much better control and lower the level of that type of risk.

4.5 Control of substances hazardous to health

The Control of Substances Hazardous to Health (COSHH) Regulations, first introduced in 1988, were the most significant piece of legislation introduced in the UK after the HASAWA; since that time, they have been subject to many amendments. Control of Substances Hazardous to Health Regulations set out the requirement for a management system to assess health risks, put in place suitable control procedures, and to monitor these controls. The COSHH regulations pre-dated the Management Regs, and were the first regulations to require employers to perform risk assessments. Thus, it brought about significant changes to health and safety practices. These regulations also established the need to monitor the health of employees. Two potential types of ill health can occur – acute and chronic. Acute conditions include asthma or fainting. They are usually of short duration with a rapid onset of symptoms following a single or short-term exposure to the causative agent. Although acute ill health can be severe, it is usually reversible. **Chronic ill health** includes cancers caused by exposure to industrial and environmental carcinogens. Chronic ill health normally develops over a long period following repeated exposures to a harmful agent. The onset of symptoms is often so gradual that the effects on health can be undetected and the condition remains undiagnosed for years. Unlike acute conditions, the effects of chronic diseases tend to be irreversible.

The COSHH regulations cover almost all hazardous substances, however, some agents are subject to separate, and more specific regulations. These agents include lead, asbestos, and sources of radiation.

Classification of hazards

A chemical substance is considered hazardous if it is classified as dangerous to health under the Chemicals (Hazard, Information and Packaging for Supply) (CHIP) Regulations 1994. The CHIP regulations place a duty of care on all suppliers when they package and transport hazardous substances. All packages must be clearly labelled with hazard information contained in the form of a Material Safety Data Sheet (MSDS). Both hazard warning labels and MSDSs are essential sources of the information required to perform a COSHH assessment. The packaging and labelling of biological substances is legislated by a number of transport regulations, and is discussed in Section 4.9.

There are four main types of chemical hazards: **irritant**, **harmful**, **corrosive**, and **toxic**, which are represented by the symbols you can see in Figure 4.7, or by recognized abbreviations. Irritants, (Xi), are non-corrosive substances that cause inflammation of the skin or respiratory

¹carcinogenic, germ cell mutagenic, toxic to reproduction
²specific target organ toxicity

FIGURE 4.7
(a) Traditional hazard symbols commonly used on chemical packages. Since 2006 these have been gradually replace worldwide by (b) the globally harmonized system (GHS) of classification and labelling of chemicals. The European Union adopted GHS at the end of 2008.

tract after repeated exposure; acetone and formaldehyde are two examples of irritant chemicals. They may also cause individuals to become sensitized or allergic to that substance, or may increase the severity of other allergies from which an individual may suffer. Many everyday substances are harmful (Xn), for example, cleaning products and paints. These substances may cause minor health effects if swallowed or inhaled, or if they penetrate the skin. Fortunately, the use of PPE (Section 4.7) is usually sufficient to be able to work safely with them. Corrosive (C) substances are usually strong acids or alkalis, which can attack or burn living tissues. Many cleaning agents contain corrosive substances. Toxic (T) substances, otherwise know as poisons, reduce or prevent the function of major organs such as the heart, liver, and kidneys. Mercury and carbon monoxide are examples of toxic substances.

Even pure chemicals may have more than one hazard classification, for example, some cleaning agents are both harmful and corrosive. Some laboratory reagents contain a number of substances of these types. Under the COSHH regulations, the controls identified and put in place must be adequate to control the most serious hazard to health. Risks associated with **Flammable** (F) substances are not covered by COSHH, but if the substance has other classifications, such as being harmful or an irritant, these risks must be considered under COSHH.

SELF-CHECK 4.8

What is the correct abbreviation for each of the classified hazards shown in Figure 4.7(a)?

Complying with COSHH

Seven basic steps must be satisfied, if an organization is to comply with the COSHH regulations:

1. Risk assessment
2. Identify precautions
3. Prevent or adequately control exposure
4. Ensure control measures are used and maintained
5. Monitor exposure of employees to hazardous substances
6. Carry out appropriate health surveillance where necessary
7. Ensure employees are informed, trained, and supervised

A COSHH risk assessment must be detailed and focus on the hazards to health from the substance or mixture of substances, and the way they are used and the environment it is used in. Thus, for example, different hazards are presented when a substance is sprayed compared with those present when it is painted onto a surface; spraying is more difficult to control, may affect a larger area, and can be inhaled far more easily, and therefore can potentially affect more people.

To identify precautions that prevent or control exposure, COSHH regulations prescribe a more specific hierarchy of control than the management regulations described in Section 4.3 that give priority to mechanical or engineering controls when a substance cannot be removed or substituted. The COSHH regulations also specifically state that PPE should be used as a control when no other measures are available, although, of course, PPE can be used in combination with other control methods. To ensure control measures are maintained, COSHH regulations include a specific requirement to ensure that engineering controls, such as safety cabinets and LEV, undergo thorough examinations at least every 14 months with written records made of them, which must be kept for at least 5 years.

Employers must monitor the exposure of employees to hazardous substances that have exposure limits assigned to them and this must be recorded. Hazardous substances have **workplace exposure limits** (WELs) set for them, which are available in EH40, a publication of the HSC. The limits are set at a level at which ill health from the substance would be unlikely to occur in the majority of workers. The limits may be short-term (15 minutes) or long-term (8 hours) depending on whether the agent usually causes acute or chronic ill health effects. General records must be kept for at least 5 years, but a record of an individual's (personal) exposure must be available for 40 years. Substances that require the monitoring of exposure usually require the employer to carry out health surveillance. In large organizations, an occupational health department provides and reviews health screening questionnaires for staff exposed to such hazards. The department may also conduct tests, such as skin checks and lung function tests, to monitor the health of staff and detect any possible ill health effects.

Control of Substances Hazardous to Health Regulations also place a duty on employers to inform employees of any risks and the relevant control measures in use. They must also train staff in how to use and maintain all control measures, and perform activities safely. There is also a duty on the employer to supervise employees, and ensure the control measures are used correctly and are well maintained.

Dangerous Substances and Explosive Atmosphere Regulations

The Dangerous Substances and Explosive Atmosphere Regulations (DSEAR) are designed to protect employees from the risks of substances likely to cause fire, create dangerous atmospheres, and/or explosions. The use, handling, and storage of flammable substances, such as industrial methylated spirits (IMS), often used in clinical laboratories, are controlled by these regulations. Flammable substances also fall under these regulations, although like any other hazards to health, they are also covered by COSHH regulations. Like all regulations, DSEAR prescribe systems to be established that allow appropriate risk assessments to be performed and which implement controls that must be monitored and reviewed.

4.6 **Fire regulations**

In 2005, a major change to fire legislation in the UK was made with the introduction of the Regulatory Reform (Fire Safety) Order. The need for fire certification of premises by the fire brigade was removed and replaced by a requirement by employers to carry out a fire risk assessment. Employers should appoint a person responsible for fire safety and this person should:

- Carry out a fire risk assessment to identify the risks and hazards
- Consider who may be especially at risk
- Eliminate or reduce the risk from fire as far as is reasonably practical
- Provide general fire precautions to deal with any residual risk
- Take additional measures to ensure fire safety where flammable or explosive materials are used or stored
- Plan procedures to deal with any emergency and document these plans
- Review the findings of risk assessments as necessary

The regulations also contain minimum standards for the provision of fire fighting equipment, and fire-related safety signs to show the location of this equipment, alarm call points, and

FIGURE 4.8
Fire triangle showing the three elements required to start a fire.

fire exits. There is also a requirement that all employees receive annual training on fire procedures, which should include training in the use of the fire fighting equipment available in the workplace.

The control of fire risks requires knowledge of how a fire starts. The initiation of a fire has three requirements: oxygen or air, fuel, and a source of heat, which are often represented as a fire triangle, which you can see in Figure 4.8. The absence or removal of any one of the three will prevent a fire starting.

SELF-CHECK 4.9

List two sources of heat and two types of fuel that are routinely found in clinical laboratories.

Fire fighting equipment, such as fire blankets and fire extinguishers, dampen fires by removing one or more of the three elements required for a fire to burn.

Fire extinguishers are predominately red, but have colour-coded labels that easily identify the different types of extinguisher. The commonest types found in laboratories are, for example, red and black labelled extinguishers, which are water- and carbon dioxide-based, respectively. Water extinguishers function by removing heat, while carbon dioxide extinguishers reduce oxygen levels. Each type of extinguisher must be used correctly: water extinguishers must not be used on electrical fires as this carries the risk of an electric shock. Figure 4.9 shows some of the different types of extinguishers and the types of fires they are able to extinguish.

Fire Extinguisher Chart

Extinguisher		Type of fire				
Colour	Type	Solids (wood, paper, cloth, etc.)	Flammable Liquids	Flammable Gases	Electrical Equipment	Cooking Oils & Fats
	Water	✓ Yes	✗ No	✗ No	✗ No	✗ No
	Foam	✓ Yes	✓ Yes	✗ No	✗ No	✓ Yes
	Dry Powder	✓ Yes	✓ Yes	✓ Yes	✓ Yes	✗ No
	Carbon Dioxide (CO_2)	✗ No	✓ Yes	✗ No	✓ Yes	✓ Yes

FIGURE 4.9
Fire extinguisher selection chart showing the type of extinguishers and their colour codes, together with their recommended uses.

4.7 **Personal protective equipment**

We looked at the regulations designed to ensure the provision and use of PPE to control risks that cannot be controlled by other means in Section 4.3. We will now look at some of the common types of PPE that you might use in a biomedical laboratory. Figure 4.10 shows a selection of the types of PPE that are used in biomedical laboratories.

Body protection

Laboratory coats are worn to protect everyday clothing and skin from exposure to hazardous substances. Depending on the nature of the hazards, laboratory coats may need to be supplemented by aprons, or the use of a disposable coat or overalls. For example, disposable coats are used in Category 3 containment laboratories. The fastenings, cuffs, and sleeves of laboratory coats should be inspected before first use and the fit must enable the person to fully close the coat. Overly long coats or sleeves are additional hazards, as parts of the coat could get trapped, or caught on equipment or furniture, causing an accident.

It may be necessary to have transport coats for staff, for use when moving between risk areas of the laboratory to relatively clean ones, such as corridors. A clean coat must always be worn when leaving a laboratory to go to a clean site or to visit clinical areas for laboratory-related purposes. This ensures that the clean area is not contaminated with substances from the laboratory transported on a used laboratory coat.

Hand and foot protection

Many different types of safety gloves are routinely used in clinical laboratories. As with all equipment, the type of glove selected must effectively control the risk; the circumstances

FIGURE 4.10
Selection of personal protective equipment (PPE) that might be used in a biomedical laboratory. Courtesy of Sperian Protection, UK.

of use must also be considered. Liquid resistant gloves can control the risks associated with handling substances where contact between them and the skin must be avoided. However, many types of glove perish upon contact with commonly handled laboratory reagents, such as xylene; therefore, careful consideration must be given to the material of the glove. Vinyl, latex, nitrile, and rubber gloves are all available. An additional consideration when selecting safety gloves is that of latex skin allergies or sensitivities. Most workplaces try to remove all latex material from use; however, some non-latex gloves do not, in some people's opinion, give the same dexterity and sensitivity of touch so they may still be used in a limited number of areas. A detailed risk assessment justifying their use should be carried out and anyone at risk clearly identified.

Safety gloves made of a suitable material may also be used to protect against sharp or rough surfaces in MHOs and when handling substances at, or working in, extremes of heat.

Gloves must also be of the correct size to ensure safe handling of the substance: loose gloves can lead to reduced grip and tight gloves to skin irritations. You can see an example of a glove selection chart, showing the type of gloves used when handling some common chemical and biological hazards in Table 4.1.

Foot protection is sometimes required to control the risk of hazardous substances falling onto feet. In most circumstances, rigid, fully enclosed shoes offer sufficient protection. In others, reinforced footwear might be necessary to protect against falling sharp objects or chemicals.

TABLE 4.1 Example of a glove selection chart showing the type of gloves to be used with some common types of chemical and biological hazards

Glove material	Uses	Recommended for:	Not recommended for:	Comments
Latex (natural rubber)	Incidental contact	Weak acids and bases, alcohols, aqueous solutions	Oils, greases, and organic solvents	Good for biological and water-based materials. Poor for organic solvents. Little chemical protection. Can puncture easily. Can cause or trigger latex allergies
Nitrile (synthetic rubber)	Incidental contact	Oils, greases, acids, caustic solutions, aliphatic solvents	Aromatic solvents, many ketones, esters, and chlorinated solvents	Good for solvents, oils, greases, some acids and bases. Clear indication of tears and breaks. Good alternative for those with latex allergies
Polyvinyl alcohol (PVA)	Specific uses	Wide range of aliphatic, aromatic ketones (except acetone), esters, ethers and chlorinated solvents	Acids, alcohols, bases, water	Good for aromatic and chlorinated solvents. Poor for water-based solutions
Polyvinyl chloride (PVC)	Specific uses	Strong acids and bases, salts and other aqueous solutions, alcohols, glycol ethers	Aliphatic, aromatic, chlorinated solvents, nitrocompounds	Good for acids, bases, oils, fats, peroxides, and amines. Good resistance to abrasions. Poor for most organic solvents

This table is for general reference only; for specific recommendations contact the glove manufacturer or MSDS.

Specialist footwear is sometimes used to control other risks, for example, using non-slip soles in hazardous floor conditions, or electrically insulated shoes to protect against electric shock, and shoes incorporating anti-fatigue properties for people that spend long periods of time standing.

Eye, head, and ear protection

Numerous types of spectacles, goggles, and face shields are available to protect the eyes and/or face from contact with hazardous substances. Here, again, the material used to construct them must be sufficient to control the risks involved. The comfort of the person required to wear the PPE must be considered, as otherwise they are unlikely to wear them consistently.

Head and ear protectors are not commonly used in laboratories. Head protection involves wearing hard hats or bump caps to protect against objects falling from above or when working in confined spaces. Ear plugs or defenders can be used to protect against noise that at some levels and frequencies damages hearing.

Respiratory protection

Filtration masks, respirators, and breathing apparatus can be used to control the hazards of airborne agents when engineering solutions, such as ventilation systems and safety cabinets cannot be used. Often these are used in non-routine procedures like conducting maintenance of the normal engineering controls or after a spillage. All masks, respirators, and breathing apparatus must be tested to ensure they fit the individual who will use them. Factors such as beards can compromise the fit and, therefore, the level of protection given will be reduced. Fume hoods do not suffer this disadvantage; however, they must be fitted with the correct air filters and checked as part of a detailed maintenance schedule.

4.8 Reporting of Injuries, Diseases and Dangerous Occurrences Regulations

The Reporting of Injuries, Diseases and Dangerous Occurrences Regulations (RIDDOR) set out the processes of reporting injuries, ill health, or other dangerous events to the HSE or, in some cases, the local authority. The aim of the regulations is to provide data on those significant events that can be analysed to reveal any national patterns and so lead to new health and safety advice and/or regulations. When serious events are reported, the HSE reviews the information it has received about the incident and may decide to inspect the workplace. The data is also used to collect information about occupational illnesses, which may lead to a better understanding of how they develop and how to control the hazards that cause them.

There are several levels of reporting that depend on the severity of the situation:

- Major injuries and dangerous occurrences
- Three-day-plus accidents
- Reportable diseases

Major injuries and dangerous occurrences include deaths, major injuries, or those requiring hospital treatment, fires, electrical shocks, and the release of biological agents.

Such events must be notified *immediately*. If an injury sustained at work prevents a person from doing all of their normal work for more than 3 days, it must be reported to the HSE on a specific form within 10 days of the incident becoming reportable. Should an employee be diagnosed with a reportable disease by a medical practitioner, the employer must report this to the HSE on the specified form. Examples of such diseases include occupational dermatitis, occupational asthma, mesothelioma, tuberculosis, and some musculoskeletal disorders.

The Health and Safety (First Aid) Regulations 1981 require an employer to provide **first aid** equipment, facilities, and personnel. The specific nature of the provision required is related to the findings of both general and COSHH risk assessments.

SELF-CHECK 4.10

How should a dangerous occurrence be reported to the HSE?

4.9 **Transport regulations**

Transporting dangerous goods is regulated by several pieces of legislation. These include, for example, the CHIP regulations we discussed in Section 4.5, the Carriage of Dangerous Goods and Use of Transportable Pressure Equipment Regulations (2004), and a European agreement know as *accord européen relative au transport international des marchandises dangereuses par-route* (ADR) of 2009. The Department for Transport publishes guidance on these regulations, and, with the HSC and Vehicle and Operators Service Agency (VOSA), is also responsible for enforcing them.

Dangerous goods include pathology samples, gas cylinders, chemicals, explosives, flammable liquids, and radioactive substances. Each specific class has defined controls for their packaging and labelling, and what must be used when they are transported.

As with all other modern regulations, risk assessment, training, and information about safe systems of work, risks, and controls is an essential part of the duties imposed. Every employer that transports dangerous goods as part of their business must have a Dangerous Goods Safety Advisor (DGSA), who must have attended a training course and passed the associated examination. The role of the DGSA is to ensure all staff involved in the processes are aware of the requirements needed to meet the regulations. Drivers who transport dangerous goods by road must carry instructions in writing, known as a **Tr**ansport **Em**ergence (**TREM**) card, detailing the substance, its category name, UN number and class, details of what to do and who to contact in an emergency, and first aid information in case of spillage of the substance being carried.

Pathology samples

Blood, urine, faeces, swabs, tissue samples, and clinical waste are infectious substances under the regulations, and assigned to class 6.2, which is further subdivided into A and B groups. The majority of samples fall into category B and may contain hazard group 1, 2, or 3 pathogens. They must be labelled: *UN3373 – Biological substance, category B*. Category A samples are defined as cultures of hazard group 3 pathogens or samples suspected of containing hazard group 4 pathogens. Packages with category A samples must be labelled: *UN2814 – infectious substance affecting humans*. Only specified carriers may transport category A samples.

Packaging and labelling of biological materials

Pathological specimens may be transported by road, rail, or air if the packaging and labelling instructions given in Packaging Instructions P650 (found in part 4 of ADR) are followed. These instructions may be summarized as follows. The primary layer (the sample container or tube) must be watertight and leak proof. There must be an absorbent material, sufficient to soak up the volume of sample, between the primary and secondary layers. The secondary layer (sample bag) must be watertight. A rigid outer packaging that has passed a 1.2-m drop test and has minimum dimensions (on one face) of 100 × 100 mm is required. Cushioning material must be inserted between the secondary layer and the outer packaging. Either the primary or the secondary packaging must be able to withstand a pressure differential of 95 kPa or more. Usually, the request form is attached to the outside of the bag to protect it from damage by any leakage. The package must be labelled with the UN3373 diamond and biohazard label as shown in Figure 4.11 below.

Samples that do not have an associated risk of infection, such as processed tissue blocks, can be transported as 'exempt human specimen', but must be packed according to the instructions contained in P650.

4.10 **Personal health and safety**

Every workplace will have its own specific health and safety instructions that you must follow at all times. The initial training you receive when you first take a job in a laboratory must always cover the hazards, risks, and controls of each process you undertake. However, you will later receive supplementary specific and mandatory training related to the regulations we have already discussed. If you believe that you have received inadequate information, you should consult a senior member of staff. However, some rules and practices are common to all laboratories, and include the following:

- A waterproof dressing must cover all cuts, abrasions, or other types of wounds to the skin especially of the hands before you start work. If you puncture your skin in an accident at work, gently encourage it to bleed under cold running water. Report the incident to a senior member of staff and seek medical advice as necessary

- You must never take personal items, such as pens, pencils, cosmetics, toiletries, and hairbrushes, into the laboratory. Cosmetics are not to be applied in the laboratory; indeed, hands must never be in contact with the mouth, face, eyes, and nose. Always wear a correctly fastened coat when in the laboratory. It must be cleaned regularly and immediately changed if it becomes contaminated

BIOLOGICAL SUBSTANCE, CATEGORY B

FIGURE 4.11
Labelling required for transporting category B specimens.

- Never take food or drink (including sweets and medication) into a laboratory nor eat or drink there. Always follow the correct laboratory procedures. Never deviate from them unless you have received alternative instructions from a senior member of staff for a specific reason. Should circumstances arise that prevent you from following the procedure, always inform a senior member of staff before continuing. Wear gloves when necessary; if this is a requirement, it will be specified in the standard operating procedure. Replace the gloves if they become contaminated or perforated. Avoid all practices that can cause splashing or the release of airborne substances unless they are performed in an appropriate containment cabinet. Report all accidents, breakages, spillages, or damaged equipment to a senior member of staff immediately

- Always wear a clean coat if visiting a clinical area. Clean coats must be stored in a separate, clearly marked area, separate from that for other laboratory coats. Remove your coat and wash your hands before leaving the laboratory

- If you become ill and consult a doctor, always inform him or her where you work, and ask them to talk to your manager if they require further details

- Lastly, *never* take any unnecessary risks in a laboratory

Occupational health

Most employers will have their own occupational health (OH) service, or access to one. The role of OH is to assess whether individuals are fit to perform the role in question *before* an offer of employment is given, and to give advice on health issues that relate to the specific area of an employee's work. Occupational health may also provide services such as vaccinations and health monitoring. Staff can also discuss concerns related to their health and welfare with them, in confidence.

Risk awareness

Some people are natural risk takers; others are naturally more cautious. It is essential that in the workplace you take a responsible attitude to risk and avoid it at all times. You should also remember that the severity of a hazard is rarely the same as the level of risk; laboratory procedures contain a wide range of controls that, if followed, greatly reduce the level of risk.

Infection control

Hand hygiene is an essential control to minimize the risk of infection. It is essential that you take great care every time you wash your hands. The seven-step technique, which is usually displayed above sinks designated for hand washing (Figure 4.12), should be followed to ensure every area of the hands, fingers, and wrists is cleaned. Great care should also be taken to dry hands thoroughly as leaving them damp can cause dermatitis. False or long nails and jewellery can harbour dirt, or infectious and hazardous material: never wear them in areas where they pose a risk.

Clothing and personal appearance

Many workplaces have a clothing and personal appearance policy. Such policies are partly intended to encourage health and safety elements, such as wearing enclosed footwear, having hair tied up, and not wearing large earrings or necklaces, but they are partly intended to ensure staff present a professional appearance. Standards of dress are subjective; hence, these

Step 1
Squeeze a small amount of sanitizer gel or soap over left palm. Dip all the fingers of the right hand into the left palm and vice versa

Step 2
Join hands palm to palm

Step 3
Rub right palm over left hand as shown and then left palm over right hand

Step 4
Interlace fingers with hands joined palm to palm

Step 5
With fingers still interlaced, rub the backs of fingers against opposing palms

Step 6
Clasp thumbs in opposing palms and wash with rotational movements

Step 7
Wash wrists with rotational movements of opposing hand

Rinse and dry thoroughly

FIGURE 4.12

Seven-step technique of hand washing. Once clean, the hands must be thoroughly dried.

policies can set clear guidelines for staff to follow. Dress standards are closely related to infection control policy as long-sleeved tops or shirts and ties can transport infectious material around the workplace. Thus, a balance between presenting a professional appearance, perhaps by wearing a tie, and infection control, must be established. A professional appearance by staff is especially important in service industries such as healthcare, and is an essential part in developing trust and positive working relationships with professionals in other departments, as well as with patients.

4.11 Universal or standard precautions

The Centers for Disease Control (CDC) in the USA first recommended universal or standard precautions 1985, in response to the risk of transmitting human immunodeficiency virus (HIV). It is not always possible to tell if a clinical sample poses an infection risk; hence, universal precautions is a system of work and controls that assume all patient-related contact carries a risk of infection.

The key points of universal precaution systems are:

- Good hand hygiene
- The wearing of PPE at all times
- The correct use and disposal of sharps, such as needles and blades
- Disinfection of body fluid spills
- Careful disposal of waste

Using universal precautions in the wider healthcare setting has been the driver for decontaminating hands after every patient interaction, irrespective of the level of infection risk from the examination or procedure. This has lead to the increased use of hand gel cleaning agents.

In a clinical laboratory, implementing universal precautions means treating all specimens as if they are extremely hazardous, and wearing gloves and possibly eye protection at all times. Wearing of gloves would even be necessary when using equipment such as telephones and personal computers. This approach would certainly simplify the use of precautions and, therefore, make it easier to follow and monitor the use of controls. However, the long-term use of PPE can lead to ill health issues, for example, wearing gloves for extended periods can lead to dermatitis.

 SUMMARY

- The Health and Safety at Work Act and the Management of Health and Safety at Work Regulations form the legal framework that ensures the safety of everyone in the workplace. They place numerous duties upon employers, but also some on employees.

- All employees have a duty to ensure their actions or omissions do not cause a risk to themselves or to the safety of others working with them. They also have a duty to use and to care for equipment as instructed, and to report defects in equipment or hazards to the health and safety officer.

- Risk assessment is the primary tool prescribed in legislation to ensure high standards of health and safety. Risk assessments identify hazards and those who are potentially in danger; they evaluate the risks associated with those hazards and suggest how the risks can be reduced to acceptable levels. All risk assessments need reviewing on a regular basis to ensure they are still adequate. Other regulations that require a more detailed form of risk assessment include COSHH, DES, Manual Handling and Fire Safety.

■ The ethical duty of everyone to ensure the safety of everybody affected by the workplace is upheld in law. Ensuring a high standard of health and safety in the workplace is also desirable because the overt costs associated with incidents and injuries are considerable; hidden costs, such as insurance, time lost in investigations, and negative publicity, are also substantial.

FURTHER READING

● HSE Books *Safe Working and the Prevention of Infection in Clinical Laboratories and Similar Facilities*. HSE Books, Bootle 2003.

Gives detailed information on health and safety specific to clinical laboratories.

● HSE Books *5 Steps to Risk Assessment*. HSE Books, Bootle, 1998. Available at: http://www.hse.gov.uk/risk/fivesteps.htm.

This is an essential guide to performing risk assessments.

● Hughes P, Ferrett E. *Introduction to Health and Safety at Work*, 4th edn. Butterworth-Heinemann, Oxford, 2009.

This is an excellent book, covering all areas of health and safety, including the legal aspects and case law.

Useful Websites

■ www.hse.gov.uk
■ www.hsebooks.co.uk
■ http://www.coshh-essentials.org.uk/

These websites are excellent sources of information about general health and safety issues and COSHH information. Many of the publications are available to download.

QUESTIONS

1. The abbreviation, **Xn** is used in chemical hazard classification to represent which of the following?

   (a) Flammable
   (b) Irritant
   (c) Harmful
   (d) Corrosive
   (e) Toxic

2. Which of the following statements are **TRUE**?

   (a) The penalties for failing to comply with HSE enforcement notices include up to 6 months' imprisonment
   (b) The act introduced to ensure workplaces were safe was The Safety Act

    (c)   Under DSE rules you are entitled to free glasses if you require them to do your job

    (d)   RSI stands for recent strain injury

    (e)   You must wear gloves when opening specimen containers to get a better grip on the lid

3. Arrange the following two lists into their most appropriate pairings.

ACD	Display screen equipment
ACOP	Health and Safety at Work Act
DSE	Personal protective equipment
HSE	Dangerous Substances and Explosive Atmosphere Regulations
WRULD	Reporting of Injuries, Diseases and Dangerous Occurrences Regulations
DSEAR	Lifting Operations and Lifting Equipment Regulations
HASAWA	Health and Safety Executive
HSC	Provision and Use of Work Equipment Regulations
IOSH	Repetitive strain injury
LOLER	Advisory Committee on Dangerous Pathogens
MHO	Work-related upper limb disorder
WELs	Approved codes of practice
PUWER	Manual handling operations
RSI	Workplace exposure limits
RIDDOR	Institute of Occupational Safety and Health
PPE	Health and Safety Commission

4. What are the five steps required to perform a general workplace risk assessment?

5. (a) What are the possible routes of entry to the body for a chemical mist? (b) Suggest possible control measures to prevent its entry.

6. What are your duties as an employee under the Health and Safety at Work Act?

7. List your duties as an employee under the Management of Health and Safety Regulations?

8. (a) What are the three elements of the fire triangle? (b) Which of these is eliminated when using a carbon dioxide fire extinguisher?

9. (a) Give an example of a mechanical control, as described by the COSHH regulations.

    (b) What are the requirements under COSHH to maintain such equipment and associated records?

10. What are the differences between a laboratory using universal precautions compared with one applying control measures based on risk assessment?

Answers to self-check and end of chapter questions are available in the Online Resource Centre accompanying this book.

 Go to www.oxfordtextbooks.co.uk/orc/glencross

Statistics and handling data

Andrew Blann

Learning objectives

After studying this chapter, you should be able to:

- Explain the concepts of quantitative and qualitative data
- List the different modes of distribution: the normal distribution and the non-normal distribution
- Describe simple statistical tests to analyse data
- Be familiar with correlation

Introduction

Biomedical science is dominated by information, almost all of which refers to the health status of an individual or groups of individuals. Many biomedical scientists are directly involved in the generation of numerical or descriptive data derived directly from human body tissues. For haematologists, the data generated may be the number of red and white blood cells in a sample of venous blood, while for biochemists the data may be the concentration of glucose in that blood. Microbiologists may collect data that tells which particular microbes are present in a sample of sputum from a patient with a lung problem, and histologists may study lung tissues from that patient, and the data they handle may determine the presence or absence of a cancer. Data may not only be at a single timepoint; in some cases we are interested in how an index (for example, cholesterol concentration) changes as someone undergoes a change in their diet and lifestyle.

The data handled by a biomedical scientist is typically generated in one of two ways: either directly by the biomedical scientist (for example, using microscopy, as in many microbiology, histology, and cytology analyses) or indirectly, through the use of some kind of automated analyser (which are used for most haematology and biochemistry tests). Immunologists and haematologists often use both analysers and microscopy. Whatever the means through which the data are collected, the value of data, and how it can be interpreted must be understood. The limitations of the techniques and what steps can be taken to ensure these techniques are delivering the correct data must also be addressed. Failure to grasp these concepts may lead to errors and possibly the loss of life.

This chapter gives a broad overview of the issues surrounding data handling and statistical analyses as it applies to biomedical science. It will explore the different ways in which data can be described and analysed, and consider issues that relate to best practice when handling data.

5.1 Types of information and how to describe them

Let us first consider why data is needed. The understanding and treatment of many human diseases and conditions relies heavily on the generation and interpretation of numerical data. For example, a person with diabetes needs to monitor and control their blood glucose concentration: they need to collect data, such as measuring blood glucose, and interpret this data. Is the value too high, too low, or within acceptable limits? In another setting, such as the study of large numbers of people, it is important to be able to understand those factors (such as high blood pressure and smoking) that predict the major health issues of a stroke or heart attack.

Often, a biomedical scientist handles more than just a single value (as in the case of a single blood glucose measurement as above), but needs to compare sets of data (that is, multiple measurements from different groups). The analysis of data in this way starts to draw upon reasonably sophisticated numerical and statistical tools. Quite often, these tools seem quite intimidating at first glance.

The nature of information

As data comes in many different forms, it is helpful to be able to characterize it in some way so that it provides an idea of how it is best to present and also perform an analysis. In biomedical science, information generally comes in one of two types:

- **Quantitative** data is that where the information is assigned a directly measurable, numerical value. Examples of this are height, weight, age, red blood cell count, temperature, serum potassium concentration, or perhaps the proportion or percentage of patients with a particular problem (such as an infection) or a particular ABO blood group. In almost all cases, the value of the data is defined not by an individual, but by an objective observation or scale, itself often derived from a machine.

- **Qualitative** data, on the other hand, is formed from information in words (often descriptive terms such as good, reasonable, occasional, frequent, moderate, severe). These items of information are generally obtained directly from people being investigated, not merely patients with a particular health problem, and use of a questionnaire is common. However, at least one observer has made the distinction of qualitative research *methods*. A major aspect of word-data of this type is that the value or relative importance of the word itself is defined by the individual and, therefore, has a high degree of subjectivity.

Despite the possible value of these two types of data, some types of information lie somewhere between the two. It is also possible for some types of qualitative data to be converted into numerical forms for rigorous analysis, that is, for it to be changed from a qualitative to a quantitative form. A weakness with qualitative data is in the subjective definition of the information. Consider pain. What one person may consider as being moderate, to another person could be severe. This gives a degree of uncertainty. However, another study may compare the number of people who use the word 'moderate' to describe their pain, with those that use the word 'severe', or even 'excruciating'. In the latter case, the assessment of pain is being evaluated in a qualitative way (i.e. the 'quality' of the pain).

Quantitative data

Broadly speaking, we can classify just about all quantitative data into one of two types: that which is categorical and that which is continuous.

Categorical data fits into one of any number of discrete boxes or categories; there are no 'in-betweens'. An example of this is the number of men and the number of women in a particular group. In the case of sex there are generally no in-betweens. Here, it is important that data lies only in one of a discrete number of categories, and sex is a good example of this. Another example is the consequence of sex: children. In real life you can't have a fraction or proportion of a child, only whole numbers.

This concept can be extended to more than two discrete groups. Just about everyone belongs to one of the four ABO blood groups of A, B, AB, and O; there are few (if any) exceptions that, if they do occur, will be so rare as to be ignored in the final analysis (unless you are an expert in blood transfusion!). Our colleagues in microbiology can say with a good degree of confidence that an organism (such as methicillin resistant *Staphylococcus aureus* – MRSA) is present or absent – there should be no in-between. Immunologists can say 'yes' or 'no' to the presence of antibodies to organisms such as a bacterium or a virus, or if the patient has high levels of certain antibodies to a particular infective agent. Histologists will tell us that a tissue either is or is not invaded by a cancer, whilst cytologists report the presence or absence of certain abnormal cells, such as in cervical cancer.

Continuous data includes factors such as height, weight, blood pressure, and just about all haematology and biochemistry results (such as the number of red blood cells and concentrations of triacylglycerols and Na^+ in the serum). The data, consisting of individual numbers, can be described by almost any figure in a given range (such as age, which can be anywhere between 0 and 120 years or so).

The presentation and analysis of continuous data is more complex than if the data is categorical. When the data is continuous we need to consider three aspects: the central point or tendency, the variance, and the distribution. These three aspects will now be explained in turn.

The central point

The most important statistical aspect of continuous data is the consideration of the central point of that particular data set. Some statisticians prefer to use the word 'tendency' (which may seem rather inexact) instead of 'point', which is not unreasonable as there are actually two such values within a single data set: the average value and the middle value. These two different central points are the **mean** and the **median**, respectively. However, some consider there to be a third point – the mode – but this is very rarely used.

- The mean point of a set of data is obtained by simply adding up all the data, and then dividing by the number of the individual data points. As such, the mean may also be seen as the 'average' number. How the mean point is obtained is described in Box 5.1.

BOX 5.1 The mean value

Consider a group of 11 data points with individual values of 25, 30, 21, 27, 35, 28, 22, 25, 29, 27, 24. These data (which may be the ages of a football team) sum to 293. Thus, the mean value is 293 divided by 11, which (to three significant figures) is 26.6.

By placing the eleven data points in order from lower to highest, i.e.

21, 22, 24, 25, 25, 27, 27, 28, 29, 30, 35

The mean value (26.6) is indeed, very close to the middle value of the whole set (the 6th, with a value of 27). The importance of 'closeness' of these two figures will be discussed further in this section.

- One way of arranging a set of numbers is to list them in rank order, i.e. from the lowest to the highest. The data point that is in the middle of the entire series is the median, and this value is arrived at by a completely different set of rules than is the mean value. This is described in Box 5.2.

BOX 5.2 The median value

Consider the data set of 21, 60, 15, 32, 19, 25, 113, 24, 29, 35, and 22. If these are arranged, or ranked, in order from the smallest (on the left) to the largest (on the right), a series is obtained:

15, 19, 21, 22, 24, **25**, 29, 32, 35, 60, 113.

In this series of 11 data points, the one in the middle is the 6th one – i.e. with 5 data points larger (greater than 25) and 5 data points smaller (less then 25). This central point is the value 25 – also known as the median (highlighted in bold). Note that it is not the mean. In this data set the mean (average) would be the sum of all the individual points (394) divided by the number of individual data points (11) to give 35.8 – a very different figure to 25. This difference is important and will be discussed later in this section.

If the data set is of an even number of data points, when there is no exact middle point, this is taken as the average of the two middle points. For example, in the series . . .

10, 11, 12, 14, 18, 26, 39, 60

... there is no one single middle point, so we must create one, which in this series is the midpoint between 14 and 18, i.e. 16.

So each set of data has both a mean and a median. Sometimes they are the same number (or are close together), but in other data sets they may be very different.

We can classify just about all quantitative data into one or two classes. What are these groupings?

What are the mean and the median of the data set: 156, 217, 199, 306, 224, 256, 178, 267, 317? A calculator will probably be useful in working out the mean, although it can be done by hand.

Variance

A second concept in getting to grips with a set of data is its **variance**, a term derived from its root word 'variety'. This word helps give us information such as the highest and lowest points in a particular group of data, and the extent to which the data clusters tightly around the central point of the mean or the median. If it is tightly clustered, the data is said to have low variance: if more diverse (spread out), it has high variance. We use one of two particular measures of variance depending on the nature of the central point (the mean or the median). When the central point of a data set is the mean, we use **standard deviation** (SD) to describe the variance as illustrated in Box 5.3. However, if the data is such that the central point is the median, then the variance is described in terms of the **inter-quartile range** (IQR) as described in Box 5.4.

The SD provides an idea of the degree to which the data is clustered close about the mean value, or is more spread out, i.e. the variance. For example, suppose a data set has a mean of 100 and a SD of 20. The SD is telling us that most of the data points (in fact, about two-thirds of them) lie within 1 SD either side of the mean, that is between 80 and 120. However, we can go further, and often find that nearly all the data points (actually, about 95% of them) are between 60 (the mean minus two SDs) and 140 (the mean plus two SDs). However, if another data set has a similar mean of 100, but a much smaller SD, such as 8, then 95% of the data points should range between 84 (100 minus 16) and 116 (100 plus 16). So a small SD tells us that the set of data is tightly clustered near to the mean, and when the SD is large, this means that the data is more spread out.

When describing the first set of data we would generally say that the mean and SD are 100 (20), although some use the notation 100 ± 20. Similarly, the second set of data can be written as 100 (8), or perhaps 100 ± 8. The SD is quite hard to obtain on a simple calculator, but is easier to obtain using a programmable calculator and, of course, on a computer loaded with a statistical software package.

Figure 5.1 illustrates variance with two different patterns. It shows two sets of data, pattern A and pattern B, which both have the same central point value of about 95. However, the two patterns have very difference variances. One of them (pattern A, lower) covers a much greater range (from about 65–120) compared with the other (pattern B, upper), which runs from 90 to about 98. So in both cases the mean is 95, but the SD of pattern A is 9, whilst the SD from pattern B is only 2.

The second method of describing variance, the inter-quartile range is derived from the data points that are one-quarter and three-quarters of the way into the complete data set when it is ranked from lowest to highest. So starting from the lowest point, and working our way up,

BOX 5.3 Variance, standard deviation, and standard error

The expressions variance, standard deviation, and standard error are worthy of further explanation. Take, for example, the data set we have already seen in Box 5.1.

21, 22, 24, 25, 25, 27, 27, 28, 29, 30, 35

The mean value is 26.6, but some numbers (such as 35) are greater than this and some (such as 21) smaller. The difference between these two numbers and the mean is 8.4 and 5.6, respectively. The variance of a data set is simply the degree to which individual data points are far from the mean, and this can be quantified. This difference (i.e. 8.4 or 5.6) is officially called the 'deviation'.

So how is the variance of this data set calculated? The first step in defining the variance is to take the square of each individual deviation (i.e. 8.4 becomes 70.54 and 5.6 becomes 31.36). Next we add up all the squares of the individual deviations, giving something called 'sum of the squares', which in this case adds up to 154.56. The number of data points in the set is called 'n', so in this case $n = 11$. To obtain the variance we divide the sum of the square by $n - 1$, hence 154.56/10 = 15.46 (rounded up).

The standard deviation is the square root of the variance (15.46) and = 3.93.

If we divide the SD by the square root of the number of data points (n) then we get another index, the standard error (SE) which is therefore: 3.93/3.31 and = 1.19.

So we can summarize our data set as:

Mean = 26.6
$n = 11$
Variance = 15.46
SD = 3.93
SE = 1.19

Thus, the variance of a data set is simply a stepping stone to the standard deviation, and similarly the standard deviation is simply a stepping stone to the standard error. Note that the standard deviation is a fixed property of a data set, whereas the standard error is dependent on the number of items of data (i.e. on 'n'), and so is not fixed but can fluctuate widely depending on the sample size. What this means in practice is that if we summarize our data set as mean with standard deviation, we get a different result than if we summarize it as mean with standard error, i.e. 26.6 (3.93) versus 26.6 (1.19). This is why, in order to minimize confusion, data sets are always described in terms of mean and SD.

The SE is a badly misused index, especially when put together with the mean in analysing differences between groups. For example, if there were twice the number of data points in the sample, we would obtain the same mean (26.6) and almost exactly the same SD (3.8). However, the SE will be very different, coming down from 1.19 with 11 sets of data to 0.82 with 22 sets of data. Nevertheless, the SE does have a place as it gives us information about the accuracy of a particular technique or method.

All this can be checked by inputting the dataset into the statistical package of your choice.

BOX 5.4 *The inter-quartile range*

Consider the data set of 15 points:

65, 43, 79, 250, 82, 49, 148, 55, 77, 40, 516, 121, 59, 47, 156

In order to obtain the median and IQR, we must unravel, or sort out, the data and place it in a rank order, from lowest on the left to highest on the right, as follows:

40, 43, 47, **49**, 55, 59, 65, <u>77</u>, 79, 82, 101, **148**, 156, 250, 516

The middle or median value is the 8th one (i.e. has 7 points above and 7 points below) is 77 and is underlined. Similarly, the middle value of the lower group of 7 (from 40 to 65) is 49 (highlighted in bold), and the middle value of the higher group (from 79 to 516) is 148 (also highlighted in bold). So the IQR is 49 to 148.

after we have reached 25% of the data points, we have the 25th percentile. Continuing to work up the data set, the half-way point (after 50% of the data points) is the 50th percentile. This is the median, as the middle of the data set has been reached. Continuing up the rank, after 75% of the data points have been passed, the 75th percentile is achieved. Finally, the highest value is the 100th percentile, as 100% of the data points will have been assessed. So unlike the SD, the IQR is relatively easy to obtain from the raw data, as is illustrated in Box 5.4.

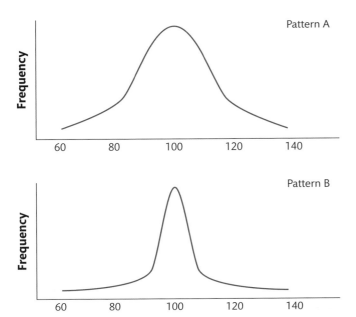

FIGURE 5.1
Illustration of the differences in variance between two different data sets that have a normal distribution and an approximately equal mean value (i.e. the 'peak' of each distribution). Despite the equivalence of a mean value, variances are different – the upper figure is more rounded or spread out, whereas the lower figure is more of a sharp peak. It is likely that the standard deviation of the upper distribution is greater than that of the lower distribution.

The IQR, like the SD, also gives us an idea of the spread of the data. We would write a summary of the data in Box 5.4 of the median and IQR as 77 (49–148):

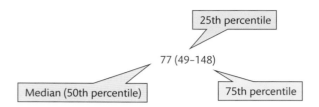

Note that the median value here (77) is certainly not in the middle of the IQR of 49–148. Clearly, 77 is much closer to 49 than it is to 148, and this fact is an important feature of this type of data.

Distribution

We bring the concepts of the central point and variance together to describe the distribution of a set of data. This is effectively what the data set 'looks' like when represented graphically. The vast majority of data in biomedical science is generally of a **normal distribution**, and includes many laboratory and physiological indices, such as haemoglobin and albumin concentrations in serum, heights, and body mass indices. In some cases, it may be described as 'bell-shaped'. An important component of data with a normal distribution is that the mean is taken to be the central point of the data set. When data is normally distributed, the mean and median are close together (such as 26.6 and 27, respectively, in the data set in Box 5.1), and the SD is relatively smaller than the mean [e.g. 100 (20)]. Figure 5.2 provides an illustration of a data set that is normally distributed.

Data that is **non-normally distributed** is less common, perhaps the best example from biochemistry being concentrations of serum triacylglycerols in a healthy population, and from physiology the duration of a pregnancy. This type of data may also be described as 'skewed'.

Example 1

FIGURE 5.2

Representation of a normal distribution. The data is presented as histograms, with the tallest (i.e. of greatest frequency, with a value of about 100 units) roughly in the middle of the entire data set. The histograms of lesser frequencies are smaller as the data sets move further from the middle towards the two polar extremes (where, on the left, the data is as low as perhaps 65 units, and on the right, where the data has a point at 120 units).

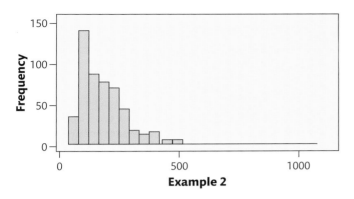

Example 2

FIGURE 5.3

Representation of a non-normal distribution. The data is presented as histograms, with the tallest (i.e. of greatest frequency, with a value of about 100 units) far to the right of the entire data set. Indeed, there is only one histogram to its left. Conversely, there are nine histograms to the right, and the frequencies fall roughly as the value of the data set rises, with a maximum to 500 units.

A characteristic of data with a non-normal distribution is that the mean and the median are always far apart (e.g. the mean of 35.8 and the median of 25 in Box 5.2), so that the average of the data set is not the middle point. Furthermore, when data is non-normally distributed, the SD is often quite large compared with the mean. For example in Box 5.3, the mean is 118, but the SD is 120. This helps confirm that the data set has a non-normal distribution. So in these cases, we use the median to define the central point and the IQR to define the variance. These points are illustrated in Figure 5.3.

Why the fuss about these two different modes of distribution? First, we need to know about distribution so we can accurately find out if there is a **statistically significant difference**, between different sets of data, such as the blood pressures of one group of people compared with those of a different group. You might think that the blood pressures of one group at, for example, 156 (25) mmHg is significantly different from that of another group, which may be 132 (21) mmHg. Indeed, this difference of about 18% seems large and so may be significantly different. However, as scientists we must apply scientific methods and ensure that any difference does not merely 'seem' significant, but actually is statistically significant.

We can establish whether or not a statistically significant difference exists by using a suitable statistical test. These generally involve complex mathematics that are best left to a computer. Nevertheless, the actual test that we would choose to use depends on the way in which the data are distributed: we would use a different test if both the data sets we seek to compare are normally distributed than we would if one or both sets of data are distributed non-normally. These tests will be introduced to you in Section 5.2.

A second occasion when we need to know about the distribution is in order to be able to interpret a particular result from a certain individual. For example, is a serum triacylglycerols result of 2.1 mmol dm^{-3} normal or abnormal? This is important as a high level may indicate a particular syndrome, the presence of a particular disease, or an increased risk of a heart attack and, if so, we should be concerned. In this case, we need to compare the value from that individual with the results of the same test from a large number of people we know to be healthy. Data from this large group make up what is called the **reference range**. It is necessary to know the distribution of triacylglycerol data in this healthy population to be able to make a judgement of whether or not a result from an individual is within or is outside this reference range.

Let us suppose that 100 healthy people provide blood for a triacylglycerol test, and that the results that come back have a non-normal distribution with 95% of the results lying between 0.5 and 2.0 mmol dm^{-3}. So under these conditions the result for the individual in question of 2.1 mmol dm^{-3} only just exceeds the top of this reference range, and so may possibly require additional investigation.

However, if the reference range has a normal distribution, it will provide different values that we would consider to be healthy, such as between 0.8 and 1.5 mmol dm^{-3}. If so, then the result from the individual patient of 2.1 mmol dm^{-3} is of greater significance because it is much higher than the top of the reference range and so may have significant clinical repercussions.

How can I tell if my data is normally or non-normally distributed?

The nature of the distribution is important because you will need this to be able to apply the correct statistical test to your data (this will be explained in the section that follows). If you chose the wrong test, you might well get an incorrect answer. Broadly speaking, there is a small number of simple rules to determine whether or not data is distributed normally or non-normally.

- If the SD is a lot smaller than the mean, then the data is more likely to have a normal distribution. We have already seen data like this, i.e. mean of 100 with a SD of 20. However, if the SD is large compared with the mean, such as mean/SD of 100/90, then the data is more likely to be distributed non-normally

- If the mean and median have similar values the data is likely to be normally distributed, e.g. mean 26.6 and median 27 (a difference of only 1.5%), as in Box 5.1. If the mean and median are far apart, such as 35.8 and 25 (a difference of 43%) as in Box 5.2, then the data is likely to be non-normally distributed

- A problem with these two methods is that they work very well only for data where the nature of the distribution (e.g. mean 34, SD 2.5) is reasonably obvious and/or you have a great deal of experience. For this reason, use of professional statistical software packages capable of determining the type of distribution is strongly advised

Once the nature of the distribution of sets of data has been determined, we can look to see if there are differences between them applying the correct statistical test. This will be explained shortly.

Qualitative data

Here, information is gathered and analysed primarily as words (single or in phrases), possibly in narrative form or with descriptive quotations. The proponents of qualitative research quite reasonably argue that not all information can or should be reduced to a number. For example, how can you reasonably quantify (place a numerical value on) someone's feelings, attitudes, or opinions, such as a belief, fear, and love? Qualitative data is often obtained from people using surveys, interviews, or observations, generally from a validated or structured questionnaire. An alternative would be to observe, and record verbal or written comments from a defined focus group.

A further distinction between the qualitative and quantitative approaches is not so much with the collected information itself, but the purpose to which that data is directed. The tradition from which quantitative research has evolved is of prediction and causal analysis, both of which are central in asking research questions. For example, we can suppose that a given group of patients will experience a change in a defined index (blood pressure, blood cholesterol) following treatment with a particular drug. In contrast, qualitative research generally does not generate numerical data that can easily answer a precise research question. Instead, the approach often aims to develop a theoretical explanation for the subject of study, the strength of which is therefore not in predicting an outcome, but in an increased understanding of the topic being investigated. Thus, the qualitative researcher may attempt to gain insights into an individual's feelings and thoughts, often within their own natural setting.

The questionnaire is often at the centre of many qualitative research projects. Considerable effort must be directed to ensure, as far as is possible, that questions are appropriate, unbiased, and cannot be misunderstood. For example, whilst 'permit' may be seen as the opposite of 'forbid', the latter word is viewed by many as pejorative, whilst 'not permit' elicits more responses. Questions themselves may be open or closed. In open questions, the respondent is allowed to use their own words and phrases, which can result in dozens of different answers that have no clear theme from only a handful of subjects.

Alternatively, the questioner may offer a dozen or more alternative words, and let the subject choose those he or she is happiest to use. In closed questions, the respondent is offered only a small number of answers, such as yes, no, and don't know, or may be asked to provide a response from a scale such as agree strongly, agree mildly, neither agree nor disagree, disagree mildly, and disagree strongly. A weakness of this latter scale of five is the temptation to sit on the fence and give the middle response, and be neutral, a position easily changed by simply removing the middle expression and reducing the number of options to four, thus forcing a choice. An additional problem with questionnaires is that respondents sometimes give a different response if re-tested, giving rise to fears of poor reproducibility, and so questionable reliability.

5.2 Analysis and interpretation of data

Once we have data, what do we do with it? As biomedical scientists we use scientific methods to understand the disease process with a view to providing a diagnosis and then monitoring treatment. For many of us, this will be in comparing the pathology result from an individual whose health status is unclear with that from a group of individuals we know to be healthy. We have already seen how we use a comparator group of healthy people to provide a reference range for serum concentration of triacylglycerols in order to determine whether or not the result from a single patient gives us cause for concern.

In other studies (possibly in research), we may compare data from a group of individuals (such as with a particular complaint, symptom, or frank disease, often referred to as the cases) with that from another group generally considered to be free of the problem (the controls – hence, a case-control study). In both types of analysis, the control group provides the normal or reference range.

One reason to be able to correctly collect and analyse data is to produce good research. A key step in research is to be able to fashion what it is you want to find out in terms of a question, and then turn it into a statement, generally called a **hypothesis**. Indeed, much of the research process is called hypothesis testing. People with an interest in heart disease may ask 'Does the blood cholesterol level in one group differ from that in another group?' This can be turned into a hypothesis statement such as 'the blood cholesterol level of one group is higher than that in a different group'. Other researchers may form their hypothesis as 'the blood pressure of a group of patients will be reduced if they regularly take a particular drug'. A third group may hypothesize that workers in one industry have more cancer than workers in a different industry.

The first two hypotheses stated above rely on data (i.e. blood cholesterol concentration and blood pressure, respectively) that is continuously variable, and this type of data will be examined first. Later in this section, we will consider the type of data that will test the third hypothesis, which is categorical (i.e. people fall into only one of a small number of discrete groups, such as those with or without a well-defined disease or condition like cancer or diabetes).

In another, perhaps more formal research setting, suppose there is interest in determining the levels of cholesterol not as above in an individual, but in a group of individuals such as workers at a particular factory or people following a particular diet. One may therefore test the hypothesis that cholesterols are significantly greater in the blood of workers at the factory compared with levels in the blood of people working in an office. Once the hypothesis has been formed, the method for testing the hypothesis needs to proceed along a number of formal steps.

- First, it is essential to find out how many people need to be recruited to make sure the findings are reliable. An appropriate degree of **power** is required in the study. If the study is underpowered and not enough people have been recruited, then the difference in levels of cholesterol between the two groups, although seeming large, may not be *significantly* large. This can lead to a false negative (i.e. we have failed to find a particular difference to be statistically significant when in fact it is indeed significant). A statistician should be consulted at an early stage to determine the number of subjects to be recruited. In this example, we should recruit 35 people in each group

- Before embarking on a study the approval of the local Ethics Committee must be secured. This body ensures that patient safety is considered and that good research practice is followed. In the United Kingdom, every hospital should have a Research Department from whom advice (and permission) should be obtained. It may also be necessary to obtain the funding to perform the blood test and cover the costs of analysis

- The third step is to recruit the people themselves, take the blood samples, and then analyse them in the laboratory

- The statistician will then be consulted once more to analyse the data. The first step will be to enter the data into a statistical package (such as *Minitab* or *SPSS*) on a computer which can then define the nature of the distribution of the data (normal or non-normal) for each group. Although it is well established that cholesterol data is distributed normally, it is nonetheless good statistical practice to check this is so

Probability

It is well established that normal, healthy adult men are generally taller than normal, healthy adult women. However, this may not be the case in all populations. Women who have grown up with high levels of growth hormone are likely to be taller than men who have grown up with low levels of growth hormone. Despite this, at the practical level, we can predict that the average height of even a small group of men is likely to be taller than the average height of a similar sized group of women. However, statisticians use the word **probability** (p) to give a more scientific and secure basis to this likelihood. In biomedical science, we are keen to establish whether or not a difference in two sets of data is genuinely due to, for example, a pathological process, or is due simply to chance.

Statisticians have developed a consensus that says that we are prepared to accept a difference as real if the probability of it being a real effect (that is, *not* due to chance) is greater than 95%. We express 95% in the decimal form as 0.95. If we're accepting that there is a 95% probability that the difference is real, then we are also accepting that there is a 5% probability that the difference could be coincidental, that is, due to chance. We express 5% in decimal form as 0.05. Hence, our requirement for p to be less than 0.05 (i.e. $p < 0.05$). It follows that if $p = 0.06$ (i.e. we have a likelihood of 94% that the difference is genuine), we do not consider this to be of sufficient reliability, and so describe it as statistically not significant.

Key Points

$p < 0.05$ – the level of probability at which we accept that the difference is real and true, and not merely due chance, co-incidence, or some other factor.

It follows that if the chance of an effect being spurious is only 1 in 10, i.e. $p = 0.1$, then this difference is not statistically significant because 0.1 is greater than 0.05. What would be even more significant would be that a difference is so large that the probability of such a difference being spurious or co-incidental in only 1 in 200 (i.e. 0.5%), which means that the chances of the effect being real are 199 chances in 200 (i.e. 99.5%). Thus, a probability of 1 in 200 gives $p = 0.005$, a result that is considerably less (indeed, 10 times less) than 0.05. Overall, the smaller the p value, the greater is the likelihood that the difference is real and is not due to chance.

SELF-CHECK 5.3

Why is it necessary to have enough power in a research study?

Analysis of data from one individual

Suppose that one of your close relatives has had a blood test for lipids and tells you that he or she has a serum total cholesterol result of 5.9 mmol dm^{-3}. Are you concerned? This depends on the age, sex, and some other health issues of your relative. However, an initial question would be to ask 'Is this result within the reference range?'

As we have discussed, the reference range is composed of results from hundreds or even thousands of people who are supposed to be healthy, such as blood donors. Using the example of cholesterol, we would expect data from a large pool of healthy people (say, 400) to have a normal distribution, with a mean of 4.5 mmol dm^{-3} and a SD of 0.5. As already described, an essential component of the relationship between the mean and SD of a normally distributed index (such as cholesterol) is that the results from 95% of those people (i.e. about 380 of them) will lie between 3.5 and 5.5 mmol dm^{-3}; this range being derived from the mean value plus two SDs and the mean value minus two SDs.

The value of 95% of the data points (as opposed to, for example, 70 or 80%) is chosen because it is most likely to provide representative and reliable information. This is why we often see mean and SD data being presented, as in this example, 4.5 ± 0.5 mmol dm^{-3}. So if 95% of the results of this population are within two SDs either side of the mean, what about the remaining 5%? There are likely to be as many at the bottom end of the scale as at the top end of the scale, and in our example it means that 10 people will have a cholesterol concentration below 3.5 mmol dm^{-3}, and another 10 will have a value greater than 5.5 mmol dm^{-3}. These people are not immediately to be considered as being in ill health. This is illustrated in Figure 5.4.

So given the above criteria, a serum cholesterol of 5.9 mmol dm^{-3} is certainly above the top of the reference range of 3.5–5.5 mmol dm^{-3}. However, is this person in ill health? This we cannot say as more details are needed before we give the individual a potential diagnosis, which

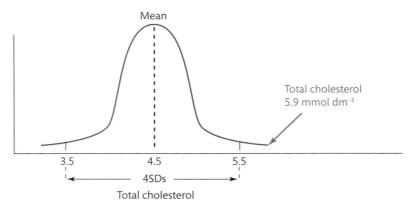

FIGURE 5.4

A representation of a normal distribution of concentrations of total cholesterol in a healthy population. The mean value is 4.5 mmol dm^{-3}. If the standard deviation of this data set is 0.5 mmol dm^{-3}, then 95% of the individual data points should lie within 4 SDs of the mean, that is, between 3.5 and 4.5 mmol dm^{-3}. It follows that a total cholesterol result of 5.9 mmol dm^{-3} is outside this 95% region. It is necessary to stress that we do not take this result of 5.9 mmol dm^{-3} to be abnormal. However, in the face of other information, such as a family history of premature cardiovascular disease, then it may prompt action.

Cross reference

You can read about hypercholesterolaemia in Chapter 9 'Abnormalities of lipid metabolism' in the companion book, *Clinical Biochemistry*.

in this case would be hypercholesterolaemia. Certainly, we feel that the individual should be made aware of this high cholesterol result, and generally instructed on the risk factors for atherosclerosis and their contribution to heart attack and stroke.

Whilst experienced biomedical scientists will become familiar with the reference range over a period of years, so that they will be able to spot an unusual result promptly, at the practical level, many blood analysers in pathology laboratories (principally in haematology and biochemistry) can be programmed to create an alarm if they consider a result to be so unusual (and possibly life-threatening) as to warrant immediate attention. This extends to other results that are outside the normal range. The results sheets that go to the ward or clinic will also have abnormal results 'flagged' – often simply marked with an asterisk, or the letters H (for high) or L (for low) to draw attention to them.

Although the cholesterol result mentioned above (5.9 mmol dm^{-3}) may be markedly raised compared with the reference range, it is generally inappropriate to say it is 'significantly' raised in the statistical sense of the word. However, it is possible to define the exact probability that a single laboratory result is outside the reference range, but this analysis is complex and is beyond the scope of this chapter. We reserve the use of expressions such as significance and probability for situations where there is a different mode of analysis, such as that of data from groups of individuals or sets of data, as we shall now examine.

Analysis of data from two groups of individuals

In looking at data from two groups of individuals, different statistical tests should be used, depending on the nature of the data and what hypothesis is being tested:

- In comparing continuously variable data from two different groups (such as certain blood lipids), the **Student's *t*-test** or the **Mann–Whitney *U*-test** would be used, depending on the distribution of the data. i.e. normal or non-normal, respectively

- If a relationship between two sets of continuously variable data within a single group of individuals is sought (such as between height and weight), then a test of **correlation** should be used

- If the data is categorical (such as the proportions of people with cancer in one group compared with another group) the **chi-squared test** is used

- **Paired tests** are used to determine any difference between two sets of data from the same individuals taken at different timepoints, or that are linked in some other manner

We will now look at each of these types of analysis in turn.

Continuously variable data from two different groups of subjects

Suppose a researcher wants to see if there is a difference in the levels of certain blood lipids (such as total cholesterol and triacylglycerols) between different groups of people, such as 30 factory workers and 30 office workers. When comparing two groups of data that are continuously variable, we use the Student's t-test or the Mann–Whitney U-test to determine if such a difference is present. Which test to use depends on the distribution of the two sets of data. If both sets of data are of normal distribution, then the Student's t-test should be used. Alternatively, if one or both sets of data are non-normally distributed, we use the Mann–Whitney U-test. Thus, a key step is to determine the distribution of the data.

It is accepted that total cholesterol is normally distributed, and so we would present the data as mean and SD. Let us say that the results from the factory workers is that the total cholesterol is 5.9 (0.7) mmol dm^{-3}, and that the results for the office workers is 5.1 (0.6) mmol dm^{-3}. By applying the Student's t-test, a p value of 0.025 is obtained. This is less than our cut-off point of $p < 0.05$, so it can be said with confidence (i.e. $p = 0.975$, or 97.5%) that the difference is *statistically* real and is not spurious.

By contrast, serum triacylglycerols have a non-normal distribution, and as such would be presented with a median and IQR. The results for the factory workers is 1.7 (1.2–2.8) mmol dm^{-3}, whilst in the office workers it is 1.1 (0.9–1.6) mmol dm^{-3}. This data would be analysed by the Mann–Whitney U-test, and gives a probability value of $p = 0.002$. This is a highly significant difference, as it can be said that the probability that the difference is real is 99.8%, whilst the probability that the difference is spurious is only 0.2%.

Therefore, there is a difference in total cholesterol levels of $p = 0.025$, and a difference in triacylglycerol concentrations of $p = 0.002$. The questions that therefore follow may include, 'why are these differences present?' and, perhaps later, 'if these differences have clinical consequences, do we need to act on either of them?'

Two sets of continuously variable data from a single group of subjects

Not only can we compare a set of data (such as serum cholesterol) from two different populations (such as in factory workers and office workers), we can also look at two different sets of data (such as serum cholesterol versus blood glucose concentrations) *within* a single population (perhaps just the office workers). Another two sets of illustrative indices are the age of an athlete and their athletic performance (such as distance they can run in a set time). Age is not the sole determinant of the performance of an athlete, and a fit older athlete with more experience and determination may well out-perform someone younger and less fit. Correlation is explained in Box 5.5.

BOX 5.5 Correlation

Consider these data from 12 people: their height, weight, age, and distance that they can run in a certain fixed time period such as 30 minutes.

Subject	Height (m)	Weight (kg)	Age (years)	Distance run (m)
1	1.56	70	43	6100
2	1.23	65	56	5200
3	1.70	72	29	6900
4	1.81	84	59	4800
5	1.46	72	34	7100
6	1.50	70	56	4800
7	1.59	66	45	4500
8	1.66	79	72	3700
9	1.70	74	56	5100
10	1.48	69	45	4800
11	1.85	88	67	4200
12	1.66	75	55	5250

By plotting the height and weight for each person in a graph we can obtain a plot such as . . .

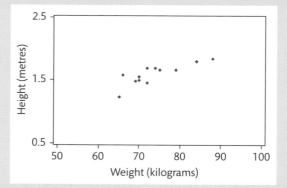

As it happens, the tallest person (1.85 m) is also the heaviest (88 kg), whilst the shortest (1.23 m) is also the lightest (65 kg). However, in between these polar extremes, the relationship is not clear-cut, although an overall trend is present. The correlation coefficient (r) gives an index of how well this data adheres to a straight line and so is a linear relationship. Because the two indices have a normal distribution, we use Pearson's method. In doing so, we obtain $r = 0.85$, which indicates a strongly positive relationship. It also happens that this is statistically significant, as $p = 0.001$.

Similarly, if we plot each person's weight against their age we can also get a graph. A quick look at the plot below also gives the impression that as someone's age increases, so does their weight. Indeed, the correlation coefficient for this data is $r = 0.56$, which would normally be a 'respectable' value. However, the probability that this data is a true reflection of the relationship between age and weight just fails to reach statistical significance as $p = 0.06$ (i.e. the likelihood of a difference being genuine is 94%).

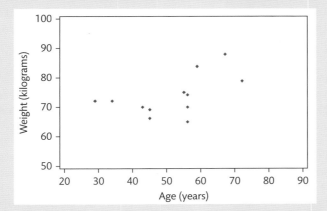

A close look at this data can provide clues as to why this seemingly good data is in fact not significant. Compared with the first height/weight plot, the data in this weight/age plot is more spread out (i.e. it has greater variance) and is not as good a straight line. A second reason why we do not have significance is because of the small numbers of data points and therefore the low level of statistical power. What this means in practice is that had we recruited an appropriate number of subjects we would have undoubtedly obtained significance with $p < 0.05$.

These first two plots indicate a positive or direct relationship between the indices, i.e. as one index increases, the other index also increases. However, we say that the relationship is *inverse* if when one index increases, the other falls. This can be illustrated by the plot of age versus the distance that the subject can run in 30 minutes.

Analysis of this data points to an excellent correlation coefficient where $r = 0.87$, slightly better than the relationship between height and weight. Indeed, the probability that this data truly does reflect a real association is highly significant, with $p < 0.001$. However, as the relationship is inverse, then we have to place a minus sign before the correlation coefficient, i.e. $r = -0.87$. If all the data points were to be found on a straight line, then the correlation coefficient would be -1. In practice, however, this is almost never found in biomedical science.

An essential aspect of correlation is to be able to define the extent to which the two indices correlate, and for this one uses a **correlation coefficient**. This is represented by the Greek letter *rho* (ρ) to define this value. A strong association between two sets of indices would be represented by a correlation coefficient where *r* is close to 1 (say, 0.92), whereas one would consider the relationship to be weak if a small *r* value of perhaps 0.15 was obtained.

We must be cautious about interpreting correlations. Just because two indices correlate strongly, does not mean that one causes the other. A crucial comment is always to recall that:

correlation does not imply causation

A suitable example of this is the relationship between height and weight. It is quite well established that, in general, tall people are heavier than short people, but it is just as well known that two people of the same height can have very different weights (and vice versa). It is also evident to most people that, in general, height correlates with weight. However, is this because the taller someone is, the heavier they become, or is it that the heavier one is, the taller they become? The former seems more likely. A better example from biomedical science is from human epidemiology and clinical studies where it is known that in a large population systolic blood pressure generally correlates very strongly with diastolic blood pressure. However, the increased systolic value does not *cause* the diastolic value to rise; and both systolic and diastolic blood pressures rise in parallel and do so independently of one another.

An amusing example of a relationship between correlation and causation occurred during the 1950s, when the rise in the import of bananas into the United Kingdom correlated strongly with an increase in the divorce rate. It could be argued that the increased consumption of bananas caused marital problems and, hence, a rise in the divorce rate or, equally, that newly single people eat more bananas once they are divorced. Both hypotheses are interesting.

As with comparing two groups of data from different populations, where we use the Student's *t*-test if the data is normally distributed, and the Mann–Whitney *U*-test if it is non-normally distributed, we have a choice of different tests of correlation. If the two sets of data have a normal distribution (such as height and weight), we use **Pearson's correlation method**. However, if one or both sets of data have a non-normal distribution (such as total cholesterol and triacylglycerols) we have to use **Spearman's correlation method**.

Analysis of continuous data obtained at two time points

So far, we have been looking at data that is generally obtained at a single time point, but it is also necessary to be able to assess whether or not there is a change in a particular index in a group of people at two time points. This type of data is called 'paired' because there are always pairs of figures (i.e. before and after), never just a single point. In clinical medicine, as part of the development of new therapeutics, we need to know how the new drug genuinely influences the biological system it is designed to act upon. For example, to be convinced that a new drug designed to lower the blood pressure actually does so, one needs data on the reduction of systolic blood pressure and/or diastolic blood pressure from the same people before they are placed on the drug, and then again later once they have been taking the drug for perhaps weeks or months. This is shown in Box 5.6.

As before, the exact statistical test used depends on the nature of the distribution of the data: if the difference between the two sets of data is normally distributed, then a **paired *t*-test**

BOX 5.6 *Data at two time points*

Suppose we wish to determine the effect of a new drug on the functioning of an organ like the kidney. One of the most useful indices to test renal function is the concentration of creatinine in the blood. We must first identify a group of patients with kidney problems who will therefore be likely to have abnormal levels of serum creatinine. We need to obtain a blood sample before the drug is issued, and then place the patients on the new drug for perhaps several months, and then take a second sample of blood. Typical results may be:

Concentration of creatinine (mmol dm^{-3})

Patient number	Before the drug	After the drug	Difference
1	150	126	−24
2	225	196	−29
3	176	168	−8
4	166	172	+6
5	145	140	−5
6	226	196	−30
7	189	168	−21
8	168	173	+5
9	173	146	−27
10	149	144	−5
11	171	155	−16
12	156	149	−7
Mean (SD)	174 (27)	161 (21)	−13 (13)
Median (IQR)	169 (151–186)	161 (144–172)	−12 (−5–26)

Cross reference

Renal function is described in Chapter 3 'Kidney disease' in the textbook *Clinical Biochemistry*.

Note that in 10 of the 12 patients, the level of creatinine has decreased. However, in two patients (numbers 4 and 8) levels have increased. Nevertheless, overall, there has been a decrease in creatinine that, assuming no other changes in the lives of the patients, implies that the drug may be active in alleviating the renal disease in these patients.

However, to be fully confident, we need to apply the correct statistical test to the data. The choice is a paired *t*-test, or the Wilcoxon test. Although the mean difference (−13) is the same as the SD (13), it is virtually the same as the median difference (−12), so we can be fairly confident that the data is normally distributed and so the test to be used would be a paired *t*-test. This test will give $p = 0.004$, which is considerably smaller than the required cut-off point for statistical significance of $p < 0.05$. In fact, this *p*-value tells use we can be 99.6% confident that the effect is real, and only 0.4% confident that the difference is due to chance.

NB. Strictly speaking, in this type of study, we should also measure creatinine in a similar group of patients over the same time period who have not been taking this particular drug. We call these subjects the control group – this is a vital consideration when designing and carrying out experiments.

is an appropriate test. However, if the difference between the two sets of data is distributed non-normally, then **Wilcoxon's test** should be used.

Paired analyses need not be linked in time – they may be linked in other ways, such as the circumference of the left arm compared with that of the right arm in the same person, or the difference in a component of the blood measured in serum compared with plasma.

Analysis of two sets of categorical data

An example of categorical data is the number of people who fall into only one of a small number of well-defined categories, such as being male or female, whether or not you smoke cigarettes, your ABO blood group, or one's marital status. In pathology, alternatives may be the presence of a certain cancer, the presence of a condition such as diabetes, or a history of heart disease.

Let us use diabetes and heart disease as an example, and form an hypothesis such as 'people with diabetes have more heart disease than people without diabetes'. In order to test this hypothesis we need to examine the presence of heart disease in two groups of people: those with diabetes and those free of diabetes. Those free of diabetes we call the control group. First, as with the example of levels of cholesterol in factory workers versus office workers, we need to find out how many people to recruit to have sufficient statistical power for our result to be reliable; this may be 100 people in each group. As with the definition of a normal or a non-normal distribution, there are statistical programs available that can tell you the exact number of people you need to recruit.

If we find that 35 of the 100 people with diabetes have heart disease, and that 20 people free of diabetes have heart disease, is this seemingly large difference in proportion (i.e. 35% versus 20%) statistically significant? The most appropriate test to use in this instance is the chi-squared test, which effectively compares the proportions of those with and without heart disease in the diabetes group and control group free of this risk factor. We summarize this data in Table 5.1.

TABLE 5.1   Number of people in each group

	Diabetics	Non-diabetics	Total
With heart disease	35	65	100
Free of heart disease	20	80	100
Total	55	145	200

The chi-squared test gives us a probability that the difference is real at $p = 0.018$ (that is, the probability that the results are down to chance is just 1.8%, making it 98.2% likely that the results are real, and are due to some genuine difference). The result of $p = 0.018$ is less than the cut-off point of 0.05 so our results support the hypothesis that diabetes is associated with excess heart disease – an established finding in human pathology.

Data from three or more groups of individuals

Analysis becomes increasingly complicated once we move away from two to three or more groups. The biggest difference is that the simple t-test and the Mann–Whitney U-test are appropriate only in the analysis of two groups. When we have three on more groups to analyse, we have to use a different test. **Analysis of variance** (ANOVA) is the correct test to use when all the groups have a normal distribution. However, if the distribution of even one of these multiple groups has a non-normal distribution, we have to use the **Kruskall–Wallis test**.

Figure 5.5 illustrates these points. The top part shows three sets of data: 'A', 'B', and 'C'. A cursory 'eyeball' view of the spread of this data gives the impression that it is normally distributed, as it indeed is, with most of the individual data points clustered around the central section of each of the data sets. So we have to use the ANOVA test to look for differences in this data, and in doing so we get $p = 0.001$. What this means in practice is that the probability of this difference being due to chance is only 0.1%. So around the other way, this means it is highly likely (to 99.9%) that there is a real difference between some of the three individual data sets. What *can't* be done is a series of different t-tests between each pair in turn, i.e. 'A' versus 'B', then 'A' versus 'C', and then 'B' versus 'C'. This may well lead to error.

The mean, median, and SD for the data sets 'A', 'B', and 'C' in Figure 5.5 are as follows:

'A':   100, 100, 15
'B':   79, 82, 17
'C':   146, 150, 17

Although we should use a formal test to confirm the distribution of the data, it is possible to get an impression of distribution by looking at the values of the mean, median, and SD, as we did in Section 5.1. Now confirm with yourself that the three sets of data all seem to have a normal distribution.

However, ANOVA does not tell us exactly where the difference is to be found. Is it between 'A' and 'B', or between 'A' and 'C', or is it between 'B' and 'C'? To do this we have to use a second test. There are several to choose from, but **Tukey's post-hoc test** is one of the most popular (another is Dunnett's test).

Using this test, we find that there is a significant difference between 'A' and 'B', between 'B' and 'C', and between 'A' and 'C', each with a probability value of $p < 0.05$.

The lower part of Figure 5.5 has three different data sets: 'X', 'Y', and 'Z'. We can again do an 'eyeball' examination of the spread of this data, but it is distributed in a different manner to groups 'A', 'B', and 'C'. In each of 'X', 'Y', and 'Z' most of the data points are clustered not in the central part of the entire set (as they are in 'A', 'B', and 'C'), but instead are skewed over to the far left of each group, suggesting that each set has a non-normal distribution, which indeed is correct.

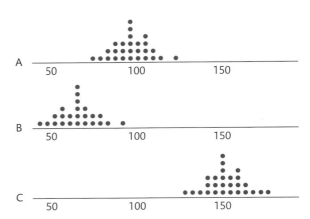

Dotplot for A–C

FIGURE 5.5

The top figure (Set A–C) shows three data sets on a common scale. As each data set has a normal distribution, the combined group must be analysed by one way analysis of variance. If significance is found (i.e. $p < 0.05$), then an additional test (such as Tukey's test) must be performed to find out where such a difference lies. In the present set, it is likely that each individual set is statistically different from the two other sets. The lower figure (Set X–Z) shows three data sets on a different scale. Unlike A–C, the patterns in X–Z are non-normal, as the bulk of the data points are skewed to the left. Because these data sets have a non-normal distribution, then the Kruskal–Wallis test must be used to determine if there is a difference between sets X, Y, and Z. However, if there is indeed a significant difference (i.e. $p < 0.05$), we cannot use Tukey's test to determine where the difference is because the latter can only be used when the data has a normal distribution. One way round this is to convert the distribution of the non-normal data set to a normal distribution, and then perform Tukey's test.

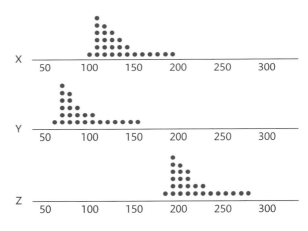

Dotplot for X–Z

SELF-CHECK 5.5

The mean, median, and SD for 'X', 'Y', and 'Z' are as follows:

'X': 150, 115, 121
'Y': 106, 85, 71
'Z': 257, 190, 176

As in the previous self-check, it is possible to get an impression of distribution by simply looking at the values of the mean, median, and SD. Again, look at the data – what factors lead you to suspect that these data have a non-normal distribution?

Just as it is inappropriate to use a series of individual t-tests when three groups of data are normally distributed (as in 'A', 'B', and 'C'), one *cannot* simply do a series of individual Mann–Whitney U-tests between 'X' and 'Y', then between 'X' and 'Z', then between 'Y' and 'Z'. Instead, the correct test to use is the Kruskall–Wallis test, which in this case again gives a high probability (i.e. $p = 0.001$) that there is a genuine difference between some of the groups. However, like the ANOVA test, the Kruskall–Wallis test does not tell us exactly which pairs of data are significantly different from each other, and to achieve this we have to use an additional test, such as Tukey's test, to which you have already been introduced.

The problem is that Tukey's test is designed to analyse data that has a normal distribution, and our data sets ('X', 'Y', and 'Z') all have a non-normal distribution. One way of getting around this is to convert data that is non-normally distributed into data that is normally distributed. This can be achieved by transforming the data into a logarithmic form that frequently (but not always) converts it to a normal distribution. This is called **logarithmic transformation**.

So once the data sets 'X', 'Y', and 'Z' data has been log transformed, and is effectively normal, an ANOVA followed by the Tukey test can be performed. In this case we find a statistically significant difference between 'X' and 'Z', and between 'Y' and 'Z' (both with a probability of $p < 0.05$), but the difference between 'X' and 'Y' was not significant.

Once more, analyses of these sorts are virtually impossible to do without a good statistical software package running on a computer.

Correlation of multiple groups

The ability to correlate three or more continuously variable sets of data at the same time is extremely complex and is very rarely required in biomedical science. Consequently, it is far beyond the scope of this chapter.

Analysis of multiple groups of categorical data

In analysing categorical data, we generally use the chi-squared test. Almost exactly the same analyses are performed when we look at three or more groups as for two groups. An example of this could be the number of children born to 293 women who live in different areas (143 live in Area One and 150 live in Area Two). This data is summarized in Table 5.2.

If we simply look at the percentages of the women in each of the three groups (no children, one child, or two or more children), we find 24, 46, and 30%, respectively, in each group for the women in Area One. Similarly, the percentages for women in Area Two are 17, 62, and 21%, respectively. A quick view of this data seems to suggest a difference – for example, 46% would generally be thought of as being considerably less than 62%. However, statisticians do not make claims simply by suggestion – a formal statistical test must be used to confirm our suspicion.

If we apply the chi-squared test to this data we get $p = 0.025$. This tells us that there is a high probability (i.e. 97.5%) that there is a meaningful difference, and only a 2.5% chance that the

TABLE 5.2 Number of children born to women in different areas

	No children	One child	Two or more children	Total number of women
Women in Area One	35	66	42	143
Women in Area Two	26	93	31	150
Total	61	159	73	293

TABLE 5.3 Comparing women with no children or with one child

	No children	One child
Women in Area One	35	66
Women in Area Two	26	93

TABLE 5.4 Comparing women with one child or with two or more children

	One child	Two or more children
Women in Area One	66	42
Women in Area Two	93	31

TABLE 5.5 Comparing women with no children or with two or more children

	No children	Two or more children
Women in Area One	35	42
Women in Area Two	26	31

differences are spurious. So it appears that women in Area One do indeed have a different distribution of the number of their children than the women in Area Two. However, as for the ANOVA and Kruskall–Wallis tests, we cannot tell exactly where this difference is. In Area One, is the group of 66 women (i.e. 46% of the whole group) with one child really different from the 93 women (62% of the whole group) with one child in Area Two? In order to do this we have to break up the data set and then perform individual chi-squared tests on each pair of data points as shown in Tables 5.3, 5.4, and 5.5.

A chi-squared analysis of the data in Table 5.3 gives a probability of $p = 0.034$. This means that the difference in proportions between the women in Area One compared with the women in Area Two is 96.6% is likely to be due to the 'Area' and not due to some other unknown factor (which would have a likelihood of 3.4%). A chi-squared analysis of the data (Table 5.4) gives a probability of $p = 0.023$ which means that the difference in proportions between the women in Area One compared with the women in Area Two is 97.7% likely to be due to the 'Area' and not due to some other unknown factor (which would have a likelihood of 2.3%). A chi-squared analysis of the data in Table 5.5 gives a probability of $p = 0.985$. This indicates that the difference in proportions is too small to be statistically meaningful. Indeed, that the likelihood of the difference being due to the difference in area is only 1.5%.

So in summary we can say that the difference in the proportions of the number of children born to 293 women living in two different areas is statistically significant to the level of $p = 0.025$ (i.e. we are 97.5% confident that the difference is genuine and only 2.5% as being due to chance). Looking at this group in more detail, we can say that this difference is due to a difference in the proportion of women with no children versus women with one child ($p = 0.023$), and also in the proportions of women with one child or with two or more children ($p = 0.034$). However, the difference in the proportion of women with no children compared with the proportion of women with two or more children was not statistically significant ($p = 0.985$).

Analysis of three or more sets of linked data

You have already seen how to analyse paired samples of plasma taken at two different time points or linked together in some other way. Similar analyses are possible for the assessment of data obtained at three or more time points or linked by three or more factors.

Consider the development of a new drug designed to reduce the numbers of platelets in the blood. It is necessary to know how long the drug takes to have an impact. Is it days, weeks, or months? Such an experiment would have to obtain blood samples on different occasions, then determine the platelet counts, and then submit the data to **two-way analysis of variance**, although this may also be known as repeated measures analysis of variance, or Friedman's ANOVA. It is described as two-way ANOVA because we need to be able to factor in the possible effects of differences (a) between patients, and also (b) between different stages of the experiment. This is described in Box 5.7.

BOX 5.7 Two-way analysis of variance

Consider these data from 10 people: the platelet counts (units: $10^9/cm^3$) have been measured before, and on three weekly occasions after the introduction of a new drug.

Although the mean platelet count falls from 324 to 312 in just a week, is this a significant drop? Note that levels then marginally rise to a mean of 315 after 2 weeks, then fall to a mean of 292 after 3 weeks. Overall, there is a significant change in the mean values with a likelihood of $p = 0.002$. However, note also the variation in results from different subjects. The average platelet counts over the whole of the experiment in subjects 5 and 9 both seem to be considerably less (256 and 265, respectively) than the averages from subjects 4 and 10 (369 and 390, respectively). Indeed, the probability that the difference between subjects is statistically significant is $p = 0.001$.

Thus it may be that any difference in the overall result may in fact be due to changes in only some of the subjects, which may not be truly representative of the entire data set. The complicated and powerful mathematics of the two-way ANOVA adjusts for any possible differences in particular groups.

Subject	Before drug use	After 1 week of the drug	After 2 weeks of the drug	After 3 weeks of the drug
1	299	278	284	256
2	314	325	311	289
3	287	281	294	271
4	379	376	366	356
5	269	252	261	241
6	312	311	335	326
7	325	329	356	313
8	378	311	278	251
9	267	275	261	258
10	412	387	404	356
Mean	324	312	315	292
SD	50	44	49	43

Data need not be linked by time in order to be analysed by this method. For example, we would use a two-way ANOVA to analyse levels of a component of the blood from the same person that has been measured in serum, in plasma anticoagulated with sodium citrate, and also in plasma anticoagulated with EDTA.

Finally, Figures 5.6 and 5.7 provide summaries and flow paths for some of the different methods of analysis we have been looking at, based on the nature of the data, and on how to present the data. Figure 5.6 refers to two sets of data, Figure 5.7 is three sets of data.

5.3 **Statistical packages**

Efficient handling and analysis of data is impossible without a good statistical package. As effective giant calculators, these packages can be purchased direct from the manufacturer, although in practice many large organizations, such as hospitals and universities will have site licences, allowing a *bona-fide* individual almost immediate access. Indeed, higher and further Education Institutions are almost certain to have a Department of Mathematics and/or Statistics from where packages, and advice, can be sourced.

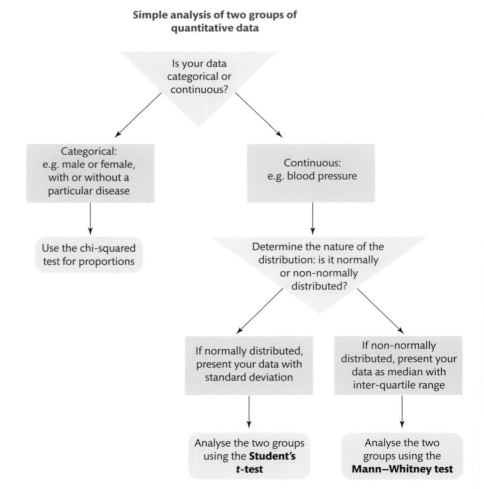

FIGURE 5.6

Flow chart for analysing two sets of data. For categorical data, use the path on the left. For continuous data, use the path on the right.

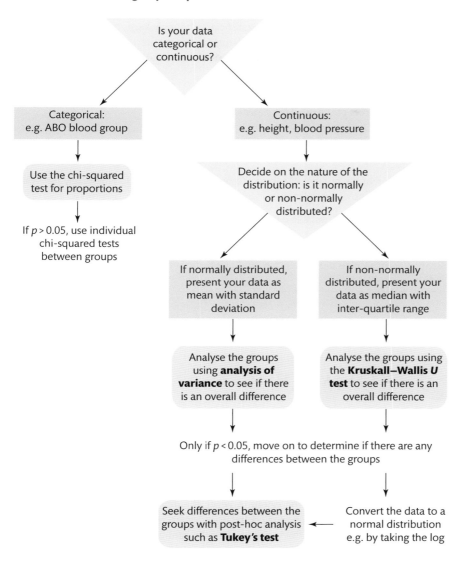

Simple analysis of three or more groups of quantitative data

Is your data categorical or continuous?

Categorical: e.g. ABO blood group

Use the chi-squared test for proportions

If $p > 0.05$, use individual chi-squared tests between groups

Continuous: e.g. height, blood pressure

Decide on the nature of the distribution: is it normally or non-normally distributed?

If normally distributed, present your data as mean with standard deviation

If non-normally distributed, present your data as median with inter-quartile range

Analyse the groups using **analysis of variance** to see if there is an overall difference

Analyse the groups using the **Kruskall–Wallis U test** to see if there is an overall difference

Only if $p < 0.05$, move on to determine if there are any differences between the groups

Seek differences between the groups with post-hoc analysis such as **Tukey's test**

Convert the data to a normal distribution e.g. by taking the log

FIGURE 5.7
Flow chart for analysing three or more sets of data. For categorical data, use the path on the left. For continuous data, use the path on the right.

Although claiming to be user-friendly, these packages are often complicated and are certainly not for the untutored. It is essential that those fresh to the subject complete a formal training programme, generally offered by their host organization (hospital, university, pharmaceutical company), although these can also be offered by commercial training organizations.

However, even the most sophisticated statistical packages are unable to counter the biggest problem in data management that is, the reliability of the data that is being entered. Even the most powerful computers and statistical packages are unable to correct data that is either incorrectly derived at source or is incorrectly entered by an operator – hence, the concept of 'garbage in – garbage out'. This is why the concepts of quality control and quality assurance are so important.

The best known statistical packages are (in no particular order), *Minitab*, *SPSS*, *SAS*, and *Stata*. All of these offer the most simple tests (such as Student's *t*-test and the chi-squared test) to the most complex analyses, but some are certainly more user-friendly to the occasional user, whilst others are designed for the full-time statistician. These packages often come with software to generate plots and figures (such as the figures in this chapter). Advice can generally be obtained from experienced users, themselves often in formal statistical units. Table 5.6 summarizes key aspects of the uses of statistical software.

TABLE 5.6 Examples of the need for statistical software

- Determination of the nature of the distribution of continuously variable data (i.e. normal or non-normal)

- Determination of the number of subjects to be recruited in order to adequately test your hypothesis (i.e. a power calculation)

- Analysis of continuously variable data (e.g. Student's *t*-test and the Mann–Whitney *U*-test)

- Analysis of categorical data (e.g. the chi-squared test)

- Generation of a correlation coefficient and graph showing the relationship between two indices

 SUMMARY

- Information can be as letters and words (i.e. qualitative) or as numbers (i.e. quantitative). Quantitative data can be categorical or continuously variable

- The central point of a continuously variable data set is the mean (the average) or the median (the middle)

- Variance is used to describe the extent to which the individual points of a data set are tightly bunched together around the central point, or are widely spread out. The standard deviation is used to quantify the variance when the central point is the mean. The inter-quartile range is used to quantify the variance when the central point is the median

- When data is normally distributed, the central point is the mean. When data is non-normally distributed, the central point is the median

- Statistically significant difference is present when a difference between two data sets is proven by a mathematical analysis to be significant

- The reference range is a group of numbers used to provide an idea of expected or desirable results

- Correlation describes the extent to which there is a relationship between two sets of continuously variable data

- The Student's *t*-test is the method for determining a difference between two sets of data that are both normally distributed. The Mann–Whitney *U*-test is the method for determining a difference between two sets of data where at least one is non-normally distributed

■ Statistical power is the extent to which your data set can be relied on to give valid results

■ Probability (abbreviated to *p*) is the extent to which your result is truly a consequence of a real effect (such as the disease process), or is different merely because of chance or co-incidence. Generally, when *p* is less than 0.05 (i.e. $p < 0.05$) a difference is statistically significant

■ The chi-squared test is used to look for a difference in the proportions of a particular index (such as ABO blood groups) between groups of individuals

■ Data at two time points can be analysed by a paired *t*-test or Wilcoxon's test

■ When looking for a relationship between two sets of data from the same individuals, we use Pearson's or Spearman's correlation methods

■ When analysing three or more sets of data, we use ANOVA or the Kruskal–Wallis test. The chi-squared test can be used to search for difference in the proportions of data from three or more groups. Two-way ANOVA is used to analyse three or more sets of data that are linked, perhaps by time

■ An appropriate statistical software package is essential to allow accurate statistical analyses. These packages can provide all the common statistical tests, including *t*-test, Mann–Whitney, and chi-squared tests. The package will also provide a correlation coefficient and tell you how much power you have.

FURTHER READING

- **Altman DG. *Practical Statistics for Medical Research*. Chapman & Hall, London, 1991.**

- **Daly F, Hand DJ, Jones MC, Lunn AD, McConway KJ. *Elements of Statistics*. The Open University/Addison Wesley, Maidenhead, 1995.**

- **Holmes D, Moody P, Dine D. *Research Methods for the Biosciences*. Oxford University Press, Oxford, 2006.**

- **Petrie A, Sabin C. *Medical Statistics at a Glance*, 2nd edn. Blackwell, Oxford, 2004.**

- **Swinscrow TDV. *Statistics at Square One*, 9th edn (revised by Campbell MY). BMJ Books, London, 1996.**

QUESTIONS

1. Calculate (to the nearest whole number) the mean and median of the data set: 25, 37, 29, 30, 43, 32 and 22.

2. A project assessed the relationship between indices A and B in 30 people, and found a correlation coefficient (*r*) of 0.36. Another project found the correlation between C and D in the same population to be $r = 0.45$. Which relationship is likely to be more significant?

3. Which test is used to search for a difference between three data sets, all of which have a normal distribution?

4. True or False: A data set with a mean of 23.4 and a standard deviation of 21.2 is likely to have a normal distribution.

5. Consider the two sets of data:

    Set A: 124, 99, 346, 150, 115, 140, 77
    Set B: 120, 179, 81, 96, 143, 118, 285

    Which of the following statements is **TRUE**?

    (a) Both sets of data are normally distributed

    (b) The median value of Set A is 124

    (c) A difference between the two sets can be determined by the chi-squared test

    (d) The two data sets can be correlated and will give a positive *r* value

    (e) The inter-quartile range of Set B is 96–179

Answers to self-check and end of chapter questions are available in the Online Resource Centre accompanying this book.

@ **Go to www.oxfordtextbooks.co.uk/orc/glencross**

6

Preparing and measuring reagents

Ian Graham

Learning objectives

After studying this chapter, you should be able to:

- Recognize the key features of top pan and analytical balances, and be able to use them to weigh materials

- Demonstrate the need for correct installation, calibration, and maintenance of balances

- Distinguish between ways of measuring and delivering volumes of liquids, and demonstrate an appropriate selection and application of volume measurement

- Recognize the key features of pipettors and demonstrate their correct use in measuring and delivering volumes

- Know why there is a need for calibration and maintenance of pipettors

- Distinguish between different ways of expressing concentrations and show how dilution affects concentration

- Demonstrate the ability to calculate concentrations, know how to prepare solutions of a particular concentration and make dilutions

- Show an awareness of correct practical procedures and health and safety requirements

Introduction

As a biomedical scientist, you will undertake much practical work where quantities of materials and solutions must be measured and delivered. You may have to dilute a blood sample with a solution such as saline, mix volumes of different solutions in a colorimetric assay or, perhaps, dilute a microbial suspension. Solutions may be bought in ready prepared or they may be prepared by yourself or others within the laboratory. In large laboratories, many routine tests will be automated and autoanalysers will make volume measurements and mix together different solutions. Although you may not have to prepare many solutions or undertake significant amounts of volume measurement and delivery, it is essential you understand how to

BOX 6.1 *The SI system and units of volume*

Europe and much of the rest of the world uses a set of units of measurement called SI units (SI stands for *Le Système International d'Unités*). Within the SI system, volume measurements are based upon the m^3, which is equivalent to 10^3 L. However, the USA and some other countries have not fully adopted SI units, and volumetric glassware and pipettes are usually calibrated in L, mL and μL. For these reasons, volumes will be quoted in both sets of units as appropriate. Table 6.1 shows the relationship between SI (based on the m^3) and non-SI (based on the L) volume measurements.

TABLE 6.1   A comparison of SI (m^3) and non-SI (L) volume measurements

SI based volume	m^3 equivalent	dm^3/L based equivalent
m^3		10^3 dm^3/L
dm^3	10^{-3} m^3	1 dm^3/ L
cm^3	10^{-6} m^3	1 cm^3/mL
mm^3	10^{-9} m^3	1 μL

Cross reference

You can read more about automation in Chapter 17 'Laboratory automation'.

undertake these activities as some assays may be performed manually. You need to be able to use accurately balances and pipettes, or other volume measurement devices and know how to make solutions, perform dilutions, and understand the effects of dilution upon concentration (see Box 6.1 for the SI system and units of volume). These are essential and fundamental skills for all biomedical scientists and the quality of your work depends upon their correct application.

6.1 **Balances and weighing**

This section will consider:

- Key features of balances and their use
- Care and maintenance
- Safe working practices

Types of balance

Cross reference

Centrifugation techniques are described in Chapter 12 'Centrifugation'.

Several types of balance are available. The simplest are beam balances, but more sophisticated top pan and analytical balances are normally used in clinical laboratories. You should be able to recognize Figure 6.1 as balances. Top pan and analytical balances are mainly used to measure a particular amount of a solid for preparing a solution, but they can also be used to weigh liquids. Simple beam balances have limited applications in laboratories. For example, they can be used in balancing a pair of centrifuge tubes that contain a suspension prior to centrifugation.

(a)   (b)

FIGURE 6.1
Photographs showing
(a) a top pan and (b) an
analytical balance.

The type of balance used determines whether **mass** or **weight** is being measured. An item is 'weighed' on a balance, but there is an important difference between mass and weight. Weight is *a force* (measured in, for example, Newtons, grams, or pounds) reflecting the effect of gravitational force at a particular location upon a mass (measured in, for example, grams or kilograms). Mass does *not* vary and is a measure of the amount of matter present. Although the two are related, the weight of a given mass will change if the gravitational force varies. A beam balance measures mass while spring balances and top pan balances measure a force, the weight of the item. Why this difference?

A beam balance consists of a beam supported at the midpoint by a near friction-free fulcrum. The sample is placed in a weighing or measuring pan at one end of the beam and one or more standard masses are placed in a scale pan at the other end to achieve equilibrium. The sum of the standard masses in the scale pan when balance is achieved gives the mass of the sample because gravity applies equally to each side of the fulcrum point. Thus, beam balances measure mass not weight. Variants of this arrangement have been introduced over the years. An asymmetric fulcrum arrangement reduces the requirement for large reference masses. For larger-scale applications, a floating platform may be provided by using a cantilever system. By comparison, spring and top pan balances are force-measuring devices, and have different forms of construction and operation that give weight measurements. Although they measure weight, the requirement for top pan and analytical balances to be calibrated means they can be used to measure mass. In practice, the terms *weight* and *weighing* have such common usage, they tend to be applied whether mass or weight is actually being measured.

SELF-CHECK 6.1

If you were a scientist on a manned lunar mission with a top pan balance as part of your equipment, would the mass and weight of an object be the same on earth and the moon or different?

Top pan and analytical balances differ in appearance to a beam balance, they include electronics, have digital displays, and are easier and more convenient to use. High levels of measuring precision are achieved by combining the application of mass changes, and a spring force in achieving the final balance.

Top pan and analytical balances

Top pan and analytical balances work in a different way from the simple mechanical beam balance. They are compact, robust, and measure to high levels of accuracy and precision.

The routine top pan balance may weigh to within 0.1 g or 0.01 g and cover a weighing range or **capacity**, up to 200 g, 1000 g, or more. They have a broad range of weighing applications, are simple to use, relatively cheap and robust if not misused. Analytical balances are a variant of the general top pan balance, measuring to within 1 mg (10^{-3} g) or less, and fully enclosed within a draught and dust shield that incorporates several sliding doors. Inevitably, these are more expensive and used for more demanding purposes where only small quantities are to be measured. Catalogues for electronic balances refer to **readability**, which is simply the number of decimal places to which that particular balance can be used. A balance with a readability of 0.1 g would be used in general applications that differ from the more demanding applications of one with a readability of 0.0001 g.

SELF-CHECK 6.2

Which one of the following sets of features would you associate with an analytical balance?

(a) Cheap to buy, high levels of readability, includes a draught and dust shield, generally used for weighing smaller quantities

(b) Relatively expensive to buy, low levels of readability, includes a draught and dust shield, generally used for weighing larger quantities

(c) Relatively expensive to buy, high levels of readability, includes a draught and dust shield, generally used for weighing smaller quantities

(d) Relatively expensive to buy, high levels of readability, lacks a draught and dust shield, used for weighing larger quantities.

Whatever the type, modern electronic balances are designed for a long service life with minimal maintenance if not abused. However, they do require regular recalibration to ensure accurate and consistent measurements.

Uses of laboratory balances

Normally, a balance would be used to weigh dry or partially hydrated solid materials to be used in the production of a solution. Section 6.4 of this chapter considers matters related to the production of solutions and the effects of dilutions. The amount of material needed depends upon both the concentration and the total volume of the solution to be prepared. Although it is relatively easy to calculate the required amount to be weighed, in a clinical laboratory a standard protocol is usually followed, with amounts already determined. Balances can also be used to measure volumes as shown in Box 6.2.

Correct installation of a balance

Top pan and analytical balances are portable. It is essential that they are correctly located and installed, since several factors can affect their use, including:

- Draughts, which are caused by open windows and doors, rising warm air from radiators, other hot air sources, and air conditioning ducts
- Sources of vibration or sudden movements
- Magnetic fields from some types of adjacent equipment
- A non-level, unstable support surface

BOX 6.2 *Measuring liquid volumes using a balance*

Balances are usually associated with the weighing of solid materials. There are occasions where it is more convenient to weigh a liquid as opposed to measuring and delivering a particular volume of that liquid. To do this a weighing container suitable for the liquid will be required and the *density* of the liquid at the ambient temperature (density is temperature dependent) needs to be known.

$$density = mass/volume$$

Therefore

$$mass = density \times volume$$

The required mass can be calculated using the density of the liquid. For example, if a volume of 25.0 cm³ is required where the liquid density is 1.06 g cm⁻³ at laboratory temperature, the calculation is:

$$1.06 \text{ g cm}^{-3} \text{ (density)} \times 25.0 \text{ cm}^3 \text{ (volume)} = 26.5 \text{ g (mass required)}$$

SELF-CHECK 6.3

Glycerol is a viscous liquid sugar alcohol with a density of 1.261 g cm⁻³. How could you use a top pan balance to measure a 15 cm³ volume of this liquid?

These are relatively easy to identify and rectify. For example, to level correctly, balances have a level indicator and several adjustable feet.

Choice of balance

The choice of an appropriate balance to use is dictated by two key factors, the weight to be measured (capacity of the balance) and the degree of accuracy of the measurement (readability of the balance). For example, if 15.624 g of glucose were required to prepare a solution for use in an assay, a 1000-g capacity balance with a readability of 0.1 g would be inappropriate. However, a 65-g capacity balance with a readability of 0.1 mg (10^{-4} g) would be a sensible choice.

Balance calibration

Some anticipation of use is required as balances may take some time to warm up and self-check. In a general purpose laboratory, it is likely that balances will be kept switched on. Once on, the balance needs to be set for the weighing units and mode of use that is required. These are generally gram units and **tare** operation, respectively. The process of taring sets the balance readout to 0 g when an empty container is placed on the pan. When material is added to the empty container the readout is therefore an indication of the total weight of material in the container. This makes the weighing simple. Prior to use, any spilt material left by a previous user must be carefully removed to avoid contamination of future samples and damage to the balance.

Key Points

Balances are often operated in tare mode so that the displayed weight only measures the material added to the weighing container, excluding the weight of the container itself.

There must be an established schedule of balance calibration in the laboratory, and it is useful to understand how it is achieved. When supplied new, a balance will have been calibrated at the factory, but regular rechecking is required to ensure that the calibration and, hence, accuracy of the instrument is not lost. This is particularly necessary when a balance has been moved due to the influences of location, temperature, and possibly risk of physical knocks. The commonest balances are top pan types, which measure weight. Their calibration

METHOD How to use a balance

- Before commencing the weighing process, you must be aware of any issues that may affect how you proceed. Relevant health and safety information needs to be read and potential problems with the material understood. This will determine how the material can be safely handled, how solid and solution containers should be labelled and how the material should be disposed if spilt. Where handling of the material may release a dust you will need to proceed with caution.

- Samples should reach ambient temperature before being weighed to avoid condensation onto the material and to reduce air currents around the weighing pan. Material that is hygroscopic (absorbs or adsorbs water from the surroundings) or that loses water rapidly should be placed in a sealed container before weighing. The same precaution will be required where volatile liquids are weighed.

- Some form of container for the material to be weighed will be required, a plastic or foil weighing boat for small amounts or possibly some form of glassware for larger amounts. A clean weighing container and spatula are essential to avoid contamination.

- The commonest mode of using a balance is with tare selected, but the procedure will vary with the model of balance. First, the empty weighing container should be placed centrally on the pan, followed by a pause until there is an indication that the balance has stabilized (the response time varies with model). The balance is then tared so that the readout is zero. Material is gradually added to the weighing container using a spatula, remembering to pause until the balance is stabilized

- before adjusting the amount of added material in order to achieve the desired weight. Many balances have a communications port that allows the attachment of a printer or computer, and the recording of data if this is required.

- The weight of the container into which the material is placed needs to be as low as possible compared with that of material to be added so as to maximize the weight change. For instance, the addition of 0.50 g of material to a 0.50 g weighing boat would result in a 0.50–1.00 g weight change. The addition of the same amount to a 100.00 g glass container would result in a change to 100.50 g, a proportionally smaller weight change.

- Electronic balances are precision instruments and the self-checking procedures, occasionally result in error codes on the digital display. Should an error code appear, its meaning can be found in the manufacturer's operation manual.

- A balance must be cleaned of spilt material after each period of use for which a brush can be left near the balance for this purpose.

- Balances require minimal cleaning and a wipe with a soft cloth moistened with a dilute mild detergent solution is usually sufficient.

- There may be circumstances where a balance is located away from the open laboratory because of some potentially harmful feature of the material handled. The reasons for this must be clearly identified near the temporary location, so that other laboratory users are not inadvertently exposed to an unnecessary risk.

involves checking against known standard calibration weights, which may be either built in, or externally supplied and placed on the weighing pan. Two common approaches to calibration are the simpler *span* calibration and more sophisticated *linearity* calibration methods.

Span calibration uses two weight values, zero and a near to or at capacity value to check the balance. The *linearity calibration* approach uses an additional value near the middle of the weighing range. This latter approach reduces the likely extent of deviation between the true and displayed weight. The procedure used will depend upon the balance make and model. Calibration checks on a regular basis are clearly essential. The Method Box 'How to use a balance' summarizes these points.

Cross reference
Health and safety issues are described in Chapter 4 'Health and safety'.

SELF-CHECK 6.4

Why is it necessary that top pan and analytical balances are correctly located and calibrated?

6.2 **Volume measurements and delivery**

This section will consider:

- The selection of appropriate procedures for measuring volumes
- Advantages and disadvantages of different approaches

When preparing solutions by dissolving a solute in a solvent, diluting a more concentrated solution, or delivering specific sample volumes, an appropriate means of volume measurement and delivery must be chosen. Consider the two scenarios given in Case Study 6.1 and think about the differing requirements. This will make you aware of the need for an appropriate level of accuracy in different situations.

CASE STUDY 6.1 *Choosing an appropriate method for volume measurement and delivery*

Cross reference
Electrophoretic methods are described in Chapter 14 'Electrophoresis'.

The purpose of this case study is to help you to think about the importance of choosing an appropriate volume measurement method for a particular situation. Details about alternatives will be given later in this section.

Consider the following scenarios, A and B.

Scenario A

You need to prepare 200 cm^3 of electrophoresis buffer for gel electrophoresis to separate serum proteins. The laboratory stores a stock solution of the buffer that is 20 times more concentrated than required.

What approach might you use to produce sufficient electrophoresis buffer of the required dilution?

Scenario B

You are setting up an assay to measure the concentration of glucose in an unknown sample. A standard (calibration) curve is required to estimate the unknown concentration. To produce an appropriate range of known glucose concentrations for this standard curve, a series of different volumes (in the range 0–1 cm³) of a stock glucose solution will be required. The volume of each glucose sample will then be made up to a total volume of 1 cm³ with buffer prior to the addition of a colour generating reagent. This will react with the glucose to generate a colour that will be measured colorimetrically and used in the production of the standard curve.

How might you approach the measurement and delivery of these volumes?

Are the specific volume measuring requirements for B similar to or different from A? You do not need to consider dilutions in detail at this point, only how you would proceed in terms of measuring and delivering the required volumes. The effects of dilutions and the approach to working out how to dilute will be considered in the last section of this chapter.

Alternatives for measuring and delivering volumes of liquid

Figure 6.2 shows a selection of equipment that is generally available for measuring and delivering volumes of liquids. Such equipment can be grouped in various ways, for example, convenience of use or volume range that can be delivered. The most useful grouping is probably according to level of **accuracy** of the delivered volume. Accuracy relates to how close the measured volume

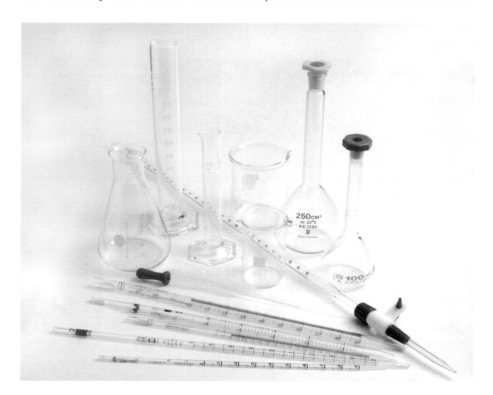

FIGURE 6.2
A selection of equipment to measure volumes of liquids.

is to the actual volume, whereas the **precision** of volume measurements refers to the closeness of repeat deliveries of (supposedly) the same measured volume. At the low end of the accuracy scale lie Pasteur pipettes (droppers) together with beakers and conical flasks that contain a volume mark. Measuring cylinders with a graduated scale provide an intermediate level of accuracy. Higher levels of accuracy are provided by various types of pipettes, burettes, volumetric flasks, and certain syringes. You may be familiar with many if not all of these, but may not have thought much about their relative strengths and weaknesses. Certain precautions need to be taken when measuring and handling certain liquids (Box 6.3). As mentioned in the Introduction, volumes can also be delivered by weighing a liquid. This approach can deliver volumes to a high level of accuracy, but lacks some convenience compared with the use of a mechanical pipettor.

Cross reference

Chapter 5 'Statistics and handling data'.

The range of volume measurement procedures commonly available will now be considered. Most emphasis will be upon mechanical pipettors (such as Gilson pipettes) as they are the most sophisticated and accurate volume measurement devices available in the routine laboratory.

Pasteur pipettes, beakers, and conical flasks

These provide a relatively simple means of measuring and delivering specific volumes, but at a low level of accuracy. The original Pasteur pipette was made of glass with a rubber teat to create a partial vacuum to draw up liquid. It is possible to make an approximate calibration of a fixed volume by drawing that volume of sample into the Pasteur pipette, marking the level on the glass. Disposable one-piece plastic Pasteur pipettes are available with several approximate volume markings on the body of the pipette. Pasteur pipettes are cheap and easy to use where approximate volumes are required, such as dispensing a few drops of a cell suspension or for completely filling wells on an enzyme-linked immunosorbent assay (ELISA) plate when a washing step is required. Where their limitations are acceptable, Pasteur pipettes are useful: they

Cross reference

You can read about ELISA in Chapter 15 'Immunological techniques'.

BOX 6.3 *Precautions to be taken when measuring and handling certain liquids*

In addition to the practical considerations related to accuracy and precision, convenience, flexibility, reliability, and the particular nature of a liquid may all influence how it is measured and handled. For instance:

- Some solutions are viscous and care must be taken to ensure complete measurement and delivery of the desired volume. This will include allowing sufficient time for the liquid to be drawn into and to drain from the measuring device

- Sampling from a cell suspension, such as blood or microorganisms requires that the suspension is first gently swirled to ensure an even distribution of cells

- Protein solutions can denature at air–water interfaces if excessively shaken, stirred, or poured, and it is necessary to keep frothing to a minimum

- Some solutions such as strong acids are highly toxic or corrosive and care has to be taken to avoid spillage and contamination. It may be appropriate to use disposable 'plasticware' where suitable

- Some organic solvents may evaporate and so exposure to air needs to be minimized by keeping them in appropriate sealed containers. Care has to be taken if using plasticware as some organic solvents can cause deterioration of the plastic.

(a)

(b)

(c)

FIGURE 6.3
(a) Correct alignment of the meniscus with the indicated calibration line. (b) and (c) show the volume incorrectly indicated.

are simple to use, quick, and the problems of cross-contamination are easily avoided with the use of disposable plastic Pasteur pipettes.

Beakers and conical flasks are available in a range of volumes with a horizontal line indicating the fill level for the quoted volume at the stated temperature. They are useful for measuring approximate volumes, but must be kept horizontal when checking the volume measurement, and are best used as holding receptacles in solution preparation and mixing.

Burettes, measuring cylinders, and volumetric flasks

Burettes and volumetric flasks are calibrated for volume measurement with high levels of accuracy at or close to room temperature. The temperature at which the calibration has been made is stated on the glassware. Measuring cylinders provide a lower level of accuracy (but higher than that for beakers and conical flasks). Whereas a volumetric flask is used for measuring a fixed volume (although a range of flask sizes are available), measuring cylinders and burettes have calibration subdivisions over a range of volume and, again, can be obtained in various sizes. For making a solution, the use of a volumetric flask containing solute would be the preferred approach, solvent being added up to the defined volume measurement line on the flask. This is both simple and convenient.

Burettes are glass cylinders with a range of volume calibrations along their length and a tap that can be opened and closed near one end. Burettes are useful for measuring and delivering small volumes (aliquots), such as in titrations where, for instance, an alkaline solution is added in small incremental volumes to a solution of acid that is stirred in a conical flask beneath the burette. Care must be taken to clean a burette both before use, by washing with the solution to be used, and also after use, usually by washing with water. The latter is necessary when using solutions such as sodium hydroxide, which can cause the burette tap to seize in its mounting.

The use of measuring cylinders, volumetric flasks, burettes, and pipettes requires *care in observing the volume measurement mark at the base of the liquid meniscus*. A meniscus is the curved surface of liquid when viewed in a narrow cylindrical space, such as the neck of a volumetric flask or the body of a pipette. The meniscus forms due to capillary action; aqueous solutions and most other liquids produce a concave meniscus in which the centre is lower than the edges. For consistent measurement, the centre point at the base of the meniscus should be observed at eye level against the appropriate volume calibration mark, as you can see in Figure 6.3. Observing the level of the meniscus is a potential source of measuring error if a consistent and appropriate approach is not adopted. When completing the filling of a volumetric flask, or, indeed, a measuring cylinder, with solvent the final small volume needed to take the solution to the correct meniscus position on the calibration mark have to be made by small, dropwise additions using a Pasteur pipette. All solutions need to be appropriately labelled, stating what they are, their concentration with any associated hazards.

SELF-CHECK 6.5

What aspects of the uses of a burette and volumetric flask could lead to inconsistencies of measurement?

Although the items mentioned above have traditionally been made of glass, plastic alternatives are becoming more common. Plasticware is safer to use as it is less likely to be damaged than glass and can also be disposable. Glassware needs to be checked before use to identify any cracked or broken items. In solution measurements and transfers between containers, care must be taken to avoid spillages and cross-contamination. Spillages must be dealt with according to local health and safety requirements.

Pipettes

Together with certain types of syringes, pipettes measure small volumes in a reproducible and accurate manner. They are commonly used to deliver volumes in the 0–20 cm^3 range, while pipettors can deliver much smaller volumes, as low as 1 μL. Pipettors are complex, precision measuring devices whose construction and operation need to be fully understood to operate correctly and reliably. They will be considered in Section 6.3.

Cross reference

Chapter 4 'Health and safety'.

Pipettes were originally made from glass, although they have increasingly been replaced by plastic alternatives. Two types are available. *Bulb* pipettes deliver a fixed volume and have a large bulbous region part way along their length with a single, fixed volume mark near the top. Although they only measure a single fixed volume, different sizes are available. There is only one calibration mark and so volume measurement mistakes are unlikely. Bulb pipettes are simple to use and potentially more accurate than a graduated pipette. *Graduated* pipettes are straight-walled and calibrated across a particular volume range such as 0–10 cm^3 in 0.1 cm^3 increments. Care must be taken to check the way the scale is presented, for example, 10–0 cm^3 or 0–10 cm^3, and whether the pipette is of the 'blowout' type or not. Blowout pipettes require that the last drop of liquid is expelled from the tip to deliver the intended volume. Non-blowout pipettes rely upon gravity to deliver the volume and allow for a small volume of liquid to remain in the tip of the pipette. This difference is not an issue if the volume to be measured and delivered is less than the total volume of the pipette. For example, a 5 cm^3 pipette could be used to measure and deliver 4.0 cm^3 with the pipette filled to the top mark (5.0 cm^3) and liquid delivered until the level had dropped to the 1.0 cm^3 mark. In this situation, it would not matter whether you used a blowout or non-blowout pipette. With both types of pipette, care has to be taken to be consistent in reading the meniscus against the graduated scale.

SELF-CHECK 6.6

What do the terms *accuracy* and *precision* mean in the context of measuring volumes with a pipette?

From a health and safety perspective, *pipettes must always be operated with a pipette filler*, which creates a partial vacuum to draw the liquid into the pipette. This avoids any chance of liquid entering the mouth as occasionally used to happen before pipette fillers became available. It is good practice to draw the liquid to just above the required mark, slowly releasing the vacuum to allow the meniscus to fall to the mark. The tip of the pipette is placed above the receiving receptacle and the pipette filler operated to release the liquid as required. As with all laboratory procedures, a correct and consistent approach to the operation of a pipette will deliver accurate and consistent results.

6.3 **Pipettors**

This section will consider:

- The design features and use of mechanical pipettors
- Optimal performance and calibration of pipettors

Pipettors are also known as *autopipettes* and sometimes *micropipettes*. They are usually adjustable volume mechanical pipettes and have become the commonest type of pipette in many laboratories. Despite their cost, they are the volume measuring tool of choice because they are highly accurate, precise, and reliable. They are sophisticated mechanical instruments and so have to be operated and handled correctly to avoid damage. Electronic versions of pipettors are also available.

In the early 1970s Warren Gilson developed an adjustable volume, high precision pipette with a volume readout display. Although many manufacturers produce pipettors, they are still often called 'Gilsons' for this reason. Pipettors are of particular value in biomedical science because they can deliver small volumes, as small as 1 μL or less, which are required in applications such as in molecular biology. They use a *disposable plastic tip* that can be autoclaved and this adds to their versatility. Why might this be an advantage? There are a number of situations where the use of a 'clean' tip may be essential, such as for work with nucleic acids where a tip contaminated with nucleases could degrade nucleic acids in the sample. Plastic tips are cheap and so avoidance of cross-contamination is easy to achieve by using different removable tips. Pipettors can be obtained in both single and multi-channel forms, different models covering various volume ranges and to different levels of accuracy. Multi-channel pipettors have a special manifold to which can be attached multiple tips (such as 8- or 12-channel types). This permits simultaneous deliveries of a particular volume of solution such as to the wells of an ELISA plate (a multi-well plate) in immunology. You can see examples of single channel and multi-channel pipettors in Figure 6.4.

Features and types of pipettors

The features of a pipettor are as follows:

- A disposable plastic tip placed by the operator on the tip holder (different pipettor models and capacities require different types of tips)

- A ceramic or metal piston within an airtight channel that can displace the air and create a vacuum, allowing liquid to be drawn into the disposable tip

- A thumbwheel that can be rotated to adjust the position of the piston and change the volume setting, which is indicated on a digital display on the side of the pipettor body

(a)

(b)

FIGURE 6.4

(a) A single and (b) multi-channel pipettors.

BOX 6.4 *Air-displacement and positive-displacement pipettors*

Air-displacement pipettors have general purpose uses for dispensing aqueous solutions and are the type you are most likely to use. You can see the general features of their construction in Figure 6.5. These include a piston that moves within the chamber of the pipettor. The piston moves down to release air, and the volume of air remaining in the chamber is inversely proportional to the volume of liquid drawn from the sample when the piston is allowed to rise. The smaller the cushion of air left within the pipettor body, the larger the sample volume. The cushion of air between the liquid being sampled and the piston prevents the piston being contaminated with liquid. In contrast, positive-displacement pipettors work more like a syringe, as you can see in Figure 6.6, which contrasts the workings of the two types. Thus, they lack an air cushion between liquid and piston, which are in direct contact with one another. Hence, the pistons must be disposable because of contamination with liquid. Positive-displacement pipettors have the advantages of being more accurate than air-displacement types because they are less subject to variations in temperature and because they do not form aerosols, which can be produced by the air-displacement types. They are mainly used to measure and dispense viscous, volatile, corrosive, or radioactive liquids.

FIGURE 6.5
Schematic showing the construction
of an air-displacement pipettor.

FIGURE 6.6
Comparison of the actions of (a) air-displacement and (b) positive-displacement
pipettors. See text for details.

TABLE 6.2 The Gilson Pipetman®
P range of pipettors

Model	Volume range/µL
P2	0.2–2
P10	1–10
P20	2–20
P100	20–100
P200	50–200
P1000	200–1000
P5000	1000–5000
P10 mL	1000–10000

Information taken from the Gilson online
brochure at http://www.gilson.com

- A moveable sleeve above the tip that allows the tip to be removed without direct contact with the operator, hence avoiding physical contact with the tip following use and possible cross-contamination with new tips

There are two different types of pipettor, *air-displacement* and *positive-displacement* (see Box 6.4). Pipettors are available in various volume ranges. For many applications, a range such as the Gilson Pipetman® P is suitable. Table 6.2 shows the volume ranges for this family of pipettors.

SELF-CHECK 6.7

Which of the following features would you *NOT* associate with a pipettor?

(a) Disposable tip

(b) Permanent or disposable piston

(c) Moveable sleeve for ejecting the tip

(d) Thumbwheel for adjusting the position of the tip

Using a pipettor

If the steps for use of a pipettor given in the the Method box 'How to set up and use a pipettor' are followed in a systematic manner, accurate and reproducible measurements will be achieved. It is worth practicing these steps a number of times to achieve consistency.

SELF-CHECK 6.8

In what ways does a pipettor differ from a pipette?

METHOD *How to set up and use a pipettor*

- Choose a pipettor appropriate to the task in hand. For a given model, it is essential to use the pipettor only within the volume range indicated. Use beyond this range can damage the pipettor mechanism.

- Organize the work area so that all materials and items of equipment to be used are conveniently placed. The work surface needs to be at a comfortable height. If necessary, this can be achieved by using a height adjustable seat. Ideally, while holding the pipettor you should be able to comfortably rest your elbow of the same arm on the work surface. This ensures the arm can be rested for some of the time in lengthy procedures whilst holding the pipettor in a near vertical position. When not in use, the pipettor must be placed vertically in an appropriate rack.

- Fit an appropriate tip onto the pipettor, applying a little force to get a good, leakproof fit. This is most readily achieved with the tips stored vertically in a tip rack. A dry tip should be wetted with solution before use. This reduces the surface tension of the inside surface of the tip and will help to minimize evaporation when the sample is drawn.

- Set the volume to be measured by rotating the adjuster. A Gilson Pipetman® P20 can deliver volumes between 2 and 20 µL. The digital display on the P20 shows three digits. The third, which is in red, represents tenths of a microlitre. Therefore, 020 represents 2.0 µL, 130 is 13.0 µL and 200 is 20.0 µL. For a P200, 095 is 95 µL and for a P1000, 085 represents 850 µL. The meaning of the three digits on the display varies with the model so instructions must be checked otherwise an incorrect volume will be delivered.

- To measure and deliver a volume of sample, the plunger is first depressed as far as the first stop (resistance will be felt) with the pipettor held away from the liquid. Next the tip should be placed below the liquid surface with the pipettor body held in a near vertical position. The plunger should be released slowly to allow the liquid to be drawn into the tip. Care must be taken to ensure that the tip remains within the liquid and that air bubbles are not drawn into the tip. After a few seconds, the pipettor should be withdrawn, keeping it near to vertical.

- To dispense the liquid, the tip is allowed to touch the side of the receiving container by slightly tilting the pipettor away from the vertical, avoiding any other liquid that may have been added already. The plunger is slowly depressed through the first stop and as far as the second stop (the final resistance). After a few seconds, keeping the plunger fully depressed, the pipettor is removed. Only when the pipettor has been removed should the plunger be released.

- The used tip can be removed by pressing the ejector button which moves a collar onto the top of the tip, loosening and releasing it. The used tip must be ejected into an appropriately labelled and suitable waste container.

Obtaining optimal performance from a pipettor

Pipettors are designed to deliver high levels of performance, but inconsistencies in results may occur if certain precautions are not taken. Thus, the mode of operation must be followed in a consistent manner. The metal components of low volume pipettors can expand if heat is transferred from the operator to the plastic handle part of the pipettor. Appropriate pipettor design can partly reduce such heating but the practice of using several pipettors on a rotational basis during prolonged use can help to overcome the potential effect of temperature changes upon the volume determining mechanism. Extended use, particularly if the pipettor is not held in the most comfortable position, may lead to repetitive strain injury. This type of discomfort can lead to the operator holding and using the pipettor in a way that is not consistent with high levels of performance. These problems can largely be overcome by using an electronic pipettor, which removes the need for the operator to exert a significant force to fill and empty the pipettor, and eject the tip. As with all experimental devices, there is a limit to the period of continuous use before operator fatigue begins to compromise performance. The work regime needs to address this issue where concentrated periods of pipettor use occur.

Inexperienced users of pipettors tend to make one or more of the following mistakes which can be overcome with practice:

- The tip may leak if it is not fitted securely to the pipettor; the fitting of the tip must be airtight
- The pipettor can be damaged if used outside the designated volume range
- Incorrect volumes may be drawn if the plunger is not depressed to the first stop or if the pipettor is raised during filling so that the tip is above the liquid level
- An incorrect volume may be delivered if the plunger is not fully depressed to the second stop or if the pipettor tip is still in contact with the delivered sample while the plunger is released
- Liquid may enter the internal mechanism and damage it, if the pipettor is held near to or horizontal when liquid is in the tip. To avoid this, pipettors should be held as close as possible to the vertical when in use

It is possible to check for consistency of approach by delivering a measured volume into a small, tared container on a balance. For water, it can be assumed that 1 µL weighs 1 mg. Even if the pipettor is miscalibrated (accuracy is compromised), successive deliveries of the same volume should still give the same measured weight (precision is maintained).

Calibrating a pipettor

In addition to routine inspections of pipettors, calibration checks are also required. An understanding of why pipettors can deteriorate can help minimize deterioration in performance, although this can be compromised in several ways. For example, misuse by the operator including incorrect volume adjustment, inappropriate handling that results in physical knocks, and dropping of the pipettor can all lead to internal damage. Clearly, the more skilled the operator, the less likely that such damage will occur. It helps if pipettors are stored in a vertical rack when not in use. The use of certain liquids results in volatile, corrosive vapours entering the internal mechanism and can damage the piston, spring, or seals that will affect the airtight seal of the pipettor and the volume that is delivered.

Regular calibration checks are necessary where high levels of accuracy and precision are required. Local laboratory procedures and practices will indicate the required frequency of pipettor servicing and calibration.

How is pipettor calibration checked? This is quite straightforward: the easiest way is to deliver a measured volume of water into a tared receptacle on an analytical balance. Ambient conditions, such as temperature and pressure together with the density of water are taken into account in converting the weight of the sample into the corresponding volume. This volume can then be compared with the volume originally selected on the pipettor. Calibration checks, as well as routine physical inspections will inform the need for maintenance, which will either be undertaken locally in the laboratory or by the manufacturer.

FIGURE 6.7
Digital volume display of a pipettor.

SELF-CHECK 6.9

Look at Figure 6.7, which shows the digital display of a pipettor. What volume would the display represent if the pipettor was from the Gilson Pipetman P range and was (a) a P20, (b) a P100, and (c) a P200?

6.4 **Preparing reagents: concentrations and dilutions**

This section will consider:

- A solution preparation checklist
- The mole (mol) and molar concentrations
- Alternative ways of expressing concentration
- Different types of dilution and their impact upon concentration

A **solution**, for instance of glucose in water, consists of a quantity of **solute** (such as glucose) dissolved in a volume of **solvent** (water in this case). The use of balances and volume measurements are the two essential techniques for the preparation of a solution and its dilution. If a set method is to be followed, where all quantities are documented, calculations need not be performed. However, do you know how to weigh an amount of glucose equivalent to 0.1 mol, can you prepare a 0.2 mol dm^{-3} sodium acetate solution, or are you able to calculate the final concentration of a 0.0005 mol dm^{-3} protein solution when 1 cm^3 is diluted to 25 cm^3? If not, then this final part of the chapter will be of help. The precautions to be observed before preparing solutions are explained in Box 6.5.

BOX 6.5 A precautions checklist when preparing solutions

Several matters need to be considered and understood before proceeding to prepare and dilute solutions:

- The health and safety assessments for all materials being used, any precautions that must be observed, procedures for appropriate labelling of solutions in storage containers, dealing with spillages, and safely disposing of waste must be fully adhered with

- Appropriate protective equipment must be used, including a fully fastened laboratory coat, together with safety glasses and gloves, as determined by the nature of the materials to be handled and procedures to be undertaken

- Solutions must never be pipetted by mouth

- The location of all relevant safety equipment in the laboratory, such as an eye wash, safety shower, first aid kit, fire extinguisher, and fire blanket must be known

- Additional precautions must be observed in some situations, such as the addition of a strong acid to water (and *not* water to the strong acid) because of the heat generated in the process. This is minimized by keeping the concentration of acid low initially and by adding it in small quantities to a much larger volume of water

- Where a generic health and safety assessment for a particular procedure is documented and followed, a check must be made to ensure that the details of what is being undertaken do not significantly differ from it. For instance, a risk assessment for preparing and using a 1×10^{-5} mol dm^{-3} solution of a slightly hazardous material would need to be modified where a more concentrated, 0.1 mol dm^{-3} solution is being prepared and used, as concentration is a key part of the risk assessment of a potentially harmful solution

Cross reference
Chapter 4 'Health and safety'.

Solutions and concentrations

The *concentration* of a solution is determined by the *quantity* of solute and the *volume* of solvent as described in Box 6.6.

Key Points

Solutions consist of a solute dissolved in a solvent; the concentration of the solution is determined by the relative proportions of solute and solvent.

The different ways in which concentration may be expressed will now be considered. The commonest form is **molarity** and this along with the concept of the **mole** will be considered in some detail followed by other common forms of concentration, **molality**, % (per cent) concentration, and parts per million (ppm). Finally, we will discuss what the effects of dilution upon concentration.

Moles, molar concentrations, and molarity

The mole is one of the base units of the SI system of units (Section 6.1). It is a measure of the *amount* of a substance. The mole is that amount of substance that contains as many elementary entities, such as photons, atoms, or molecules, as there are atoms in 0.012 kg (12 g) of carbon (^{12}C). This value is called **Avogadro's number** and is equal to 6.022×10^{23}, and is a constant for a mole of any substance. Although a molecule of water (M_r, relative molecular mass, 18) is small compared with a molecule of haemoglobin (M_r 64,500) (see below for an explanation of M_r), a mole of each would contain this same number of individual molecules. Of course, the mass of a mole of these two different molecules would be vastly different.

This numerical nature of the mole is of little help when measuring a particular amount of a substance as counting individual molecules is not a practical proposition! Instead, it is much more convenient to use **molar mass,** which is the mass of *all* the elementary entities in one mole of a substance and has units of g mol^{-1}. For glucose the value is 180.16 g mol^{-1}. A numerically equivalent alternative to molar mass is M_r, which makes reference to the mass of ^{12}C. For an atom the corresponding term is relative atomic mass (A_r):

$$M_r = \text{(mass of one molecule of substance)}/\text{(1/12 mass of one molecule of } ^{12}C)$$

BOX 6.6 Concentrations of molecules and ions

The numerical nature of the mole means that a solution contains a particularly large number of individual solute molecules. Aqueous sodium chloride solutions contain sodium and chloride ions (1 NaCl → 1 Na$^+$ + 1 Cl$^-$). A 0.04 mol dm^{-3} aqueous solution of sodium chloride would have a concentration of 0.04 mol dm^{-3} for each of the individual Na$^+$ and Cl$^-$ ions, but 0.08 mol dm^{-3} for the *combined* concentrations of both.

BOX 6.7 *Molecular size and the dalton*

The unit called the **dalton** (Da) is sometimes used to express a mass equal to one-twelfth of that for ^{12}C. It is often adopted in biochemistry and molecular biology texts when referring to the sizes of macromolecules. A protein with M_r of 45,000 could be described as being 45,000 Da or 45 kDa in size.

SELF-CHECK 6.10

Distinguish between a mole, molar mass, and relative molecular mass.

As M_r is a ratio of masses, it has no units, but may be expressed in **daltons** as described in Box 6.7. M_r is equivalent to the now outdated term *molecular weight*, although the latter term is still found on many reagent bottle labels, many textbooks, and is still used by many biomedical scientists. To obtain one mole of glucose, you need to weigh 180.16 g of it; that is, obtain a mass equivalent to the value of its M_r.

SELF-CHECK 6.11

The amino acid glycine has the formula $C_2H_5NO_2$. How much glycine would you need to weigh to obtain 0.8 moles?

Method Box 'Preparation of solutions' gives practical hints on how to prepare, in this example, a solution of glucose.

How are concentrations expressed using the mole?

A one *molar* solution contains one *mole* of solute in a total solution volume of one dm^3 or *litre*. This forms the basis of the molarity form of concentration. Technically, this is not in accordance with the SI system of units as the SI unit of volume is the m^3. However, as the m^3 is 1000 times larger than a dm^3 (which is equivalent to a litre), the use of molarity is numerically more convenient.

The number of moles in a particular volume of a solution can be calculated by applying the following equation:

$$\text{Number of moles (mol)} = \text{molarity (mol dm}^{-3}) \times \text{volume (dm}^3)$$

Key Points

Molarity is the most common representation of concentration and is the number of moles of solute in 1 L of solution.

SELF-CHECK 6.12

What is the molar concentration of a 6 mg cm^{-3} solution of lactose ($M_r = 360.31$)? How many moles would be in 50 cm^3 of this solution?

METHOD *Preparation of solutions*

How could you prepare 1 dm³ of a 1 mol dm⁻³ aqueous solution of glucose?

You would:

- Find out the M_r value for glucose from the reagent bottle label or from data sheets ($M_r = 180.16$)
- Weigh this amount using an appropriate balance then place the glucose in a 1 dm³ volumetric flask
- Add distilled water to the mark on the volumetric flask
- Mix well

However, you do not always prepare one dm³ volumes of solutions, most likely much less than this. How could you prepare 100 cm³ of a 0.2 mol dm⁻³ solution of glucose? It is easiest to split the calculation into several simple, logical steps:

- 1 dm³ of a 1 mol dm⁻³ solution will contain 180.16 g of glucose
- 1 dm³ of a 0.2 mol dm⁻³ solution will contain 180.16 × 0.2 g glucose, which = 36.032 g
- If a volume of 100 cm³ is needed, the amount of glucose to be weighed is:

$$100 \text{ cm}^3 \times (36.032 \text{ g}/1000 \text{ cm}^3) = 3.603 \text{ g}$$

Although 3.603 g is equivalent to 0.02 mol glucose (mol = mass/M_r), the concentration is 0.2 mol dm⁻³ because the 0.02 mol glucose is dissolved in a 100 cm³ (0.1 dm³) volume. Molarity expresses concentration as though the volume is 1 dm³ (1 L).

$$\text{Molarity (mol dm}^{-3}) = (\text{mass}/M_r) \text{ dm}^{-3}$$

Other common representations of concentration

Molarity is the commonest way of expressing concentrations and is therefore the one you are most likely to be familiar with. However, there are several other ways in which concentration may be expressed, three of which we will now consider.

Molality is a form of concentration where the mass of the solvent is used, rather than the volume of solution, as used in molarity. For example, a 0.1 molal solution would contain 0.1 mol solute/kg solvent. For a dilute aqueous solution where the mass of the solute is relatively low compared with the volume of solvent and where water has a density of 1 g cm⁻³, molal and molar concentrations will be similar in value. With greater amounts of solute, the values will differ as the addition of solute will affect the final volume of solution compared to the volume of solvent used. However, using molality can be convenient because the only measurements are those of mass, unlike molarity, which involves both mass and volume measurements.

SELF-CHECK 6.13

Would you expect any differences between the molal and molar concentrations of 1 dm³ volumes of aqueous solution containing (a) 0.1 mg of solute and (b) 50 g of solute?

Per cent concentrations are expressed as parts of solute per 100 parts solution. For example, % w/v (*mass–volume or weight–volume percentage*) means mass parts of solute per 100 volume parts of the resulting solution. Thus, a 10% w/v solution of glucose contains 10 g glucose per 100 cm³ of the glucose solution. *Mass* percentage (% w/w) expresses the concentration as solute mass parts per 100 total mass parts. It is also known as weight percentage or weight–weight percentage. This could be used in the preparation of an aqueous solution of an alcohol. For instance, 25 g alcohol plus 75 g water would make a 25 % w/w solution. Finally, % v/v (*volume–volume* percentage) expresses the concentration as volume parts of solute per 100 volume parts of solution.

Convert a 0.1 mol dm^{-3} concentration of glucose into a mass–volume percentage (%w/v).

Parts per million (ppm), like per cent concentrations, is a non-SI expression of concentration. Similar forms of expression include parts per thousand. You will have come across ppm if you have carried out flame photometry or atomic absorption spectroscopy work as these techniques detect relatively low concentrations of analytes. A concentration of 1 ppm means 1 g solute per 10^6 g of solution or the aqueous solution volume equivalent, per 10^6 cm^3. This is the same as 1 µg per g or 1 µg per cm^3.

Per cent concentrations and parts per million are useful ways of expressing concentrations where the M_r of the analyte in question is unknown.

Dilutions

This final part of the chapter will consider how to dilute a concentrated solution (often called a *stock solution*) to make other less concentrated solutions of known value. Where assays and other procedures that require a range of solutions are undertaken on a regular basis, it is convenient to store pre-prepared concentrated stock solutions for subsequent dilution as and when required. This is, of course, only possible where the solution is stable and unaffected by long-term storage, and the dilutions are performed in an accurate and precise manner.

Preparing diluted samples from a stock solution

We will consider the preparation of diluted samples from a stock solution using the practical example given in Case study 6.2.

Using the approach in Case study 6.2 will allow you to handle any situation that involves dilutions. The equation $c_1v_1 = c_2v_2$ also allows you to prepare a particular dilution, where the required final concentration is known because v_1, the volume of stock solution needed for the preparation of the final volume of solution, v_2, can be calculated.

Key Points

The equation $c_1v_1 = c_2v_2$ facilitates the calculation of the effect of a dilution and how to produce a specific dilution from a stock solution.

It is helpful to remember that for any dilution the following points apply:

- The final concentration will be less than that of the stock solution because the number of moles of solute taken in the measured volume of stock solution is now dispersed in a larger volume

- The extent of the dilution will determine the concentration change, so the greater the dilution the lower the concentration of the final solution

- The effect of a dilution can be calculated using the expression $c_1v_1 = c_2v_2$ (Case Study 6.2)

- The accuracy of a dilution will be influenced by the choice of methods for measuring and delivering the required volumes

- Information about a dilution needs to reference both the initial volume to be used and the final volume. For example, a 20-fold dilution would involve adding nineteen volumes of

CASE STUDY 6.2 *Quantitatively diluting a stock solution*

A common form of assay is one where a range of known concentrations of a solution is prepared and a standard curve (calibration curve) produced. The curve can then be used to estimate the concentration of an unknown sample. A property of the solute (such as absorbance due to a coloured compound) that varies with concentration is measured. The Beer–Lambert law provides the basis for a linear relationship between absorbance and concentration.

Cross reference

You can read more about the Beer–Lambert law in Chapter 11 'Spectroscopy'.

Usually solutions are colourless and some form of colorimetric assay will be needed to form a coloured derivative of the solute, whose absorbance can then be measured using a colorimeter or spectrophotometer. For instance, there are several reducing sugar-based colorimetric assays for the detection of glucose. The Somogyi–Nelson method utilizes an alkaline copper solution that reacts with glucose resulting in the conversion of oxidized $Cu(II)^{2+}$ to reduced $Cu(I)^{+}$. The latter reacts with arsenomolybdate to form a blue coloured complex. The absorbance of the blue solution can be measured.

A simple assay to assess the concentration of haemoglobin in a sample will be used to illustrate the effects of dilution of a stock solution. Look at Table 6.3 which has a blank row and requires your input of the concentration values. The table shows a series of dilutions from a stock solution of haemoglobin which is, of course, coloured. You should be able to see that the dilutions have been achieved by aliqoting a range of volumes of the haemoglobin solution between 0 and 3.00 cm^3 and adding an appropriate volume of saline to each to give a total volume of 3.00 cm^3 in each case. The stock haemoglobin solution has a concentration of 2.00 mg cm^{-3}. You need to work through the following information so as to be able to complete the Table.

How might you proceed to fill in the concentration values? Tubes 1, 5, and 7 should pose little difficulty. Tube 1 lacks haemoglobin and so the concentration of its contents will be zero. At the other end of the range, tube 7 contains undiluted haemoglobin and the concentration will therefore be the same as the stock solution (2.00 mg cm^{-3}). Tube 5 has a mixture of equal volumes of haemoglobin and saline, so the concentration in this tube has halved. How can the other concentrations be determined? The extent of dilution will need to be calculated by comparing the initial

TABLE 6.3 Preparation of a range of haemoglobin (Hb) concentrations for the construction of a standard curve. Stock [Hb] = 2.00 mg cm^{-3}. The final row in the table is for you to complete

Tubes	1	2	3	4	5	6	7
vol Hb/cm^3	0.00	0.25	0.50	1.00	1.50	2.25	3.00
vol saline/cm^3	3.00	2.75	2.50	2.00	1.50	0.75	0.00
[Hb]/mg cm^{-3}							

volume of haemoglobin with the final volume of haemoglobin plus saline. This is readily achieved by using the equation:

$$c_1 v_1 = c_2 v_2$$

where c_1 = concentration of stock solution; v_1 = volume of stock solution used; c_2 = concentration of the diluted solution; and v_2 = total volume of solution after dilution.

It is essential that both c_1 and c_2 have the same units of concentration. Similarly, v_1 and v_2 must have the same units of volume.

Applying this equation to tube 6:

$c_1 = 2.00$ mg cm^{-3}
$v_1 = 2.25$ cm^3
c_2 = unknown concentration
$v_2 = 3.00$ cm^3 $(2.25 + 0.75)$
N.B. v_2 will be the same value for all tubes.

By rearranging the above equation, the unknown concentration c_2 can be obtained:

$c_2 = (c_1 \times v_1)/v_2$
$\quad = (2.00$ mg cm$^{-3} \times 2.25$ cm$^3)/3.00$ cm^3
$\quad = 1.50$ mg cm^{-3}

By repeating this approach you can work out the remaining concentrations and complete the last row of Table 6.3.

TABLE 6.4 **Cumulative effects of a two-fold serial dilution**

Dilution step	Dilution
1	2 ×
2	4 ×
3	8 ×
4	16 ×
5	32 ×
6	64 ×
7	128 ×
8	256 ×
9	512 ×
10	1024 ×

diluent to 1 volume of the stock solution to give 20 volumes of diluted solution (this could be called a 1 in 20 dilution)

SELF-CHECK 6.15

What is the concentration of the coenzyme NADH in a dehydrogenase enzyme assay in which 0.20 cm^3 of stock 150 mmol dm^{-3} NADH solution is mixed with several other solutions to give a final volume of 3.00 cm^3?

Specific types of dilution

There are certain situations where a particular pattern of dilution (a *dilution series*) is required. These may be a linear or a non-linear series. Where a standard curve is being constructed, equal or near to equal increments in the dilution series will ensure that there is an even spread of data points. The range of concentrations in your completed version of Table 6.3 is of this type. Sometimes adjustments to the pattern of dilutions are necessary so as to spread evenly the reciprocals of substrate concentrations, as in the Lineweaver–Burk enzyme kinetics plot. There are, however, circumstances in which a wider range of dilutions using non-linear incremental changes are required. This may, for example, reflect an uncertainty of knowledge about the density of cells in a cell suspension or concerning the level of biological activity in a sample, such as may be encountered with the ELISA. In these situations, some form of serial dilution is undertaken.

A *two-fold* or *log$_2$* dilution is where successive dilutions halve the concentration. This is achieved by measuring a volume of sample and mixing it with an equal volume of diluent and repeating this '1 + 1' approach for each successive dilution. Table 6.4 illustrates the effect of this type of dilution. You can see that after only a few cycles of dilution, the concentration falls rapidly.

Cross reference

See Chapter 15 'Immunological techniques'.

A *ten-fold* or log_{10} dilution occurs when each successive dilution reduces the concentration ten fold. This is accomplished by measuring one volume of solution and adding nine volumes of diluent to it: a '1 + 9' approach for each successive dilution. After several cycles of serial dilutions, the low concentration can become difficult to comprehend easily in a table of data. Ten-fold dilutions make graphical representation of concentration data on a linear scale rather meaningless. Table 6.5 shows how logarithmic values can be used to represent such data. The log values could be used on a linear graph scale as an alternative to using logarithmic graph paper.

SELF-CHECK 6.16

How many cycles of dilution will achieve a minimum of a 1,000-fold dilution using (a) a two-fold and (b) a ten-fold dilution series?

TABLE 6.5 **The effect of ten-fold dilution steps upon antigen (Ag) concentration in an ELISA. The stock [Ag] is 1 mg cm^{-3}. A log$_{10}$ value change of 1 represents a 10× change in the concentration.**

Dilution step	[Ag]/mg cm^{-3}	log^{10} [Ag]
1	0.1	−1
2	0.01	−2
3	0.001	−3
4	0.0001	−4
5	0.00001	−5

SUMMARY

- The use of balances for weighing and pipettors and other volume measurement methods for volume delivery are key techniques in the production of solutions and their dilution

- Top pan and analytical balances measure weight; with calibration they can be used to measure a required mass

- For correct operation, balances must be appropriately sited and calibrated

- The choice of method for measuring and delivering a volume of liquid depends upon the level of required accuracy and precision

- Burettes, pipettes, and volumetric flasks provide high levels of accuracy if used correctly

- Pipettors are the volume measurement tool of choice for highly accurate and precise work. They use disposable tips for convenience, require calibration and, being precision instruments, must be used with care

- Pipettors must be used in a correct and consistent manner to achieve accurate and precise performance; to avoid operator fatigue and discomfort in extended use, a comfortable posture must be adopted

- Solutions are prepared at particular concentrations of solute in solvent, most frequently expressed as molarity, but also in other forms such as molality

- The production of solutions of a particular molarity requires an understanding of the mole and how the relative proportions of moles of solute and volume of solvent affect the concentration of the solution

- Dilutions increase the proportion of solvent compared to solute and, therefore, reduce concentration; the application of the equation $c_1v_1 = c_2v_2$ simplifies the calculation of dilutions and concentrations

- Dilution series can be linear, such as in standard curves or non-linear such as successive two-fold and ten-fold dilutions

 # FURTHER READING

- **Graham I (1983) Difficulties encountered by biology students in understanding and applying the mole concept.** *Journal of Biological Education*, 1983; **17**: 339–342.

  This article addresses some of the common areas of confusion and misunderstanding in applying the mole concept and considering concentrations and dilutions.

- **Reed R, Holmes D, Weyers J, Jones A.** *Practical Skills in Biomolecular Sciences* **3rd edn. Pearson Education Limited, Harlow, 2007.**

  Chapters 22–24 of this comprehensive basic laboratory skills textbook complement much of this chapter.

Useful Websites

- **http://www.fisher.co.uk/phpfiles/techzone/question.php?questionid=181** (accessed on 25 January 2008). This web page provides useful tips on choosing the right pipettor and tip, recommendations for correct use, and a link to a fuller article on liquid handling.

- **http://www.gilson.com/Products/product.asp?pID=67** (accessed on 25 January 2008). This web page introduces the Gilson Pipetman® P range of pipettors, one of the most useful ranges of pipettor, and also links to illustrated PDF documents on the use of pipettors.

 # QUESTIONS

1. Indicate which of the following statements are **TRUE** or **FALSE**. Justify your answer.

   (a) It does not matter what size of pipette or pipettor is used to measure a particular volume of solution

   (b) When purchasing a balance for a laboratory, it is necessary to obtain the most sophisticated model that funds permit

   (c) Dilutions for the production of a standard curve should be prepared using equipment with high levels of accuracy and precision

2.   An acid has a M_r of 192 and is used in the preparation of various aqueous solutions. What are the concentrations in mol dm^{-3} of the following solutions?

(a)   100 cm^3 contains 2 g of acid

(b)   200 cm^3 contains 4 g

(c)   5 cm^3 contains 0.05 g

(d)   2.5 dm^3 contains 45 g

(e)   10 g of the acid is diluted in 1 dm^3, which is then diluted 25-fold.

3.   (a)   Which of the following represents a 10 parts per million concentration of an aqueous solution?

I   10 g per dm^3

II   1 mg per 100 g

III   100 µg per kg

IV   1 µg per mL

(b)   Assuming that (a) (I) is a solution of glucose, calculate (I) the molar and (II) mass–volume percentage concentrations.

4.   There has been an electrical fault in your section of the laboratory and you have to temporarily relocate the analytical balance that you are using. List the key issues that you would need to address before being able to operate the balance in the new location.

5.   Construct a table with the column headings cost, simplicity of construction, convenience of use, and level of accuracy. Under each heading, use a simple scoring system ranging from low (—) to high (++) for a Pasteur pipette, beaker, measuring cylinder, pipette, and pipettor. This exercise will help you to summarize the key features of different items of volume measuring equipment. Looking at your table, do you think that 'you get what you pay for'?

6.   What are the advantages and disadvantages of using molarity and mass percentage (% w/w) forms of concentration?

7.   Thinking about how it is achieved, what is the main precaution that must be taken when preparing a two-fold dilution series?

Answers to self-check and end of chapter questions are available in the Online Resource Centre accompanying this book.

Go to www.oxfordtextbooks.co.uk/orc/glencross

7

Samples and sample collection

Joyce Overfield

Learning objectives

After studying this chapter, you should be able to:

- Identify major factors that contribute to 'sample quality'
- Outline guidelines and procedures for the collection of samples
- Describe the procedures for processing a range of blood samples
- Discuss sampling of histological specimens
- Describe the key steps in obtaining a sample for microbiology testing
- Describe how samples should be handled during the pre-analytical stages

Introduction

Blood and other tissues, and body fluids can be tested in clinical laboratories to aid in preventing, diagnosing, and treating diseases and disorders. A variety of analytical techniques are used, which are being continually developed and improved in accuracy and ease of use by biomedical scientists. Samples for clinical testing include whole blood, plasma, serum, urine, and faeces for chemical and cellular analyses; tissues, and **curettings**, which are scrapings of tissues, for histological examination to identify pathological changes; swabs may be taken from various parts of the body for microbiological culture. Examples of body fluids that may be tested for infection include blood, urine, cerebrospinal, bronchial, pleural, synovial, and bone marrow.

Samples obtained from patients must be placed in suitable containers, transported to the clinical laboratory, and stored correctly, if clinical test results from them are to be accurate. Most clinical laboratories, including those within the National Health Service in the UK, have standard operating procedures (SOPs), which specify the manner in which samples must be treated. These procedures must always be adhered to.

This chapter will describe the range of samples and the procedures for their collection for analyses. It is not possible here to include sample preparation for every clinical test available and if you have particular interests, the further reading at the end of the chapter is wider ranging.

SELF-CHECK 7.1

Describe the key factors that contribute to pre-analytical quality assurance of clinical samples.

7.1 **Blood**

Blood is the most commonly tested type of clinical sample because of its accessibility and the wealth of information that can be obtained by analysing its many different constituents. A clinician or general medical practitioner can request that a range of different blood tests be performed in the clinical diagnostic laboratories by simply ticking or writing on a request form, an example of which you can see in Figure 7.1.

Only a limited number of tests are performed on whole blood; most are done using plasma or serum. Blood tests are performed for a number of clinical reasons, some of which you can see listed in Table 7.1. These include, for example, to assist in the diagnosis of diseases or conditions, or in screening tests, such as testing a cohort of individuals for the presence of human immunodeficiency virus. Appropriate testing can also eliminate the possibility of a particular condition even if the patient's symptoms have suggested its possibility. Testing of samples of blood from patients is also used to monitor the activity and severity of some clinical conditions. A blood test may help to see if the patient is responding to treatment. For example, a case of iron deficiency anaemia treated with oral iron supplements should begin to show an increase in the concentration of haemoglobin in the blood. Samples of blood are also required to test the effectiveness of specific organs, such as liver and kidney function, and of the effects of medical or surgical treatment on these organs.

Naturally, a blood sample from a patient is also needed to establish his or her blood group in cases where a blood transfusion is required.

FIGURE 7.1

Example of a request form for laboratory tests. See also Table 17.3.

TABLE 7.1 Examples of tests performed on blood samples

[1]A full blood count (UK) or complete blood count (US). This repertoire of tests checks for anaemia, and other conditions affecting erythrocytes, leucocytes, and platelets

[2]Blood chemistry profiles

[2]Kidney function (urea, Na^+, K^+, creatinine)

[2]Liver function tests

[2]Hormone (thyroxine, thyroid stimulating hormone) concentrations

[2]Blood glucose (sugar) concentrations

[3]Blood clotting tests

[4]Tests for inflammation

[2]Blood lipids and cholesterol concentrations

[2]Monitoring the doses of specific medications

[5]Immunology or serology tests such as testing for antibodies to specific viruses and bacteria

[3]Blood grouping and screening for blood group antibodies

> **Cross reference**
> [1] See the *Haematology* book of this series; [2] see the *Clinical Biochemistry* book of this series; [3] see the *Transfusion and Transplantation Science* book of this series; [4] see the *Histopathology* book of this series; [5] see the *Immunology* book of this series.

Blood sampling procedures

The healthcare worker responsible for drawing blood samples may be a specialist **phlebotomist**, a clinician, or other healthcare worker trained in blood sampling techniques.

Samples of blood may be peripheral or venous depending on whether the sample is obtained by pricking a finger or by venepuncture. Finger prick blood is used mostly for point of care tests and is only acceptable providing a free flow of blood can be obtained. However, for laboratory analyses, it is more common to obtain blood by venepuncture. Blood can be drawn from a vein in one of several ways depending on the age of the patient and clinical tests required. Usually, the antecubital vein, which is located in the crook of the forearm, is used as a source (Figure 7.2). Occasionally, when veins are difficult to locate or have been used many times, blood samples may be taken from the veins on the back of the hand.

In the UK, USA, and Australia most blood samples are collected using commercial vacuum tube systems called **vacutainers**. Under certain circumstances, a syringe may be used, often with a butterfly needle, which is a short needle with an attached plastic **cannula**; the name is derived from the characteristically shaped plastic flanges that aid handling as you can see in Figure 7.3(b).

Antecubital vein

FIGURE 7.2
Schematic illustration of a venepuncture, indicating the position of the patient's arm and antecubital vein.

(a)

(b)

FIGURE 7.3
(a) Selection of conventional syringe needles and (b) a butterfly needle. Note the characteristically shaped plastic handling vanes that give the needle its name.

However, in most parts of the developing world, needles and syringes continue to be the commonest method of obtaining venous blood.

Key Points

All blood is a potential source of infection and poses a risk to the phlebotomist; in particular, blood may carry human immunodeficiency or hepatitis viruses. Thus suitable precautions must be taken at all times.

Cross reference
Chapter 4 'Health and safety'.

Irrespective of whether blood is collected by vacuum venepuncture or using a syringe and needle, a number of general procedural rules are the same. First, any dietary requirements and other appropriate preparatory instructions should be noted before performing investigations. Some blood samples must be taken without a tourniquet (Figure 7.2) so it may be necessary to check with the laboratory for the appropriate procedure. Blood samples must not be contaminated by patient infusions such as a saline drip or blood transfusion.

The correct type of sample tube (see later) as specified by the **pathology** laboratory must always be used. If a blood sample is added to the wrong type of tube by mistake, it should never be decanted into a second (correct) tube. Blood collection tubes must not be shaken as this can cause **haemolysis**, which is the breakdown of the membrane surrounding the erythrocytes and leakage of their contents. If this does occur, it is likely to invalidate subsequent test results; hence, tubes should be mixed gently by inversion. Finally, it is essential to avoid over- or under-filling tubes that contain an anticoagulant.

Venous blood sampling using vacutainers

Vacuum tubes were first marketed by the U.S. Company, Becton Dickinson Limited under the trade name *Vacutainer tubes*. You can see a range of vacutainer tubes in Figure 7.4. Many companies now sell vacuum tubes as the patent for this device is in the public domain.

FIGURE 7.4
Range of blood collection
tubes with colour-coded lids.
Each type of tube is used for
different clinical tests.
See also Tables 7.2 and 17.2
and Figure 12.14.

A vacutainer consists of a plastic hub, a hypodermic needle, and a vacuum tube. It is based on the use of *evacuated* tubes so that venous blood automatically aspirates through the needle inserted into the tube. Vacutainer systems incorporate a plastic cap known as a multiple sample sleeve, which fits over the posterior end of the needle. This mechanism allows multiple vacuum tubes to be attached and removed in turn from a single needle, allowing numerous blood samples to be obtained from a single venepuncture. This multiple sample sleeve also prevents blood from leaking onto both the phlebotomist and the patient. When using the vacuum tube system, the needle pierces the top of the sample tube and will potentially come into contact with the additives in the tube (see below). As it is hollow, the needle can carry some additive into the next tube and contaminate it. The additive most likely to cause trouble is ethylenediamine-tetraaceticacid (EDTA). This will affect coagulation time assays and may interfere with some clinical biochemistry tests by chelating and removing divalent metal ions, for example, Ca^{2+}, and can affect the determination of serum K^+ by reducing erythrocyte integrity. Thus, blood samples for plain tubes should be drawn first and those to be added to EDTA-containing tubes last.

Standard operating procedures for venepuncture samples must state that the healthcare worker accurately identifies a patient by verbal confirmation, explain the reason(s) for the procedure, and obtain the patient's consent. The patient's arm is examined to identify the most easily accessible vein for the venepuncture. It is advisable not to take blood from an arm containing a drip or transfusion as this will contaminate the sample. A tourniquet is placed in position on the upper arm without applying pressure (to allow for patient comfort), and the arm positioned comfortably on a support. Alcohol gel or hand washing must be performed between each patient and latex-free safety gloves may be worn if necessary. The tourniquet is tightened to facilitate the distension of the vein. The vein may also be gently palpated.

The vacutainer needle is now attached to the holder and its protective shield removed. The needle is inserted into the distended vein. The vacutainer tubes are gently, but firmly pushed in the appropriate order into the needle holder while keeping the holder steady with the other hand. Once filled, all tubes are gently inverted as they are removed from the holder, ensuring that the blood and anticoagulant are mixed to prevent clotting.

When all tubes have been filled, the tourniquet is released and removed. The final tube should be removed from the vacutainer holder before removing the needle from the vein as this practice reduces the risk of a **haematoma**.

The needle is withdrawn and pressure applied through a piece of gauze to prevent haemorrhage from the puncture wound. The patient should be asked to maintain pressure on the needle site for approximately 5 minutes to aid wound sealing.

Venous blood sampling using a needle and syringe

A sterile, plastic syringe fitted with a suitable needle may be used to extract venous blood from a patient. Needles for venepuncture are available in a variety of sizes, which are described by their gauge (g) number. The most commonly used needles are:

- 21g (green top)
- 22g (black top)
- 21g (green label) butterfly
- 23g (blue label) butterfly
- 25g (orange label) butterfly

The last is only used in paediatric blood sampling or other extreme cases where the vein is particularly small. A 19g (brown top) needle has a larger bore and should be used in cases where a large volume of blood is required, for example up to 100 cm^3.

A needle and syringe may be more suitable for obtaining samples from elderly patients, or those with difficult or inaccessible veins. Given that syringes are manually operated and the amount of suction applied can easily be controlled, their use is particularly suitable with patients that have small veins, which collapse under the suction of an evacuated tube. In children or other circumstances where the quantity of blood gained may be limited, it is useful to know how much blood has been obtained before distributing it amongst the various tubes. Needles and syringe are also used for collecting blood for microbiological examination (Section 7.5). In all cases, the tube contents must be mixed by inverting gently a number of times immediately upon adding the blood to prevent clotting occurring in the tube.

Once used, the syringe and needle should be disposed of as a whole unit, that is, the needle must not be disconnected, into a bin dedicated to sharp items. Nor must the protective sheath be threaded back onto the needle before discarding it. Needle-stick injuries pose a risk of infection to all healthcare workers because of the exposure to the patient's blood. It is therefore essential to follow good practice. Patients may be identified as 'high risk', that is already diagnosed with infections, for example with HIV or hepatitis, or they may be carriers and unaware of their status. Box 7.1 describes the precautions necessary when obtaining blood from a patient known to be of high risk.

Cross reference
You can read more about the features on health and safety in Chapter 4 'Health and safety'.

Key Points

Irrespective of whether blood is collected by vacuum venepuncture or using a syringe and needle, it must be added to the type of tube appropriate to the clinical tests required.

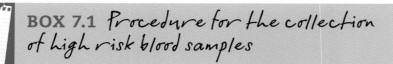

BOX 7.1 *Procedure for the collection of high risk blood samples*

- Samples must be collected by the clinician requesting the test or by an experienced and fully trained phlebotomist

- Plastic disposable apron and disposable gloves must be worn. It is recommended that suitable eye protection such as goggles or safety spectacles be worn, but a face mask is unnecessary

- A waterproof disposable sheet must be placed under the patient's arm

- The patient's skin must be cleaned and the needle inserted in the usual way using a vacutainer or syringe

- Any spillages must be cleaned with freshly made 3% hypochlorite solution

- Used swabs must be discarded into a disposal bin specifically designated for sharp items

- Gloves and apron must be discarded into a plastic sack for incineration (colour-coded yellow in the UK)

- Hands must be washed with soap and water from elbow taps and dried with disposable paper towels

- 'Danger of infection' labels need to be attached to request forms and samples, which must be placed in plastic sample bags and sealed

Blood containers

Once collected, the sample of blood can be added to one of several different containers depending on the tests required. Some tests must be performed using serum and thus require the prior clotting of the blood. In contrast, tests on plasma require the addition of anticoagulants to the blood sample to prevent its clotting. If you are unclear, Box 7.2 describes the differences between plasma and serum. If the concentration of glucose in the blood is being measured, then the blood is added to a container containing a preservative. Sample bottles with specifically coloured lids (see, for example, Table 7.2) indicating the additive present and the test profiles for which they can be used have been developed to help phlebotomists in adding blood samples to the appropriate container.

Anticoagulants and other additives

The tubes in which venous or capillary blood is transported to the clinical laboratory may contain a variety of additives or none at all. It is essential to know which anticoagulant or additive is required by the laboratory for particular tests. Table 7.2 describes anticoagulants, additives, and their use.

For most haematology assays, particularly full blood counts (FBC), blood group determinations, antibody screening, and transfusion compatibility tests, whole blood needs to be mixed with the anticoagulant EDTA, which acts by chelating and removing Ca^{2+} from solution. This prevents clotting, as Ca^{2+} is essential to the clotting cascade that forms fibrin. However, if blood clotting tests are required, then trisodium citrate is traditionally used as the anticoagulant of choice.

BOX 7.2 What is the difference between serum and plasma?

Plasma is formed from blood samples that have been prevented from forming a clot by adding anticoagulants; the erythrocytes, leucocytes, and platelets are then allowed to settle or are separated by centrifugation, leaving the plasma as a separate upper layer in the tube. Portions of plasma can then be removed for analytical tests. Serum, however, is obtained from whole blood that has been allowed to form a clot. The clot is formed by the conversion of soluble fibrinogen to a fibrin gel. The gel consists of a mesh of protein fibres that traps erythrocytes, leucocytes, and platelets. If the clot is allowed to settle or the tube is centrifuged, the serum forms a layer above it. Samples of serum can then be easily removed with a pipette or in a clinical analyser for testing or can be stored in a refrigerator or freezer. Whole blood samples cannot be stored in this way as the erythrocytes would haemolyse. In addition, many of the analytes estimated in clinical laboratories are present in the serum and using whole blood is not satisfactory for their analyses. Thus, plasma contains fibrinogen, which is absent in serum.

TABLE 7.2 Commonly used coloured containers and anticoagulants. See also Tables 17.2 and 17.3.

Additive	Optimum concentration	Mode of action	Lid colour code[1]	Examples of clinical tests
[2]Ethylenediamine tetraacetic acid	1.2 mg anhydrous salt per cm^3 of blood	Chelates and removes Ca^{2+}	Lilac	Full blood count Sickle cell tests Glandular fever tests
[2]Ethylenediamine tetraacetic acid	As above	As above	Pink	Blood group and compatibility tests Antibody screening
[2]Trisodium citrate	1 volume 0.109 mol dm^{-3} trisodium citrate per 9 volumes of blood	As above	Blue	Coagulation (clotting) tests, e.g. prothrombin time, activated partial thromboplastin time
[2]Trisodium citrate	1 volume 0.109 mol dm^{-3} trisodium citrate per 4 volumes of blood	As above	Black	Erythrocyte sedimentation rate (ESR)
[2]Lithium heparin, [2]Heparin	15 ($\pm$ 2.5) international units per cm^3 blood.	Neutralizes thrombin by inhibiting several clotting factors	Green	Blood glucose, Ca^{2+}
[3]Fluoride oxalate		Fl^- prevents pre-analytical breakdown of glucose (glycolysis). Oxalate binds Ca^{2+}	Grey	Blood glucose
[3]Gel clot activator	Gel substance	Accelerates clotting of blood	Red/brown/yellow	Tests requiring serum

[1]Paediatric sample bottles have different coloured tops; the anticoagulant is stated on the label.
[2]Anticoagulant.
[3]Not an anticoagulant.

It is essential to add appropriate volumes of blood samples to the tubes as under or overfilling means the proportion of blood to anticoagulant will be incorrect, which will decrease the accuracy of test results. This is especially the case when coagulation tests are to be performed. Some assays require whole blood and in those cases, tubes containing lithium heparin are used.

The majority of clinical biochemistry tests are performed on serum and so a plain tube or a tube containing a clotting accelerator (known as a gel tube, see Figure 12.14) are used. These are usually a plastic or glass tube with kaolin-coated plastic granulate or a polymerized acrylamide resin or gel. The clotting accelerator can interfere with some clinical assays and so a plain tube is recommended for certain tests, in which the blood must be allowed to clot and the clot to retract naturally, leaving the serum freely available. If serum is removed before the clot has retracted, it may continue to clot and form a fibrin gel. The advantage of adding a clotting accelerator is that the rapid formation of the clot means tests can be performed immediately after centrifugation.

Labelling of tubes

It is essential that all tubes are legibly and accurately labelled with information that allows subsequent matching of the patient and sample by the laboratory. This information may include the patient's family and given names, date of birth, NHS/CHI number, hospital number, gender, source, date, and time of sample collection. Quality assurance in laboratory professional practice begins with sample collection. Thus, it is essential that healthcare professionals have an understanding of the sampling procedures and the correct labelling of these samples. Each discipline in biomedical science will have particular sample labelling requirements which are fully explained in the companion volumes in this book series, but the bare minimum of information is outlined in Table 7.3.

Once appropriately labelled, the sample tubes are placed in a plastic bag which is then sealed. The patient should be checked for bleeding from the venepuncture and pressure continued until the bleeding stops if necessary. The needle site may be covered using a fresh piece of folded gauze and micropore tape. Finally, hands must be washed with soap and water after removing gloves, or washed immediately if they have been contaminated with blood.

TABLE 7.3 Labelling of sample request form

Labelling of specimen containers	Labelling of request form	
Full name of patient	Name of patient – forename and surname	
Date of birth and hospital number	Date of birth and hospital number	
	Clinical details – including any information likely to aid the clinician or other healthcare workers	
Name of consultant	Name of consultant	
Ward or clinic	Ward or clinic	

SELF-CHECK 7.2

What are the minimum identification details to be included on the label of a sample?

Factors in blood sampling that may affect accuracy of results

Cross reference

You can read about the principles of centrifugation in Chapter 12 'Centrifugation'.

To perform tests on serum or plasma obtained from a sample of blood, the blood must first be centrifuged. Centrifugation must be performed at the appropriate speed and length of time to isolate the serum or plasma from the blood. It is essential that cellular components do not leak their contents into the serum or plasma because this contamination will affect the concentrations of some analytes.

Blood must be centrifuged within a suitable time of sampling, normally 4 hours post-venepuncture. It is essential to prevent haemolysis as this will affect many biochemical and haematological tests. High concentrations of blood lipids also affect the accuracy of many clinical tests. This is known as **lipaemia**, and is due to a high fat diet or a lipid abnormality. Figure 7.5 shows the levels of haemolysis and lipaemia in centrifuged blood samples.

Serum must be removed from erythrocytes if the sample is to be stored overnight, to ensure accurate test results, particularly for concentrations of K^+ and glucose. Samples for coagulation tests need to be centrifuged and tested as soon as possible as coagulation factors are relatively labile. Alternatively, the plasma may be removed and frozen prior to analysis. When freezing samples, the permitted length of storage may depend on the temperature or the substance being analysed. The normal recommended temperature is $-20°C$.

FIGURE 7.5

Levels of (a) haemolysis and (b) lipaemia in centrifuged blood samples. The black line gives an indication of the opacities of the supernatants. LDH, lactate dehydrogenase.

Arterial blood sampling for blood gas analysis

Arterial blood samples are required when investigating conditions such as diabetes and lung diseases. These can affect the functioning of kidneys or lungs, and result in an imbalance in blood pH, carbon dioxide, and oxygen contents, and/or electrolyte concentrations. Arterial blood transports oxygen to the tissues and venous blood takes waste to the lungs and kidneys, therefore, the gas concentrations and pH differ between these two types of blood vessels. To estimate blood gases, blood is sampled from the radial artery, although in the case of newborn infants may be taken from the umbilical cord. The radial artery is located on the inside of the wrist, just beneath the thumb as shown in Figure 7.6, where a pulse is found. In difficult cases an artery in the elbow (brachial) or the groin (femoral) may be used. An arterial blood sample must be obtained by a clinician who is trained in this specific procedure. Blood is drawn into a syringe, which contains the anticoagulant heparin. A small amount of heparin is expelled from the syringe before taking the sample; adequate heparin will remain in the 0.2 cm^3 dead space of the syringe barrel and needle. A minimum of 3 cm^3 of blood is required to avoid excessive dilution by the heparin. Samples containing air bubbles must be discarded because this will cause gas equilibration between the air and the blood, thus lowering the pressure of arterial CO_2 and increasing the pressure of arterial O_2. After sampling, pressure must continue to be applied to the artery for at least 5 minutes to ensure bleeding has stopped.

The sample is retained within the syringe, which is mixed gently by inversion, and sealed so that exposure to air cannot occur. The syringe must be transported to the laboratory for testing as soon as possible and should be kept at $4\,^{\circ}C$ if any delay is expected, up to a maximum of 1 hour.

7.2 **Urine**

The diagnosis of many disorders is aided by collecting urine and analysing its metabolites and chemicals. Urine may be collected as single or random samples, or as the total volume of urine voided over a 24-hour period (Table 7.4). Random or early morning samples of urine may be tested for the presence of a variety of substances using a urine dipstick (Figure 7.7). The glucose tolerance test used in diagnosing diabetes mellitus requires three samples of urine, which are collected at specified intervals and analysed for their glucose concentrations.

Radial artery

FIGURE 7.6
Schematic showing the sampling of arterial blood from the radial artery.

TABLE 7.4 Types of urine samples and their corresponding analytes or tests

Type of urine collection	Analyte/test in urine
Random urine collection	Bence-Jones protein Reducing substances Indicans Osmolarity Porphyrins (protect from light) Pregnancy testing
24-hour collection *without* preservative	Creatinine Electrolytes Protein Urate Citrate
24-hour collection *with* preservative	Ca^{2+} Oxalate Phosphate 5-Hydroxyindole acetic acid Catecholamines

Collection of 24-hour urine samples

Twenty-four-hour urine samples give the most accurate estimates of the chosen analyte. However, it is essential to obtain a collection of the total urine output over an *accurately* timed 24 hours. Thus, careful attention to detail is essential. The patient is supplied with a special plastic container, with capacity of up to 2 dm³. If urine output in the 24 hours exceeds this, a second container is provided. The container should not be emptied or rinsed as it is chemically cleaned, and in some cases may contain a preservative such as hydrochloric or acetic acid. Care must be taken to avoid spillages. It is usual to start collecting urine on rising in the morning. The patient should initially empty their bladder and discard the urine, noting the exact time. In the next 24 hours, all urine voided should be included in the collection and added to the container. On the following day, at the same time as the collection started on the previous day, the bladder is emptied and all of this urine is collected into the container. Urine collection is then complete.

The container must be clearly labelled with the patient's name and the dates and times of collection. Additional instructions may be required for patients whose urine is to be tested for catecholamine levels (see Box 7.3).

BOX 7.3 *Foods and drugs that affect urinary catecholamine concentrations*

Many foods and drugs can disturb the biochemistry of blood and produce metabolites that are detectable in samples of urine. The concentrations of catecholamines in urine are particularly affected. Thus, the following foods and drugs should *not* be taken for 48 hours before and for 24 hours during, the collection of a urine sample:

- *Foods:* bananas, fruit, coffee, chocolate, vanilla, and flavourings.

- *Drugs:* aspirin, monoamine oxidase inhibitors, phenothiazines, imipramine, labetalol, guanethidine, reserpine, levedopa, tetracycline, or alpha methyldopa.

FIGURE 7.7
Renal dipstick used for clinical testing of urine.

Like blood, urine may also be collected for microbiological analyses (Section 7.5).

7.3 Fluids other than blood and urine

Apart from blood and urine, samples of a number of other body fluids are collected for clinical testing. These include:

- Cerebrospinal fluid
- Bronchial fluid
- Pleural fluid
- Bone marrow

A minimum of 1 cm^3 of these samples are collected by invasive procedures, which have some degree of risk to the patient and must therefore be performed carefully by correctly trained healthcare workers. Often, second samples tend to be unobtainable. Once collected, the samples are added to sterile universal bottles. Some of these are mainly tested for the presence of microorganisms and so we will discuss them in Section 7.5; others may be subjected to other clinical biochemical or immunological analyses.

Cerebrospinal fluid

Samples of cerebrospinal fluid (CSF) are required to investigate infections of the central nervous system (Section 7.5) or cases of suspected subarachnoid haemorrhage. The fluid is removed using a needle from the spinal canal in the lower back, in a procedure called a lumbar puncture (Figure 7.8). Three samples of CSF are needed from patients with suspected subarachnoid haemorrhages.

The samples should be investigated by the microbiology laboratory to determine the numbers of cells and types of microorganisms present. In addition, the clinical biochemistry laboratory may, for example, determine the concentration of protein in the sample.

Bronchial fluid

A fibre optic bronchoscope is used to obtain fluid from the bronchioles of the lungs, as you can see in Figure 7.9. The tip of the bronchoscope is introduced into the main stem bronchus. The bronchus is usually rinsed with a saline solution. Suction is then applied to the main stem bronchus and other segments of the bronchioles to remove fluid. The bronchial fluid may be

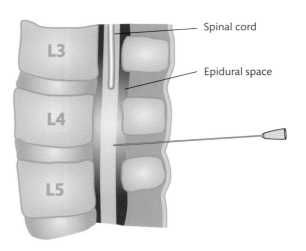

FIGURE 7.8
Simplified illustration of the sites and technique used to perform a lumbar puncture. Note how the needle is inserted between the vertebrae allowing CSF to be collected.

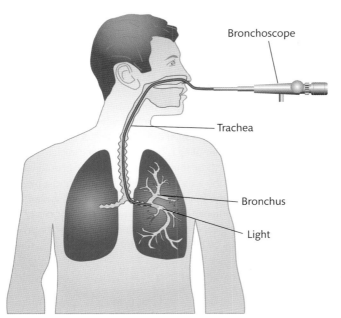

FIGURE 7.9
Schematic illustration of a bronchoscopy, a procedure that allows the respiratory system to be directly viewed. Samples of bronchial fluid can be removed as described in the main text.

filtered through sterile gauze before being added to a sterile container for transport to the laboratory. The sample may be centrifuged to obtain cells for cytological investigations (Section 7.4), or cultured to detect and identify any microorganisms which it may contain.

Pleural fluid

Fluid that collects between the lungs and the rib cage is known as a pleural effusion. The fluid may be sampled as illustrated in Figure 7.10, to investigate for disease. The patient is prepared by disinfecting the skin on the patient's back, which is then numbed using a local anaesthetic. A needle attached to an empty syringe is inserted into the fluid pocket, which is located approximately 3 cm below the surface. The syringe is filled or attached to soft plastic tubing to

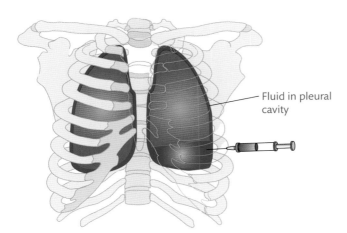

FIGURE 7.10
Schematic illustration of the removal of a sample of pleural fluid.

remove the pleural fluid into a bag or jar. The total sample, or a portion of it, is labelled with the patient's details and transported to the laboratory for clinical investigation.

Synovial fluid

Synovial fluid cushions and lubricates the mobile joints of the body where it is found in small quantities in the spaces contained within the synovial membranes. Analysis of synovial fluid may identify diseases that affect the structure and function of the joints, including osteoarthritis, rheumatoid arthritis, and gout. Synovial fluid sampling is performed by a clinician using a sterile syringe and needle in a procedure called arthrocentesis, which you can see outlined in Figure 7.11. A local anaesthetic is used to numb the area surrounding the joint and a needle is then gently inserted into the space between the bones, and a sample of fluid sucked out into a syringe. The fluid is placed in a sterile container and transported to the laboratory for analysis. Microscopical (Section 7.4), microbiological culturing (Section 7.5), and biochemical tests, such as those for uric acid and glucose may be performed on the sample.

Bone marrow

Samples of bone marrow are used diagnostically to observe changes to cellular morphology, and for clinical and immunological investigations. Samples are obtained by aspiration or by trephine biopsy. Bone marrow may be sampled from a variety of sites but the commonest are the posterior iliac crest and sternum. As you can see in Figure 7.12, aspiration involves inserting a hollow needle, to which a syringe is attached, into the marrow. The marrow is then sucked into the syringe and spread directly onto glass microscopical slides for analysis or into a sample pot to be used for cytogenetic investigations, culturing the cells present, biochemical analyses, and a variety of other specific investigations.

A trephine biopsy is performed using a trephine needle, known as a Jamshidi bone trephine needle. The sample comprises a core of solid bone, which includes the marrow. This is placed into a solution of formalin for fixation and examined by the histology laboratory.

Cross reference

Gout is described in Chapter 4 'Hyperuricaemia and gout' in *Clinical Biochemistry*.

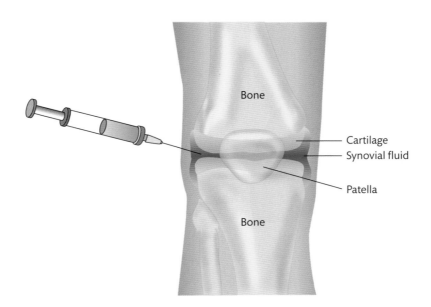

FIGURE 7.11

Schematic illustration of the removal of a sample of synovial fluid, in this case from the knee joint, using a needle and syringe.

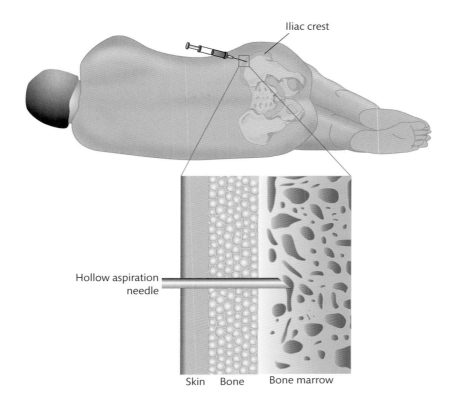

Iliac crest

Hollow aspiration needle

Skin   Bone   Bone marrow

FIGURE 7.12
Schematic illustration of the sites and technique used to obtain a sample of bone marrow from an iliac bone using a robust intra-osseous needle and syringe.

7.4 Cytopathology and histopathology samples

Cytology is the study of cells and histology that of tissues. Cytopathology and histopathology are the branches of pathology that study and diagnose diseases at the level of the cell and tissue, respectively. The study of cytological and histological abnormalities requires samples of tissues and cells.

In general, the majority of histopathology samples are transported to the laboratory in plastic pots of varying sizes or glass 'universal' bottles containing a fixative (Table 7.5). In the UK, 4% formaldehyde (as formalin or formal saline) is used to fix the majority of samples. The ideal amount of fixative is 10 times the volume of the sample. Failure to use adequate amounts will potentially result in a loss of tissue and cellular morphology, and could adversely affect subsequent staining and molecular techniques that may be required for diagnosis.

Cells

Cytological samples are normally provided unfixed, as cell suspensions in transport media or as smears of cells on glass slides. The latter are fixed using alcoholic spray fixatives. The commonest cytological samples are taken from the cervix by swabbing (Section 7.5) and are used in smear testing to detect cervical cancer at an early, treatable stage (see Figure 8.2).

TABLE 7.5 Fixative solutions for histology samples

Fixative	Constituents	Use
Formal saline	100 dm³ 40% formaldehyde 9 g NaCl 900 dm³ tap water	General preparation of paraffin embedded tissue sections from e.g. liver, kidney
Alcoholic formalin	100 dm³ 40% formaldehyde 900 dm³ 95% ethanol	Specific cases, as advised by the clinical laboratory
Bouin's fluid	75 dm³ saturated aqueous picric acid 25 dm³ 40% formaldehyde 5 dm³ glacial acetic acid	Biopsies of testicular, prostate, and other tissues
Glutaraldehyde	5 dm³ glutaraldehyde 95 dm³ 0.1 mol dm⁻³ cacodylate buffer, pH 7.4	Ultrastructural studies

Fine needle aspirates

A clinical sample can be taken from superficial sites such as breast, lymph nodes, thyroid, and salivary glands, and subcutaneous lesions in general by aspiration using a fine needle. Aspirations can usually be performed in an outpatient department immediately by a pathologist or on in-patients in their ward. Aspirations of deep-seated sites may be undertaken using needles of the correct length and using some form of image guidance, such as ultrasound or X-rays. If a pathologist is not available or the clinician wishes to perform the aspiration then a member of the cytopathology laboratory may assist in the spreading, fixation, and transport of the sample. Aspirations are generally considered to be cytopathological samples, unless larger aggregations of cells, akin to tissues, are removed. A diagnosis can normally be reached the same day, allowing for more rapid determination of a treatment programme for the patient.

Tissues

Post-mortem and post-operative samples may be major parts of bodies, portions of organs, or small pieces of tissue taken as biopsies. Sections cut from pieces of tissue are processed and stained on glass slides in a variety of ways to allow them to be visualized using a microscope. Samples of tissue obtained during surgery may be rapidly frozen using liquid nitrogen, which prevents ice crystals forming and disrupting the morphology of the tissues. Thus, intra-operative samples should be sent immediately to the laboratory, without fixation. Sections of the frozen sample can then be cut and stained for microscopic observation. Sections prepared in this way can aid a rapid diagnosis and guide the surgeon during an operation. Samples may also be obtained from patients by curettage, which are essentially scrapings of tissues.

Biopsies such as skin or renal tissue samples, which require testing by immunofluorescence, must be sent to the clinical laboratory as soon as possible in normal or physiological saline solution (9 g dm⁻³). These are then frozen and sections of tissue prepared. Removing biopsies of muscle tissues may require specific techniques that can be obtained by direct discussion with the laboratory. Testicular and prostatic biopsies are recommended to be placed in Bouin's fixative (see Table 7.5). Lymph nodes and other samples on which microbiological or biochemical examination is required must be divided and a portion placed in a sterile container. Lymph nodes removed from patients with suspected lymphoma should be fresh and intact and sent immediately to the histopathology laboratory.

Cross reference
Chapter 8 'Microscopy'.

Cross reference
You can read more about immunofluorescence in Chapter 15 'Immunological techniques'.

Tissues may also be sampled to investigate specific enzyme activities, such as that of ATPase; activities of this change in different types of muscle fibres in some muscle disorders. In such circumstances, the sections are fixed *following* the investigative technique to ensure that enzyme activity is not compromised. Muscle biopsies may have specific requirements, which can be obtained after direct discussion with the laboratory.

Transport of cytological and histological samples

Samples of cells and tissues for cytological and histological investigations are placed in pots with tightly closing lids to prevent leakage. The pots are enclosed in plastic bags for transport to the laboratory. Samples must be accompanied with an appropriate request form in a manner that keeps the form and sample together, but avoids contamination of the form. As with all samples, it is essential to complete the request form, as incorrect or incomplete forms may result in a delay in results. Any fresh samples must be sent to the laboratory as quickly as possible to avoid deterioration of these samples.

When the sample is received in the laboratory, it must be matched to its request form, ensuring all details are legible and correct. Due to the diagnostic nature of many histological or cytological samples, it is important that full clinical details are provided on any previous medical history, biopsies, or treatments. Failure to provide this information may result in the sample being processed in a sub-optimal way.

Key Points

Samples for histology or cytology are often unique and it is not possible to take a repeat sample if something goes wrong

---

SELF-CHECK 7.3

List three ways in which a sample of tissue can be obtained for histological examination.

Storing cytological and histological samples

Cytological and histological samples are often stored for longer periods than samples for other disciplines, being kept (as a minimum) until a definite diagnosis has been established. The resulting prepared specimens are also stored for long periods of time, which presents its own problems.

Storage of samples and specimens is now governed by the requirements of the various human tissue acts and is the subject of professional guidance from the Institute of Biomedical Science and the Royal College of Pathologists.

Cross reference
More material about the human tissue acts and blood regulations is given in Chapter 19 'Quality assurance and management'.

7.5 Samples for microbiology testing

Microbiology laboratories of hospitals investigate samples for viruses, bacteria, and fungal infections. Virology laboratories offer routine and specialist tests for the presence of viruses, including the detection of those causing hepatitis, acquired immunodeficiency syndrome (human immunodeficiency virus, HIV), German measles (rubella) and, despite being caused by a bacterium, often syphilis. The bacteriology section of the microbiology laboratory deals with the isolation, identification, and antibiotic sensitivity testing of clinically significant bacterial isolates obtained from a range of patient samples. These include blood, urine, sputum, and other body fluids, faeces, swabs from for example wounds, eyes, nose, throat, vagina, and groin. In addition, scrapings and nail clippings are also cultured for microbiological growths. Fungal infections are investigated by the mycology section of the microbiological laboratory.

In all cases, the healthcare worker must ensure that the sample is correctly labelled with the patient's identifying details, and that the appropriate details, including medical and treatment history, are provided on the request form.

Blood

Samples of blood for microbiological culture in aerobic and anaerobic conditions are obtained using a needle and syringe (Section 7.1) to keep the environment sterile. Bottles used for the blood cultures do not fit in the standard vacuum tube holders, although it is possible to obtain blood culture bottles with long narrow necks and which do fit, or to obtain adaptors that allow the culture bottles to be used as a vacuum tube. When blood cultures are collected using a winged blood collection set, air in its tubing means the phlebotomist must collect the blood sample for aerobic culture first. A blood sample for anaerobic blood culture can then be collected once the air has been flushed out of the tube.

It is essential that the collecting procedure is performed in as sterile a manner as possible to prevent skin **commensals** contaminating the samples. To reduce this danger, an abrasive swab is used to prepare the site by removing upper layers of dead skin cells along with their contaminating bacteria. Povidone-iodine has traditionally been used to clean the skin, prior to sampling, but in the UK a 2% chlorhexidine in 70% ethanol or isopropyl alcohol solution is preferred and time must be allowed for it to dry. The tops of all blood containers should also be disinfected using a similar solution. The bottles are then incubated in specialized incubators for 24 hours. This step allows the very small numbers of bacteria (potentially only one or two organisms) to multiply sufficiently for identification and antibiotic resistance testing. Modern blood culture bottles have an indicator in their base that changes colour in the presence of bacterial growth and can be read automatically by machine. The bar-coded stickers found on these bottles must not be removed as they are used for identification by the automated systems.

Urine

Urinary tract infections are common in certain groups, such as the elderly, immunocompromised patients, or pregnant women. The aim is to collect urine from the patient with minimal contamination so that any pathogenic infection can be detected by culturing and identifying bacteria present in the sample. Patients need to be given clear instructions as to how to collect the urine sample. Obviously, if they require assistance this must be given. Samples of urine should be collected midstream (MSU) or from a urinary catheter (see below). Early morning urine samples contain the highest numbers of bacteria, as the urine has been 'incubating' in the bladder overnight. The genitals should be thoroughly washed with soap and water, and the first part of the urine stream discarded into the toilet or bedpan. The urine sample is then collected into a sterile utensil such as a foil bowl or sample pot, by intercepting a continuous stream of urine. Sterile plastic screw-capped bottles are used for all samples. Sample pots for MSU contain boric acid as this keeps bacterial numbers constant for at least 24 hours.

If the urine sample is to be used to investigate the possible presence of mycobacteria that cause tuberculosis (TB) then complete (that is *not* midstream) early morning urine samples need to be collected over 3 days. If testing is to be for *Chlamydia trachomatis*, at least 15–20 cm³ of the **first voided urine** (FVU) of the day is required.

In all cases, a request form should accompany the urine sample, indicating the type of sample and any patient history of antibiotic treatment. If there is to be a delay in transporting the sample to the laboratory, the sample should be stored refrigerated, but never in a food refrigerator!

Collecting urine with a catheter

Sometimes, for example, following an operation or when the patient cannot urinate for themselves, the urine is drained directly from the bladder through a catheter (catheter sample urine, CSU) into a drainage bag. Urine samples should not be collected directly from the drainage bag, but taken from the sampling port by clamping the drainage tube immediately below the sampling port and waiting for 10 minutes until enough urine has accumulated. Hands must be washed prior to cleansing the sampling port with a medi-swab. A needle and syringe is then used to collect approximately 10 cm³ of urine from the sample port and transfer it to a sterile pot. The sample pot should be labelled with the patient's details and sent to the microbiological laboratory for appropriate testing.

Collection of sputum

The aim when collecting sputum is to obtain samples of deep respiratory secretions that are not contaminated with upper respiratory tract bacteria. Sample collection begins by explaining the procedure to the patient, encouraging them to breathe deeply, and on exhalation to cough directly into a sterile screw-capped container. Microbiological culturing of sputum samples should use freshly collected specimens. If sputum is being sampled to investigate for the possible presence of TB-causing mycobacteria, at least three separately collected samples should be tested over three consecutive days.

All samples of sputum should be labelled and sent to the microbiological laboratory in a sample bag separate from those for other types of samples, for example blood.

Other fluids

Other fluids collected for microbial analysis include CSF, bronchial, pleura, and ascitic. These fluids are obtained by invasive techniques and are from normally sterile sites, as we described in Section 7.3. Samples must be added to sterile containers. Care must be taken by healthcare workers to avoid contaminating the sample with normal microbial flora. If the patient has started antimicrobial therapy prior to sampling, this must be noted. In many cases, the results of culturing these fluids to detect microorganisms present are particularly useful clinically. Therefore, samples should be tested immediately by the laboratory.

The skin must be thoroughly disinfected prior to taking samples of CSF by lumbar puncture (Figure 7.8). In addition to patient identifier on the sample bottles, the bottles need to be labelled in the order the samples were collected Three samples are collected; the first two are used for biochemical analyses. The third should be sent to the microbiological laboratory for examination by Gram staining, appropriate culturing of the bacteria present, which depends on the findings of the Gram stain, and counting any cells that may be present in the sample.

Ascites is the accumulation of fluid in the abdominal cavity in a number of pathological conditions, such as cirrhosis of the liver, heart failure, certain cancers, and infections, for example, pancreatitis, chronic hepatitis, and TB. Fluid may be removed by draining (Figure 7.13) or during surgery, and subjected to relevant clinical testing.

Sweat testing involves the collection of a sample of body sweat to determine its concentrations of Na^+ and Cl^-. This has been the most widely used diagnostic test for cystic fibrosis. Sweat collection requires specific training as a 'good' sample depends on care and skill in avoiding evaporation and contamination, both of which lead to falsely increased concentrations of Na^+ and Cl^-.

Collecting a sample for a sweat test uses a small, painless electric current to help draw sweat to the surface of the skin, from where it can be collected for analysis. The sweat may be drawn from the thigh or the forearm, depending on the age and size of the patient. The area is prepared by washing and drying, after which two metal electrodes are attached and fastened with straps. Two gauze pads, one soaked in saline or hydrogen-carbonate and the other in pilocarpine, a drug that stimulates sweating, are placed under the electrodes. The electric current is applied to the skin for five to 10 minutes to carry the pilocarpine into the skin where it then stimulates the sweat glands. Once the sweat is being produced the electrodes are removed, and the skin is washed with distilled water and redried. A dry piece of filter paper is taped to the area where the pilocarpine was applied. The filter paper, known as a sweat patch, is covered with wax or a sheet of plastic to prevent evaporation. After 20–40 minutes, the plastic is removed and the filter paper is placed in a container, which is sealed and labelled with the patient's details for transport to the laboratory.

Collection of faeces

The microbiological examination of faeces is complex. A full clinical history must be taken from the patient, including foods eaten because of the possibility of food poisoning, any travel to foreign countries, any medication being taken, as well as the more usual basic information.

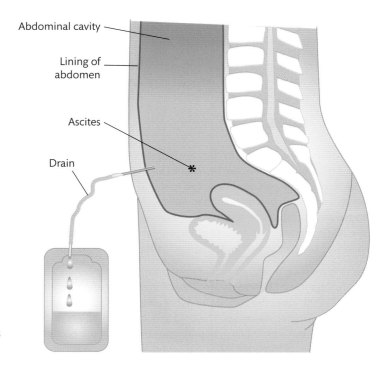

FIGURE 7.13
Schematic illustration of the collection of ascites fluid by drainage.

A sample of faeces is collected using a spatula. A piece the size of a large pea is adequate. The sample, usually called stool, is transferred to a plastic screw-cap faeces container. If the faecal sample is liquid, the pot is filled to about one-third. Hand-washing and the wearing of safety gloves are essential. The presence of blood in faecal sample, known as **occult blood,** can be detected using specialized cards as indicate in Figure 7.14. The faecal material is smeared on the relevant portion of the card, which is then sent to the laboratory for analysis. When testing for occult blood, patients must not eat any of the following for 3 days prior to the test: red meat, black pudding, liver, kidneys, fish with dark meat (salmon, tuna, sardines, mackerel), cauliflower, horseradish, tomatoes, radishes, melon, bananas, soya beans, alcohol, iron tablets, aspirin, and ascorbic acid.

Collection of samples using sterile swabs

Simple sterile swabs consist of a plastic stick ending with a pad made from cotton, Dacron, or alginate, which is usually contained in a plastic sleeve containing a transport material to preserve the sample or swabbing (Figure 7.15). Swabbings usually consist of body tissues and fluids for the detection of microbiological **pathogens**. They are transported from the patient to the microbiological laboratory in media contained in the corresponding sleeve.

(a)   (b)

FIGURE 7.14
(a) Photograph of an open and closed occult blood card. (b) Schematic illustration of a faecal sample being added to an occult blood card. Note three cards are provided for collecting samples on different occasions.

FIGURE 7.15
Photograph of a sterile swab and its container, which contains transport medium.

The commonest areas sampled by swabbing are:

- High vagina
- Nose and ears
- Throat
- Conjunctiva (eye)
- General wounds

Obtaining a high vaginal swab requires a speculum being placed into the vagina to dilate its walls. The sterile plastic envelope containing the swab is discarded. The swab is removed from its sleeve and inserted as high as possible into the vagina and gently rotated. Any visible discharge is particularly sampled. The swab is returned to its plastic sleeve, which contains the transport medium and sent to the laboratory for suitable analyses. Plain swabs are suitable for the detection of all vaginal pathogens except *Chlamydia* or *Herpes*, which require the use of specialized types of swabs.

Nasal swabs are obtained by moistening the swab in sterile saline, and then rubbing and rotating it against the anterior nasal hairs. The swab is then placed into its sleeve, where it is moistened in its transport medium. Ear swabs are collected by placing a sterile swab into the outer ear cavity, and rotating it to collect any pus or discharge. The swab is then placed into the sterile sleeve, labelled with the patient's details, including whether the sample is from the left or right ear, and transported to the microbiological laboratory.

Swabbing the throat involves depressing the tongue with a wooden spatula. The swab is then moved to the back of the throat and the area around both tonsils rubbed using gentle rotary movements. The swab is then placed in its plastic sleeve containing transport medium.

Microbial infections of the eyes can lead to a discharge of pus or other fluid. Conjunctival (eye) swabbing involves thoroughly soaking the swab in the exudate by rubbing it across the lower eyelid from the inner to outer corners. It is necessary that the patient remains still during the procedure. The swab is then added to its sleeve with transport medium.

Wound swabs should be obtained before wounds have been cleaned, patients bathed, or antibiotic treatment commenced or changed. Swabbing should only be applied directly to an infected site to avoid contaminating undamaged skin or mucous membranes. The swab should be rotated in areas showing pus and then placed in transport media. Both the swab and its request form must be appropriately labelled; particularly relevant information regarding antibiotic therapy and the site of the wound.

It is essential to detect the presence of methicillin-resistant *Staphyloccocus aureus* (MRSA) in or near wounds. When MRSA has previously been shown to be present in a patient, this must be stated in the clinical details and MRSA tests asked for on the request form. This allows the laboratory to carry out additional tests to check for MRSA at the same time as routine culturing is being performed. Patients may need screening for the presence of MRSA prior to their admission to hospital. In such cases, swabbings from the throat, nose, axillae, and groin are required.

SELF-CHECK 7.4

List the key steps in obtaining a wound swab for microbial culture.

Samples for investigating fungal pathogens

Samples for mycology testing should be collected into special collection envelopes supplied by the laboratory (Figure 7.16). As with other samples, they are sent with a request form including details of the patient's clinical symptoms and history.

Material from skin lesions is collected by scraping material from their outer edges, usually with the edge of a glass microscope slide or scalpel blade. The edge of the lesion is the part most likely to contain viable fungal material. Scrapings from the scalp are also obtained in this way, but should include hair stubs. Hairs may be plucked from the scalp, but cut hairs are unsuitable for testing as fungal infections are usually below the surface of the scalp.

Nail clippings for mycological investigation should be taken from the discoloured or brittle parts of the nail, and cut back as far as possible from the free edge as the growth of some fungi are restricted to the lower parts. Scrapings can also be taken from under the nails to supplement clippings; however, even if fungi are present in the nail clippings, they often fail to grow on culturing.

Cross reference
You can read more about medical microbiology in the companion book of that name.

High risk, hazardous microbiology samples

If it is thought that there is a likelihood of group III pathogens (Box 7.4), such as tubercle bacilli and hepatitis B and C viruses, being present in clinical samples, then the samples must be placed in a plastic sample bag, which is then enclosed in a second sample bag attached to the request form. The request form and sample must be marked with a *Danger of Infection* label.

Cross reference
Chapter 4 'Health and safety'.

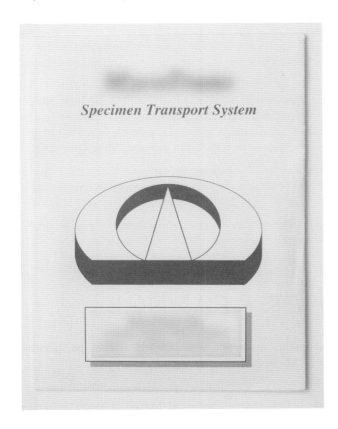

Specimen Transport System

FIGURE 7.16
Photograph of an envelope for collecting samples for mycological testing.

BOX 7.4 Categories of pathogenic microorganisms

The Advisory Committee for Dangerous Pathogens has divided pathogenic organisms into four groups, according to their level of danger. Group I are the least dangerous, IV the most pathogenic. *Mycobacterium tuberculosis*, one of the two pathogenic organisms that cause tuberculosis is a group III pathogen.

SUMMARY

- The quality assurance in obtaining samples for testing by clinical laboratories includes correctly identifying patients, obtaining samples using correct procedures, meticulous labelling of samples and requests forms, adhering to health and safety procedures, and being aware of standard operating procedures for sample processing

- Procedures for obtaining and processing blood samples require selection of the appropriate container. An anticoagulant or additive is necessary for some laboratory analyses. Venepuncture procedures may employ a needle and syringe or needle and vacutainer. All blood samples should be free from haemolysis. Plasma or serum samples are obtained by centrifuging blood samples

- Samples for histological examination may be obtained as tissue biopsies. Sections cut from pieces of tissue are processed and stained on glass slides, and visualized using a microscope. Samples of tissue obtained during surgery may be rapidly frozen using liquid nitrogen. In some situations, samples may be obtained by aspiration using a fine needle

- Key steps in obtaining a range of samples for microbiology testing including the relevance of sterility and appropriate use of transport media in which to transport samples to the laboratory

- Urine samples for biochemical analysis may be collected randomly or over a 24-hour period. Containers for the latter may require the inclusion of a preservative

FURTHER READING

- **Ageneye F. Pre-analytical quality assurance: a biomedical science perspective.** *The Biomedical Scientist*, 2007; **51**: 86–87.

  A short paper in a professional journal from the Institute of Biomedical Science.

- **Bowen RA, Hortin GL, Csako G, Otañez OH, Remaley AT. Impact of blood collection devices on clinical chemistry assays.** *Clinical Biochemistry*, 2010; **43**: 1–2, 4–25.

● **Ford M. (editor) Medical Microbiology.** Oxford University Press, Oxford, 2010.

A companion book in this series.

● **Haverstick DM, Brill LB, Scott MG, Bruns DE. Preanalytical variables in measurement of free (ionized) calcium in lithium heparin-containing blood collection tubes.** *Clinica Chimica Acta* 2009; **403**: 1–2, 102–104.

Two articles that provide details of the types of blood collection tubes and discuss their effects on blood chemistry analytes.

● **Jacobs DS, DeMott WR, Oxley DK.** *Laboratory Test Handbook*, 2nd edn. Lexi-Comp, Inc, Ohio, 2002.

An excellent pocket-sized reference book (1348 pp.) containing sample details for all tests imaginable, clinical reference ranges and much more.

● **Pitt SJ, Cunningham JM.** *An Introduction to Biomedical Science in Professional and Clinical Practice.* Wiley-Blackwell, Chichester, 2009.

This textbook covers many aspects of laboratory practice, including an introduction to the organization of sample analyses by clinical pathology laboratories.

● **Polack B. Evaluation of blood collection tubes.** *Thrombosis Research*, 2007;**119:** 133–134.

A journal article focusing on types of tubes for blood samples required for coagulation and thrombosis investigations.

Useful Websites

■ Human Tissue Authority. The Human Tissue Act. Available at: **http://www.hta.gov.uk/ legislationpoliciesandcodesofpractice/legislation/humantissueact.cfm**

■ Epsom & St Helier NHS Services. Available at **http://www.epsom-sthelier.nhs.uk/ our-services/a-to-z-of-services/diagnostics-and-pharmacy/pathology/chemical-pathology/** (accessed 26 December 2009). A detailed National Health Service (UK) website that describes specimen reception for pathology samples.

■ Lab Tests Online-UK. Available at: **http://www.labtestsonline.org.uk/**. A peer reviewed public resource on clinical laboratory testing with a patient-centred approach. This website was established with the aid of a grant from the Health Foundation, and is maintained by the Royal College of Pathologists and the Department of Health, UK.

■ The retention and storage of pathological records and specimens. Available at: **http:// www.rcpath.org/resources/pdf/g031retentionstorageaugust09.pdf**

 QUESTIONS

1.   Trisodium citrate is used as an anticoagulant for which of the following?

   (a)   Catecholamine tests

   (b)   Coagulation tests

   (c)   Investigating suspected cholera cases

   (d)   Investigating suspected Chlamydia cases

   (e)   Clinical chemistry testing

2.   Which of the following statements are **TRUE**?

(a)   The commonest type of clinical sample analysed is blood

(b)   A sample of serum does not contain fibrinogen

(c)   Using an insufficient amount of fixative in preparing histopathology samples results in a loss of cell morphology

(d)   A bone marrow biopsy is called a trephine biopsy

(e)   Patients providing faecal samples for occult blood tests must avoid eating bananas

3.   List the common additives used in blood collection tubes, their modes of action, and the clinical tests with which they are compatible.

4.   Explain why clinical samples must be appropriately centrifuged and stored prior to clinical analysis.

5.   What instructions would you give to a patient whose urine is to be tested for its concentrations of catecholamines?

Answers to self-check and end of chapter questions are available in the Online Resource Centre accompanying this book.

 Go to www.oxfordtextbooks.co.uk/orc/glencross

Microscopy

Tony Sims

Learning objectives

After studying this chapter, you should be able to:

- Define magnification, contrast, and resolution
- Explain the principles of image formation
- Identify the various components of a microscope and describe their functions
- Describe the major types of microscopes used in clinical laboratories
- Discuss the applications of the various forms of light and electron microscopy used in clinical laboratories
- Set up a microscope with optimum illumination

Introduction

A **microscope** is an instrument that forms an enlarged image of an object that would normally be too small to be seen with the naked eye as shown in Figure 8.1 (a) and (b). **Microscopy** is the use of a microscope to examine and analyse such objects. Microscopes that use a single lens are called simple microscopes; those with more than one are compound microscopes.

Microscopes are perhaps the most widely used instruments in biomedical science. They have contributed greatly to our knowledge and understanding of pathological processes, and are used in all branches of biomedical science. There are several different types of microscope, each of which has its own specific applications. *Light* microscopes are used to look at cells and tissues. They are used in many areas of biomedical science and form the work horses of histological examinations or in screening programmes, such as that for cervical cancer (Figure 8.2). The *electron* microscope, with its vastly increased **magnification** and **resolution**, is used to visualize virus particles, explore the structures of bacteria, and observe more fully the sub-cellular components seen in both normal and diseased cells and tissues.

In this chapter we will describe how the various types of microscope are constructed and work, and how they can be applied to diagnostic biomedical practice.

(a)

(b)

FIGURE 8.1

Microscopical examinations of (a) blood smear and (b) the bacterium *Staphylococcus aureous*. Courtesy of M Hoult, School of Healthcare Science, Manchester Metropolitan University, UK.

FIGURE 8.2

Light micrograph of a cervical smear showing pink stained normal squamous cells from the superficial layer of the cervix. The blue stained cells are from the cervical wall immediately below that are squamous cells; their comparatively large nuclei indicate that they are abnormal and the patient is at risk of developing cancer of the cervix. Courtesy of Manchester Cytology Centre, Manchester Royal Infirmary, UK.

8.1 **Microscopy and image formation**

Three main features limit the use of any microscope:

1. Magnification
2. Contrast
3. Resolution

The magnification of a specimen given by any microscope is simply the apparent size of the image formed divided by its real size. We will discuss how a microscope produces a magnified image in Section 8.3. The *contrast* of an image is the difference between the darkest and lightest coloured areas, which in microscopy reflects differences in the densities of different parts of the structure viewed. In general, a biological specimen will show poor contrast and will therefore be difficult to see unless the contrast between different parts of it are increased by differential staining. Specimens for light microscopy are generally stained with coloured dyes, which are only absorbed by some areas of the specimen so boosting the contrast, as you can see in Figure 8.3. You can read on outline of how clinical specimens are prepared for light microscopy in Box 8.1.

The contrast of specimens for electron microscopy is increased by staining specimens with the salts of heavy metals, such as lead, silver, or gold. Even for specimens with excellent contrast, the magnification any microscope can give is limited by the resolution, which we will discuss below.

Follicular cell

*

Follicle containing
thyroglobulin

FIGURE 8.3
Histological structure of the thyroid gland.
**Note how staining of the different portions has
increased the contrast of the image. Courtesy of
Dr A.L. Bell, University of New England, College
of Osteopathic Medicine, USA.**

BOX 8.1 *Preparing samples for light microscopy*

In general, tissue samples are transported to the laboratory in containers holding a fixative (Table 8.1) to preserve structure. In the UK, 4% formaldehyde (known as formalin) is the most commonly used fixative. Ideally, about a 10-fold volume of fixative to sample is required. Failure to use adequate amounts will potentially result in a loss of tissue and cellular morphology, and could adversely affect subsequent staining and molecular techniques that may be required for diagnosis. Large tissue samples may be sliced open to facilitate optimal fixation.

Once fixed, samples of tissue for histopathological examination must be processed to allow thin (2–3 µm thick) sections to be produced for microscopical study. For routine processing into paraffin wax, tissues are exposed to a number of reagents, for example, transferring the samples through a series of solutions with increasing concentrations of alcohol, to remove fully their water content. The alcohol is then replaced by an intermediate (or link) reagent that is not miscible with water, but allows molten wax to penetrate the sample block. In the UK, USA, and Australia, this is routinely undertaken on automated, fully enclosed tissue processors. In developing countries, this can still be undertaken manually; however, the nature of the solutions used to process the tissue does pose a health and safety risk to the laboratory worker.

Sections are cut from the wax-embedded tissue using a microtome; these can also be automated to produce sections of consistent and reproducible thickness.

Situations may arise that require the production of sections for rapid diagnosis or investigation of materials. For example, samples of tissue obtained during surgery may be rapidly frozen using liquid nitrogen (−196°C), which prevents ice crystals forming that would disrupt the architecture of the tissue. Sections of the sample are cut and stained

Cross reference

You can read more about health and safety in Chapter 4.

TABLE 8.1   Fixative solutions for histology samples

Fixative	Constituents	Use
Buffered formalin	100 dm³ 40% formaldehyde 9g NaCl 900 dm³ tap water	General preparation of paraffin-embedded tissue sections from e.g. liver, kidney
Alcoholic formalin	100 dm³ 40% formaldehyde 900 dm³ 95% ethanol	Specific cases, as advised by the clinical laboratory
Bouin's fluid	75 dm³ saturated aqueous picric acid 25 dm³ 40% formaldehyde 5 dm³ glacial acetic acid	Biopsies of testicular, prostate, and other tissues
Glutaraldehyde	100% glutaraldehyde	Ultrastructural studies

for microscopic observation to provide a rapid diagnosis and guide to the surgeon during the operation.

In some instances, for example, renal biopsies, it may be necessary to produce sections of less than 2 μm thickness. Often, this requires processing the tissue sample into resin, rather than wax. Sections are cut on an automated microtome using a knife blade strengthened, for example, by coating its cutting edge with tungsten carbide or diamond. Using the more solid medium of resin allows thinner sections to be cut, but it does suffer the disadvantage of absorbing many of the stains used to heighten contrast. Consequently, interpreting these sections can be more difficult. Resin is also used to embed small pieces of tissues that require examination using an electron microscope (see Box 8.3).

Once the tissue has been processed and sectioned, it must be stained to provide contrast between the different tissue and cellular components. There is a multitude of different methods that may be utilized to highlight different tissue components, but the routine stain used is haematoxylin and eosin (H&E). Such stains highlight the general architecture of tissues and determine if a pathological process is occurring, or if any further techniques are required to reach a definitive diagnosis.

Samples received in histopathology laboratories are varied: it is not unusual to receive samples of bone or other mineralized materials. These cannot be processed straight to paraffin wax and sectioned using a microtome because the mineralized areas are too tough for normal microtome knives to cut. However, the mineral content of the fixed tissue can be reduced by exposing it to **decalcifiying agents**. For samples with relatively small levels of mineralization, gentle chelating agents such as ethylaminediaminetetraacetic acid (EDTA) may be used. For heavily mineralized tissues, such as bone, harsher agents such as formic or hydrochloric acid must be applied. Over-exposure to the decalcification agents can be detrimental to tissue architecture and have a negative impact on the staining properties of the tissue. In particular, antigenic sites may be destroyed, leading to false negative results in subsequent investigations. Thus, samples should be examined regularly using for example X-rays, to determine the optimal extent of decalcification. The tissue should then be thoroughly rinsed and returned to fixative solution. In samples where the degree of mineralization is crucial to diagnosis, as for example when investigating osteoporosis, decalcifying agents cannot be used and tissues should be processed into resin.

In cytopathology, the cells of the sample must be made available on a slide for viewing. This can be achieved in different ways. The sample may be smeared onto the slide, for example, from a swab or a spatula. A specialized centrifuge called a cytospin may be used, to force the cells out of suspension (be it urine, transport medium) and onto the slide. If a sample has very few cells in relation to its volume, it is first subjected to conventional centrifugation to concentrate the cells. The pellet of cells is then resuspended in a relatively small volume, which is subjected to the **cytocentrifuge**.

In the UK, many cervical samples (Figure 8.2) are received as part of the national cervical screening programme. These samples are delivered to the cytology laboratory in transport fluid or fixative and the cells are separated by using membrane filtration or density gradient systems that deposit the cells onto glass slides. As with histopathological sections, once the cells are fixed on the slide, they must be stained to increase contrast for visualization by light microscopy.

Cross reference

Chapter 7 'Samples and sample collection' describes how swabs can be used to obtain clinical samples.

Key Points

All microscopes used in biomedical science are likely to be compound in nature. That is they consist of two main elements, called objective and eyepiece lenses. By using lenses in this way, vastly improved magnifications can be achieved over the use of a single lens.

Resolution

Resolution is a measure of the clarity or acuteness of an image. It is the smallest distance that can be distinguished between two points and objects smaller than the resolution cannot be observed. The resolution of the unaided eye is approximately 200 μm.

SELF-CHECK 8.1

Define what is meant by the term *resolution*.

Resolution is a limiting feature of all microscopes, as simply magnifying an image will not always provide more information. As magnification increases, resolution or the resolving power of a microscope must also improve. Consider, for example, a digital image that is enlarged to the point of pixellation: clearly, the clarity of the image has been decreased despite the increase in magnification and is known as *empty magnification*.

In mathematical terms, resolution can be represented by the formula:

$$\mathbf{R} = 0.61\lambda / NA$$

Where **R** is resolution, λ is the **wavelength** of the radiation or light and NA is the **numerical aperture** of the lens.

Numerical aperture is a quantitative representation of a lens' light (or radiation) gathering capacity and is represented by the formula:

$$\mathbf{NA} = \mathbf{n} \sin \alpha$$

Where **n** is the **refractive index** of the medium between the glass coverslip (or the prepared specimen) and the front of the objective lens, and α is the angle between the outermost ray of light (or radiation) that enters the front of the objective lens and the optical axis of the objective lens, as shown in Figure 8.4. If the medium between the coverslip and objective is air then the maximum theoretical numerical aperture is 1.0. However, it is not possible to achieve this, as the lens would need to lie directly on top of the coverslip. If, however, we substitute oil or water for air then the theoretical numerical aperture is increased to 1.3–1.5. In practice, the theoretical values cannot be attained and in air the best possible value of numerical aperture is approximately 0.95 and in oil approximately 1.4. Thus, the resolution of a light microscope is approximately half (0.61/1.4) of the wavelength of visible light.

SELF-CHECK 8.2

Calculate the approximate resolution obtained with a microscope fitted with an oil immersion objective lens and using light of wavelength 600 nm.

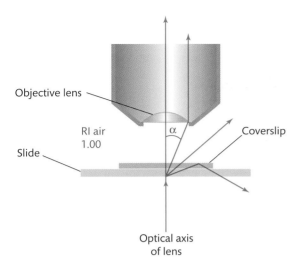

FIGURE 8.4
Schematic illustrating the angle of the light path when calculating the numerical aperture of a lens.

Light and waveforms

Light emitted from a single source can be regarded as a series of waves travelling in straight lines in all directions. The length or periodicity of these waves is termed the 'wavelength' and gives the light its colour. White light consists of a spectrum of light of different wavelengths (and different colours) ranging from short ultraviolet through to long infrared light.

When light passes through air it travels in a straight line. If, however, it strikes a different medium such as glass, the light path is both retarded and refracted, or deviated at the air/glass interface. Light of different wavelengths is refracted to variable amounts. Refraction is best seen when light passes from air into a glass prism and out again into air and is shown in Figure 11.7(c).

Monochromatic light consists of light of a single wavelength whose amplitude determines the brightness of light. The higher the amplitude of the wave, the brighter the light will be. It is necessary to understand these two properties of the waveform of light to appreciate the effects they have on any image seen in a microscope (Figure 8.5).

Cross reference
Chapter 11 'Spectroscopy', gives a fuller description of electromagnetic radiation.

Image formation and lens defects

When parallel light rays pass through a **lens**, they are refracted by the lens and brought to a common focus by it forming an image. This focus is called the **focal point**. The distance between the centre of the lens and the focal point is called the **focal length**. However, lenses can suffer from a number of defects or aberrations, which may prevent light being brought to

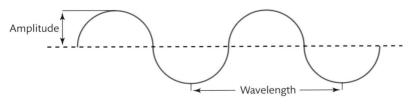

FIGURE 8.5
Schematic illustrating the definitions of amplitude and wavelength.

this common focus. This is because at its edge, a lens will also act as a prism due to its shape and the greater curvature of the lens which will exacerbate any aberrations.

These defects include:

- Chromatic aberration
- Spherical aberration
- Astigmatism

Chromatic aberration

Chromatic aberration is the inability of a lens to bring light of different wavelengths to a common focal point resulting in a blurred image with multiple coloured fringes. It is caused by different wavelengths being refracted to different degrees, as seen in a simple prism, as lenses also act as prisms. Thus, light rays of shorter wavelengths are brought to different focal points than ones of longer wavelengths, as you can see in Figure 8.6.

Key Points

Chromatic aberration is the inability of a lens to bring electromagnetic rays of different wavelengths to a common focal point.

Spherical aberration

Spherical aberration is the inability of a lens to bring light passing through the periphery of the lens to the same point as light passing through the central part of the lens (Figure 8.7). This is caused by the lens refracting the light to different degrees, which depends upon the curvature of the lens and the angle at which light enters the lens.

Key Points

Spherical aberration is the inability of a lens to bring electromagnetic rays of different wavelengths to a common focus.

Both spherical and chromatic aberration result in images that are significantly reduced in quality. These aberrations can both be corrected by using different glass additives and lenses in combination that have differing amounts of curvature and different shapes.

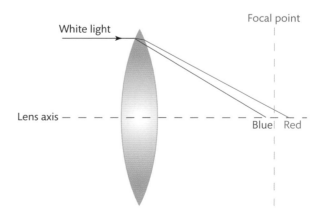

FIGURE 8.6
Schematic illustration of chromatic abberation in a glass lens.

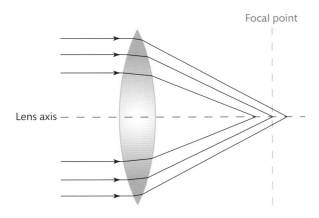

Focal point

Lens axis

FIGURE 8.7
Schematic illustration of spherical abberation in a glass lens.

SELF-CHECK 8.3

How can chromatic aberration be corrected?

Astigmatism

Astigmatism results when a lens is unable to bring light passing through one part of a lens to the same focal point as light passing through another part. This results in a distorted blurred image and is best illustrated when viewing a lattice. Astigmatism is perhaps the least important lens aberration since correction of either chromatic or spherical aberration usually results in the elimination of astigmatism.

> ### Key Points
>
> **Astigmatism is the inability of a lens to bring electromagnetic rays passing through different parts of the lens to a common focal point.**

8.2 Components of a microscope

The basic construction of a compound light microscope has changed little since the mid-nineteenth century. All consist of a system of lenses and their associated components, which are used to form a magnified image of the specimen. However, since the 1800s, significant improvements have been made in the quality of lenses, resulting in greatly improved images (Figure 8.8). Much emphasis has also been placed on the ergonomic design of microscopes to provide the user with the most comfortable working position.

Light source

All microscopes require a reliable light source. In early designs, this source came from the sun and a mirror was used to reflect its rays into the microscope condenser. As optics improved, this form of illumination became inadequate, since it was not possible to adjust the brightness. Tungsten filament lamps fitted with a variable rheostat that allows the brightness to be adjusted have proven to be a reliable mechanism. However, tungsten filament bulbs generate considerable heat and may need cooling to be used safely. Consequently, most modern

(a)

(b)

Eye

Eyepiece

Objective lens

Specimen

Specimen stage

Condenser

Mirror

Light source

FIGURE 8.8
(a) Modern compound light microscope. (b) Schematic illustrating the lightpath of a compound microscope.

microscopes now use low voltage halogen light sources. These provide a more intense light and a brighter image without the heat associated with tungsten filament light sources.

Fluorescent microscopes require a light source capable of emitting a high level of ultraviolet light, which neither tungsten nor halogen lamps are capable of producing. For this purpose the best light source is a high-pressure mercury vapour lamp. This lamp, however, requires a specialist lamp holder and power supply and generates a significant amount of heat that needs to be dissipated. Care must also be exercised regarding health and safety issues (Health and Safety Box 'The use and disposal of high-pressure mercury vapour lamps').

HEALTH & SAFETY

The use and disposal of high-pressure mercury vapour lamps

The mercury vapour lamp contains mercury gas at high pressure within a glass housing. The intensity of light emission from these lamps diminishes with use and, in addition, the glass becomes increasingly unstable and prone to explode. To ensure that the lamp operates optimally, manufacturers incorporate an elapsed timer to record how long the lamp has been in use and recommend its replacement and correct disposal after a known length of use. Disposal must be controlled to ensure mercury does not enter the environment.

It is also essential that after the lamp has been switched off, it is not re-used until it has cooled down. Failure to observe this cooling period may result in the lamp exploding. Manufacturers now incorporate switching timers into the lamp power supply that prevent the lamp being re-used until an adequate cooling period has elapsed.

Cross reference

Health and safety issues are discussed in Chapter 4.

Illumination of specimens

Effective microscopy relies upon the correct illumination of specimens. This is done by using the condenser or *sub-stage condenser* as it is sometimes called, to focus the light produced by the lamp into a cone onto the specimen giving it a maximum and even illumination. In its simplest form, the sub-stage condenser consists of one or two lenses with an adjustment that allows vertical focusing of the light to compensate for the varying thickness of microscope slides.

For best illumination, it is necessary to be able to accurately align the axis of the condenser with the objective lens, and to be able to adjust the size and brightness of the cone of light hitting the specimen. The condenser is fitted with adjustable screws to move the condenser in the x–y axes to allow the light cone to be centred, and with an iris diaphragm to adjust the diameter of light cone.

It is possible to set the condenser in one of two ways to optimally illuminate the specimen:

- Köhler illumination
- Critical illumination

Each method has its advantages and applications within biomedical science.

Köhler illumination requires the condenser to be set in such a way that light reaching the specimen evenly illuminates a wide area. This form of illumination is particularly useful when recording photographic images to ensure even illumination across the whole field. **Critical illumination** is the focusing of the light source at a point directly onto the specimen using the sub-stage condenser. It is an effective way of ensuring that the maximum level of illumination reaches the specimen. This method is particularly necessary when using high magnification objective lenses that have short working distances, but when combined with low magnification objective lenses result in uneven illumination.

You can read an outline of how to set up a microscope for optimal illumination in the Method Box 'Setting up of microscope for optical illumination'.

METHOD *Setting up of microscope for optimal illumination*

- Whatever type of microscope is being used, the basic principles of setting it up for critical or Köhler illumination are essentially the same. First, adjust the condenser to as high a position as possible and then focus on a sample using a low power objective lens (normally 10× or 4×). Secondly, close the field diaphragm in the condenser and sharply focus the diaphragm onto the image of the sample using the condenser focus controls.

- The field diaphragm can be considered to represent the illumination axis and should be adjusted using the x and y condenser adjustment screws. The condenser lens system focuses as much light onto the sample and the lamp is focused onto the plane of the field diaphragm of the condenser. This is termed Köhler illumination. Now open the field diaphragm until it is just outside the field of view.

- If the microscope does not have a condenser fitted with a field diaphragm, it is necessary to place a sharply pointed object, for example a pin, onto the glass of the coverslip and focus the illumination onto this point by raising or lowering the condenser. This is termed critical illumination.

Specimen stage

The specimen stage is the rigid flat surface immediately above the condenser on which the specimen sits. Most stages incorporate sample clamps attached to mechanical manipulators that allow accurate movement of the specimen. **Vernier scales** are incorporated into the mechanism, which means that when a point of interest in the specimen is observed, its Vernier values can be recorded and, therefore, the position is known and returned to later with ease.

Stand and tube

The stand of the microscope is the sturdy, rigid, and vibration-free base to which all the other components are attached. It incorporates the coarse and fine focus adjustment mechanisms of the microscope that allows for fine vertical movement of either the stage or the objective lens depending on the manufacturer.

Above the stage the stand has a *nosepiece* for holding between three and five *objective lenses* of different magnifications. The nosepiece rotates between set points allowing the appropriate objective lens to be moved into the light path. Above the nosepiece is the observation tube that houses the *eyepiece* lens. There are several types of observation tube: the simplest consists of a single eyepiece lens, but more usually consists of a binocular or trinocular attachment. A trinocular tube allows a camera to be attached for simultaneous viewing and recording of the image. The distance from the front of the objective lens and the eyepiece is termed the tube length. The length of the tube has a significant impact on the final magnification as described in Section 8.3.

Recent developments in lens construction have resulted in the production of 'infinity corrected' lenses. Here, the focused image leaving the lens is not brought to a focal point, but rather remains parallel at the same focal plane to infinity with no change in magnification. This development has allowed manufactures to vary the tube length without altering the magnification allowing them to develop and design more ergonomic stands and multi-headed microscopes.

Key Points

Microscopes are able to have multiple viewing heads (up to 20) that allow all observers to see exactly the same image. This facility is particularly useful for teaching or demonstration purposes.

Objective lens

The objective lens collects the light from the specimen and forms a **real image** of the specimen. It is perhaps the most important component of the microscope in that it contributes most to the ultimate quality of the image. There are numerous types of objective lenses; each have specific uses and qualities. The selection of an objective lens is determined by the function to which the microscope is to be put.

There are three main types of objective lenses:

1. Acromatic

2. Semiapochromatic or fluorite

3. Apochromatic

All lenses have other features that contribute to their effective use and objective lenses are no exception. Two features of note are *depth of field* and *working distance*. Depth of field is the amount of a specimen in focus at any one time. Generally lower power lenses have a greater depth of field and higher power lenses have a smaller depth of field so require constant refocusing. Working distance is the distance between the coverslip and the bottom of the objective lens and again lower power lenses have a longer working distance than higher power ones.

Achromatic

Achromatic lenses are the simplest types of corrected lens available. They correct for chromatic aberration in two wavelengths and for spherical aberration for only one wavelength. Achromatic lenses have a relatively low numerical aperture, which allows a good working distance and depth of field. When using this type of lens, the quality of the image is best in the lens axis. Blurring occurs at the periphery of the image, which makes achromatic lenses unsuitable for colour photomicrography.

Semiapochromatic

Semiapochromatic or fluorite lenses incorporate the mineral, fluorite into the glass of the lens. This allows lenses to be constructed with improved numerical apertures and increased image resolution. These lenses still only compensate for chromatic aberration in two colours and spherical aberration is still present, however, they provide superior performance over simple achromatic lenses.

Apochromatic

Apochromatic lenses are the highest quality lenses available. They correct for three colours, and do not exhibit spherical aberration or astigmatism. These lenses provide the highest numerical aperture and, consequently, the highest resolution. The working distances of these lenses are usually short but this can be used to advantage by the addition of high refractive index immersion oil or water between the specimen and the objective lens. As we described in Section 8.1, this has the effect of increasing numerical aperture and thus increasing the potential resolving power of the lens. Using apochromatic lenses means it is possible to achieve numerical apertures in excess of 1.4 with full colour correction.

Eyepiece

The eyepiece is the last lens system through which the image passes. It magnifies the image formed by the objective lens and tube stand and allows a **virtual focused image** to fall on the eye of the observer. Some eyepieces are capable of accommodating a range-measuring graticule. Graticules are glass discs onto which are etched micrometer scales, grids, or points, and which can be used for quantitative and accurate measurements of images. Modern microscopes are fitted with wide field, flat field, and possibly high focal point eyepieces especially designed for spectacle users. In choosing the eyepiece, it is necessary to match its quality to that of the objective lens. It is inappropriate to use the best quality apochromatic objective lens with the simplest of eyepiece lenses.

8.3 Magnification

The magnification produced by a *lens* can be defined by using the distance from the object to the lens and that from the lens to the image, using the formula:

Magnification = Real image distance/Specimen focal distance

The real image distance or optical tube length has been standardized in microscopes to 160 mm. The focal distance is the length from the optical centre of the lens to its focal point.

You can see from the formula above that the smaller the focal distance, the larger the magnification obtained. The surface curvature of the lens can be increased to shorten the focal distance, but maximum curvature is limited to that of a spherical lens. By adding multiple lenses it is possible to significantly increase magnification: in contemporary microscopes both objective and eyepieces lenses are constructed from numerous lenses of differing curvatures and diameters.

The total magnification obtained using a compound microscope when the microscope has a standard optical tube length of 160 mm can be obtained by simply multiplying the magnification values of the objective and eyepiece lenses.

For non-standard optical tube lengths the formula in Box 8.2 must be used.

8.4 Dark field microscopy

Conventional bright field illumination will in itself provide only minimal contrast (Section 8.1) and unstained biological samples are virtually invisible since they have insufficient variations in optical densities. To overcome this problem, it is conventional to introduce contrast into specimens by staining them with coloured dyes. However, when it is necessary to examine samples in the living state, this 'normal' preparation may result in death and disruption of the sample so alternative methods of illuminating the specimen are needed to prevent this.

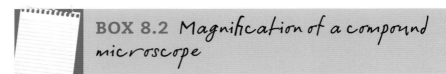

BOX 8.2 *Magnification of a compound microscope*

The final magnification produced by a compound microscope can be expressed as **objective magnification** × **tube lens factor** × **eyepiece magnification**:

Tube lens factor = Optical tube length/Objective focal length

Example: Magnification using a 10× objective lens with a tube lens factor of 1.25 and an 8× eyepiece

Magnification = 10 × 1.25 × 8 = 100 ×

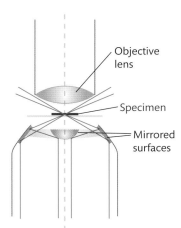

FIGURE 8.9
Schematic outline of lightpath in dark field illumination.

The simplest method of introducing contrast into unstained samples is to use oblique light. In doing so, the sample is illuminated by light from an angle that has the effect of increasing refraction while creating a decrease in direct illumination. Unstained samples thus appear bright against a dark background. The dark field condenser was developed to take advantage of this effect. This consists of a parabolic mirror placed in the substage condenser that only allows light to reach the specimen stage at an acute angle as you can see illustrated in Figure 8.9. When a specimen is not present on the microscope stage, light does not pass through to the objective lens. When a sample is placed on the stage, the small variations in refraction created by the specimen are amplified to make the sample appear bright against the dark background.

Dark field condensers are occasionally used in conjunction with the transmission fluorescence microscope and this application will be described more fully in Section 8.7.

8.5 **Phase contrast microscopy**

As you learned in Section 8.1, light is a waveform. When two rays of light originate from the same source they are said to be **coherent**, because they are in **phase** with each other and their amplitudes coincide. As a consequence, they are able to combine to form a ray with an amplitude twice the height of the original rays resulting in a brighter one. If, however, the two rays become out of phase then they are able to interfere with one another. If the rays were out of phase with one another by half of the wavelength, then this would result in the extinction of the ray. This phenomenon of interference is utilized in the phase contrast microscope. You can see an outline of its operation in Figure 8.10.

The sample is illuminated with a cone of light produced by an annular stop, which consists of a ring positioned in the bright field substage condenser. The cone of light passes through the sample, which is positioned at its focal point. Light leaving the sample enters the objective lens of the microscope.

Objective lenses in phase contrast microscopes have a **phase plate**, which consists of a disc of glass with a ring etched into it that corresponds exactly to the shape and size of the annulus. The depth of the etching is critical: it must be of such a depth that it retards the light rays passing through the full thickness of the phase plate by ¼ wavelength when compared with the light passing through the etched ring. This retardation, while resulting in interference, does not in itself produce the desired increase in contrast. The sample is responsible for producing this effect.

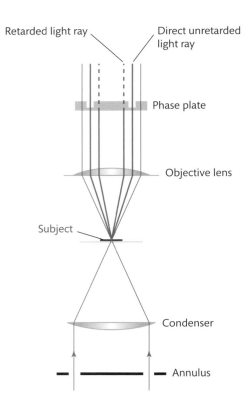

FIGURE 8.10
Schematic illustration of the lightpath in a phase contrast microscope.

Tissues will also retard light by approximately ¼ of a wavelength; hence, the combined retardation from the phase plate and the specimen produces the total interference.

When light passes through a sample, some of it is scattered and passes into the objective lens in the normal way and passes to the eye of the observer through the non-etched part of the phase plate. When the unaltered rays are focused with the retarded rays they combine to form the real image of the specimen. Subtle changes in refractive index within the sample are seen as varying degrees of brightness in the image against a dark background.

When setting a microscope for phase contrast it is essential to ensure that the annulus and the phase plate are exactly aligned. It is usual to use an auxiliary microscope to view the back focal plane of the phase contrast microscope, which allows the operator to view the phase plate and use adjusting screws fitted in the condenser to align it with the annulus with great accuracy. In practice each objective lens and phase plate has a matching annulus fitted into a rotating disc set into the base of the substage condenser. Each annulus can be adjusted and set individually to match its objective. If properly adjusted, a good phase contrast microscope should be able to distinguish differences in refractive index of less than 5% easily, producing images of high contrast without the use of staining.

Phase contrast microscopy is used extensively in biomedical science when it is necessary to examine unstained tissue samples, such as examining cell deposits from urine samples from patients suffering from urinary tract infections. Phase contrast microscopy is used to examine and identify the cellular component of such samples.

Cross reference

You can read more about urinary infections in Chapter 3 'Kidney disease', in the companion text, *Clinical Biochemistry* and in *Medical Microbiology*.

SELF-CHECK 8.4

In a phase contrast microscope, by how much does the phase plate retard the rays of light?

8.6 **Polarization microscopy**

As described in Section 8.1, light emitted from a single source can be regarded as electromagnetic waves whose amplitude represents brightness and the wavelength its colour. This waveform is emitted and vibrates in all directions. **Polarized light**, however, only vibrates in a single direction. Certain crystals are capable of producing plane polarized light by splitting incident rays entering the crystal into two components, which are refracted by the crystal in two different planes. This phenomenon is termed **birefringence**.

If we place two similar polarizing crystals so that their planes of polarization are parallel to each other, then any light entering the first crystal would be capable of passing through the second. However, rotating the second crystal through an angle of 90° would result in the extinction of light and none would exit from the second crystal. This phenomenon is exploited in the polarization microscope, which is used to examine and measure substances that are capable of polarizing light.

The polarizing microscope has a **polarizer** filter situated in the substage condenser. This illuminates the specimen with plane polarized light. A second polarizing filter or **analyser** is situated below the eyepiece and is capable of being rotated through 180°. When using the polarizing microscope, the two filters are orientated so that their planes of polarization are at a right angle resulting in light extinction. However, any specimen that contains a birefringent substance will rotate the illuminating plane polarized light so that it is able to pass through the analyser and appear bright against a dark background. This form of microscopy is particularly useful in the examination of fluids that may contain crystals, such as urine and joint fluids.

> **Cross reference**
> You can read about the deposition of urate in gout patients in Chapter 4 'Hyperuricaemia and gout', in the companion volume, *Clinical Biochemistry*.

8.7 **Fluorescence microscopy**

The use of shorter wavelength light in the ultraviolet range is outside the visible spectrum so its use to improve resolution is limited. However, certain substances when illuminated with high energy light of a short wavelength (that is ultraviolet) will re-emit light of a lower energy of longer wavelength that is within the visible spectrum. This is called fluorescence. The use of fluorescent microscopes allows us to visualize small quantities of biological components within tissue samples since they appear bright against a dark background.

Some substances are auto-fluorescent or exhibit **primary fluorescence**, (that is they fluoresce spontaneously under UV light). Other substances do not fluoresce but can have fluorescent dyes called fluorochromes attached to them, which will produce a **secondary fluorescence** when stimulated with ultraviolet light. Examples of substances that exhibit auto-fluorescence are the vitamin As and some porphyrins. A number of dyes which exhibit fluorescence can be used to demonstrate the presence of specific cellular components in specimens.

> **Cross reference**
> Fluorescence is also discussed in Chapter 11 'Spectroscopy'.

The light source in all fluorescent microscopes must be capable of producing a high proportion of light rays in the ultraviolet range. Halogen lamps produce significant amounts, but almost all of it is in the ultraviolet blue range and so is of limited value. The commonest light sources in use are high pressure mercury vapour or Xenon arc lamps. Both are highly efficient, but requiring specific power supplies and appropriate lamp housings.

There are two types of fluorescence microscope, which differ in their manner of illuminating the specimen. In *transmission* fluorescence microscopes, the light source is situated beneath the specimen as in a conventional light microscope. In *incident* light fluorescence microscopes, however, the specimen is illuminated from above and the viewer sees visible light transmitted from the specimen against a black background.

Transmitted light fluorescence

To gain the highest performance from the fluorescence microscope it is necessary to use filters that are able to transmit the specific wavelength of the excited light, while removing unnecessary and unwanted wavelengths that may reduce specificity. A transmission fluorescence microscope has two types of filters (Figure 8.11): an **exciter filter** placed in the light path between the lamp and the condenser and a **barrier filter** situated in the body tube before the eyepiece. The filters must be specifically manufactured to correspond to each type of fluorochrome in use.

The exciter filter allows the maximum amount of ultraviolet light of an appropriate wavelength to excite the specimen while filtering out the levels of visible and inappropriate ultraviolet light from the illumination. This ensures that the maximum level of fluorescence is emitted from the sample. The light entering the objective lens is a combination of unscattered ultraviolet light and the longer wavelength visible light produced by the fluorochrome(s). It would be dangerous for the observer if this light were to fall directly onto the eye and may cause blindness. This is prevented by the barrier filter that removes all of the unused ultraviolet light and allows only the fluorescent visible light to be transmitted to the eye, thus protecting the observer. Unfortunately, to produce a filter capable of both blocking ultraviolet light and allowing transmission of specific wavelength visible light, means the overall

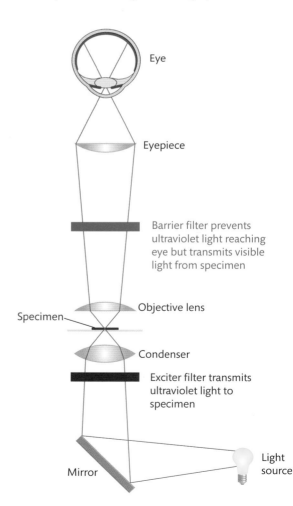

Eye

Eyepiece

Barrier filter prevents ultraviolet light reaching eye but transmits visible light from specimen

Objective lens

Specimen

Condenser

Exciter filter transmits ultraviolet light to specimen

Mirror

Light source

FIGURE 8.11
Outline of a transmission light fluorescence microscope.

level of transmission is lowered and significantly reduces the intensity and brightness of any fluorescence produced.

An alternative to the use of a barrier filter is to use a dark ground condenser. This ensures that the only light reaching the eye of the observer is that emitted by the specimen since unscattered, indirect ultraviolet light does not reach the eye. This has the added advantage of forming a dark contrasting background that gives an apparent increase in the brightness of the fluorescent light.

SELF-CHECK 8.5

Which light has the shorter wavelength: ultraviolet or blue? (You might find it helpful to consult Figure 11.2.)

Incident light fluorescence

Since the fluorescent light from a fluorochrome is emitted in all directions it is possible to illuminate the specimen in a fluoresence microscope from above. This is called *epi-illumination*. Some of the excitation light from the fluorochrome is directed back into the objective for observation by the user.

In operation, the exciting light passes from the lamp and through the excitation filter to a dichroic mirror or prism, as you can see in Figure 8.12, which filters out all the unwanted light, only allowing ultraviolet light of a specific wavelength to pass through. The resultant ultraviolet light passes to the specimen where the fluorochrome is excited to generate visible light. The visible light passes into the objective and then to the observer through the same dichroic prism, which is able to transmit visible light in the opposite direction to the exciter light, and to the eye of the observer. The image therefore appears brightly coloured against a dark background.

Incident light fluorescence microscopy has significant advantages over transmitted light fluorescence. The reduced intensity of light resulting from the use of barrier filters is eliminated, a substage condenser is not required and the level of light reaching the specimen is considerably increased giving a much brighter image. In addition, since the fluorescence microscope uses light that is produced from sources above the specimen it is possible to simultaneously use other forms of transmission microscopy. This allows both morphological and fluorescent images of the specimen to be seen simultaneously. This is of particular value when additional staining of specimens may mask the fluorescence produced. In this case a phase contrast or dark ground condenser can be used to introduce additional contrast.

Uses of fluorescence microscopy

Both transmitted and incident light fluorescence microscopy are used extensively in the study of immunological disorders. The technique of **immunofluorescence** is used to identify specific immune responses in disease. These methods rely on specific interactions and binding between antibodies labelled with fluorescent dyes (fluorochromes) and antigens present on cell surfaces. Once bound, the visualization of the fluorescence label confirms the presence of the antigen and also identifies its specific location within the tissue (Figure 8.13).

Cross reference

You can read more about immunofluorescence in Chapter 15 'Immunological techniques' and the accompanying volume, *Immunology*.

SELF-CHECK 8.6

How does an incident light fluorescence microscope differ from a transmitted light fluorescence one?

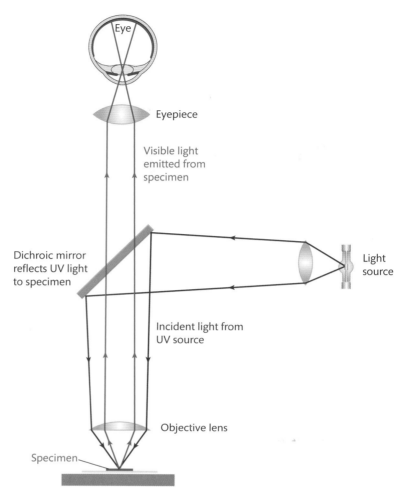

FIGURE 8.12
Outline of an incident light fluorescence microscope.

8.8 **Confocal microscopy**

In all forms of microscope, the final image is affected by the thickness of the specimen being examined because of the relatively small depth of focus. This is a particular problem in fluorescence microscopy where the fluorescent light emitted from the specimen is derived from the full thickness of the specimen resulting in blurring of the image caused by those portions of the specimen that are above and below the focal plane. This blurring is minimized in the confocal microscope.

Confocal microscopes use a narrow beam of light from a laser, thus, the specimen is illuminated with coherent light of high intensity. The beam of laser light is scanned through the specimen using a system of galvanometer mirrors that oscillate at adjustable speeds to divert the laser beam to produce a scan. Reflected light produced is re-focused by a confocal aperture or pinhole which only allows light from a specific focal plane to fall onto a photomultiplier and produce a digital signal. A computer then processes this signal to produce an image that has a great depth of field giving the specimen a three-dimensional appearance. If the specimen is

FIGURE 8.13
Incident light fluorescence micrograph of a section of breast tumour stained using fluorescent *in situ* hybridization (FISH) to demonstrate amplification of the *HER2* gene on chromosome 17. This method is used as a prognostic marker for treatment with the drug herceptin. See also Figure 16.26 and its associated text.

successively scanned through a range of focal planes, the images at each level can be electronically stored and used by the computer to produce a full three-dimensional image of it.

To-date, confocal microscopy has found limited application in diagnostic biomedical science. It is, however, used extensively in scientific research (Figure 8.14) since biological material can be examined in three dimensions without the need for extensive sample preparation.

SELF-CHECK 8.7

How is the quality of an image improved using confocal microscopy?

8.9 **Inverted microscopes**

When examining living material, for example certain types of cells grown in tissue culture, it is necessary to use an *inverted* microscope. This allows the bottom surface of a culture flask, where the cells grow and adhere, to be examined with minimal disturbance to the specimen. The inverted microscope is constructed so that the specimen stage has a large flat surface capable of accommodating tissue culture flasks or Petri dishes (Figure 8.15). The objective lens and condenser are designed to work at a much longer working distances than those of conventional microscopes to allow for the thickness of the culture vessel. Inverted microscopes

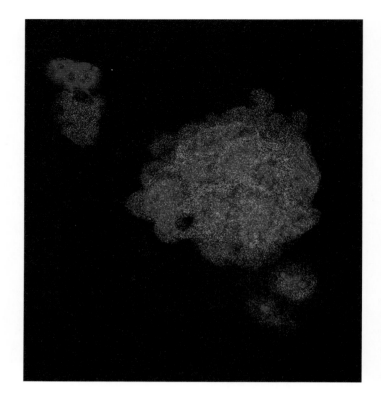

FIGURE 8.14

Confocal image of a group of stem cells. The green areas are parts of the plasma membranes and the purple portions are nuclei. Courtesy of Dr Q. Wang, School of Healthcare Science, Manchester Metropolitan University, UK.

FIGURE 8.15

Myoblasts growing in culture viewed using an inverted microscope. Courtesy of Dr Q. Wang, School of Healthcare Science, Manchester Metropolitan University, UK.

also have the facility to utilize phase contrast or dark field illumination and fluorescence giving the instrument maximum flexibility.

8.10 Electron microscopy

In Section 8.1 we noted that the resolution of a microscope depends largely upon the wavelength of the illuminating light; using visible light limits resolution to about 200 nm. Using radiation of short wavelengths is a way of increasing the resolution of a microscope.

This feature is exploited in the electron microscope. Like light, moving electrons are part of the electromagnetic spectrum; the faster they move the shorter their wavelength.

Modern electron microscopes have resolutions that are less than 0.2 nm when used with biological specimens, which is 1000× the resolution of a light microscope.

Cross reference

The electromagnetic spectrum is described in more detail in Chapter 11 'Spectroscopy'.

Components of the electron microscope

Electron microscopes were first developed in the 1920s. Their components are analogous to those of the conventional light microscope, as you can see if you compare Figure 8.16 with 8.8.

Electron microscopes have a source of electrons, commonly called an *electron gun*, a condenser system that collects the electrons and focuses them on the specimen, a system of

FIGURE 8.16
(a) Modern transmission electron microscope (JEOL JEM-ARM200F). Courtesy of JEOL Ltd.
(b) Cross-section diagram of a transmission electron microscope. See text for details.

objective lenses to magnify the image formed and a projector lens system to focus the image onto a phosphorescent screen, thus making it visible. Glass cannot focus beams of electrons, but they can be focused using electromagnetic lenses.

The electron beam in an electron microscope is energetic and if they were allowed to travel far in air, collisions with its molecules would soon dissipate the beam. Thus, to produce a sufficiently long electron path within the microscope a high vacuum is required. Electromagnetic lenses generate a large amount of heat so a suitable cooling system is also required.

The electron gun

The electron gun of an electron microscope consists of a tungsten filament bent into a V-shape, which is heated by a low voltage direct current. This causes a cloud of electrons to be released from the tip of the V. The number of electrons emitted from the filament can be varied by altering the size of the current passing through the filament.

A *beam* of moving electrons is formed by a Wehnelt cap and a positively charged (anode) plate, as shown in Figure 8.17. The Wehnelt cap is made of metal and encases the filament. It is maintained at a potential difference of about 20 V negative to the filament resulting in an attraction to the electrons. A hole in the cap directs the electrons towards the **anode plate**, which is maintained at earth potential. The potential difference between the filament and the anode determines the **acceleration voltage** and in most electron microscopes can be varied between 20,000 and 100,000 V.

For maximum efficiency and the production of a suitable electron beam, it is essential that the filament is heated to the correct temperature, often referred to as the saturation point, that the Wehnelt cap is accurately aligned, and that the distance between the filament and cap is critically adjustable.

Electromagnetic lenses

Electromagnetic lenses consist of a coil of wire encased in soft iron. They have a pole-piece opening at the point through which the electron beam passes (Figure 8.18). When a direct

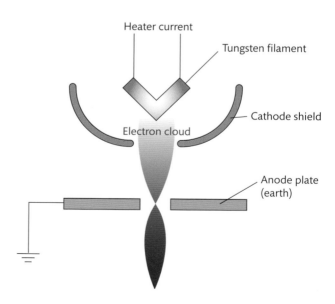

FIGURE 8.17
Schematic illustration of an electron gun. See main text for details.

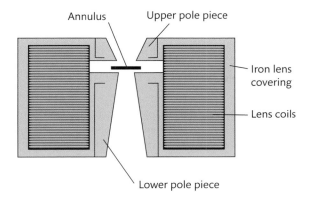

Annulus   Upper pole piece

Iron lens covering

Lens coils

Lower pole piece

FIGURE 8.18
Schematic showing a cross-section of an electromagnetic lens.
See main text for details.

current is applied through the wire it induces a strong magnetic field through the axis of the coil. Since electrons are electromagnetic radiation, an electron beam passing through the pole-piece opening is affected by the magnetic field in the metal coil, which allows the beam to be focused. Varying the strength of the current passing through the coil we can vary the degree by which the electrons are focused allowing the beam to be convergent or divergent. A number of such lenses are built into the column of the electron microscope.

As with all lenses, electromagnetic ones suffer from several aberrations. As you learned earlier in Section 8.1, optical lenses suffer from chromatic and spherical aberrations, as well as astigmatism. Electromagnetic lenses suffer similar problems; however, the ways in which these defects are overcome differ.

To recap: chromatic aberration is the inability of a lens to bring electromagnetic rays of different wavelengths to a common focal point. Chromatic aberration in electromagnetic lenses is overcome by using a beam of monochromatic electrons. Manufacturers go to great lengths to ensure the high tension voltage between the filament and the anode plate within the electron gun is stable, which ensures a constant accelerating voltage and thus a monochromatic beam of electrons. A further potential source of chromatic aberration in the electron microscope is the energetic interactions between the electron beam and the specimen under examination. This reduces the energy of the electrons and so changes their wavelength to produce a potential polychromatic beam. This form of chromatic aberration can be minimized by using higher acceleration voltages and preparing extremely thin sections of specimens for examination.

Spherical aberration is caused by the inability of a lens to bring electromagnetic waves to a common focal point. it is not possible to correct this type of aberration in electromagnetic lenses because the magnetic field they produce varies across their diameter, and the physical dimensions of the lenses prevents the production of compound lens. However, spherical aberration can be overcome by using only very small apertures in the centres of the lenses, which only allow electrons at the central axis of the lens to pass through and illuminate the specimen. Electrons at the periphery of the lenses are prevented from doing this. Given electrons travelling at the central axis of the lens exhibit the least spherical aberration, this approach minimizes the defect.

Astigmatism, the inability of a lens to bring electromagnetic rays from a single source to a common focus, is caused by the asymmetry of the electromagnetic field within the lens. It is usually caused by defects in pole piece manufacture or, more commonly, contamination on the pole piece. This results in electrons passing through different parts of the lens being focused to different degrees. Naturally, contamination on the pole pieces must be minimized to reduce astigmatism effects. This can be achieved by ensuring that the microscope column is kept scrupulously clean and by using an anticontamination trap or cold finger comprising a small metal plate cooled by liquid nitrogen situated close to the specimen. Thus, any contaminating vapour liberated from the specimen during irradiation condenses on this cold plate, rather than the lens pole pieces.

It is not possible to remove the asymmetries within lenses that lead to astigmatism, but it is possible to compensate for them by surrounding the lens with a number of small secondary electromagnets. Individual adjustments of these so-called astigmatism coils make it possible to adjust the degree of astigmatism within the lens and to compensate for the defect.

Vacuum systems

Electrons do not travel far in air since they collide and react with the gas molecules. The distance that an electron can travel without collision is called the mean free path; in modern electron microscopes this needs to be in excess of 2 m. To achieve this distance, it is necessary to create a very high vacuum in the region of 1.33×10^{-2} to 1.33×10^{-3} Pa (10^{-4} to 10^{-5} mm of mercury) within the column of the microscope. This high vacuum also has the effect of reducing oxidation of the tungsten filament and thus extending its life. In addition, it also helps to maintain a clean environment within the column and reduce contamination of the lens pole pieces and specimens.

Several types of vacuum pumps are employed, for example, oil rotary pump and oil diffusion pumps, which are typically used in combination.

Oil rotary pumps have a simple mechanism consisting of a series of vanes that rotate in an oil bath. This mechanism traps gas within a chamber forcing gas molecules out of the column. They are usually employed in the initial stages of evacuating the microscope column. Oil rotary pumps are reliable and have the advantage that they can start to operate at atmospheric pressure. They do, however, have the disadvantage that they can only produce a vacuum of approximately 1.33 Pa (10^{-2} mm of mercury) so they must be used with other types of vacuum pump.

Oil diffusion pumps operate by heating inert oil to its boiling point. The resultant oil vapour is trapped within a series of cones with downward-pointing pipes that form a jet of oil vapour directed towards a cooled surface where the oil condenses and traps gas from the surrounding atmosphere. They are efficient pumps; capable of forming a vacuum of approximately 1.33×10^{-3} Pa (10^{-5} mm mercury) and are silent in operation. However, they do not operate from atmospheric pressure and require an initial vacuum of 1.33–0.133 Pa (10^{-2} to 10^{-3} mm mercury) to operate effectively. The assistance of a rotary pump, which is mounted in series, is therefore useful to obtain this degree of vacuum before the diffusion pump begins to operate. It is essential that diffusion pumps are not exposed to vacuums of less than 1.33 Pa (10^{-2} mm mercury) when the oil is at operating temperature, since this may cause back streaming of the oil vapour passing the cooling plates allowing it to enter the microscope column causing considerable contamination. Manufacturers prevent this from happening by placing a pressure sensitive baffle plate at the top of the pump, which automatically closes if the vacuum fails or is insufficient.

SELF-CHECK 8.8

Why is it not possible to operate an oil diffusion pump from atmospheric pressure?

Cooling system

Electromagnetic lenses generate considerable heat as a by-product of their action. This heat can produce instabilities in the action of the lenses and can also lead to the thermal expansion of metal components around the lenses that can adversely affect the formation of a well-focused image. To prevent these undesired effects, lenses are water cooled. In addition, oil diffusion pumps also require cooling for them to operate. This is simply achieved by passing water pipes around and through the electromagnetic lenses and the pump, through which chilled water is circulated maintaining the temperature at 18–20 °C.

Viewing screen and recording images

Electrons are invisible to the human eye due to their very short wavelengths. Thus, to see the image of a specimen produced in an electron microscope it is necessary to allow the focused electron beam to hit a phosphorescent screen placed in the beam path. The electrons react with the **phosphor**, which emits light. The intensity of this light is proportional to the number of electrons striking the phosphor and thus a visible image of the specimen is formed.

The image formed on the phosphorescent screen is transient being only present when illuminating the specimen. The traditional method of making permanent reproducible records of the image was to form a photographic record. Fortunately, photographic emulsions are extremely sensitive to electrons and are capable of producing high resolution copies of an image. Microscope cameras have to be mounted within the column and the vacuum system, since it is necessary for the electron to fall directly onto the photographic emulsion. They can be situated beneath the phosphorescent viewing screen, which is moved out of the path of the beam to allow exposure of the film or placed above the screen immediately beneath the projector lens and moved into the beam path when a photograph is required.

Photographic film contains a large quantity of moisture, which if exposed to the high vacuum environment of the electron microscope column would result in considerable contamination and deterioration of the vacuum. It is necessary to dry the film prior to placing it into the column and most manufacturers incorporate a desiccation chamber within the microscope that can be maintained under vacuum by the rotary oil pump. Most manufacturers also incorporate an air lock into the camera, which facilitates film change without the need to bring the whole column back to atmospheric pressure.

With the rapid development of digital imaging systems, manufacturers of electron microscopes have been quick to realize the advantages of these types of imaging systems. Digital images are readily accessible, easy to store electronically, and can be transferred to multiple viewers simultaneously by email. They are also of high quality and are the medium of choice for scientific publications.

Digital cameras can be incorporated into the electron microscope in the same way as conventional film cameras: either situated beneath the phosphorescent screen or placed in the column beneath the projector lens. The image electrons are allowed to fall directly onto a small phosphorescent screen placed in the beam and the resultant image is directed to the camera.

To take advantage of increased magnifications and resolution possibilities of the electron microscope requires specimens to be processed in a particular way as described in Box 8.3.

BOX 8.3 *Preparing samples for electron microscopy*

Sample preparation for transmission electron microscopy (TEM) is designed to take advantage of the increased magnification and resolution that TEM offers over light microscopy (Box 8.1). Unfortunately, the majority of the reagents and solutions used in preparing samples for EM present some form of risk to the user. Hence, sample preparation must only be undertaken when wearing appropriate personal protective equipment and within a fume cupboard.

Electron beams are destructive to tissues, so the preparation of samples must result in specimens that can withstand being bombarded with electrons in a vacuum. This is normally achieved by using extremely potent fixatives, such as glutaraldehyde and embedding the sample in hard epoxy resin, which is capable of being thinly sectioned.

Preparation of slices of tissue or suspensions of cells for electron microscopy begins by fixing them in a strong cross-linking reagent, such as glutaraldehyde or osmium tetroxide, to minimize the effects of autolysis and microbial degradation. This type of fixation also produces samples that are able to withstand the subsequent processing steps. The samples are then dehydrated using a graded series of concentrations of ethanol, ending with 100% strength, before being impregnated in liquid epoxy resin. The samples are then baked to set the resin. Once embedded, sections approximately 1 μm thick are metochromatically stained with, for example, toluidine blue and examined using light microscopy, to initially identify any areas of particular interest. Once such areas are exposed, ultra-thin sections of only 50–90 nm are produced using an ultramicrotome. The sections are mounted on support grids, which typically are about 3 mm in diameter and made of copper. The holes in the grid allow the beam of electrons to pass though the sections, while the lattice supports the section. Prior to viewing, the sections are stained using a variety of heavy metal salts including those of lead, uranium, or less often gold, silver, or platinum. Once stained, the grids may then be placed in the electron microscope for visualization of the areas of interest, such as those shown in Figures 8.19–8.22.

Cross reference

Chapter 4 'Health and safety'. You can read more about electron microscopy and sample preparation in the accompanying volume – *Histopathology* – in this series.

Application of transmission electron microscopy in pathology

There are several pathological processes which can only be examined using electron microscopy because the cellular changes are so small that they are beyond the resolution of light microscopes (Figure 8.19). This is especially true in the case of some renal diseases. There are changes to the glomerular basement membrane and to the epithelial cell foot processes which can only be seen using an electron microscope, which you can see in Figure 8.20 (a) and (b).

The electron microscope is also used extensively in tumour diagnosis. Particularly where there is poor cell differentiation, and it is difficult to see specific features and markers at the resolution of light microscopy. An example of this occurs in the differential diagnosis of malignant melanoma where the quantity of melanin pigment within the tumour cells is so small that conventional staining and immunocytochemical demonstrations appear negative. With the electron microscope it is possible to see very small numbers of melanosomes and even the non-pigmented pre-melanosomes within the cytoplasm of melanoma cells. The development of specific tumour markers and the use of immunocytochemistry have, however, led to less reliance on electron microscopy in tumour diagnosis.

Cross reference

You can learn about the glomerular basement membrane and epithelial cell foot processes of the kidney in Chapter 3 'Kidney disease' of the companion textbook, *Clinical Biochemistry* and *Histopathology*.

Cross reference

Immunocytochemistry is described in Chapter 15 'Immunological techniques'.

FIGURE 8.19

Electron micrograph of amyloid fibres within the wall of a small blood vessel from a patient with chronic inflammatory disease. Note the characteristic fibrillary nature of this abnormal protein. Electron microscopy is the definitive method of choice for the positive identification of amyloid.

(a)   (b)

FIGURE 8.20

(a) Electron micrograph of renal glomerular basement membrane from a patient suffering from membranous glomerular nephritis. (b) Note how its appearance differs in structure from that seen in a healthy person. Courtesy of Dr J. C. Jeanette, Department of Pathology and Laboratory Medicine, University of North Carolina, USA.

Viral particles are far too small to be seen with a light microscope, but are clearly visible using electron microscopy (Figure 8.21). Early identification and diagnosis of viral infections is sometimes essential if the patient is to be diagnosed quickly and receive prompt, effective treatment. This is particularly the case in patients infected with human immunodeficiency virus (HIV) and suffering from acquired immunodeficiency syndrome (AIDS) where their immune

FIGURE 8.21
Electron micrograph of influenza viral particles. Courtesy of H. Cotterill, Manchester Royal Infirmary, UK.

response is severely compromised. Electron microscopic examination of suitable samples can be used to identify viral particles by their characteristic morphology and offers a simple and quick method of sample analysis. You can see an electron micrograph of HIV particles escaping from an infected cell in Figure 8.22.

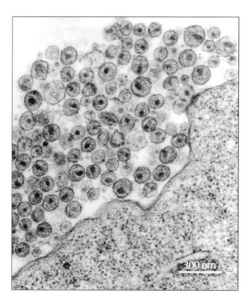

FIGURE 8.22
Electron micrograph of human immunodeficiency viral particles being released from the surface of a cell. Courtesy of H. Cotterill, Manchester Royal Infirmary, UK.

SUMMARY

- Microscopes are an essential part of laboratory testing and diagnosis
- Microscopes are used in many different disciplines
- Various types of microscopy exist and are used for different purposes
- Microscopes must be set up and adjusted correctly to ensure optimum visualization of specimens
- When significant increases in magnification or resolution are required, electron microscopy must be used instead of light microscopy

FURTHER READING

- Bradbury HSM. *Introduction to Light Microscopy*, Royal Microscopical Handbook. BIOS Scientific Publishers, London, 1998

- Rogers K. *Complete Book of the Microscope*. Usborne Publishers Ltd. London, UK, 2006

- Pawley JB. *Handbook of Biological Confocal Microscopy*. Springer, New York, 2006.

- Bozzola J, Russell LD. *Electron Microscopy*, 2nd edn. Jones and Bartlett, Boston, 1998.

- Ghadially F. *Diagnostic Ultrastructural Pathology A Self Evaluation and Self Teaching Manual* Hodder Arnold Publishers, London, 1998.

- Cheville N. *Ultrastructural Pathology*, 2nd edn. Wiley Blackwell, Oxford, 2009

QUESTIONS

1. If the diameter of a virus particle appears to be 5 mm when magnified 100,000×, its 'real' diameter is:

   (a) 1.0 μm

   (b) 0.005 mm

   (c) 50 nm

   (d) 500 μm

   (e) 5 μm

2. Indicate which of the following statements are **TRUE**.

   (a) Specimens for light microscopes are thicker than those for electron microscopy

   (b) Electron microscopes have a greater resolving power than light microscopes

   (c) Moving electrons have shorter wavelengths and therefore give poorer resolution than light rays (d)　Electromagnetic lenses can focus beams of electrons as well as light rays

(e)　Chromatic aberration, spherical aberration, and astigmatism can affect both glass and electromagnetic lenses

3.　(a) Describe what is meant by chromatic aberration. (b) How is this defect in a microscope lens corrected?

4.　Describe how you would set up a light microscope illumination system to give Köhler illumination.

5.　What two main factors affect the resolution of a microscope?

6.　Why is it necessary for the column of an electron microscope to be maintained under very high vacuum?

7.　Describe with the aid of a diagram, how the incident light fluorescence microscope works and can be used in the biomedical science laboratory. You may find it helpful to read relevant parts of Chapters 15 'Immunological techniques' and 16 'Molecular biology techniques'.

8.　What is the total magnification of a microscope using a 40× objective lens with a focal length of 2 mm and an eyepiece with magnification of 10×?

Answers to self-check and end of chapter questions are available in the Online Resource Centre accompanying this book.

Go to www.oxfordtextbooks.co.uk/orc/glencross

9

Electrochemistry

Peter Robinson

Learning objectives

After studying this chapter, you should be able to:

- Define key terms related to electrochemical analyses
- Describe the reactions that occur in a typical electrochemical half cell
- List common substances that may be measured in a clinical laboratory using electrochemical analytical devices
- Describe the Nernst equation and outline its use in potentiometric analysis
- Describe the underlying principles of amperometric electrochemical devices
- Outline the principles and clinical applications of ion selective field effect transistors
- Define and describe the underlying principles of operation of biosensors
- Discuss the development of biosensor technology

Introduction

It is likely that a pH electrode was the first piece of analytical equipment that you used when you first entered a science laboratory. Like so much modern technology, you could use such a device without ever really understanding how it works. However, working within a clinical laboratory you may well have to use several, possibly dozens, of different electrode devices to measure a wide range of **analytes** (the term generally used for the substances being measured) in clinical samples. Ion selective electrodes, oxygen electrodes, carbon dioxide electrodes, and various biosensors are now found in most clinical settings. It is useful therefore to have sufficient background knowledge to be able to understand the potential and limitations of such devices. The recognition that such electrodes fit into a small number of 'families' should enable you to develop experience of using one electrode and help you understand how other 'close relatives' might also function. Learning a few fundamental principles can therefore save hours of having to 'rote learn' the specifics of each device in isolation. In addition, some knowledge of the principles on which these devices are based is particularly useful when, as can occur, an electrode appears to be malfunctioning.

In many ways electrochemical analysis is ideally suited for use in the clinical laboratory. Indeed, many techniques are also the basis of the point of care testing devices described in Chapter 18.

Once set up and calibrated, electrochemical devices are simple to use, such that after even the most basic of training you will be able to measure numerous analytes in clinical samples with high accuracy and reliability. Because electrochemical techniques do not rely on any colour formation they are highly suited to measurement of samples with a high background colour, in particular blood. Finally, unlike most other analytical techniques, electrochemical techniques use few reagents. Once an electrode and meter have been purchased the actual running cost per sample is generally low, and this may be of great importance to the financial competitiveness of a clinical laboratory.

Cross reference
You can read about automated clinical laboratory instruments in Chapter 17 'Laboratory automation'.

The aims of this chapter are to outline the likely uses of these electrode devices within a clinical laboratory and to describe their fundamental principles of operation. General advice on the care and maintenance of electrodes is also included. However, you are always advised to consult the manufacturer for specific information regarding such matters, as electrodes from different manufacturers may have some quite specific requirements.

9.1 Basic concepts and definitions

Electrochemistry is the study of the chemical changes produced by electrical currents and also the production of electrical currents by chemical reactions. For more than 200 years electrochemical techniques have developed through advances in chemistry, physics, electronics and, more recently, biology. The sheer number of electrochemical techniques can be bewildering, and while there may be fundamental differences between some groups of techniques, it is often the case that numerous techniques differ from each other in only minor detail. Sadly, even the naming of electrochemical analytical techniques has been applied with little or no semblance of a systematic approach. As a consequence, the nomenclature surrounding the subject can be quite confusing and, at times, plainly inconsistent!

It is therefore necessary to go back to the basics of electrochemistry before jumping into a detailed description of the devices and techniques themselves.

(a)

What is an electrical current?

(b)

FIGURE 9.1
(a) Conventional and (b) electron flow notation for an electric current showing the apparent inconsistency in direction of current flow.

An electrical current is simply a movement (flow) of electric charge. However, one fundamental inconsistency that might well lead to confusion relates to something as simple as the actual direction of the flow itself. This inconsistency dates back to the mid 18th century when Benjamin Franklin was studying static electricity. He found that when a wool cloth was rubbed over a paraffin wax block the two materials subsequently attracted each other. Franklin considered that the wool cloth was in some way removing an 'invisible fluid' from the wax block, hence the type of charge that was associated with rubbed wax became known as 'negative' (because it had a deficiency of the 'fluid'), whilst that associated with the wool became known as 'positive' (because it had an excess of the 'fluid'). This terminology was widely adopted such that when electric batteries were later developed, it was natural to assign the direction of the flow of current to be from the positive to the negative pole. Over a century later, however, J.J. Thomson showed that cathode rays were negatively charged particles, which he called 'corpuscles', and claimed that atoms were built from these corpuscles. The electron had been discovered and it was suddenly realized that it was electrons that carried the current in metal wires. Thus, electrons must be moving in the opposite direction to that of the 'conventional' current. Sadly, it was much too late to change Franklin's naming convention and today it must be accepted that current can be considered to move in either of these two ways. Hence, both of the conventions shown in Figure 9.1 are acceptable.

BOX 9.1 *Ohm's Law*

Ohm's Law can be rearranged to describe the current, voltage, or resistance. The easiest way to remember these relationships is to use a **VIR triangle** as shown in Figure 9.2. To use this, simply use your finger to hide the value that you want to calculate. The remaining two values then show you how to do the calculation.

FIGURE 9.2
VIR triangle used to help calculate Ohm's Law.

Ohm's Law

A flow of an electrical current only occurs when there is a *potential difference* between two points within a circuit. It is also influenced by *resistive forces* present within the system. These relationships are usually described by **Ohm's Law** (Box 9.1), which states that within an electrical circuit the current passing between any two points of a conductor is directly proportional to the potential difference, that is voltage drop or voltage, between the two points but is inversely proportional to the resistance between them. In mathematical terms, Ohm's Law may be written as:

$$I = V/R$$

where I is the current in amperes (often abbreviated to amps, A), V is the potential difference in volts (V), and R is the resistance measured in Ohms (Ω).

SELF-CHECK 9.1

Calculate the voltage required to force a current of 0.2 A through a wire with a resistance of 1200 Ω.

9.2 Principles of electrochemical techniques

Whenever you study electrochemical reactions you will come across the concept of an electrochemical cell that is made up of two electrochemical **half-cells**. Each half-cell contains an electrode, which at its simplest is just a metal wire, that is surrounded by an electrolyte. The two half-cells are joined by a shared electrolyte or by a relatively concentrated solution of different ions, often saturated or 4 mol dm^{-3} KCl or KNO$_3$, called a **salt bridge**.

Consider a relatively simple half-cell consisting of a copper wire dipping into a solution of $CuSO_4$. Within this system, **oxidation** may occur so that the copper electrode may lose copper ions to the solution, leaving behind electrons delocalized on the metal surface:

$$Cu_{(electrode)} \rightarrow Cu^{2+}_{(aqueous)} + 2\ e^-_{(electrode)}$$

This leaves the electrode negatively charged with respect to the surrounding solution. Alternatively, the conditions may favour a **reduction**, where copper ions in solution combine with electrons at the metal surface to deposit copper atoms:

$$Cu^{2+}_{(aqueous)} + 2\ e^-_{(electrode)} \rightarrow Cu_{(electrode)}$$

This produces a deficit of electrons in the copper electrode making it positively charged with respect to the surrounding solution. However, in both cases, a spontaneous potential difference called the **electrode potential** is established across the electrode/electrolyte interface. Hence, when two half-cells are combined, a flow of electrons, that is an electrical current, may occur. Consider, for example the Daniell cell shown in Figure 9.3, which consists of zinc and copper wires immersed in solutions of $ZnSO_4$ and $CuSO_4$ respectively. If the Zn^{2+} and Cu^{2+} concentrations are approximately equal, then reactions will result in the oxidation of the zinc metal in one half-cell to leave the zinc electrode negatively charged, and the reduction of the copper ions in the other half-cell, which results in the copper electrode becoming positively charged. If the two electrodes are then connected by a wire, electrons will flow from the zinc anode to the copper cathode. The Daniell cell can therefore produce an electrical current and is thus a simple battery.

The salt bridge connecting the two solutions of the half-cells in Figure 9.3 completes the circuit and allows migration of ions between the two compartments, but prevents gross mixing of the two solutions. The salt bridge might be as simple as a strip of filter paper soaked with a relatively inert electrolyte such as KCl, or a U-shaped glass tube filled with the same electrolyte held within an agar gel.

Schematic representation of electrochemical cells

In 1953 the International Union of Pure and Applied Chemistry (IUPAC) developed a standard notation to represent electrochemical cells that is still accepted today. The IUPAC notation for the Daniell cell shown in Figure 9.3 may be written as follows:

$$- Zn \mid Zn^{2+} \mid\mid Cu^{2+} \mid Cu^+$$

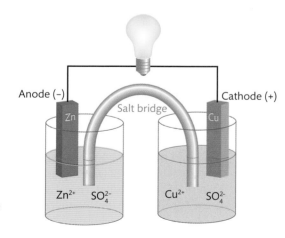

FIGURE 9.3

Daniell cell. See text for details. Note, this is a Galvanic cell and, therefore, the anode and cathode are negatively and positively charged respectively, which is the opposite of the electrodes in simple electrophoresis (Chapter 14).

where the single vertical lines represent a phase boundary (electrode/electrolyte interface) and the double vertical lines the salt bridge. Note that according to this notation the negative electrode (**anode**) is placed to the left, the positive electrode (**cathode**) is placed to the right, and electrons would therefore flow from left-to-right in the connecting wire.

Classification of electrochemical techniques

Electrochemical cells can be categorized as **galvanic** or **electrolytic**. A galvanic cell is one where the reactions occur spontaneously at the electrodes when they are connected externally by a conductor. In an electrolytic cell, however, the electrochemical reactions are not spontaneous and only occur if an external voltage is applied across the two electrodes. A modern rechargeable battery is, in essence, an electrolytic cell when it is being recharged.

> A **cation** (pronounced cat-ion) is an ion that has fewer electrons in its electron shells than protons in its nucleus. It thus carries a net positive charge and will move towards the negative electrode during electrolysis.
>
> An **anion** (pronounced an-ion) is an ion that has more electrons than protons. Thus it has a net negative charge and will move towards the positive electrode during electrolysis. You might be able to remember that it is the anion that is negatively charged by the mnemonic 'a-n(egative)-ion'

SELF-CHECK 9.2

State whether the Daniell cell shown in Figure 9.3 is a galvanic or electrolytic cell.

Electrochemical analytical techniques may be conveniently grouped according to whether their electrochemistry is galvanic (spontaneous) or electrolytic (non-spontaneous). **Potentiometry** involves the use of galvanic cells, and the measurement of the spontaneous potential of an electrode with minimal current flow. The commonest applications of potentiometry are the pH electrode and ion selective electrodes described in Section 9.3. In contrast, **voltammetry** (including amperometry) uses electrolytic cells and the measurement of the current passing through an electrode when an external potential is applied. By far their commonest application is the Clark oxygen electrode. This device is described in Section 9.4 and relies on the reduction of dioxygen at a platinum cathode to give a current that is proportional to the oxygen concentration in the solution. Many biosensors also incorporate amperometric detectors, as described in Section 9.5.

9.3 **Potentiometric techniques**

In potentiometric analysis, an electrode, often called the **working** or **sensing electrode**, is used that generates a spontaneous potential at its surface with the analyte, that is, the solution being measured. However, only potential *differences* can be measured, hence, this cell needs to be connected to a second cell that produces a standard potential so that the *difference* between the two can be evaluated. This second, so called **reference electrode**, is a necessary component of all pH and ion selective electrodes.

The behaviour of potentiometric electrodes is generally described by the **Nernst equation**, where the total potential, E (in volts) developed between the working and reference electrode is:

$$E = E_{constant} + 2.303 \frac{RT}{nF} \log_{10} A$$

or simplified as:

$$E = E_{constant} + S/n \log_{10} C$$

where $E_{constant}$ is a constant potential that depends mainly on the reference electrode; 2.303 RT/F (=S) is the Nernst factor or slope (where R is the universal gas constant, T is the temperature in

BOX 9.2 Activity

Activity is a useful chemical concept. Consider an individual ion in solution. The ability of this ion to participate in any chemical reaction will be reduced if it is surrounded by other ions, which 'shield' it. This is more likely to occur when there are more such ions in solution. Activity therefore is a true measure of the ion's ability to affect chemical equilibria and reaction rates; it is often said to be the *effective* concentration of the ion in solution. In most biological situations, where the concentrations of ions are rather low, the values of the concentration and activity are equal. At higher concentrations, however, the activity may become significantly less than the concentration.

Kelvin and F is the Faraday constant); n is the number of charges on the ion; A is the activity of the ion; and C is the concentration of the ion.

Potentiometric electrodes actually measure the **activity** of an ion in solution (Box 9.2). In most biomedical situations the concentrations of ions are low and activity is synonymous with concentration.

The Nernst equation shows that the responses of potentiometric electrodes will depend both on the temperature and the number of charges carried by the ion. At 25°C the Nernst factor (2.303 RT/F) becomes 0.059 and thus the equation becomes:

$$E = E_{constant} + 0.059/n \log_{10} C$$

This means that there will be a 59 mV change in potential for a 10-fold change in the concentration of monovalent ion, such as H^+, Na^+, K^+, and Cl^-.

A potentiometric electrode is said to have Nernstian characteristics if it obeys Nernst's law. If the changes in potential relative to concentration are less than the theoretical values, it may indicate an electrode malfunction or interference from other ions.

SELF-CHECK 9.3

Given that the universal gas constant is 8.314 J mol^{-1} per K, T is the temperature in Kelvin (=273 + T°C) and that the Faraday constant is 96,485 coulomb mol^{-1}, calculate the change in potential in mV for a ten-fold change in the concentration of H^+ when the temperature of the system is 35°C.

pH electrode

Virtually all textbooks, including this one, describe the pH electrode separately from ion selective electrodes (ISEs) even though it is simply one specific type of ISE that responds to H^+. However, it is much commoner in laboratories than other ISEs and more likely to be used by students and thus tends to get a more detailed coverage by authors.

In 1906, Max Cremer found that a thin glass membrane separating the two electrodes of a galvanic cell could produce a potential that was responsive to changes in the concentration of H^+. Three years later, what we would recognize as a glass electrode was constructed, and studied by Fritz Haber and Zygmunt Klemensiewicz. However, this device did not become popular until the 1930s, when reliable amplifiers became available to measure the output from the electrodes. Glass electrodes are still the most convenient and accurate way of determining pH and the pH electrode/pH meter is a basic item in any science laboratory.

What most people refer to as a glass electrode, as you can see in Figure 9.4 (a), is technically a device, not an electrode. It consists of a thin soft glass membrane, which gives the electrode its name, and an electrically insulating tubular body made of hard glass or epoxy. This structure encloses an internal electrolyte solution (usually 0.1 mol dm^{-3} HCl) and a silver/silver chloride **internal reference electrode**. The silver/silver chloride internal reference electrode is connected to the pH meter. A potential difference must, however, be measured between two points. Hence a second **external reference electrode** is required, which also needs to be immersed in the same test solution. Frequently, this second electrode is made up of a second silver/silver chloride wire surrounded by a solution of 4 mol dm^{-3} KCl saturated with AgCl. The external reference electrode can be a separate probe as shown in Figure 9.4 (b), although it is commoner to find this external reference electrode built into the glass electrode in a concentric double barrel arrangement, commonly referred to as a **combination electrode** as you can see in Figure 9.4 (c).

Whichever reference electrode is being used, it must also form electrical continuity with the test solution. In many electrodes, the KCl of the reference electrode is an aqueous solution that flows out through a small hole in the electrode body to form a liquid junction. To minimize the loss of KCl, the junction is packed with ceramic, fritted, or fibrous material, although *some* flow

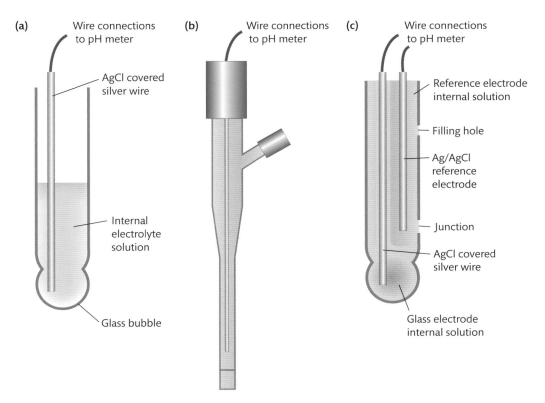

FIGURE 9.4
(a) A pH 'electrode', (b) a Ag/AgCl external reference electrode, and (c) a combination pH electrode. See text for explanations.

of KCl, however small, always occurs and therefore the pH electrode will slightly contaminate the test solution. For most analytical applications this contamination is not significant. The KCl solution lost from the electrode must be replaced. This can be done by refilling through a small hole, high up on the electrode body. Refilling electrodes adds to their maintenance costs and makes them less suited to situations where portability is a factor. To overcome such problems, the internal KCl solution may be gelled, which helps slow the loss. However, as the composition of the internal solution cannot then be restored by refilling, such **gel electrodes** generally have a shorter operational lifetime, although they are easier to use and maintain.

In situations where contamination of the test solution by the KCl of the reference electrode *is* an issue, or where the test solution contains materials that might diffuse into the external reference electrode and affect the electrochemistry, a specially designed reference electrode called the **double junction reference electrode** may be used. This device may also be selected simply to prolong the operational lifetime of the electrode. In a double junction electrode, an additional chamber between the reference electrode and the external solution is introduced. This slows the loss of KCl to the test solution and presents a further diffusional barrier to prevent the inward diffusion of any contaminating materials. Double junction electrodes are, however, more difficult to construct and, hence, are more expensive.

Modern pH electrodes are robust and reliable devices. However, you might be interested to know why the glass membrane found in these devices generates a potential when separating two solutions. Not all glasses possess this property and it has taken considerable research to optimize the composition of the glass to allow interference-free detection of H^+. Soda-silica glasses, such as Corning 015 glass, consisting of 22% Na_2O, 6% CaO, and 72% SiO_2, are effective for measuring pH. Hydrogen ions themselves *do not* pass through the glass membrane of the electrode. Rather, within such glass, the molecular network of SiO_2 contains both Na^+ and Ca^{2+} coordinated with oxygen that are able to exchange with H^+ from the test solution and thereby generate a potential across the glass membrane. The glass membrane therefore acts rather like a battery whose voltage depends on the concentration of H^+ ions in the test solution in which it is immersed. The size of the generated potential (E) is given by the equation:

$$E = 2.303 \ (RT/F) \ \log_{10} [H^+]_i/[H^+]_o$$

where $[H^+]_i$ and $[H^+]_o$ are the molar concentration of H^+ inside and outside the glass membrane, respectively. In practice, the internal H^+ concentration is fixed at 10^{-1} mol dm^{-3} since the electrode contains 0.1 mol dm^{-3} HCl. Hence, the electrode is responsive only to the concentration of H^+ in the test solution in which it is immersed.

Within the pH electrode, this spontaneously generated potential must be measured against that produced by a reference electrode. In practice, however, other potentials are also generated within the system. The so-called *asymmetry potential* is poorly understood, but is present across a glass membrane even when the H^+ concentration is the same on both sides. The liquid junction of the reference electrode may also give rise to a potential because the K^+ and Cl^- it contains may diffuse out through the junction at different rates. However, these additional potentials can be taken into account, such that at 25°C the overall potential of a glass pH electrode measured against a reference electrode will be 59 mV for a 10-fold change in the activity of H^+. Rather conveniently, since pH itself is $-\log10 \ [H^+]$, it follows that the 59 mV change in potential actually represents each division of the pH scale.

Electrode selectivity and interference

Glass pH electrodes are particularly useful in the clinical laboratory because they demonstrate little interference from the other components that may be found in a wide range of clinical samples.

One exception to this is the interference by monovalent cations, such as Na^+ and K^+. These ions behave in similar ways to H^+ at the glass surface and generate comparable potentials. Under acidic conditions, where the concentration of H^+ in solution is high, the interference is likely to be negligible. At alkaline pH, where there are much lower concentrations of H^+ in solution, interference can be more obvious and the measured pH is lower than the true pH of the test solution. This effect is often called the *alkaline error* or sometimes the *sodium error*, although not only Na^+ interferes in this way.

The careful selection of the glass composition is crucial to minimize such interferences, as it is the composition of the glass that is solely responsible for the selectivity. Most commercial pH electrodes have selectivities that are sufficient under all, but hyperalkaline conditions and below pH 12 any interference effects should be negligible even in modestly priced devices. Detailed information about selectivity is available from electrode manufacturers. It is possible to take advantage of this interference: by selecting a glass that has *high* response to Na^+ it is possible to construct a Na^+ selective electrode, as described later.

Cross reference
Measuring Na^+ and K^+ in clinical samples is described in Chapters 1 and 2 of *Clinical Biochemistry*.

Electrode storage and cleaning

The thin glass membrane of a pH electrode is fragile. Care must be taken not to break or scratch it, or to cause a build-up of static charge by rubbing it. Many pH electrodes have a plastic casing surrounding the glass electrode to protect them from damage. Even so, the temptation to use the pH electrode as a stirrer must be resisted: this is a common cause of breakage in many laboratories!

It is essential that the outer layer of glass in the pH electrode remains hydrated, so the electrode is normally immersed in a solution at all times. Most manufacturers now supply specific storage solutions for their electrodes. If these are not available, then the electrode should *never* be stored in distilled or deionized water, as this will cause ions to leach out of the glass bulb and render the electrode useless. It is acceptable to immerse combination electrodes in a buffer solution, those of pH 4 to 7 are generally recommended, for storage between frequent measurements. For longer-term storage it is generally recommended to keep the electrode immersed in 4 mol dm^{-3} KCl.

The electrode should be washed thoroughly with distilled water after each use. As the electrode is susceptible to contamination or dirt, it is recommended that it is cleaned perhaps every 1–2 months depending on the extent and condition of use. General cleaning can involve the use of a mild detergent. Any scale deposit on the electrode surface can be removed with a concentrated acid, for example 6 mol dm^{-3} HCl. Biological materials can be removed using a pepsin solution, and oily deposits with acetone or methylated spirits. However, the manufacturer's instructions should always be consulted before using any of these methods.

If an electrode has been allowed to dry out and has lost activity, it is worth contacting the manufacturer first, but a rejuvenating procedure might include soaking it in concentrated hydrochloric or nitric acid, or even boiling the electrode. Such procedures are not guaranteed to be successful and should only be used as a last resort.

pH meter

The glass pH electrodes developed in the early part of the 20th century were not used on a large scale because of various technical difficulties. The main problem was caused by the large internal resistance of the glass electrodes themselves (typically 50–500 MΩ). This necessitated the use of sensitive and expensive galvanometers, which at the time were only available in a small number of well equipped research laboratories. However, in the mid-1930s Arnold Beckman, an assistant professor of chemistry at the California Institute of Technology, was trying to develop a technique (and technology) to make quick and accurate measurements of the acidity of lemon juice. Beckman's innovative approach was to use a high-gain amplifier made using two vacuum tube valves. The amplified current could then be read with much more reliable and cheaper voltmeters. In 1934, Beckman filed a patent, not actually for the entire pH meter, but rather for the amplifier component alone. Some time later he put the amplifier, a measuring electrode, and a data meter together in a compact walnut case, which he called an *acidimeter* (Figure 9.5 (a)). Today we recognize this device as a pH electrode and meter, modern versions of which you can see in Figure 9.5 (b).

When using a pH electrode and meter, calibration usually involves the use of two buffered solutions of widely differing pH. It is increasingly common to purchase such buffer solutions ready-made from the electrode suppliers. These are often colour-coded to make the calibration process easier: the commonest buffers used are pH 4.0 (red), pH 7.0 (yellow), and pH 10.0 (blue). Calibrating most meters initially involves the pH 7 buffer. With the electrode in the buffer solution and after sufficient time to equilibrate, the 'calibrate' dial on the meter should be adjusted to give an output of pH 7.0. The electrode is then removed, rinsed with distilled water, and blotted dry before being placed in a pH 4 buffer (if the test solution is expected to be acidic) or a pH 10 buffer (if the test solution is expected to be basic). This second point is adjusted using the 'slope' control on the pH meter. Once the pH electrode has been calibrated it can simply be rinsed with distilled water and blotted dry before being immersed in the test

(a)

(b)

FIGURE 9.5

(a) The first commercial pH meter and probe, called the *acidometer*, developed by Arnold Beckman in the 1930s for the measurement of the acidity of lemon juice. Photograph supplied courtesy of Beckman Coulter. (b) Contemporary modern pH electrode and meter.

solution for a rapid and accurate estimation of pH (Box 9.3). Between measurements, the electrode should be rinsed with distilled water and blotted dry of excess liquid.

Should commercial buffer standards not be available you can make standards yourself from recipes available in for example the Merck Index or the US Pharmacopeia. Typical standard buffers might include:

- 0.05 mol dm^{-3} potassium hydrogen phthalate (pH 4.01)
- 0.025 mol dm^{-3} disodium hydrogen phosphate + 0.025 mol dm^{-3} potassium dihydrogen phosphate (pH 6.86)
- 0.01 mol dm^{-3} sodium borate (pH 9.18)

Glass pH electrodes are affected by temperature (see the Nernst equation, Section 9.3). Each division of the pH scale will, for example, represent a potential of 54 mV at 0°C, 59 mV at 25°C, and 62 mV at 37°C. Temperature will also affect the thermal characteristics of the electrode and the test solution such that, in practice, the pH electrode will have an **isopotential point** where its potential is 0 and where temperature will not have any effect on the potential. This is often designed to be near pH 7. The further the pH is from this isopotential point, the greater will be the effect of temperature and the more necessary it will be to apply temperature compensation. Most pH meters have a temperature compensation control that needs to be set to the temperature of the calibrating buffers and test solution. More advanced instruments have a temperature probe that is dipped into the calibrating and test solutions alongside the measuring electrode. This ensures automatic correction of any variability resulting from solutions having differing temperatures.

 BOX 9.3 *Expressing mean pH values*

Consider the case of measuring the pH of a urine specimen with a pH meter. Measurement of three samples of urine gives results of 5.5, 6.0, and 6.5. You then calculate the mean pH of the urine to be pH 6.0. But is this value appropriate?

This is a remarkably common situation, yet the simple analysis as described above is significantly flawed. pH is a logarithmic scale; hence, in terms of H$^+$ concentration the three pH values are not equally spaced. Converting the individual pH values into H$^+$ concentrations gives:

pH 5.5 = 3.16 × 10^{-6} mol dm^{-3}

pH 6.0 = 1.00 × 10^{-6} mol dm^{-3}

pH 6.5 = 0.32 × 10^{-6} mol dm^{-3}

The arithmetic mean of these concentrations yields a H$^+$ concentration of 1.49 × 10^{-6} mol dm^{-3}, which equates to a pH of 5.8. This difference (5.8 versus 6.0) might be small enough to ignore, but when pH readings are more widely spread the effects are significant. Try Self-check question 9.4 to see this for yourself.

You should note that many papers are published that report a simple arithmetic mean of pH value. This may have been used either pragmatically (the pH values are very close together and hence errors are small) or through the author's own ignorance. Beware!

Given three solutions of pH 5, 7, and 9, respectively, you might expect the mean pH to be pH 7. However, calculate the mean as described in Box 9.3. What is the mean value?

Severinghaus CO_2 electrode

The need to measure the carbon dioxide concentration in clinical samples, especially blood, developed out of necessity during the 1950s' polio epidemics in Europe and the USA. In the USA alone, nearly 60,000 cases of polio were reported in 1952 and over 20,000 children were left with mild to disabling paralysis. Such paralysis often produced respiratory difficulties, which were treated with the use of the negative pressure ventilator, the 'iron lung'. During such treatment it was necessary to measure blood gases accurately and quickly to assess the effectiveness of the therapy.

In 1954, Richard Stow, working at the Ohio State University in Columbus, developed an electrode capable of measuring the partial pressure (concentration) of carbon dioxide or PCO_2 in solution. The device was essentially a modified glass pH electrode. Stow realized that carbon dioxide freely penetrated rubber and that it also acidified water. Hence, he constructed a glass pH electrode and surrounded this with a thin film of distilled water held in place by a thin rubber membrane (actually, in this instance a section of rubber glove). When this device was dipped into a sample of blood, the carbon dioxide from the blood diffused through the membrane and acidified the water surrounding the pH electrode, resulting in a measurable drop in pH that was proportional to the PCO_2. Stow published and presented his findings, but his device was rather insensitive and unstable, and was never produced commercially. It was, however soon modified by John Severinghaus who added a solution of $NaHCO_3$ and $NaCl$ between the membrane and the electrode. Severinghaus also preferred a Teflon (polytetrafluoroethylene) membrane, rather than Stow's rubber membrane. In this device, carbon dioxide from the blood sample diffuses across the membrane into the $NaHCO_3$ and $NaCl$ solution, and disturbs the carbonic acid/hydrogen carbonate equilibrium to produce H^+:

$$CO_2 + H_2O \rightleftharpoons H_2CO_3 \rightleftharpoons H^+ + HCO_3^-$$

The system is allowed to achieve equilibrium, at which time the pH is measured. This is then compared with a calibration curve of pH against known PCO_2 values.

Severinghaus's modifications doubled the sensitivity of the original device, enhanced its stability, and improved the response time. In 1958, Severinghaus described a device incorporating both his PCO_2 electrode and a Clark O_2 electrode, which you can read about in Section 9.4, to form a combined blood gas analyser. Commercial manufacture of the Severinghaus PCO_2 electrode began in 1959 and such electrodes are still found in PCO_2 analysers in clinical laboratories today.

Cross reference

Measuring HCO_3^-, PCO_2, and pH in clinical samples are described in Chapter 6 of *Clinical Biochemistry*.

The Severinghaus electrode is a *gas-sensing electrode* and as it is based on the pH electrode, the electrical output is a *logarithmic function* of PCO_2. At 37°C an output of 62 mV is expected for a 10-fold change in PCO_2. The response is also temperature dependent since temperature affects the solubility of dissolved gases, the equilibrium constants within the HCO_3^- layer, and the Nernstian characteristics of the pH electrode itself. It is therefore usual to maintain the temperature of such devices at a constant 37°C. For the most accurate analyses, it may also be necessary to correct for the temperature of the patient who provided the samples, especially if the patient is known to be either febrile or hypothermic.

Ion selective electrodes

The glass pH electrode is just one of a wide range of ISEs available to the biomedical scientist. Note at the outset that these devices are called *ion selective* electrodes, not ion specific electrodes. All of these devices *do* suffer from some interference, that is, responsiveness to other ions, although in many clinical applications this interference may well be minimal. Within the clinical laboratory you may well need to use Na^+, K^+, and Cl^- ISEs to measure the electrolyte balances of samples. Lithium ion ISEs may be used to measure concentrations of Li^+ in samples of serum in cases where this is being used to treat psychiatric illness.

In general ISEs can be divided into three main groups:

- Modified glass electrodes
- Polymer membrane electrodes
- Solid-state electrodes

Modified glass electrodes incorporate a membrane of specifically modified glass, as their name implies. Polymer membrane electrodes have an ion exchange material in an inert matrix, such as PVC, polyethylene, or silicone rubber, while solid-state electrodes incorporate a membrane consisting of a single crystal of a sparsely soluble salt or a heterogeneous membrane in which the salt is incorporated in an inert binding agent.

The interfering effect of Na^+ on pH measurements was discussed earlier. We noted how the chemical nature of the glass used within a pH electrode is designed to minimize this effect. However, to measure Na^+ in clinical samples, a glass electrode with a *high* Na^+ responsiveness and with an inner reference solution of a fixed Na^+ activity are required. In early Na^+ ISEs, NAS 11–18 glass, comprising 11% Na_2O, 18% Al_2O_3, and 71% SiO_2 was found to be effective, although this has now generally been superseded by a range of lithium-based glasses (Li_2O), sometimes referred to as LAS glasses. Unfortunately, both NAS and LAS glasses are more responsive to Ag^+ and H^+ than they are to Na^+. Fortunately, Ag^+ is rarely encountered in clinical samples and such interference can generally be ignored. In some clinical applications, H^+ interference may also be of little concern. Consider, for example, the analysis of Na^+ in blood where in healthy individuals the plasma Na^+ concentration is maintained within the range 0.135–0.145 mol dm^{-3}. The pH of such samples would be expected to be close to pH 7.4, which equates to a H^+ concentration of 4×10^{-8} mol dm^{-3}, at which concentration little observable interference would be expected. However, in other applications, the interfering effects of H^+ may well be significant. Standardizing the H^+ concentration of all samples and standards to ensure that the electrode is responding to Na^+, rather than to H^+ interference must be considered.

While it is possible to produce a glass electrode that is responsive to K^+ using, for example, NAS 27-4 glass, a more common approach is to use electrodes constructed by incorporating specific ion exchange materials, such as valinomycin, into a polymer membrane. Molecules of the cyclic antibiotic valinomycin have a hexagonal ring structure with an internal space of almost exactly the same diameter as K^+. They can therefore form complexes with the K^+ and preferentially conduct them across a membrane. Similarly, Li^+ ISEs can be constructed from Li^+ selective carriers, such as crown ethers. These are capable of transporting Li^+ across the membrane and releasing the ion, in preference to the many other cations that might be present in a sample.

Various sold-state membranes are also available. Chloride ISEs are constructed using a crystalline membrane that incorporates AgCl in direct contact with the test solution. This membrane slowly dissolves and Cl^- from the electrode diffuses into the test solution to leave behind a surplus of Ag^+. This generates an electrical potential that is proportional to the Cl^- in the test solution. Such electrodes are quite robust and generally have faster response times than glass-based electrodes. However, since such devices rely on the slow dissolution of the membrane itself, they have a finite operational lifetime.

With all ISEs, we have emphasized that the electrode responds to the *activity* of the ion in solution. If the instrument is calibrated with a standard of known concentration then, provided the ionic strengths of the solutions are similar, the concentration of the test solution will be recorded. To ensure that the ionic strengths are similar, an **ionic strength adjustor** can be added to the sample and standards. Ionic strength adjustors contain high concentrations of ions and sometimes pH adjustors, and decomplexing agents or agents to remove chemical species that interfere with the measurement of the ion of interest. If some of the ion is not free in solution, but exists in a complex or as an insoluble precipitate, then ion selective electrodes will give a lower value than methods such as atomic absorption spectrophotometry that detect all of the ions present. The ISE results may, however, be more applicable because it is often the *free ions* that are responsible for any clinical or biological effects.

The response of all ion selective electrodes is logarithmic and is described by the Nernst equation. Ten-fold changes in ion activity give equal increments on the meter scale. Ion selective electrodes are not intrinsically sensitive and generally show a linear response range 10^{-1} to 10^{-5} mol dm^{-3}. In practice, the minimum detection limit of most commercial devices is 10^{-6} mol dm^{-3} at best. As with pH electrodes, the potential produced from an ISE is temperature dependent, except at the isopotential point which varies with the type of electrode. It is therefore necessary to have some form of temperature control and/or compensation. All ISEs also require a reference electrode to generate a steady potential against which the varying potential of the working electrode is compared. If either K^+ or Cl^- is measured, then a double junction reference electrode is needed to prevent contamination of the sample by the internal solution of the reference electrode.

Many ions can be measured directly or indirectly by ISEs. One form of indirect measurement is the use of the electrode as the endpoint indicator of a titration. The electrode can be sensitive to the ionic species being determined or to the titrant ion. Such titrations may be 10 times more accurate than direct measurements, since they only require the accurate measurements of a *change* in potential, rather than the absolute value of the potential itself.

Ion selective field effect transistors

A range of ion selective electrodes may be constructed by modifying field effect transistors to respond to specific ions. Such **ion selective field effect transistors** (ISFETs) are available as probe type electrodes and also as microelectrodes that are suitable for implantation and advanced experimental studies. It is likely that these devices will have an increasing clinical value due to their robustness and because they can be stored dry, unlike conventional pH electrodes and ISEs.

The electrical transistors found in nearly all electrical devices are **metal oxide semiconductor field effect transistors** (MOSFETs). Within such semiconductor devices, as you can see in Figure 9.6 (a), a channel of semiconductor material enables a current to flow from one electrode, called the *source*, to a second electrode, the *drain*. The rate of flow of electrons, that is the *conductivity* of the semiconductor, is changed by altering the electrical diameter of the channel. This is the function of the *gate* electrode, a metallic electrode that is insulated by a layer of silicon dioxide from the channel itself, such that it can only influence the source-to-drain current electrostatically. Within such a device, a small change in potential difference between the gate and the body of the semiconductor is able to change the conductivity of the channel and produce a relatively large change in the current from the source to the drain.

FIGURE 9.6
**Schematics showing the structures of
(a) metal oxide semiconductor field
effect transistor (MOSFET) and (b) an ion
selective field effect transistor (ISFET).**

In this way, the transistor acts as an *amplifier*. This is somewhat like a small tap controlling the flow of water through a pipe: a small turn of the tap can dramatically increase or decrease the flow of water.

As you can see in Figure 9.6 (b), the source and drain within an ISFET are constructed in an identical way to the MOSFET. However, the metallic gate of the MOSFET is replaced by an insulating coating capable of interacting with the analyte in the test solution to generate a potential. In a pH selective ISFET, this may simply be a layer of silicon dioxide. The source and drain leads, as well as semiconductor itself, are encapsulated so only the gate area is open for contact with the test liquid. The device also incorporates a reference electrode, so that when the device is immersed in a test liquid, an electrical circuit is set up between the reference electrode and gate surface enabling the H^+ concentration in solution to influence the channel and, hence, the current passing from source to drain.

Ion selective field transistors may be further modified by changing the chemical nature of the gate insulator or by adding an ion sensing membrane layer over the gate insulator. Using such techniques, multi-function ISFETs capable of measuring pH, Na^+, K^+, and Ca^{2+} are formed. It is likely that such devices will become increasingly common for the analysis of clinical samples. ion selective field effect transistors also make eminently suitable **transducers** for incorporation into biosensors, as described in Section 9.5.

9.4 **Voltammetric techniques**

In all voltammetric techniques, an external voltage greater than that of the spontaneous potential of the electrochemical cell is imposed on the system to force the electrochemical reaction to occur. In simple direct current voltammetry, if an electrode is placed into a solution of an electro-reducible molecule and the potential of the electrode is made progressively more negative a point is soon reached when reduction of the soluble species will begin at the electrode surface and a small current will flow. A graph of current against voltage is called a voltammogram, which you can see in Figure 9.7, The point at which the current begins to flow is E_1. As the electrode potential becomes ever more negative, the current increases dramatically as the molecular species is reduced still further. Eventually, a potential is reached where the molecular species is being reduced at the electrode surface as quickly as it can diffuse towards that surface. At this point, E_2 in Figure 9.7, a *maximum or limiting* current is produced and further increases in potential will not increase the current beyond this value.

The voltammagram shown in Figure 9.7 illustrates two notable properties:

1. The midpoint of the wave, called the half-wave potential ($E_{1/2}$), is characteristic for each molecular species reduced

2. The value of the limiting current is directly proportional to the concentration of this molecular species in solution

These two features enable voltammetric techniques to be used both to identify, as well as quantify **electroactive** materials in solution.

More advanced forms of voltammetry, such as pulse voltammetry, differential pulse voltammetry, and alternating current voltammetry all use more complex voltage changes. Stripping voltammetry allows a preconcentration step to be introduced into the analysis. These advanced techniques are more sensitive than direct current voltammetry, although because of this complexity they rarely find their way into routine laboratory analyses.

Voltammetry is closely related to **polarography** and, indeed, some authors use these terms interchangeably. However, the term polarography should only be used when a dropping mercury electrode is employed as the working electrode. This continuously renews the working electrode surface and ensures that the analysis is not affected by residual material from previous samples contaminating the electrode. **Amperometry** is a much simpler technique. It involves the reduction or oxidation of an electroactive molecular species at a constant applied potential, usually sufficient to generate the limiting current such that the current flowing is proportional to the concentration of the analyte in the test solution.

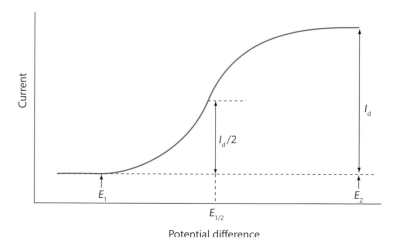

FIGURE 9.7
Typical direct current voltammogram.

Clark O_2 electrode

In 1949 Leland Clark working at the Fels Research Institute, Yellow Springs, Ohio, devised and developed an efficient bubble oxygenator for use in the oxygenation of blood during cardiac surgery. This invention necessitated the development of a device to measure the partial pressure (concentration) of oxygen (PO_2) in the blood of patients undergoing such surgery. Clark duly obliged and in 1956 described a compact oxygen probe that could measure oxygen concentrations in aqueous solutions. The Clark electrode is still widely used today. You can see an example of the device in Figure 9.8. It is an amperometric device that consists of a platinum cathode and a silver anode, both of which are immersed in a solution of saturated KCl. The two electrodes are separated from the test solution by an oxygen permeable Teflon membrane.

When a potential difference of –0.6 V is applied across the electrodes such that the platinum cathode is made negative with respect to the silver anode, electrons are generated at the anode, which reduce oxygen at the cathode.

At the silver anode (positive), silver chloride is formed with the liberation of electrons:

$$4\,Ag + 4\,Cl^- \rightarrow 4\,AgCl + 4\,e^-$$

In contrast, electrons are used at the platinum cathode (negative) to form water, consuming oxygen in the process:

$$O_2 + 4\,H^+ + 4\,e^- \rightarrow 2\,H_2O$$

Since oxygen is reduced at the cathode, the concentration of oxygen in the solution of KCl falls, which therefore acts as a sink, so that further oxygen diffuses into the solution from the sample through the oxygen permeable Teflon membrane. Since the rate of diffusion of oxygen through the membrane is the limiting step in the reduction process, the current produced by the electrode is directly proportional to the oxygen concentration in the sample. Clark electrodes generally require the sample be stirred or mixed such that gradients of PO_2 are not set up within the sample itself, leading to erroneous (lower) PO_2 values being measured.

Clark electrodes need to be maintained at a constant temperature since the solubility of oxygen in solution is temperature dependent. Similarly, atmospheric pressure also affects solubility. Published solubility tables generally quote dissolved oxygen in pure water, but salting out effects resulting from dissolved solutes may mean that the oxygen saturation values in clinical samples are below those quoted in such tables.

There are many variants of the Clark electrode. It is common to find such electrodes in combined blood gas analysers alongside both Severinghaus and pH electrodes to measure blood PO_2, PCO_2, and pH. Modified oxygen electrodes small enough to be inserted into blood vessels have been developed, but frequently this is avoided because of the dangers of infection or of forming a blood clot. Often it is considered preferable to remove small samples of blood from a warmed earlobe or a finger-tip and measure the oxygen content of the blood using a small probe-type electrode.

A type of Clark oxygen electrode commonly found in many teaching laboratories is the Rank oxygen electrode (produced by Rank Bros Ltd, UK). This is a versatile instrument and useful for performing enzyme assays in which oxygen depletion or generation can be used as measures of the rate of reaction. Glucose oxidase, amino acid oxidase, and catalase are enzymes whose properties can be studied in this way.

Following its development, it was recognized that the Clark electrode could also be used to measure hydrogen peroxide in solution simply by reversing the potential of the electrodes. If a potential difference of +0.7 V is applied across the electrodes then the platinum anode is

FIGURE 9.8
Clark oxygen electrode.

BOX 9.4 *What is the charge of an anode?*

You can see that in Section 9.2 the *anode* of a galvanic cell, such as a Daniell cell (Figure 9.3), is negatively charged, whereas in Section 9.4 the anode of a voltammetric cell is positively charged. This is not a mistake, simply the result of rather confusing inconsistency. In both cases, however, the anode is the electrode at which *oxidation* events occur and where electrons therefore become liberated. Similarly, *reduction* events always take place at the *cathode*.

In this respect, some electrochemical cells differ from the electrophoresis apparatus discussed in Chapter 14 'Electrophoresis', where the anode is *always* positively charged and the cathode *always* negative.

made positive with respect to the silver cathode and electrons are generated at the anode and used at the cathode as follows (Box 9.4). Thus, at the platinum anode (now positive), hydrogen peroxide is consumed in an electron releasing reaction:

$$H_2O_2 \rightarrow O_2 + 2H^+ + 2e^-$$

and at the now negative silver cathode, electrons are used to reduce silver ions:

$$2AgCl + 2e^- \rightarrow 2Ag + 2Cl^-$$

The flow of current is therefore proportional to the concentration of hydrogen peroxide in solution. Measuring the concentration of hydrogen peroxide is particularly useful in designing many biosensors (Section 9.5). Indeed, approximately 65% of all biosensors incorporate some form of peroxide sensitive transducer.

9.5 Biosensors

Biosensors may be defined as self-contained integrated devices that incorporate a biological recognition element combined with an electrochemical transducer capable of converting its biological or biochemical signal or response into a quantifiable electrical signal. In the majority of biosensors, the biological component is an **immobilized enzyme** that is attached to a membrane or embedded within a structure that holds it in intimate contact with the transducer. This generally allows the enzyme to be re-used and so reduces operating costs.

Much of the technological development of biosensors has been stimulated by the need to determine the concentrations of glucose in clinical samples. In 2000 the World Health Organization estimated the number of people suffering from diabetes worldwide to be 171 million, and predicted this figure would double by 2030. Thus, many companies have made significant investments in research and development programmes that have led to the production of a wide variety of biosensor devices. Within clinical laboratories, a range of high throughput laboratory analysers are available to assist in the routine analysis of blood glucose and the diagnosis of diabetes. Small hand-held devices are also readily available which enable diabetic patients to regularly monitor their own blood glucose concentrations (Figure 9.9) and so make any necessary changes in diet or influence their decision to administer insulin as appropriate. Indeed, the *availability* of such portable instruments has been a major influence in the trend towards point-of-care testing within the healthcare sector.

Cross reference

Remember, you can learn more about point of care testing in Chapter 18.

Biosensor design and operation

In 1962 Leland Clark (of oxygen electrode fame, as described in Section 9.4) first used the term 'enzyme electrode' to describe a device in which a traditional electrode could be modified to respond to other materials by the inclusion of a nearby layer of enzyme. Clark's ideas became commercial reality in 1975 with the successful launch of the Yellow Springs Instruments model 23A glucose analyser. This device incorporated glucose oxidase, which catalyses the reaction:

$$\text{Glucose} + O_2 \rightarrow \text{gluconic acid} + H_2O_2$$

The device also contained an amperometric peroxide-sensitive electrode (which we described in Section 9.4) to measure the amount of H_2O_2 produced by the reaction. Within the device, the rate of formation of H_2O_2 is a measure of the rate of reaction between glucose and oxygen, which, in turn, depends on the concentration of glucose in solution. Thus, the device can be used to estimate the concentration of glucose in samples.

In enzyme-catalysed reactions, the relationship between substrate concentration, in this case glucose, and reaction rate is not a simple linear one but is described by the **Michaelis–Menten equation**. This is true for the glucose oxidase found within a biosensor. The rate of reaction is only directly proportional to the concentration of glucose when the glucose concentration is low relative to that of the enzyme. However, at higher concentrations of glucose, the concentration of enzyme is the limiting factor, and the change in rate becomes hyperbolic. You can see the Michaelis–Menten curve in Figure 9.10. We can engineer a more linear relationship over a greater range of glucose concentrations by ensuring that the enzyme is enclosed behind or within a membrane through which the glucose must diffuse before it reacts with the enzyme. The effect of the membrane is thus to keep the glucose concentration low compared with that of the enzyme. In this way, the system is diffusionally, rather than kinetically, limited and the response of the biosensor is then more linearly related to the concentration of glucose in solution.

FIGURE 9.9
FreeStyle Lite blood glucose meter from Abbott Diabetes Care. It is about the size of a credit card and weighs only 40g. The meter features a large and easy to read LED display with a backlight for night use. The photograph shows a test strip inserted into the meter, which is discarded after use. Photograph supplied courtesy of Abbott Diabetes Care.

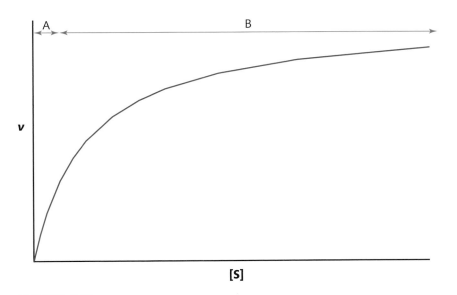

FIGURE 9.10
Representative Michaelis–Menten curve. Note where the concentration of substrate [S] is low relative to that of the enzyme (A), the rate of reaction (v) is proportional to the concentration. However, the concentration of enzyme becomes the limiting factor at higher [S], as you can see at B.

The sensing probe of a device of this type comprises a peroxide-sensitive amperometric electrode covered by a membrane comprising three layers. The first layer is made of porous polycarbonate and limits the diffusion of glucose into the second layer, which contains immobilized glucose oxidase, thereby preventing the reaction from becoming kinetically limited. The innermost, third layer is of cellulose acetate that allows only small molecules, such as hydrogen peroxide to reach the electrode. This layer eliminates many larger electroactive compounds that could interfere with the amperometric measurement.

The YSI model 23A has now been replaced by the more advanced YSI model 2300 for clinical uses and the YSI model 2700 for food analysis. As you can see in Figure 9.11, such instruments possess a carousel that enables batches of samples to be run, with each sample requiring only 100 seconds for analysis. These instruments can measure the glucose content of whole blood, plasma, or serum, and require only 25 μl of sample per analysis. The membrane-bound glucose oxidase typically needs replacing only every two or three weeks, thereby reducing the cost of analysis. These systems also offer advanced data handling and data storage facilities.

As shown in Table 9.1, these instruments can be modified to analyse a wide variety of substances of biological interest simply through the incorporation of other oxidase enzymes in the membrane. Where a single oxidase enzyme is not available for a particular analyte, analysis can be accomplished by co-immobilizing multiple enzymes that act sequentially.

Small hand-held instruments have been developed to satisfy the demands of individuals who wish to measure their own blood glucose concentration. In many of these instruments, the catalyst and transducer are more intimately linked on the sensor surface. The Medisense Exactech Glucose Meter, for example, was launched in 1986 and soon became the world's

FIGURE 9.11
Model 2300 STAT Plus glucose and lactate analyser. The figure shows the 24-position turntable, which allows batches of samples to be analysed. The buffer reservoir and waste container are both conveniently contained within the analyser. Photograph supplied courtesy of Yellow Springs Instruments Life Sciences.

TABLE 9.1 Compositions of enzyme membranes available for analysers with H_2O_2 measuring probes as transducers

Analyte	Enzyme	Reaction catalysed
Alcohol	Alcohol oxidase	Ethanol + O_2 → Acetaldehyde + H_2O_2
Glucose	Glucose oxidase	β-D-Glucose + O_2 → gluconic acid + H_2O_2
Lactic Acid	Lactate oxidase	L-lactate + O_2 → pyruvate + H_2O_2
Lactose	Galactose oxidase	Lactose + O_2 → galactose dialdehyde derivative + H_2O_2
Starch	Amyloglucosidase Glucose oxidase	Starch + H_2O → β-D-glucose β-D-glucose + O_2 → gluconic acid + H_2O_2
Sucrose	Invertase Mutarotase Glucose oxidase	Sucrose + H_2O → α-D-glucose + fructose α-D-glucose → β-D-glucose β-D-glucose + O_2 → H_2O_2 + gluconic acid + H_2O_2

best selling biosensor. The initial device was the size and shape of a pen, and used disposable electrode strips. This was followed by a credit card style meter in 1989. These devices, again, relied on glucose oxidase as the biological component, but did not measure reaction rate from the production and detection of H_2O_2. Rather, they relied on the direct measurement of the rate of electron flow from the glucose to the electrode surface. The reactions that occur within this device may be summarized as follows:

$$\text{Glucose} + \text{GOx-FAD} \rightarrow \text{gluconic acid} + \text{GOx-FADH}_2$$

$$\text{GOx-FADH}_2 + \text{Mediator}_{(oxidized)} \rightarrow \text{GOx-FAD} + \text{Mediator}_{(reduced)}$$

and at the electrode surface:

$$\text{Mediator}_{(reduced)} \rightarrow \text{Mediator}_{(oxidized)} + e^-$$

where GOx-FAD represents the FAD redox centre of glucose oxidase in its oxidized form, and GOx-FADH$_2$ the reduced form. Within the Medisense device, the **mediator** is ferrocene, although other manufacturers have produced similar devices using alternative mediator technologies. In essence, electrons are stripped from the glucose and then pass from the enzyme to the mediator, which then donates them to the working electrode surface resulting in the generation of an electrical current directly proportional to the rate of oxidation of glucose and, hence, proportional to the glucose concentration in the sample. As well as a working and reference electrode, later test strips also incorporated a third electrode, known as a background compensation electrode. This is not coated with glucose oxidase and thus measured the non-specific current generated by interfering materials present in the sample. The current generated by this electrode is subtracted from that of the working electrode to produce a response due to glucose alone. Unlike instruments of the Yellow Springs type discussed earlier, these devices do not have any requirement for oxygen. This is advantageous, as the concentration of oxygen in blood can vary, and in some cases may become limiting, such that the glucose measurement is affected.

Medisense, whose only product was their blood-glucose meter, was bought by Abbott Diagnostics in 1996 and Abbott-branded devices continued to use and develop this technology for some time. In 1999, however, Therasense marketed a glucose meter that represented the next generation of sensing technology, and integrated the enzyme even more closely with the electrode. Originally developed by Adam Heller at the University of Texas in the 1990s, **wired enzyme electrodes** did not rely on a *soluble* mediator, such as the ferrocene used in the Medisense meter.

Rather, the enzyme is immobilized in an osmium-based polyvinyl imidazole hydrogel in which the electroactive osmium groups are firmly held in place within the polymer, and cannot participate in molecular diffusion. Thus, the electrons are not passed from enzyme to electrode through a soluble mediator that moves by diffusion, but by a series of fixed electroactive osmium centres that shuttle the electrons onwards, often called *electron hopping*.

The FreeStyle Lite meter produced by Abbott Diabetes Care and shown in Figure 9.9 incorporates wired enzyme technology. Devices of this type are highly amenable to miniaturization: the Abbott device shown in Figure 9.9 is only the size of a credit card. Its size is largely limited by the dimensions required for a display that can comfortably be read by a diabetic patient, who might well have some visual impairment due to their condition. The electrochemistry is incorporated into a test strip only a few centimetres in length that is inserted into the meter as you can see in Figure 9.9. A sample of blood is obtained using a sterile lancet or similar pricking device and placed on the test strip. Only 0.3 µL blood is needed for analysis, equating to a drop the size of a pin head. Once loaded with sample, the device is able to measure the glucose concentration in approximately 5 seconds and can store the results of up to 400 analyses for comparative purposes. After a single use the test strip is discarded.

You might wonder why such glucose test strips are used once only when the enzyme is immobilized and could possibly be re-used. Re-using the enzyme would require that the user or the device itself be effectively washed between samples to ensure that results are not affected by materials remaining from previous samples. Given that these devices are developed to be used with little training by the general public and in a discrete fashion, it is clear that avoiding such washing procedures by incorporating a single use test strip has great appeal. Further, the test strip is the only part of the device that is exposed to the blood sample and this can be discarded immediately after use. Thus, it is easier to maintain hygiene and prevent possible contamination from blood borne diseases such as acquired immune deficiency syndrome (AIDS) and hepatitis. Even so, when using these meters, the test strips together with the lancets used to obtain the sample should be disposed of in accordance with relevant Infection Control and Health and Safety guidelines. These may vary slightly from region to region, but in general lancets and similar pricking devices need to be disposed of in a properly designed sharps containers (available from hospitals and pharmacists), while the test strips should be disposed of as clinical waste.

Cross reference

Health and safety are discussed in Chapter 4.

Cross reference

Measuring renal function is described in Chapter 3 of *Clinical Biochemistry*.

Single measurement devices, as described above, therefore play a key role in both the diagnosis and subsequent management of diabetes. Similar devices are available for the analysis of a number of other clinically important analytes, for example glucose and lactate. Concentrations of lactate in blood may be measured using biosensors to detect the high levels of lactate that typically accompany severe sepsis or septic shock, whilst creatinine and urea biosensors are used to monitor renal function.

Continuous measuring devices are also becoming available and may well revolutionize the control of certain disease conditions. With respect to diabetes, devices such as the FreeStyle Navigator system from Abbott use the same wired enzyme technology as described earlier, but now incorporate this technology into a tiny filament about the diameter of a thin hypodermic needle. This is inserted approximately 5 mm under the skin to measure the glucose level in the interstitial fluid that flows between the cells. The device is designed to remain in place for up to five days, during which time it can measure glucose concentration every minute. A wireless transmitter sends the glucose readings to a separate receiver anywhere within a 10-foot range, and this can then issue an early-warning alarm to alert the user to a falling or rising glucose level in time for appropriate action to be taken to avoid a hypoglycaemic or hyperglycaemic episode.

All electrochemical techniques, ranging from the earliest pH electrodes to the most advanced biosensors, share common appeal to the biomedical scientist. Electrochemical devices are often small and their portability is an advantage to those who wish to make analyses at the point of care. Such devices are also quite robust and cheap, relative to other analytical technologies, and require limited use of 'wet reagents' in their operation. The simplicity of the devices often means that they can be used with little training, and their selectivity and sensitivity are sufficient for many, although certainly not all, clinical purposes. For these reasons, electrochemical devices are frequently considered to be one of the real workhorses of the clinical laboratory.

SUMMARY

- All electrochemical techniques involve the use of two electrodes and the measurement of the potential difference between them (potentiometry) or the current that flows between them (voltammetry)

- pH, PCO_2, and ion selective electrodes are all potentiometric devices

- Potentiometry involves the generation of a spontaneous potential that is measured against that produced by a reference electrode as a potential difference

- The behaviour of these devices is described by the Nernst equation

- The minimum detection limit of most ion selective electrodes is 10^{-6} mol dm^{-3} at best

- Ion selective field effect transistors can be used to construct miniaturized solid-state potentiometric devices

- In voltammetric techniques, electrochemical reactions are not spontaneous, but forced to occur by an imposed external voltage

- These advanced techniques can be used to both identify and quantify analytes

- The oxygen electrode has widespread applications in medicine and clinical analysis

- The hydrogen peroxide electrode is used in about 65% of biosensors

- Biosensors incorporate a biological component for specific recognition, together with a transducer capable of generating an electrical output

- Biosensors are widely used in the analysis of glucose concentrations for the diagnosis and management of diabetes but also widespread applications both in the clinical laboratory and in point-of-care testing

FURTHER READING

- **Anonymous Classification and nomenclature of electroanalytical techniques (rules approved 1975).** *Pure and Applied Chemistry*, 1976; **45**: 81–97.

  A well-accepted scheme of classifying electrochemical techniques, though currently being updated by IUPAC.

- Pretsch E. **The new wave of ion-selective electrodes**, *Trends in Analytical Chemistry*, 2007; **26**: 46–61.

  An interesting review revealing the significantly enhanced sensitivity and selectivity of modern ISEs.

- Luong JHT, Male KB, Glennon JD. **Biosensor technology: technology push versus market pull**. *Biotechnology Advances*, 2008; **26**: 492–500.

  A review that describes the commercial aspects relevant to the production of and demand for biosensors. The review also describes the current commercial activities of most of the major companies producing such devices.

- Newman JD, Turner APF. **Home blood glucose biosensors: a commercial perspective**. *Biosensors and Bioelectronics*, 2005; **20**: 2435–2453.

  A thorough review of the commercial development of glucose biosensors, though missing some of the most recent developments in continuous monitoring.

- Hönes J, Muller P, Surridge N. **The technology behind glucose meters: test strips**. *Diabetes Technology & Therapeutics*, 2008; **10**: S10–S26.

  A useful article describing the technology behind a wide range of commercially available devices.

- Brett CMA, Brett AMO. *Electroanalysis*. Oxford Chemistry Primers Series, Number 64. Oxford University Press, Oxford, 1998.

  Concise and well written introductory text providing a good 'next step' in your reading around the subject. Particularly useful in its coverage of more advanced voltammetric techniques.

- Cooper J, Cass AEG. (eds) *Biosensors*, Practical Approach Series. Oxford University Press, Oxford, 2004.

  A second edition of the classic1990 text that explains how to implement diverse techniques, such as protein engineering, optical and electrochemical instrumentation, and numerical modelling in the context of producing biosensors.

Useful Website

- **http://www.freepatentsonline.com**. A free and searchable database of patents. Very useful for investigating recent developments in sensors and similar devices.

Useful Journals

- *The Analyst*
- *Biosensors and Bioelectronics*

 QUESTIONS

1. If an electrode is described as an ISE, what does the acronym ISE mean?

    (a) Ion specific electrode

    (b) Ion selective electrode

    (c) Ion sensitive electrode

(d)   Ion sensing electrode

(e)   Ion standard electrode

2. Which of the following occur within an electrochemical analytical device?

(a)   The anode always carries a negative charge

(b)   The anode always carries a positive charge

(c)   The anode is the electrode at which the oxidation reactions occur

(d)   The anode is the electrode at which the reduction reactions occur

(e)   None of the above

3. Which of the following statements are **TRUE** or **FALSE**?

(a)   All biosensors contain an enzyme able to recognize the analyte and a transducer that converts the response into an electrical signal.

(b)   In simple direct current voltammetry, the limiting current is the highest one that will flow when an analyte is being reduced at the electrode surface.

4. Given that the total potential, E (in volts) developed between the working and reference electrodes of an ISE is given by the Nernst equation:

$$E = E_{constant} + 2.303\ RT/nF \log_{10} C$$

(a) Calculate the potential developed by a pH electrode and a Na^+ ISE at 37°C. (b) Is the accuracy of analysis of these two devices likely to be the same?

5. While your main interest in electrochemistry might be for the analyses of clinical samples, such devices do have many other uses. List some other analytical situations in which these devices might be used, and consider why electrochemical techniques might be favoured in these particular situations.

6. Section 9.3 gives general information on pH electrodes. Having read this section, locate the manufacturers' instructions for the pH electrodes in your laboratory. (a) What type of electrodes are you using, for example, combination electrodes, gel electrodes, double junction reference electrodes? (b) Do they all have the same mode of operation, that is, calibration, cleaning, and storage procedures?

7. Section 9.5 describes the YSI glucose biosensor in which glucose is measured using the enzyme glucose oxidase and a peroxide sensitive electrode as the transducer. Suggest two other types of electrode transducers that could in theory be used to measure the rate of this reaction and form the basis of a glucose biosensor. Why is a device measuring peroxide preferable to either of these alternative instruments?

Answers to self-check and end of chapter questions are available in the Online Resource Centre accompanying this book.

@ **Go to www.oxfordtextbooks.co.uk/orc/glencross**

10

Radioactivity and radiation

John Gaffney and Mark Slevin

Learning objectives

After studying this chapter, you should be able to:

■ Describe radioactivity in terms of atomic structure

■ Discriminate between the main types of radiation

■ Demonstrate how the rate of decay of radionuclides can be calculated

■ Describe the methods available for the measurement of radioactivity

■ Explain the safety issues involved when handling and disposing of radionuclides

■ Identify uses of radioisotopes in biomedical science

Introduction

Radioactivity is the spontaneous emission of particles and/or rays or waves of energy from the unstable nuclei of some atomic species. Such atoms are called **radioisotopes**, and as they decay they form more stable atoms. The rays or waves, and the particles are called **radiation**. There are three forms of radiation: alpha- (α-) and beta- (β-) particles, and gamma (γ-) radiation. It is possible to detect and measure all three types of radiation.

Radiation is potentially harmful and so precautions must to be observed when using and handling radioisotopes. However, it is also of use in biological and medical research and in the diagnosis and treatment of some types of clinical conditions.

10.1 Atomic structure

The atomic nucleus contains most of the mass of the atom in the form of protons (p), and neutrons (n). The mass of a proton is 1.67×10^{-27} kg. This is approximately the same as that of the neutron; however, they differ in that protons are positively charged while neutrons are electrically neutral. The nucleus or **nuclide** can be thought of as being surrounded by a cloud of negatively charged electrons. The number of electrons is the same as that of the protons so an atom is electrically neutral. Electrons are responsible for the chemical properties of the element. They occur only at specific energy levels called **orbitals**, which increase in energy with the distance of the orbital from the nucleus. Electrons can move between different orbitals if they absorb or emit energy. The lowest energy state and therefore the most stable for an electron in an atom is called the **ground state**. If an electron gains energy by interacting with radiation, the electron may be excited or energized sufficiently to move to a higher energy orbital; from there it can fall back to its original orbital with the emission of light or heat.

Nuclides can be represented in the abbreviated form, $_Z^A X$, where X represents the symbol of the chemical element, A is the **mass number**, which is numerically equal to the sum of the numbers of n and p, and Z is the **atomic number**, that is the number of p in the nucleus. For example, $_6^{12}C$ describes a carbon nucleus that contains 6p and 6n and therefore has an atomic mass of 12 units. Although this form of carbon is generally written as ^{12}C, it is pronounced C-12. The atomic number is a characteristic value for each element.

> **Cross reference**
>
> You can read more about the interactions of electrons and radiation in Chapter 11 'Spectroscopy'.

SELF-CHECK 10.1

Draw a table listing the atomic mass, and the number of protons, electrons, and neutrons in the following elements: ^1H, $_3^7Li$, $_{30}^{65}Zn$, $_{46}^{106}Pd$.

Isotopes

In any given nuclide, the number of protons is the same as the number of electrons. However, although all the atoms of an element have the same atomic number, the number of neutrons, and therefore the mass number, can vary forming **isotopes**. For example, carbon has isotopes with mass numbers ^{10}C, ^{11}C, ^{12}C, ^{13}C, ^{14}C, and ^{15}C. These contain between four and nine neutrons, but all have six protons. Hydrogen occurs as one of three isotopes: hydrogen (^1H), deuterium (^2H), and tritium (^3H); all contain a single proton, but 0, 1, or 2 neutrons, respectively. Approximately one-third of the known elements exist as different isotopes. Many nuclides are unstable and with time will decay to a more stable form with the emission of energy, that is the nuclides are radioactive. However, isotopic species have variable stabilities, but not all are radioactive.

SELF-CHECK 10.2

Naturally occurring chlorine (Cl) is a mixture of 75% ^{35}Cl and 25% ^{37}Cl. What would be the observed atomic weight of chlorine?

Nuclear stability

Two forces operate in the nucleus: the short-range, attractive strong nuclear force between neutrons and between protons, and Coulomb (electrical) repulsive forces between the positively charged protons. The protons and neutrons are held in a stable state in the nucleus by the binding energy of the strong nuclear force. Thus, the stability of the nucleus is a balance between these attractive and repulsive forces and depends on the numbers of protons and neutrons. The most stable nuclei, those of low atomic number, have equal numbers of these particles. A straight line relationship occurs between the numbers of protons and that of neutrons up to atomic number 26 (that of iron (Fe)); which is the most stable element. You can see this in Figure 10.1.

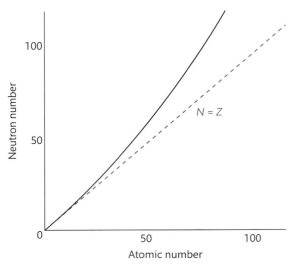

FIGURE 10.1
A graph of the atomic number (*Z*) plotted against the numbers of neutrons (*N*) for the elements of the periodic table. The hypothetical curve where *N* equals *Z* is a straight line (red), but naturally occurring elements deviate from this (blue) because many have more neutrons than protons.

An unstable nucleus with an excess of protons or neutrons, emits radiation until a stable ratio occurs. However, elements of atomic number greater than 30 require an excess of neutrons for stability. Indeed, all nuclei that contain 84 (that of polonium) or more protons are unstable and decay. Although many isotopes are found naturally, others, especially ones used in medical procedures, are produced artificially by bombarding lighter nuclei with radiation in nuclear reactors or in cyclotrons.

10.2 **Types of radioactive decay**

Radiation is energy travelling in the form of electromagnetic waves or as a stream of particles.

Radioactive decay involves the disintegration of atomic nuclei. The **radiation** released by unstable nuclides until they reach a stable state can be particulate (small particles) or photons of electromagnetic radiation called:

- Alpha (α)-particles
- Beta (β)-particles and
- Gamma (γ) rays

Although lighter elements may decay and emit radiation in a single step, heavier ones, with atomic numbers greater than 70, undergo a series of changes before reaching a stable state and release α-particles in the process.

α **Radiation**

An α particle is a helium nucleus consisting of two protons and two neutrons ($^4_2\mathrm{He}^{2+}$). Thus, the release of an α-particle (α-decay) decreases the atomic number and the mass of the nucleus

TABLE 10.1 Summary of the types of radiation emitted by unstable nuclei

Name	Symbol	Nature	Range of energies/MeV*	Range of penetration in air/m	Protective shielding
Alpha particle	$^4_2\alpha$	Helium ion	4–8	0.025–0.080	Not needed
Electron	$^0_{+1}\beta^-$	Energetic and fast moving electrons	0.01–3	0.15–6	Plastic
Positron	$^0_{-1}\beta^+$	Energetic positively charged electrons	0.01–3	0.15–6	Plastic
Gamma ray	γ	Electromagnetic radiation	0.03–3	1.3–13	Lead

*1 MeV = 1.6×10^{-13} J.

by 2 and 4 units, respectively. For example, ^{226}radium decays to ^{222}radon by α-emission; however, the latter is also radioactive and can decay into further products:

$$^{226}_{88}\text{Ra} \rightarrow {}^{222}_{86}\text{Rn} + {}^4_2\text{He}^+ \rightarrow {}^{216}_{84}\text{Po} + {}^4_2\text{He}^+ \rightarrow {}^{212}_{82}\text{Pb} \rightarrow \rightarrow {}^{208}_{82}\text{Pb}$$

SELF-CHECK 10.3

Uranium (U) decays to thorium (Th) by α-emission. Insert the appropriate numerical values to complete the following equation.

$$^{238}_{?}\text{U} \rightarrow {}^{?}_{90}\text{Th} + {}^4_2\text{He}$$

Decay by α-particle emission does not have many applications in biology. α-Particles have limited penetration in air (Table 10.1) but they are toxic if the radionuclide is ingested due to their high energy. For example, the presence of ^{226}Ra at internal sites in the body may cause bone cancer.

β Radiation

Nuclei that are unstable due to an excessive number of neutrons may lose energy when a neutron changes to a proton with the emission of a β-particle.

$$^1_0\text{n} \rightarrow {}^1_1\text{p} + {}^0_{-1}\beta^-$$

The atomic number increases by 1 but the atomic mass does not change. A *positron* is emitted if a nucleus has more protons than neutrons. Thus, when a proton changes to a neutron there will be a decrease in atomic number of 1.

$$^1_1\text{p} \rightarrow {}^1_0\text{n} + {}^0_{+1}\beta^+$$

A negatively charged β-particle (β⁻) is an electron and a positively charged β-particle (β⁺) is a positron. These particles have the same mass, but opposite charges.

In these reactions, the sums of the subscripts and charges are balanced. For instance, the carbon isotope, ^{14}C decays by β⁻ emission to ^{13}N:

$$^{14}_6\text{C} \rightarrow {}^{14}_7\text{N} + {}^0_{-1}\beta^-$$

while ^{33}Cl can decay to ^{34}S by positron emission:

$$^{33}_{17}\text{Cl} \rightarrow {}^{33}_{16}\text{S} + {}^0_{+1}\beta^+$$

The general equation for β-decay of a neutron-rich isotope is:

$$^A_Z X \rightarrow\ ^A_? Y + \ ^0_{-1}e$$

Insert the appropriate symbol to replace the question mark(?).

γ Rays

γ Rays are electromagnetic energy released in association with the rearrangements that occur in a nucleus after it has emitted an α- or β-particle. γ-Rays, and X-rays, are emitted if electrons that overlap with a nucleus combine with a proton to form a neutron. When this happens, energy in the form of γ-and X-rays (see Box 10.1) can be emitted. Again, this will decrease the atomic number of the nuclide by 1, but leave the atomic mass unchanged. You can see an example of this in the way an isotope of iodine, ^{125}I, decays:

$$^{125}_{53}I \rightarrow\ ^{125}_{52}Te + \text{X-rays} + \gamma$$

The newly formed element, tellurium (Te) is deficient in an electron. However, a rearrangement of its outer orbital electrons is able to fill the vacancy and occurs with the accompanying emission of a photon of γ- or X-rays.

BOX 10.1 *Discovery of radiation and radioactivity*

Wilhelm Röntgen discovered radiation in 1895 while he experimented with the discharge of electricity in a vacuum. He used a glass tube, which contained two metal plates connected to a battery. When he evacuated the tube and switched on the electric flow, the negative charged plate or cathode glowed. Röntgen later showed that the glow was due to the emission of energy produced when electrons in the plate jumped to a higher orbital and fell back to their original level. Some of the electrons can be dislodged from the plate and interact with atoms in the glass wall. In interacting, the electrons slow down and emit radiation which Röntgen called X-rays because their nature was unknown.

Henri Becquerel discovered radioactivity in 1895 shortly after that of X-rays. He sprinkled a uranium salt onto a photographic plate, which he kept in the dark. When the plate was developed, he noticed dark areas on the plate that corresponded to the position of the uranium. Initially he thought this was due to X-rays interacting with silver atoms on the photographic plate. Later, however, it was found that the 'Becquerel rays' produced from uranium salts were different to X-rays as they ionized gases, which could be used to measure the activity of the source. In 1897, Pierre and Marie Curie repeated Becquerel's experiments and showed the intensity of the energy released was proportional to the amount of uranium. They called this energy radioactivity. In attempting to isolate a pure emitter of radioactivity from uranium and thorium ores, they discovered a new element, polonium that had a high activity.

Ernest Rutherford and Martha Thompson working in Manchester and Cambridge investigated the phenomena of radiation in more detail and recognized two types of radiation emitted from uranium. These were called alpha (α) and beta (β) radiation respectively.

γ-Radiation consists of high energy, short wavelength electromagnetic photons, which are emitted during α and β decay. It has great penetration, for example, it will pass through 15 cm of steel, but its cytotoxicity is about the same as that of X-rays. X-rays are penetrating electromagnetic radiation emitted when the inner orbital electrons of an atom are excited and release energy in the same energy range as γ-rays (0.010–10 meV), but it is of non-nuclear origin,

Table 10.1 summarizes some of the major features of these types of α, β, and γ radiations.

10.3 **Rate of radioactive decay**

The rate of radioactive decay is a measure of how fast the conversion of an unstable radioisotope decaying to a stable form with the emission of energy occurs. Each radionuclide has a different but characteristic rate of decay. The rate is unusual in that it is not affected by chemical or physical parameters such as concentration, temperature, or pressure. If a graph of the number of atoms (N) present at set time intervals (t) is drawn, it gives an exponential curve as you can see in Figure 10.2.

A simple relationship can be derived from this curve, that is $-dN/dt$ is proportional to **N** or:

$$-dN/dt = \lambda N$$

The term, $-dN/dt$ is the rate of decay for the nuclide, and is commonly called the **activity**, while λ is the decay constant, which differs for each radionuclide. The relationship between these terms given above can be converted to a second equation:

$$\ln N_1/N_o = -\lambda t$$

The symbol 'ln' represents the natural logarithm, N_o and N_1 are the number of radioactive atoms present initially and after time t, respectively. Natural logarithms can be converted to logarithms to the base 10 by multiplying by 2.303. It is usual to calculate and quote the rate of decay for a radioisotope as its **half-life** ($t_{1/2}$), which is the time taken for half of the atoms in a sample to decay. Manipulating the decay equation gives the following relationship:

$$t_{1/2} = 0.693/\lambda$$

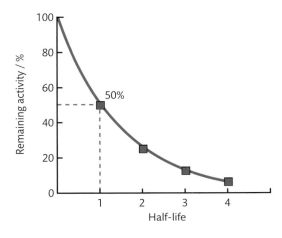

FIGURE 10.2
A graph of half-life of an isotope against the amount of radioactivity shows that radioactive decay follows an exponential curve. At each half-life the amount of radioactivity has decreased by half.

You can see in Figure 10.2, that after one half-life, only 50% of the original atoms remain, after two half-lives, 25% remain and so on.

The half-lives of common elements used in biology and medical science vary from hours to years as you can see in the examples given in Table 10.2. Unfortunately, the half-lives of the biologically important elements ^{15}O and ^{13}N are 2.03 and 10.0 minutes, respectively. These are so short, that the isotopes are of little practical value. Instead, non-radioactive isotopes, for example ^{15}N, that can be detected and quantified by mass spectrometry are used.

Cross reference
You can read about mass spectrometry in the following chapter 'Spectroscopy'.

SELF-CHECK 10.5

The half-life of ^{32}P is 14.2 days. Use the appropriate equation to calculate the corresponding value of λ.

10.4 Units associated with radioactivity

A number of units are associated with radioactivity. These include the:

- Curie
- Becquerel
- electron volt

A **Curie** (Ci) is the amount of a radionuclide that gives 3.7×10^{10} disintegrations per second (dps). This pre SI unit has been superseded by an SI unit called the **Becquerel** (Bq), which is defined as 1 disintegration per second (dps) although disintegrations per minute are also used. The Bq is such a small unit that only multiples of it are commonly used.

TABLE 10.2 Half-lives of radionuclides commonly used in biology

Isotope	Half-life	Type of decay	Examples of uses
^3H	12.26 years	β^-	Suitable for radiolabelling organic molecules with high specific activity for metabolic and enzyme studies
^{14}C	5760 years	β^-	Suitable for labelling organic molecules at different positions in the molecule for investigating metabolism
^{22}Na	2.58 years		Transport studies
^{32}P	14.3 days	β^-	Labelling proteins and nucleotides (see Chapter 16)
^{35}S	87.2 days	β^-	Labelling proteins (see Chapter 16)
^{51}Cr	27.7 days	Electron capture	Haematological investigations (see the companion textbook, *Haematology*)
^{59}Fe	60 days	β^-	Haematological investigations (see the companion textbook, *Haematology*)
^{99m}Tc	6.0 hours	Isomeric transition	Diagnosis of disease (see the companion textbook, *Biology of Disease*)
^{125}I	60 days	Electron capture	Labelling proteins (see Chapter 16)
^{131}I	8.04 days	β^-	Thyroid function (see the companion textbook, *Clinical Biochemistry*)

SELF-CHECK 10.6

Convert 1 µCi into the equivalent number of kBq.

The **specific activity** of a radioisotope is the amount of radioactivity per unit mass, which can be expressed in cps or dpm per µmol of the substance. However, other combinations of units are possible:

Specific activity = radioactivity (Bq, cpm, dpm etc)/amount of substance (mol, g, etc)

In experiments using radioisotopes, the radiolabelled compound is often diluted with its unlabelled form, called a carrier, before use. This changes the specific activity of the compound. However, there are a number of advantages to using compounds with a high specific activity, including:

• Only a relatively small amount of labelled compound need be used

• Interference in metabolic pathways is minimized

• Normal physiological conditions can be maintained

• The products of metabolic reactions can be isolated at high specific activity

SELF-CHECK 10.7

A sample of [^{14}C]-glucose (M_r of glucose is 180) was diluted with unlabelled glucose to give a concentration of 2 µg cm^{-3} and activity of 1 µCi. Calculate the specific activity of the preparation in µCi/mol.

The electron volt (eV) is the energy gained by an electron in accelerating through a potential difference of one volt. Electron volts are used when describing the energy released during radioactive decay. It is, however, an extremely small amount of energy, only 1.6×10^{-19} J, and for practical purposes keV (10^3 eV) or MeV (10^6 eV) are used.

We have summarized these units in Table 10.3.

10.5 **Interaction of radiation with matter**

Radiation interacts with matter in ways that are dependent on the energy of the radiation. Kinetic energy of motion is contained in α- and β-particles, while γ rays contain energy in the form of electromagnetic radiation. All α-particles and γ-radiation from a particular nuclide are

TABLE 10.3 **Terms and units used in describing radionuclides**

Unit	Abbreviation or symbol	Definition of unit
Counts per minute (or second)	cpm or cps	Measured rate of decay
Disintegrations per minute (or second)	dpm or dps	Real rate of decay
Curie (µCi)	Ci	3.7×10^{10} dps (Ci $\times 10^{-6}$)
Becquerel	Bq	1 dps
Electron volt	eV	Energy equal to 1.6×10^{-19} J

Cross reference

You can read more about electromagnetic radiation in Chapter 11 'Spectroscopy'.

of the same energy, and are in the range of 2–9 MeV and keV to MeV, respectively. The energy of β-particles from a given nuclide is more variable, with a range up to a maximum value, E_{max}. Their *mean* energy is 1/3 E_{max}. β-Particles vary in their energies because the amount of energy available is shared between the β-particle and a second particle, either a neutrino or an antineutrino. If decay is by β+ emission, a neutrino is also released. Decay by β- emission is associated with the release of an antineutrino. In both cases, the energy is distributed between the β-particle and the neutrino or antineutrino in a random manner. Hence, the energy possessed by a β-particle differs at each disintegration.

Excitation and ionization

When radiation interacts with matter, it can lead to excitation and ionization of electrons. If the electron gains sufficient energy, it will be lost from the atom or molecule: a process called ionization. Ionization results in the formation of an **ion pair**, consisting of a positively charged ion and a free electron. Particulate radiation, α- and β-particles, cause excitation and ionization because they directly interact with electrons. Depending on the energy of the particles, they may interact with many atoms before losing energy. For example, the particles lose about 35 eV of energy for every ionization, therefore an α-particle of 3.5 MeV will produce about 10^5 ion pairs. The effect of γ-radiation depends on its energy. At low energy, it may be unable to remove an electron and so expends its energy as heat. At higher energies, electrons will be excited or removed.

SELF-CHECK 10.8

Suggest what may happen to the free electron generated by ionization.

10.6 Detection and measurement of radioactivity

Detecting and measuring the amounts of radioactivity relies on its abilities to excite electrons and ionize atoms and molecules as described in Section 10.5. Sources of α-particles are rarely used in biomedical science, and methods to detect β-particles and γ rays are emphasized. Three methods are available to do this, based on the ionization of a gas (gaseous detection), or of a liquid (liquid scintillation counting), or the interaction of the radiation with a photographic emulsion (autoradiography).

Gaseous detectors

Gaseous detectors consist of a gas-filled chamber that contains two electrodes across which a voltage is applied. When ionizing radiation interacts with the gas it generates ion pairs. The electrons move to the positive electrode and the positive ions to the negative plate generating a current. The magnitude of the current is proportional to the numbers of ion pairs, which, in turn, is proportional to intensity of the source of radiation. A commonly used gaseous detector is the Geiger counter, whose typical appearance can be seen in Figure 10.3. The Geiger counter can detect about 99% of the β-particles from a source, but cannot detect γ-radiation.

Meter

Cathode surface

Glass window

Anode

Wire mesh

Glass bead

FIGURE 10.3
A Geiger–Muller detector.

Scintillation counters

In a scintillation counter, the radiation interacts with chemicals called **fluors**, which give off fluo-rescent energy when they interact with radiation. If the energy of an electron in its ground state is increased to an excited, higher energy level, then the subsequent decay back to the ground state is associated with the emission of light. This phenomenon is called fluorescence. Hence, the interaction of the radiation with the fluor can be detected as flashes of fluorescent light.

Liquid scintillation counting is used to detect isotopes that decay by β emission and can be used to detect the isotopes commonly used in biomedical science, such as those of carbon, phosphorous, and hydrogen. The sample is dissolved in a liquid containing the fluors called the **scintillant**. The flashes of light produced by interactions of the β-particles with the fluor are detected by photomultiplier tubes as you can see in outline in Figure 10.4. The photomul-tiplier converts the flashes of light into an electronic signal, whose magnitude is proportional to the number of β-particles detected. Scintillation counters can separate the flashes of light produced according to the energy spectrum of the original radionuclide. This allows the simul-taneous counting of two different isotopes in a single sample. For example, many biological compounds can be labelled using both ^{14}C and ^3H for use in biomedical research.

Solid scintillation counters are also available, but are only used for penetrating radiation, for exam-ple, isotopes of iodine and involve a crystal of fluorescent material held close to the sample.

A drawback to liquid scintillation counting is **quenching**, where the flash of light produced by the fluor is absorbed by a contaminant before it can be detected. This leads to undercount-ing of the amount of radionuclide in the sample. This problem is encountered in biomedical science research since organic solvents, such as alcohols or chloroform, which are efficient quenchers, are often used to extract biological compounds from tissues. Liquid scintillation counters have quench correction facilities that use standards of known activity, thus their counting allows the quenching of the sample to be estimated and corrected for.

Autoradiography

Radioactively labelled compounds separated by, for example thin layer chromatography or gel electrophoresis, can be detected by exposing the paper or gel to a photographic film.

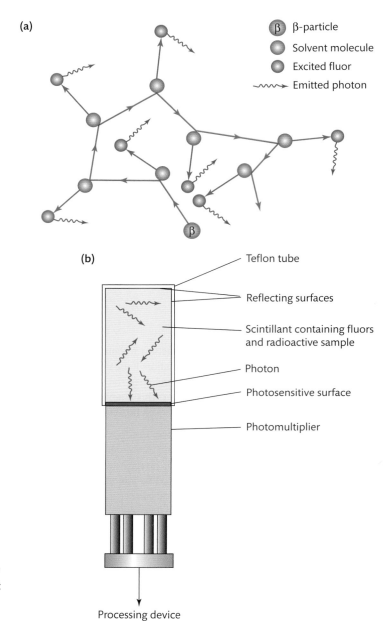

FIGURE 10.4

(a) The use of fluors to detect radioactive decay by stimulating the production of flashes of light (photons). (b) The detection of radiation in a scintillation counter.

The radiation released by the radionuclide interacts with silver grains in the film in a similar manner to that of light during photography, reducing silver halide crystals to metallic silver. Thus, an image showing the distribution of radioactivity in the paper or gel is produced. Once visible on the photographic plate, the corresponding materials in the paper or gel must then be identified. A disadvantage of autoradiography is the sample must contain enough energy to be able to interact with the film; if the sample is of low specific activity it may require weeks or months to produce a detectable image.

Autoradiography can also be performed on whole animal sections to investigate the distribution of a radiolabelled drug in the organism. Following exposure, the film will show an image of the section where the radioactivity is concentrated in tissues and organs. Generally, only weak β-emitting isotopes, such as ^{3}H, ^{14}C, ^{35}S, are used in localizing materials in tissues and cells using autoradiography. Their low energy and therefore limited penetration means they produce a sharp image on the photographic films.

Cross reference
You can read more about autoradiography in Box 13.1.

Limitations in detecting radiation

Whatever the type of detector, it will not be 100% efficient because not all disintegrations of a radioactive nuclide are detectable. In addition, some of the emitted radiation may be absorbed by contaminants, such as organic chemicals like chloroform or acids, and so are missed by the counter. Thus, only a proportion of the disintegrations can be detected and these will be recorded as counts per second (cps) or counts per minute (cpm); often the efficiency of liquid scintillation counters is only 50 to 70%. However, if the efficiency of the counting process can be determined, it is possible to convert the recorded cpm to disintegrations per minute (dpm). For example, if a counter detects 50% of the known disintegrations in a standard, then the counts obtained for the samples will have to be doubled.

SELF-CHECK 10.9

A scintillation counter has a counting efficiency of 82% for ^{14}C and gave a reading of 45 000 cpm for a sample. Calculate the true value in dpm.

10.7 Safety precautions when using radioisotopes

Radiation absorbed by tissues causes ionization and the formation of free radicals leading to cell damage, mutations in DNA and may possibly lead to cancer. However, there is no simple relationship between the amount of radiation emitted by a source and the degree of cell damage. The energy of the radiation, the amount of it absorbed by the body and the susceptibility of some tissues to radiation damages all need to be taken into account.

Absorbed dose

The absorbed dose is the energy deposited in tissues when they interact with radiation. The unit of absorbed dose is the **gray** (Gy), which is the absorption of 1.0 J of energy per kilogram of tissue. The pre-SI unit of absorbed dose was the **rad** (from *radiation absorbed dose*), which is the dose of radiation that gives an energy absorption of 1×10^{-2} J kg^{-1}. A single Gy is equal to 100 rad. However, the damage caused by identical absorbed doses varies with the type of radiation. Thus, to assess different radiations it is necessary to compare how much ionization they produce when air is exposed to their ionizing effects. Some measuring devices still recorded exposure in the pre-SI unit, the **roentgen** (R), which is equal to the quantity of ionizing radiation that will produce one electrostatic unit of electricity in one cubic centimetre of dry air at 0°C and standard atmospheric pressure. This is equivalent to the amount of radionuclide that produces 1.61×10^{15} ion pairs kg^{-1}. The SI unit of exposure is the **Coulomb kg^{-1}** (C kg^{-1}), which again is the electric charge released when radiation comes into contact with a set volume of air. A coulomb is a unit of electrical charge equal to the charge transferred by a current of 1 ampere s^{-1}. One R is equal to 258 C kg^{-1}.

Effective dose

Additional factors need to be considered when assessing damage in biological tissues. For example, the **linear energy transfer** (LET) is the rate of ionization along the path the particle takes through the tissue. The biological damage caused by α-particles is greater than a β-particle or photon of γ-radiation of the same energies because of the relative efficiencies with which energy is transferred to the tissue. The damage caused is related to the **relative biological effectiveness** (RBE) of the radiation, which is defined as:

$$RBE = \text{dose of X-rays that produces a given effect/dose of radiation required to produce same effect}$$

Typically, RBE values are 1 for high energy β-particles (emitted from the nuclides commonly used in biological experiments), 2 for low energy β-particles and 20 for α-particles. The **effective dose of radiation** is the absorbed dose $\times$ RBE and is measured in **Roentgen equivalent man** (rem) or **Sieverts** (Sv). The rem is the amount of radiation that gives a dose in humans equivalent to that of 1 rad of X-rays. It is not an SI unit and has been replaced by the Sv, which is the amount of radiation that gives a dose in humans equivalent to 1 Gy of X-rays.

See Health and Safety Box 'Health & safety in relation to radiation' for safety precautions.

Cross reference
You can read more about health and safety in Chapter 4.

HEALTH & SAFETY

In relation to radiation

Safety precautions

In the United Kingdom, the Ionizing Radiations Regulations 1999 (IRR99) established the regulatory framework for exposure to ionizing radiations and sets an annual limit of exposure of 20 mSv. Some organs of the body are more susceptible to radiation and, therefore, total organ dose equivalents are used. For example, the lens of the eye is subject to a low limit of 6 mSv.

Handling of radioisotopes

Radioisotopes must normally be used only if an alternative is unavailable and the minimum quantity should be used. A laboratory will establish local rules for the safe handling of radioisotopes. A radiation safety advisor will be responsible for overseeing these. To limit exposure, contact with radiation should be kept to a minimum.

The intensity of radiation decreases with the distance from its source. Thus, any technique that maintains distance should be used, for example using tongs to hold samples. Shields will also reduce exposure. Perspex will reduce exposure to β- and γ-emitters. Lead shields may be required for more penetrating types of radiation.

To limit internal exposure, the entry of the radionuclide into the body should be prevented by protecting the skin, and preventing its swallowing and inhalation. The normal laboratory practices of wearing gloves and a laboratory coat must always be followed. A photographic film badge needs to be worn and developed regularly at set time intervals to assess exposure. All work involving radioisotopes must be carried out in a fume hood. If a radioactive source enters through the skin it may pass directly into the bloodstream. If it is ingested its fate will depend on the route of entry and the chemistry of the particular radionuclide. If inhaled the nuclide may be swallowed, trapped in the trachea, or enter the lungs and subsequently the blood stream. Once in the body, many radionuclides concentrate in one organ, the critical organ, for example calcium and phosphorus isotopes accumulate in bone, others, such as ^3H and ^{22}Na distribute evenly in body tissues.

10.8 Clinical applications of radioactivity

In general, there are few routine applications of radioisotopes in the medical laboratory. There are, however, many applications in nuclear medicine, a term used to describe the use of radioactive compounds to perform diagnostic imaging or tests and in treating some diseases. Some of the radionuclides and their uses in nuclear medicine are listed in Table 10.4 (see Box 10.2 for more on radiopharmaceuticals).

TABLE 10.4 Radionuclides and their applications in nuclear medicine

Radionuclide	Clinical application
^{11}C	Brain imaging
^{18}F	Brain imaging when attached to glucose
^{133}Xe	Cerebral blood imaging
^{67}Ga	Tumour imaging
^{201}Tl	Cardiac muscle
^{123}I	Study of thyroid function
^{82}Rb	Myocardial perfusion

Radioisotope imaging

The extent of penetration of α- and β-particles in tissues is small making them unsuitable for imaging. However γ rays with energies greater than 50 keV are ideal for use in a number of clinical applications. A range of radioisotope imaging techniques that rely on detecting

BOX 10.2 Radiopharmaceuticals

Radiopharmaceuticals are radionuclide-labelled agents that distribute in the body. They provide information on physiological and metabolic functioning of organs in the body. This is in contrast to other imaging techniques, which provide anatomical information. The distribution of radiopharmaceuticals in the body is determined by the method of administration, the circulation, and the metabolism in the organ. Imaging is based on detection of photons in the form of X- or γ rays emitted from the source. An image is constructed from the detected photons by using a collimator, which only accepts photons from a narrowly defined region of the body.

The preferred characteristics of a radiopharmaceutical are:

- It should have a half-life as close as possible to the time of the test to avoid excess exposure and to be eliminated from the body within the half-life

- The nuclide should be a γ-emitter with an energy in the range of 50–300 keV.

- It should accumulate only at the site of interest

- It should be simple to prepare and incorporation of radioactivity should not alter the biological properties of the radiopharmaceutical γ rays are available. Calcium does not have a radioactive isotope suitable for bone scans but a radioactive isotope of technetium (^{99}Tc), is a suitable replacement. If 740 MBq (20 mCi) of ^{99}Tc-methylene diphosphanate is administered intravenously it diffuses into the extracellular fluid surrounding the bone and becomes fixed into the solid phase of the bone by exchanging with Ca^{2+}. The sites of incorporation of the isotope in the skeleton can be detected using a scanning gamma camera, which detects γ rays emitted by the nuclide and builds up a three-dimensional image of the targeted organ following repeated scans of the area. The concentration of ^{99}Tc is highest in those areas of increased bone activity and blood flow, at, for example the sites of bone fractures (Figure 10.5) or cancer metastases, providing a sensitive method of identifying such regions.

Myocardial perfusion scanning is a technique for visualizing blood vessels in the wall of the heart. It is applied to patients with angina (chest pain as a result of exercise) or with unexplained chest pain that may be due to partial blockage of arteries or to visualize partial blockages in an artery caused by plaques. Damage caused by a myocardial infarction (a heart attack) may also be evaluated by the technique. Again, commercial preparations of ^{99}Tc or thallium are injected into a vein of the arm or hand. Scanning of the patient with a gamma camera shows the areas where the radionuclide has accumulated because damaged areas have a decreased flow of blood and do not absorb as much of the isotope. Thus, a stress myocardial perfusion scan assesses the flow of blood to the heart muscle or myocardium when it is subjected to stress by exercise or medication. The technique can identify those areas of the myocardium that have decreased blood flow using a gamma camera to take pictures of the heart.

Positron emission tomography (PET) utilizes nuclides that decay by positron emission. The positron collides with an electron producing two photons of γ rays, which travel collinearly in opposite directions. A gamma camera, which contains a scintillator connected to photomultiplier tubes, is used to detect the photons. The γ rays are focused and the data collected can be used to construct a three-dimensional image of the tissue. However, the mechanisms by which this is done are complex and need not be discussed here. Positron emission tomography is particularly useful for producing images of the brain. Glucose is the major energy source for the brain where it is degraded by glycolysis. However, the modified form of glucose, 2-deoxyglucose is retained in the brain. If this compound is labelled with a γ-emitter, such as ^{18}F-2-deoxyglucose or ^{11}C-2-deoxyglucose, then areas of metabolic activity in the brain can be visualized, as you can see for the relative activities in a normal brain and that of a sufferer of Alzheimer's disease in Figure 10.6 (a, b).

SELF-CHECK 10.10

Are there any potential hazards in radioisotope imaging?

FIGURE 10.5
The presence of two fractures in the pelvis detected using radioimaging with ^{99}Tc.

(a)

(b)

FIGURE 10.6
Positron emission topography (PET) images showing metabolic activity in the brains of (a) a healthy person and (b) a patient with Alzheimer's disease. Courtesy of the Alzheimer's Disease Education and Referral Centre, National Institute on Aging, USA.

Applications in haematology

Radioisotopes have haematological roles, in particular for determining blood volume and erythrocyte survival times.

The disorder polycythaemia vera (also called polycythaemia rubra vera) is a myeloproliferative disorder in which the abnormal bone marrow produces an excess of erythrocytes. As a result, the blood is more viscous than normal and has haemocrit values (the proportion of blood occupied by erthrocytes) greater than 54% (reference range 40.7–50.3%) in men and 49% in women (reference range 36.1–44.3%). In some patients, the leukocyte and platelets may also be increased in number and morphologically abnormal. However, the spleen may also be abnormally enlarged, leading to an apparent increase in blood volume and making the haemocrit value falsely normal. Thus, a diagnosis of polycythaemia vera requires a definitive demonstration of an increase in the mass of erythrocytes. This can be measured by labelling the erythrocytes with ^{51}Cr. The assay involves withdrawing a blood sample (20–22 cm^3) from the patient and mixing it with ^{51}Cr-sodium chromate of known radioactivity. The sample is then kept at room temperature for 35 minutes to equilibrate. The blood is separated by centrifugation, the plasma discarded, and the erythrocytes washed with isotonic saline before being suspended in 12 cm^3 of isotonic saline. A sample of the blood is retained as a standard and the labelled blood cells injected back into the patient. After 10 minutes, a new blood sample is withdrawn and the packed cell volume (PCV) determined. The erythrocytes are then lysed and the released radiation measured. The erythrocyte volume can be calculated as the total cpm injected divided by the cpm per cm^3 of erythrocytes. Erythrocyte masses greater than 36 cm^3 kg^{-1} in men (reference range 28.3 ± 2.8 cm^3 kg^{-1}) and 32 cm^3 kg^{-1} in women (reference range 25.4 ± 2.6 cm^3 kg^{-1}) are abnormal.

Erythrocyte survival studies are used to assess the presence and severity of haemolytic anaemia. Haemolysis may occur by one of two mechanisms: intravascular or extravascular. Intravascular mechanisms include complement fixation, trauma, or other extrinsic factors. It is associated with conditions such as glucose 6-phosphate dehydrogenase deficiency, thrombotic thrombocytopenic purpura, disseminated intravascular coagulation, and paroxysmal nocturnal haemoglobinuria. Extravascular mechanisms are more common and here erythrocytes are removed from the circulation by the mononuclear phagocytic system because they are intrinsically defective or because of the presence of bound immunoglobulins on their surfaces.

Cross reference
You can read more about haematology in the companion textbook, *Haematology*.

Normally, erythrocytes circulate for about 120 days. This is much shorter in haemolytic anae-mia and bone marrow activities cannot compensate for their increased loss. Erythrocyte or red cell survival is tested by labelling the cells with ^{51}Cr and determining the blood volume as described above for polycythaemia vera using 10 cm^3 samples of blood removed at 24- and 48-hour intervals for up to 14 days.

Radioimmunoassays

Radioimmunoassay (RIA) was a commonly used technique for quantifying a number of clini-cally important biochemicals, such as hormones, drugs, and immunoglobulins, although its use is now much more restricted. Any material that can act as an antigen and stimulate the production of a complementary antibody can be detected using a suitable RIA. Thus, RIAs combine specificity, due to the recognition of the antigen by the antibody, with the sensitivity of radioisotope techniques. The technique is based on preparing a radiolabelled form of the material to be measured, which is used as an antigen. For example, tyrosine residues in protein antigens can be labelled with ^{125}I. Non-protein antigens can be labelled using ^{3}H.

You can see an outline of how RIAs are performed in Figure 10.7 (a). Increasing amounts of com-bined radiolabelled and unlabelled antigen are added to fixed quantities of antibody. The anti-body is unable to discriminate between the labelled and unlabelled forms of the antigen hence the amount of labelled antigen that binds to the antibody will decrease as the amount of unla-belled increases. The antibody–antigen complexes are then separated from the free antigen in one of a number of ways. For example, as shown in Figure 10.7 (a) a second antibody that binds to the first can be added to the solutions and this precipitates the antibody–antigen complexes leaving free antigen in solution. The radioactivity of each is then determined. A calibration curve is produced by adding the labelled antigen to a known range of concentrations of unlabelled antigen at the same time as the test on the sample is performed. This can be used to determine the concentration of the analyte in the unknown sample as shown in Figure 10.7 (b).

Immunoradiometric assays (IRMAs) are more sensitive and specific than RIAs in measur-ing the concentrations of small M_r peptides, and steroid hormones. The basis of IRMAs is to immobilize the antibody to the ligand to be detected, for example a hormone, on insoluble polystyrene beads. The antigen binds to the immobilized bead and the unbound antigen can be removed by washing. The bound antigen is detected and quantified by adding a second antibody labelled with ^{125}I to the bound antigen. The beads are washed again but this time to remove the excess ^{125}I-antibody. The radioactivity of the beads is counted and this is now a measure of the amount of antigen (hormone) that was present in the sample. A standard curve must also be produced using known quantities of the hormone, which is used to relate the radioactivity of the hormone bound to the beads to its concentration in the sample.

Non-isotopic methods utilizing chemiluminescence immunoassays, fluorescence immu-noassays, and enzyme-linked immunosorbent assays (ELISAs) have largely replaced RIA for most tests. Enzyme-linked immunosorbent assays use an antibody or an antigen linked to an enzyme and so combine the specificity of the antibody–antigen reaction with the sensitivity of enzyme assays. The sensitivity of ELISAs may be greater than RIA but more significantly, they are quicker to perform and do not require the use of radioisotopes. However, RIAs are still of use in diagnosing the condition systemic lupus erythematosus (SLE). Most patients suffering from SLE produce antibodies against double stranded DNA (dsDNA) and, at some time dur-ing their illness, these can be detected in their sera. The most commonly used clinical tests to detect the anti-dsDNA antibodies of SLE are the so-called Farr RIA and an ELISA. Some evalua-tions of both methods indicate that the Farr RIA-based method is superior to the ELISA dsDNA

Cross reference
You can read more about ELISA in Chapter 16 'Molecular biology techniques'.

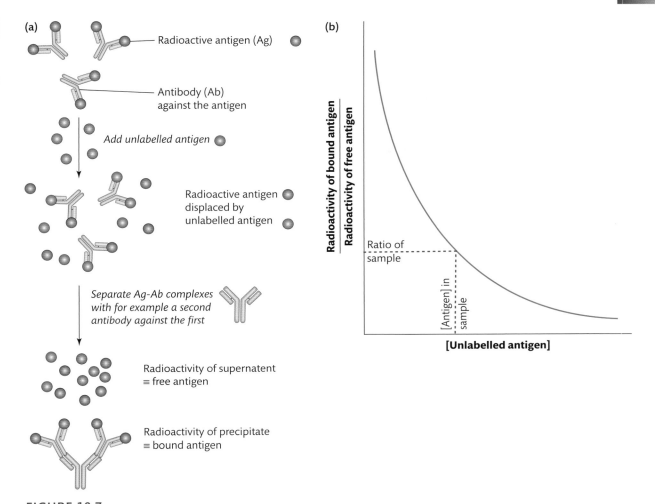

FIGURE 10.7
(a) Outline of the procedure for radioimmunoassay (RIA). See text for details. (b) Use of a RIA calibration graph to determine the concentration of an antigen in a patients' sample.

test in diagnosing (identifying) patients with SLE. For this reason, some clinical biochemistry laboratories still use it, although in most it has been replaced by the ELISA method, because of the general advantages given above.

SELF-CHECK 10.11

List two advantages of replacing RIA by ELISA.

Radiotherapy

Radiotherapy is the medical use of ionizing radiation as part of the treatment of certain cancers. Where applicable, it is used as the primary treatment to kill cancer cells, either alone or in combination with surgery or chemotherapy. It is also used as a palliative treatment to relieve symptoms, if the cancer is untreatable. The basis of the therapy is that radiation interacts with oxygen or water molecules forming free radicals that damage DNA. Since tumour cells have a higher rate of cell division than normal tissues, they are more susceptible to damage by free radicals. However, a major problem encountered in radiotherapy, is that the poorly developed blood systems of cancers means they are short of oxygen, which reduces its effectiveness.

Cells differ in their sensitivity to radiation. For example, 99% of bone marrow cells are killed by a dose of 4 Gy, but a dose of 16 Gy is required to kill 99% of thyroid cells. Similarly, cancer cells also show a range of sensitivities to radiation and so the dose has to tailored to the particular type of tumour. Leukaemias and lymphomas are very sensitive, epithelial tumours are moderately so, requiring doses in the range of 60–70 Gy. Renal tumours are insensitive. A typical dose applied to solid epithelial tumours is in the range 60–80 Gy and for lymphomas it is 20–40 Gy.

Radiotherapy is applied *externally* or *internally*. In either case, the dose of radiation used is limited because it also damages healthy tissues. The adverse effects of external treatment is reduced by irradiating the tumour with several beams of X-rays focused from different directions. The total dose of X-rays applied is fractionated into multiple doses, which are given over several weeks. Typically, doses of 2 Gy are used daily over a period of several weeks or months; a regimen known as continuous hyperfractionated radiotherapy (CHART). The advantage of dividing the dose in this way is that healthy tissue can repair itself but tumour cells are less efficient at self-repair, and so recover less well between the doses. In internal therapy, the radioisotope is placed near to or inside the tumour for a short period, a procedure known as brachytherapy. For example, intracavitary radiotherapy involves the insertion of ^{137}Caesium into a body cavity using an applicator. This form of treatment is used in treating cancers of the vagina, cervix, and uterus. Alternatively, thin radioactive wires may be inserted directly into the tumour, as, for example, in treating prostate cancers. Internal therapy may also involve giving the patient a radioactive liquid either orally or intravenously.

SUMMARY

- Radiation is the energy released when an unstable atom decays to a stable form

- Radiation is emitted as α- or β-particles or as γ-radiation

- Radiation is detected by gaseous or liquid scintillation counters or by autoradiography

- Radiation is potentially harmful and precautions must be taken in its handling and uses

- Radiation has applications in clinical imaging, diagnosis and in treating some forms of cancer, and in biomedical research

FURTHER READING

- **Hendee WE, Ritenor ER. *Medical Imaging Physics*. John Wiley and Sons, Harlow, 2003.**

  Describes the methods and theory of the application of radiobiology to imaging.

- Hogle WP. **The state of the art in radiation therapy.** *Seminars in Oncology Nursing*, 2006; **22**: 212–220.

  This paper describes the methods used to administer radiation and the effects of treatment.

- O'Neill P, Wardman P. **Radiation chemistry comes before radiation biology.** *International Journal of Radiation Biology*, 2009; **85**: 9–25.

  This article illustrates contributions of radiation chemistry to radiation biology.

- Sharp PF, Gemmell HG, Murray AD. *Practical Nuclear Medicine.* Springer-Verlag, Berlin, 2005.

  A detailed study of the theory and practice of nuclear medicine.

- Steele G. (ed.) *Basic Clinical Radiobiology.* Oxford University Press, Oxford, 2002.

  A collection of articles describing the science of radiobiology and its applications.

- Wigg D. **Radiation: facts, fallacies and phobias.** *Australasian Radiology*, 2007; **51**: 21–25.

  A discussion of the use of radiation in diagnosis and treatment, and the benefits and costs of the use of radiation.

QUESTIONS

1. Indicate whether the following statements are **TRUE** or **FALSE**:

   (a) In any given nuclide, the number of electrons is the same as the number of neutrons

   (b) Nuclei become more unstable the bigger the difference in numbers of protons and neutrons

   (c) The decay of radium ($^{226}_{88}$Ra) to radon ($^{222}_{86}$Rn) is an example of α-decay

   (d) The unit of absorbed energy from radiation is the gray

   (e) The gray is defined as the absorbed dose when 1J is deposited per gram of tissue

   (f) The linear energy transfer (LET) is the degree of ionization caused by the interaction of the radionuclide with tissue

   (g) Values for LET are in the range of 20 for β-particles and 1 for α-particles

2. Insert the correct term where a gap appears in the following text. Choose from the following list of alternatives, which may appear more than once: *radioisotopes; radiation, α-particle (x 2), β-particle, γ rays, 4, helium, electron, neutron, proton, mass, half-life, 5760 years, Becquerel, 1, 2, 1 disintegration per second, energy, 1 volt, electron volt.*

   Unstable atomic species called _____ can decay into more stable species by the emission of _____ . The energy emitted can be carried as an _____ or a _____ or as _____. An _____ has a mass of _____ units and is a _____ ion. Release of an α-particle decreases the atomic mass of the radioisotope by 4 mass units and the atomic number by _____. A β-particle is an _____ produced when a _____ decays to a _____ . In the process the _____ of the radionuclide remains unchanged, but the atomic number increases by _____ unit. The time taken for half of the radioactive sample to decay is called the _____ and varies from 8.04 days for ^{131}I to _____ _____ for ^{14}C. The SI unit of radioactivity is the _____, which is defined as _____ _____ _____ _____. The _____ _____ is the unit used to express the energy released during radioactive decay. It is defined as the _____ gained by an electron accelerated through a potential of _____ _____.

3.   Define the following terms; Becquerel, Curie, rad.

4.   The half-life of ^{32}P is 14.2 days. How long would it take a preparation of the isotope with an initial activity of 21 000 dpm to decay to 1000 dpm?

5.   What precautions should you take when carrying out an experiment to label a protein with ^{32}P?

6.   In an experiment labelling a sugar with ^3H the sample was counted in a liquid scintillation counter and a reading of 32 600 cpm was obtained. Using a standard source of ^3H known to contain 555 000 dpm, a reading of 425 000 cpm was obtained. (a) What is the efficiency of counting and (b) express the activity of ^3H as dpm?

7.   A sample of protein M_r 64 000 was labelled with ^{14}C. After its purification it was found to have a specific activity of 1.7×10^7 dpm μg^{-1}. What is the specific activity of the purified protein in Ci mol^{-1}?

8.   Explain the cytotoxic effect of radiation. Why would you expect cell damage to increase with the energy of radiation?

Answers to self-check and end of chapter questions are available in the Online Resource Centre accompanying this book.

Go to www.oxfordtextbooks.co.uk/orc/glencross

11

Spectroscopy

Qiuyu Wang, Helen Montgomery, Nessar Ahmed, and Chris Smith

Learning objectives

After studying this chapter, you should be able to:

- Define the term spectroscopy
- Describe the electromagnetic spectrum
- Discuss the general theory of spectroscopy
- List the major spectroscopic techniques
- Discuss some of the uses of spectroscopy in the biomedical sciences

Introduction

Spectroscopy is the study of the interactions between electromagnetic radiation and matter. Atoms, molecules, and ions can scatter, absorb, or emit electromagnetic radiation in ways that depend upon the chemical composition and structure of the substance. Thus, the pattern of absorption or emission is characteristic for a given substance. This means that spectroscopy can be used as an analytical technique to identify, which elements are present in a sample, determine how much of an analyte is present, and even be used to elucidate the structures of molecules. It is a non-invasive technique and samples are not destroyed during spectroscopic analyses. These characteristics give spectroscopic analysis a broad range of applications in the biomedical and biological sciences.

11.1 Radiation and the electromagnetic spectrum

Radiation is the emission or transfer of energy (*E*) as electromagnetic waves or particles. The particles are often called **photons**. Electromagnetic radiation, such as light, is propagated

Cross reference
You can read more about radiation in Chapter 10 'Radioactivity and radiation'.

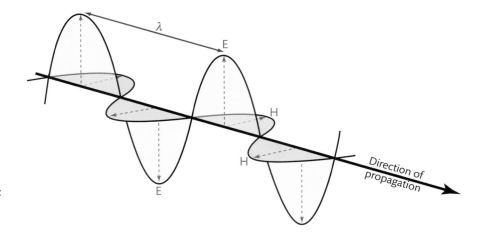

FIGURE 11.1
Propagation of electromagnetic
radiation. See main text for
explanation.

through space as electric and magnetic fields, which are mutually perpendicular to each other.
You can see this expressed diagrammatically in Figure 11.1.

Two characteristic features of any sort of wave are its wavelength (λ) and frequency (v). You
can see in Figure 11.1 that the wavelength is the distance between any two adjacent peaks.
The frequency is the number of peaks passing a fixed point in a given time, which is normally
measured in seconds (s). Wavelength and frequency are interrelated by the velocity of light (c)
in the following way:

$$\lambda = c/v$$

The energy (E) of any type of electromagnetic radiation is described as:

$$E = hc/\lambda = hv$$

where h is Planck's constant. The use of these simple equations means that the frequency,
wavelength and energy of electromagnetic waves can be easily interconverted although care
must be taken over units (see Box 11.1).

Thus, the energy of a wave is directly proportional to its frequency: the higher the frequency,
the greater the energy. However, the energy is inversely proportional to its wavelength, the
shorter the wavelength, the greater the energy. The correlation between wavelength and

BOX 11.1 *Spectroscopy and units*

Wavelengths (λ), frequencies (v), and the Planck constant (h) can all be expressed in one
of a number of units. Wavelengths are usually given in nanometres (nm), but some older
texts use the angstrom (Å), which is equal to 10^{-10} m or 0.1 nm. Frequencies are normally
given in cycles per second or hertz (Hz), which is reciprocal seconds (s^{-1}). However, frequencies may also be described as a wave number (v'), which is often expressed as reciprocal centimetres (cm^{-1}) simply because the numbers become easier to manage. Planck's
constant has a value of 6.63×10^{-34} J s^{-1}. Again, it can be expressed in alternative units
including 6.63×10^{-27} erg s^{-1} and 4.135×10^{-15} eV s^{-1} (eV, an electron volt, is the energy
acquired by a single electron accelerating through a potential difference of 1 volt and is
equal to about 1.6×10^{-19} J).

The velocity of light (c) is 2.998×10^8 m s^{-1}.

FIGURE 11.2
A portion of the electromagnetic radiation spectrum.

energy resembles the energy of a spring. Think of squeezing a spring, the more you compress it, which gives it energy, the greater its tendency to jump back to its original, resting length.

The full range of electromagnetic radiation is often called the **electromagnetic radiation spectrum**. It is often subdivided or classified in to a number of types of radiations depending upon their wavelengths (or frequencies) as shown in Figure 11.2.

11.2 **Interactions between radiation and matter**

When radiation of a particular wavelength interacts with matter, the radiation may be *scattered* or it may be *absorbed*. Scattering, often referred to as Rayleigh scattering, occurs when a photon collides with an atom or molecule, and its direction of propagation is changed. However, its wavelength and frequency, and therefore energy, are not usually altered. Turbidity and nephelometry depend on the scattering of light and are briefly described in Section 11.6. Light scattering can be used to determine the M_r of macromolecules in solution or the sizes of larger suspended particles, such as those of viruses. This technique has been superseded by simpler, more convenient methods, such as gel electrophoresis and analytical gel filtration or by using analytical microultracentrifuges.

The scattering of radiation is also the basis of a number of techniques used to investigate the structures of particles of clinical and biological interest. These techniques include X-ray diffraction, electron microscopy, and neutron scattering. A discussion of these techniques is, however, outside the scope of this chapter.

Cross reference
You can read about gel electrophoresis, analytical gel filtration, and analytical microultracentrifuges in Chapters 8, 12, 13, and 14 respectively.

If a photon of radiation is absorbed by matter, energy is transferred from it to the atom or molecule. This means that the electronic state of the atoms or molecules is affected; their energy is increased from a ground state to an excited one and the atom or molecule is described as being **excited**. The atom, molecule, or even part of a molecule that becomes excited by absorption is often called a **chromophore**. The excitation energy is usually dissipated as heat when the excited atom or molecule collides with, for example, a solvent molecule. However, some molecules can lose the energy by emitting a photon of lower energy in a process called photoluminescence. If the photon is emitted virtually instantaneously, this is called **fluorescence**. Phosphorescence differs from fluorescence in that the phosphorescent material continues to emit light after removal of the source of excitation, although this slowly diminishes. By contrast, fluorescent materials fluoresce only while excited.

SELF-CHECK 11.1

In fluorescence, which of the photons, absorbed or emitted, has the shorter wavelength?

11.3 Radiation, particles, and quanta

Electromagnetic radiation is often difficult to describe in general everyday terms. This is because it cannot be described solely in terms of the properties of waves, but neither does it possess only the properties of particles. Instead it has attributes of both. Some of the ways in which radiation interacts with matter, for example, the scattering of light, are most easily described by assuming it is a wave. However, other aspects, such as the absorption of light, are more easily envisaged if radiation is regarded as a particle.

The frequency of radiation is related to its energy content by Planck's constant (6.63×10^{-34} J s^{-1}), as described in Section 11.1. The introduction of Planck's constant means that a specific frequency or wavelength has a discrete and fixed amount of associated energy. Equally, when an atom or molecule absorb or emits energy, the amount of energy absorbed or released can only be of discrete or fixed amounts. These 'amounts' are called quanta (singular quantum); a quantum is the smallest individual quantity of energy. Hence, quanta occur in discrete, specific amounts, and not as a continuous range of values. Furthermore, atoms and molecules can possess only specific amounts of energy or be at certain 'energy levels'; so they too are quantized.

For a photon to be absorbed, its energy must exactly match the difference in energy between two energy levels in an atom or molecule. The discrete energy levels (quanta) that are associated with atoms and molecules are usually illustrated using an energy-level diagram like the one in Figure 11.3. If you examine this figure you can see that the ground state is indicated by the lower red line (I). However, in an excited state, the energy of an atom or molecule has increased by a fixed amount or quantum. One such excited state is represented by the upper red line (II). You can also see that within each of the electronic states [the ground (I) and an excited (II)], there are different possible vibrational energies, which are represented by the thinner blue coloured lines. Rotational energies also occur at values between those for vibrations. Our diagram shows only a limited number of these as short green coloured lines; otherwise it would become too cluttered!

A change between any two energy levels is called a **transition**. The vertical arrow in Figure 11.3 shows one of the many possible electronic transitions. You should note that a transition can occur only if the atom or molecule absorbs an amount of energy exactly equal to that between the two relevant states. This is indicated on the energy axis as ΔE. However, transitions can occur not only between electronic states, but also between the various vibrational and

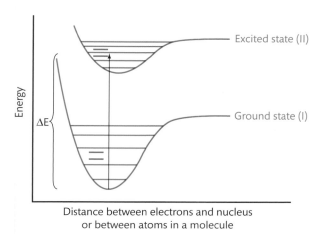

Energy

ΔE{

Distance between electrons and nucleus
or between atoms in a molecule

Excited state (II)

Ground state (I)

FIGURE 11.3

Schematic energy-level diagram from the ground state (I) to the first excited state (II). The arrow shows a transition from the ground state to the third vibrational level of the first excited state, which requires an input of energy equal to ΔE.

rotational states. The sizes of the energy changes decrease from electronic through vibrational to rotational and therefore the wavelengths of the photons required to promote them increase.

Do the frequencies of photons (or electromagnetic radiation) increase or decrease with the order of the following transitions: rotational, vibrational, and electronic?

When an excited molecule returns to its ground state, energy is lost and the transition has been from a higher to lower energy level. The energy can be lost as photons, as in fluorescence or phosphorescence (Section 11.7). When radiation is not emitted, however, the transition is called non-radiative decay. Except in the cases of fluorescence or phosphorescence, radiation is only emitted by materials at temperatures that are so high they would normally degrade molecules of clinical or biological interest. Hence, in the biomedical sciences, spectrophotometry is virtually synonymous with measuring the *absorption* of radiation. Figure 11.2 indicates the relative energies associated with different wavelengths. Each of these is used with different types of spectroscopy. All of these different types of spectroscopy have different uses in clinical practice.

These include determining the absorbance, sometimes called optical density, of a substance or its absorption spectra. We will discuss both of these analytical approaches in the following sections.

11.4 **Absorbance**

Absorbance is a means of quantifying how much light of a particular wavelength is absorbed by a sample. In contrast, absorption spectra show the fraction of electromagnetic radiation absorbed by a material over a *range* of different wavelengths, frequencies, or wave numbers (see Box 11.1).

Spectroscopic techniques can be used to make quantitative measurements. For example, to determine the amount of a substance (the analyte) that is present in a sample. The absorbance of an analyte is proportional to its concentration (c) in a manner described by the Beer–Lambert law. This law describes the absorbance of **monochromatic** light, which is one of a single wavelength, as it passes through a *dilute* solution. The law states that the absorbance is directly proportional to both the concentration of the solution and also to the length of the path that the light takes through that solution.

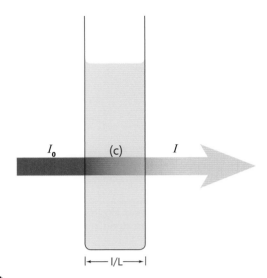

I_0 (c) I

|←— l/L —→|

FIGURE 11.4
The absorbance of monochromatic light
in accordance with Beer–Lamberts law.

$$A = \varepsilon cl \text{ or } acL$$

You can see an illustration of the law in Figure 11.4. Light of a specific wavelength and intensity I_0 is entering a sample of concentration c with a path length l or L. Light leaving it has an intensity of I. If the sample does not absorb or scatter any of the light, then I will equal I_0. If, however, some of the light is absorbed by the sample then I will be less intense than I_0, such that the transmitted light (T) is equal to I/I_0. This value is usually multiplied by 100 and expressed as a percentage value. The *logarithm* of $1/T$ or I_0/I ($\log (I_0/I)$) is the absorbance (A).

The Beer–Lambert law is summarized by the expression:

$$\log (I_0/I) = A = \varepsilon cl$$

where ε is often referred to as the molar extinction coefficient.

Absorbances are logarithmic values and so do not have units. In traditional, non-SI units, l is expressed in cm and the concentration in mol L^{-1} or M. This gives ε units of $M^{-1} cm^{-1}$. However, if SI units are used, then the equivalent Beer–Lambert law expression becomes:

$$\log (I_0/I) = A = acL$$

where the constant (a) is called the absorptivity, c is in mol m^{-3} and the path length (L) is in m. This gives absorptivity units of $m^2 mol^{-1}$. Whatever system of units is used, the generally, observed range of transmittance is from 0 to 100%. Absorbance values then vary from 0 to 2.

The absorbance of a given substance will vary according to the wavelength of the light to which it is exposed. When using absorbance for quantitative analysis, however, the absorbance of a solution is normally determined at the wavelength at which the material being investigated absorbs most strongly, that is where the absorbance is at its greatest. This wavelength is often designated as A_λ. At a single wavelength, the absorbances of dilute solutions of a given analyte are directly proportional to the concentration of analyte. This is illustrated in Figure 11.5: note how the graph is only a straight line for lower concentrations in accordance with the Beer–Lambert law.

Absorbances are usually measured with the solution contained in specialized transparent containers called **cuvettes**, which normally have a standard path length of 1 cm or 0.01 m. This length can be assumed unless you are told otherwise. Disposable cuvettes made of polymethacrylate and polystyrene are commercially available for use at wavelengths of

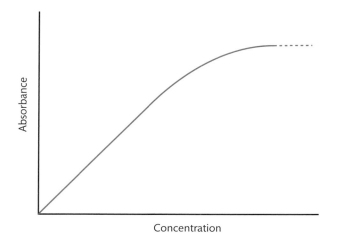

FIGURE 11.5
Graphical illustration of
the Beer–Lambert Law.
See text for details.

280–800 and 350–800 nm, respectively. They therefore suffer the disadvantage that they cannot be used to determine the absorbances of solutions of nucleic acids, which have λ_{max} at about 260 nm, and perform poorly with protein solutions that absorb maximally at about 280 nm. Reusable glass, quartz, and fused silica cuvettes are also manufactured. Again, glass cuvettes cannot be used in the ultraviolet range; their lowest usable wavelength is 340 nm. Cuvettes of quartz and fused silica are the most versatile and can be used throughout the ultraviolet and visible ranges that is from about 200–800 nm.

Given the relationship between absorbance and concentration and because many chemicals absorb light maximally at specific wavelengths, measuring the absorbance of a solution of analyte is extremely useful, since it provide a means of determining its concentrations in patient samples. In theory, if ε (or a) is known for a compound, then its concentration in a sample can be determined simply by measuring its absorbance under the same conditions for which ε (or a) was found. In practice, however, it is more usual to determine the absorbance of that sample and also for a known concentration or a series of concentrations of the analyte. The unknown concentration of an analyte can then be calculated from the ratios of the absorbances given that the concentration of the standard is known. Alternatively, a graph or standard curve of absorbance against the known concentration can be plotted from which the unknown concentration can be obtained (Figure 11.6).

SELF-CHECK 11.3

Calculate the concentration of the analyte X, given that a 10-fold dilution of the specimen gives an absorbance of 0.458 and a standard solution of the same analyte of concentration 50 µg cm^{-3} had an absorption of 0.625.

Spectrophotometers and colorimeters

Two types of spectrometers are typically encountered. These are single beam and double beam instruments. Both types have a visible light source and either type may have a deuterium lamp that produces ultraviolet light so that absorbance at shorter wavelengths can also be measured. You can see outline structures of each type of spectrophotometer in Figure 11.7 (a) and (b). In single beam instruments, a **monochromator** separates the beam into its component wavelengths. Monochromators can be prisms or diffraction gratings [Figure 11.7 (c)], but either type allow a monochromatic beam of relevant wavelength to pass through the

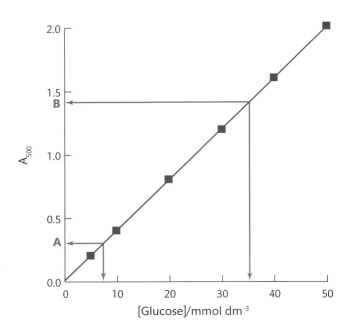

FIGURE 11.6

Standard or calibration curve, relating the concentration of glucose to the absorbance at a wavelength of 500 nm using a glucose oxidase-based test. Two serum samples, A and B were analysed for glucose using the method at the same time. Sample A was from a healthy individual and B from a poorly controlled diabetic. The absorbances obtained for A and B were 0.30 and 1.40, which correspond to concentrations of 7.0 and 35.0 mmol dm^{-3}, respectively.

sample in a cuvette. The amount of light transmitted is measured to give the absorbance of the solution. Single beam instruments must be 'zeroed' using a reference solution ('the blank') in the cuvette. This solution contains all the components of the sample other than the analyte. The reference is then replaced with the sample and its absorbance noted. The same cuvette must be used for both solutions to avoid absorbance errors arising from variations in the transparencies of different containers.

In double beam instruments, a monochromatic beam is, again, selected for use. However, it is split into two beams of equal intensity using a half-mirrored device. One of these, called the reference beam, passes through a cuvette containing only the solvent. The other half-beam passes through a second cuvette containing a solution of the sample compound being studied. The intensities of these light beams are determined and electronically compared. The intensity of the reference beam is automatically subtracted from that of the sample to give its absorbance.

Immunoassays, which we describe more fully in Chapter 15, are often performed in specialized containers that have 96 wells. These are generally referred to as 96-well plates. A single assay is performed in each well. When the assays are completed and colour has developed, the absorbance of each well in the plate is measured using a specialized spectrophotometer called a plate reader.

The term *colorimetry* is often used when the absorbance of visible, and therefore coloured, light is used. In this case, rather than using a spectrophotometer, which produces light with a limited range of wavelengths, a colorimeter is used. Colorimeters do not produce light of specific wavelengths but use a series of filters that produce narrow bands of light in the visible spectrum.

Clinical applications of measuring absorbances

Spectrophotometers are incorporated in automated analysers in hospital laboratories and used to measure a wide range of analytes colorimetrically. For example, serum creatinine concentrations are determined using the Jaffe's reaction. Creatinine is a breakdown product of

(a) Slit · Slit · Monochromator · Sample · Detector · Output device 0.24 · Light source

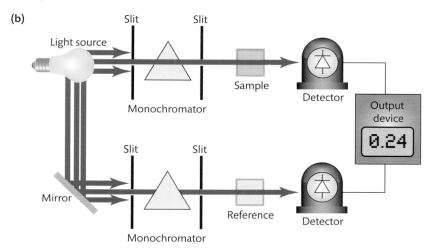

(b) Light source · Slit · Slit · Monochromator · Sample · Detector · Output device 0.24 · Mirror · Slit · Slit · Monochromator · Reference · Detector

(c) Slit · Slit

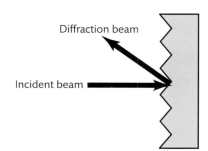

Diffraction beam · Incident beam

FIGURE 11.7
(a) Single and (b) double beam spectrophotometers. (c) Shows the use of a prism and a diffraction grating as monochromators. See text for details.

creatine phosphate in muscle tissues and its concentration in serum is used to assess kidney function because it is produced at a fairly constant rate in the body and is freely excreted by the kidneys. Reference ranges for serum creatinine are typically 60–110 µmol dm^{-3} in men and 45–90 µmol dm^{-3} in women. If the filtering mechanism in the kidneys declines, for example, due to loss of nephrons as in renal failure, then concentrations of serum creatinine will increase above these ranges in proportion to the decline in function.

Creatinine reacts with picric acid in an alkaline solution to give an orange coloured product. After an incubation of 15 minutes at room temperature to allow for colour development, the absorbance is measured at a wavelength of 490–500 nm using the spectrophotometer. The absorbance produced is proportional to the concentration of creatinine in the sample of serum.

Cross reference
You can read more about automation in Chapter 17 'Laboratory automation'. Chapters 1–3 of the companion text, *Clinical Biochemistry*, will also provide useful material.

11.5 **Absorption spectra**

An **absorption spectrum** is a graph of absorbance against wavelength and shows the fraction of incident electromagnetic radiation absorbed by a material over a range of *different* wavelengths, frequencies, or wave numbers (see Box 11.1). In a relatively short time, a suitable spectrophotometer can automatically determine absorbances, as described in the previous section, but at a number of different wavelengths.

Different regions of the electromagnetic spectrum can be used to investigate different features of an atom or molecule. Look at Figure 11.2. You can see that the higher energies of ultraviolet (usually 200–400 nm) and visible lights (approximately 400–800 nm), collectively often abbreviated to UV–Vis or 'standard' absorption spectroscopy, are associated with electronic transformations. Infrared spectroscopy uses radiation of 2500 to 16,000 nm, whose lower energies promote vibrational transitions. Rotational transitions can occur with the longer wavelengths. Microwave radiation (about 10–300 mm wavelengths) is used in nuclear magnetic resonance spectroscopy (Section 11.8). Thus, by using electromagnetic radiation of different wavelengths it is possible to investigate different aspects of the same substance, and gradually build up a picture of its composition and structure. In essence, the absorption spectrum for a compound constitutes a type of 'fingerprint' for it, which can be useful in identifying a specific analyte in a sample. Thus, absorption spectra, particularly those associated with longer wavelengths of the infrared and microwave regions, can be used to help identify specific analytes in samples. We will now consider a number of these techniques in turn.

Ultraviolet–visible absorption spectrophotometry

Many compounds absorb electromagnetic radiation of specific wavelengths. Many colourless compounds of clinical or biological interest, such as proteins, nucleic acids, purines, pyrimidines, and some drugs, vitamins, and amino acids absorb non-visible, ultraviolet light (Figure 11.2). This is of help in identifying those wavelengths most appropriate to use in measuring absorbances to determine the concentration of an analyte (Figure 11.8).

SELF-CHECK 11.4

A partial absorption spectrum of oxyhaemoglobin is given in Figure 11.8 (a). What would be suitable wavelength(s) to measure its concentration?

Infrared spectroscopy

Infrared (IR) spectroscopy is a useful analytical tool in identifying or partially identifying an analyte. In principle, it is not different from UV–Vis spectroscopy. However, the energies of photons in the IR region of the spectrum vary from approximately 0.25 to about 3.755 J mol^{-1}. Unlike ultraviolet and visible light this is insufficient to produce electronic transitions. Rather, the absorption of lower energy IR light excites covalently bonded atoms and organic functional groups in molecules resulting in vibrational and rotational transitions. Such excitation is possible because the covalent bonds in molecules do not behave like rigid, non-elastic structures, but more like springs that can be stretched and bent.

Thus molecules of virtually all organic compounds can show a large number of vibrational motions, which are characteristic of the atoms or groups of atoms present. This means infrared

(a)

(b)

FIGURE 11.8

Ultraviolet–visible absorption spectra of (a) deoxyhaemoglobin and oxyhaemoglobin and (b) for a number of vitamins.

spectroscopy is especially useful for analysing organic compounds because the different functional chemical groups present absorb IR at characteristic wavelengths. We have listed the characteristic absorption wave number values of some functional groups in Table 11.1. Thus, many of the functional groups in the molecules of an analyte can be relatively easily identified.

Absorption by these groups gives rise to seemingly rather complex spectra when compared with those of the UV–Vis range. You can see what we mean if you examine Figures 11.8 and 11.9. Look at the relative simplicity of the UV–Vis spectrum of the complex protein haemoglobin shown in Figure 11.8 (a) compared with the complex IR spectra of the structurally much simpler compounds shown in Figure 11.9. These spectra are typical of IR absorption spectra in that the absorption is expressed as percent transmittance and wavelengths are given as wave numbers (Box 11.1). However, as we noted above, the very complexity of IR spectra means

TABLE 11.1 Values of some infrared radiation wavelengths characteristically absorbed by some chemical groups

Bond	Wave no.r/cm^{-1}	Chemical group/notes
Single bonds		
C–H	2850–3000	Saturated alkane; presence too numerous to be of diagnostic value
C–C	variable	Presence too numerous to be of diagnostic value
C–N; C–O	1000–1400	Overlap makes assigning signals difficult
C–Cl	700–800	
O–H	3200–3550	Alcohols and phenols; hydrogen bonding will broaden peaks
N–H	3100–3500	Amines; primary amines give several signals, secondary give one peak, tertiary no peaks
P–H	2275–2440	Phosphorus compounds
Double bonds		
C = C	1600–1675	Often a weak signal
C = N	1630–1690	Overlaps with other signals
C = O	1715–1750	Esters
C = O	1680–1740	Aldehydes
C = O	1665–1725	Ketones
C = O	1670–1720	Carboxylic acids
C = O	1630–1695	Amides
N = O	1300–1370 1450–1570	Nitro compounds
P = O	1140–1300	Phosphorus compounds
Triple bonds		
C ≡ C	2120–2260	Alkynes; weak bands
C ≡ N	2210–2260	Nitriles

they provide a 'fingerprint' of an analyte that can be extremely useful in identifying, or at least partially identifying, what it is.

Infrared spectra can be obtained from gaseous, liquid, or solid samples. Glass cuvettes cannot be used since glass absorbs IR light. Instead, samples are usually examined as thin films sandwiched between two polished salt plates, which are transparent to IR light. Liquids can be applied as films directly between the plates. Solid samples can be examined following their incorporation into thin potassium bromide disks prepared under high pressure. They can be mixed with a solvent and ground to a fine paste, which is spread between the plates. In some cases IR spectra can be obtained directly from finely powdered preparations of samples, although the resolution is sometimes poorer than using salt plates (Figure 11.10 (a) and (b)).

(a)

(b)

FIGURE 11.9
Infrared spectra of (a) retinoic acid and (b) paracetamol.

SELF-CHECK 11.5

Identify the chemical group that is the one most likely to have produced the trough at 1690 cm^{-1} in the infrared absorption spectrum of retinoic acid given in Figure 11.9. Table 11.1 will be of use!

Macromolecules, such as proteins, can dissolve in volatile solvents; the resulting solution can then be spread as a thin film on one of the plates. Evaporation of the solvent then leaves a thin film of protein molecules on the plate for analysis. Unfortunately, organic solvents tend to denature proteins.

If a solvent is used to dissolve solids, care must be exercised in its choice; the solvents chosen should not absorb IR light in regions of the spectrum that may be of analytical use. Hence, water cannot be used because it absorbs strongly in the 1600 and 3400 cm^{-1} regions, also it is normally present in high concentrations in samples and would dissolve the salt plates. Commonly used solvents include carbon tetrachloride, chloroform, and tetrachloroethene.

FIGURE 11.10
Infrared spectra of (a) ibuprofen in KBr plates and (b) obtained directly using a powdered tablet.

Raman spectroscopy

The difficulties associated with obtaining IR spectra in water can be overcome using Raman spectroscopy, which also probes the vibrational energy levels of atoms and molecules. You can see an outline of a Raman spectrophotometer in Figure 11.11.

The incident, monochromatic, light used in Raman spectroscopy is chosen so that its wavelength is not absorbed by the sample; hence, most of it is simply transmitted. However, a relatively small proportion of the light will be scattered and this can be viewed at an angle of 90° to the sample. It was noted in Section 11.6, that scattering does not in general alter the frequency of the exciting light. However, if an exceedingly small proportion of the energy of an exciting photon is transferred to a sample molecule, it can lead to a transition from one vibrational state to another. Hence, a small proportion of the scattered light will have its frequency

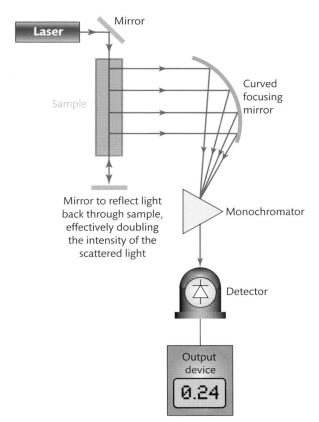

FIGURE 11.11
Outline of a Raman spectrophotometer. See main text for an explanation of its operation.

decreased due to its involvement in the vibrational transitions. The net result of such changes in frequencies means that a series of vibrational transitions can be observed as a collection of spectral lines in the scattered light collected at 90° to the sample. These spectral lines form the Raman spectrum.

Given that the information gained in Raman spectroscopy is from vibrational transitions, it is similar to that obtained by IR spectroscopy. However, it differs in that the information is independent of the frequency used to excite the atoms/molecules. Thus, any wavelength can be used to produce a Raman spectrum. However, because only a small proportion of the light is absorbed by vibrational transitions, a high intensity light source is needed; typically a laser is used. Thus, Raman spectroscopy is carried out with a dedicated, specialized spectrophotometer. The sample must also be highly concentrated. In practice, a 2% solution is usually used, which can lead to problems with molecules aggregating or precipitating.

Clinical applications of infrared and Raman spectroscopy

Infrared and Raman spectroscopies, often coupled with gas liquid chromatography (GLC, Chapter 13), have been used to study drugs such as the various derivatives of penicillin and amphetamines. Amphetamines are a group of structurally similar molecules that are difficult to differentiate. However, amphetamines in samples can be identified using differences in their infrared spectra using GLC–IR. Infrared spectroscopy is used to monitor pollutants in the environment, particularly harmful chemicals in the air. These include measurement of toxic gases, such as carbon monoxide in garages and formaldehyde produced during manufacture of plastics. There is considerable interest in the use of infrared spectroscopy as a non-invasive

Cross reference
You can read about
chromatography in Chapter 13.

technique to determine the concentrations of glucose in samples of blood from diabetic patients. The blood glucose is quantified by exposing a region of the skin to infrared radiation and analysing the resultant spectrum.

11.6 **Light scattering methods**

The two main experimental techniques encountered in the clinical laboratory based on the scattering of light are **turbidity** and **nephelometry**.

Turbidity is the degree of opacity or cloudiness of a fluid caused by particles suspended in it, which causes a beam of light passing through the suspension to scatter. An Increase in turbidity reduces the amount of light that can penetrate the suspension, i.e. its transmittance is decreased. The amount of light that is scattered depends on a number of factors including the size and shape of the particles, and their concentration. Turbidimetry is the measurement of turbidity. It relies on the use of a spectrophotometer, which can be used to determine the concentration of the suspension of particles in a turbid sample. The particles are often bacterial cells or precipitated protein. A beam of light is passed through the suspension. Some of the light is absorbed, reflected, and scattered. However, the rest is transmitted through the solution and measured by the spectrophotometer and used to determine the concentration of particles. Turbidity (Tb) can be defined mathematically as:

$$I = I_0 e^{-lT} \text{ or, more simply, } Tb = 2.303 \times 1/l \times \log (I_0/I)$$

where I_0 and I are the intensities of incident light and transmitted beams of light, and I is the length of the light path. This relationship only holds for relatively finely dispersed suspensions of absorbances less than 0.3–0.5.

Turbidimetric measurements have some clinical uses. For example, they can be used to determine the number of bacterial cells in microbial suspensions. In some clinical disorders, for example nephrotic syndrome, proteins are lost in the urine. Determining their presence and concentration therefore has medical value. The urine samples are treated with sulphosalicylic acid, which causes any protein present to precipitate as a fine dispersion. The amount of precipitation is measured by its turbidity and is proportional to the protein concentration. The technique is simple and rapid, but has the disadvantages of lacking sensitivity and precision.

SELF-CHECK 11.6

A suspension of *Eschericia coli* cells of dry weight 0.4 g dm⁻³ give an absorbance value of 0.8. Calculate its corresponding turbidity (Tb) value.

Nephelometry (from the Greek nephele meaning cloud) is a technique in which the intensity of light scattered by a suspension is measured at an angle to the incident beam of light, rather than using the transmitted light to determine the concentration of suspended particles. Thus, unlike turbidity, nephelometry is not measured in a spectrophotometer, but uses a nephelometer (Figure 11.12). Nephelometry resembles turbidity, but has the advantage of being more sensitive.

The amount of scattered light depends on the size of the particles and their concentration. Thus, for example, the concentration of protein in clinical samples can be determined by comparison with known concentrations. One clinical application of nephelometry is to measure the concentration of rheumatoid factor (RF) in samples of serum. Rheumatoid arthritis is an extremely painful and debilitating chronic condition, which is characterized by symmetrical

Sample

90°
scattering
angle

Light source

Detector

Output
device

0.24

FIGURE 11.12
Schematic illustrating measurement using nephelometry. See text for an outline of the method.

Cross reference
You can read more about the clinical uses of turbidity and nephelometry in Chapter 15 'Immunological techniques'.

arthritis, radiological changes to the bone and the presence of autoantibodies, the RF, in the plasma of patients. Rheumatoid factor are immunoglobulins that recognizes and bind to antigenic sites at specific positions of IgG antibodies. The presence of high levels of RF in patients is associated with a poor prognosis.

11.7 Fluorescence and fluorimetry

Fluorescence was introduced in Section 11.2. To recap briefly, when most molecules absorb electromagnetic radiation, the electrons move to a higher energy level before falling back to their ground state, generally releasing energy in the form of heat in doing so. However, in some molecules only part of the energy is lost as heat and the rest is emitted as fluorescent light. The wavelength of this light (λ_{em}) is longer than that used to excite the molecule (λ_{ex}) because as we noted above, the excited molecules lose some of the absorbed energy as heat. The difference between the two wavelengths is known as the Stokes shift.

A fluorimeter or fluorescence spectrophotometer is an instrument used to measure the fluorescence of a given substance. Its components are outlined in Figure 11.13. The source of light produces high energy radiation of short wavelength. This passes through the first monochromator, which can be used to select an excitation beam of appropriate wavelength that passes through the cuvette. Following its excitation, the sample in the cuvette emits fluorescent radiation in all directions. The second monochromator, which is placed at right angles to the cuvette, only allows emitted light that is of the fluorescent wavelength to fall onto the detector and be measured. The detector produces an electrical signal that is proportional to the intensity of the fluorescent light, which in turn is proportional to the concentration of the sample.

Fluorescence has two principal advantages over the use of visible or UV spectrophotometry. First, it is more sensitive, indeed this can be up to a 1000-fold greater. Secondly, it is also more specific as two wavelengths are used to produce a signal not one as in spectrophotometry. However, there are also a number of disadvantages to using fluorimetric techniques. They are sensitive to changes in pH and the presence of contaminating molecules in the solution. Furthermore, not all compounds generate fluorescence, although it may be possible to couple non-fluorescent compounds with those that do exhibit fluorescence. For example, proteins can be reacted with fluorescamine to produce a fluorescently labelled complex, which allows

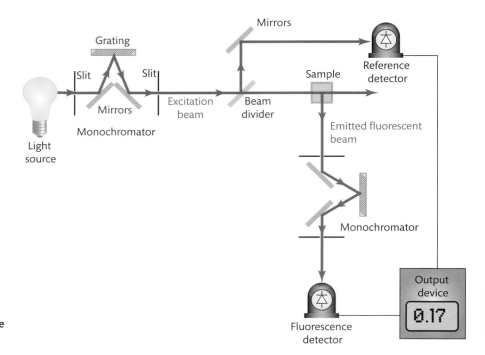

FIGURE 11.13
Outline of a fluorimeter.
The main text describes
the operational basis of the
instrument.

Cross reference

Chapter 15 'Immunological
techniques'.

the concentrations of minute amounts of them to be quantified. Fluorescent groups can also be coupled to immunoglobulins and this is the basis of immunofluorescence assays, which are described in Chapter 15.

Sometimes, the emitted fluorescence is less than that expected. This phenomenon is called **quenching** and occurs when substances in the sample absorb the emitted fluorescence or when molecules such as oxygen in the sample interfere with energy transfer.

Fluorimetry can be used to measure the concentration of phenylalanine in samples of serum from newborn infants in screening for phenylketonuria (PKU). This condition is characterized by excessive concentrations of phenylalanine in the plasma and is associated with mental retardation in untreated children. Phenylalanine is reacted with ninhydrin in the presence of the dipeptide leucyl-alanine to give a product whose fluorescence is proportional to the concentration of phenylalanine. The method is calibrated using known standards of phenylalanine.

11.8 Nuclear magnetic resonance spectroscopy

Nuclear magnetic resonance (NMR) spectroscopy provides information on the position of specific atoms within a molecule by exploiting the magnetic properties of certain nuclei. Nuclei with odd masses or odd atomic numbers, for example ^{1}H and ^{13}C, have a magnetic moment, also called spin or angular moment. The chemical environment that is the surrounding atoms, of these types of nuclei can be determined using NMR spectroscopy. This information can then be used to identify chemical functional groups in a molecule, identify some compounds in samples and even construct models of the three-dimensional structures of compounds. Most NMR studies in the biological and biomedical sciences exploit the spin of ^{1}H (^{1}H-NMR) or ^{13}C (^{13}C-NMR), since these elements are common in organic molecules.

However, a number of other magnetic isotopes are suitable for NMR studies. You can see some examples in Table 11.2.

One isotope of carbon, ^{12}C, is used extensively in biological investigations despite being nonmagnetic. Can it be used in NMR spectroscopy?

Nuclei can be regarded as moving charged particles and so have associated magnetic fields; in effect they resemble simple bar magnets. In the absence of an external magnetic field, the spins of the atoms are randomly orientated (Figure 11.14 (a)). However, when a magnetic field is applied the atoms become orientated in a direction that is parallel to that of the applied field (Figure 11.14 (b)). The spins, however, can be aligned with the field (α orientation) or 'point' in the opposite direction (β orientation).

Those nuclei with their spins aligned with the field have a lower energy than those with the opposite orientation; they therefore predominate. However, the energy difference between the two orientations is not large and an equilibrium distribution occurs in which the α:β ratio is approximately 1.00001:1.00000. You can see this situation illustrated diagrammatically in Figure 11.15 (a). Equilibrium is disrupted if pulses of radio frequency (RF) electromagnetic radiation of a certain frequency are absorbed by a nucleus. Specifically, the absorption of the RF radiation by nuclei with low-energy spins (α) changes their orientation to that of the higher-energy spin (β) state. The absorptions are often called **resonances** because the

TABLE 11.2 Examples of isotopes with odd masses or odd atomic numbers with magnetic properties making them amenable to analysis by nuclear magnetic resonance

Nuclei
^{1}H, ^{2}H
^{13}C
^{14}N, ^{15}N
^{17}O
^{19}F
^{23}Na
^{25}Mg
^{31}P
^{33}S
^{39}K

(a)

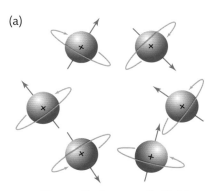

No applied magnetic field

(b)

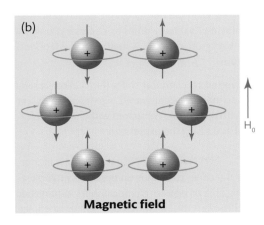

H_0

Magnetic field

FIGURE 11.14
(a) The spins of atoms are randomly orientated unless (b) they are subjected to an applied magnetic field (H_0).

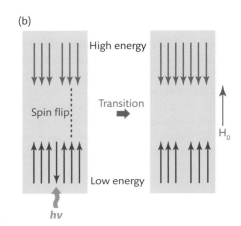

FIGURE 11.15
(a) Shows the non-random alignment of nuclei in the magnetic field H_0, using arrows to represent the direction of spin. (b) Pulses of radiation in the RF region of the electromagnetic spectrum are used to reverse the alignment of nuclear spins from the low energy spin aligned state (α) to the opposite higher energy spin state (β). The energy required for this transition depends on the strength of the applied magnetic field.

frequency of the RF radiation matches that at which the nuclei spin. The exact resonance or frequency of any given magnetic nucleus depends upon its exact chemical environment. This enables NMR spectroscopy to characterize the structures of chemical compounds. When the nuclei revert back to the equilibrium distribution, that is when those nuclei that have been excited to a high-energy spin state relax to their original low-energy state, RF radiation is emitted, and can be detected and measured. Nuclear magnetic resonance spectra are graphs of intensity against the RF applied.

Figure 11.16 shows the basic components of an NMR spectrophotometer. The sample is placed in the magnetic field and excited by pulses in the RF input circuit. The realignments of the magnetic nuclei in the sample induces a radio signal in the output circuit, which generates a complex output signal. The exciting pulse is repeated as many times as necessary to allow the signal output to be distinguished from any background noise. A mathematical (Fourier) analysis of the output produces the NMR spectrum.

In practice, the resonance frequencies are converted to chemical shifts (δ) to allow results obtained in different experimental conditions to be easily compared. Chemical shifts are defined as the resonance frequency associated with the sample compared with that of a reference compound whose chemical shift is defined as 0, using the expression:

$$\text{Chemical shift } (\delta) = (V_{signal} - V_{reference}) \times 10^6 / V_{reference}$$

You can see from this equation that the scale of the chemical shift is unitless. The multiplication by 10^6 is simply for the convenience of being able to express chemical shifts as parts per million (ppm). Tetramethylsilane [TMS, $Si(CH_3)_4$] is often the reference compound of choice

FIGURE 11.16
Basic arrangement of a nuclear magnetic resonance (NMR) spectrophotometer.

although other substances, for examples 3-trimethylsilyl 1-propane sulphonate (DSS), are used. Whatever the standard, its nature must be specified. You can see examples of NMR spectra in Figure 11.17 (a) and (b).

The frequencies and duration of the applied RF pulses may be varied in an enormous number of combinations allowing different molecular properties of the sample to be investigated. The exact frequency emitted by each nucleus depends upon its molecular environment. For example, protons in methyl groups ($-CH_3$) have chemical shifts of about 1–2 ppm, while those in aromatic rings have values of approximately 7–8 ppm. Thus, the frequencies differ for each nucleus unless they are chemically equivalent and in identical molecular environments. One-dimensional NMR spectra, like those in Figure 11.17, give valuable information about the structures of the molecules since each of the chemical shifts arises from a particular chemical functional group (Figure 11.18). Also, like the spectra obtained from IR spectroscopy (Section 11.5) the pattern of signals can act like a 'fingerprint' and help identify a particular sample.

SELF-CHECK 11.8

How many peaks would you expect to see in NMR spectra arising from the hydrogen atoms of (a) methyl bromide (CH_3Br) and (b) methyl ethanoate (CH_3COOCH_3)? Explain your answer.

In theory, it should be possible to obtain chemical shifts for all the hydrogen atoms present in a molecule and, therefore, obtain a one-dimensional NMR spectrum for any molecule. In practice, however, this is not possible for complex molecules such as those of polypeptides and nucleic acids. These molecules contain many hydrogen atoms and therefore the differences in their chemical shifts are smaller than the resolving power of the technique. In other

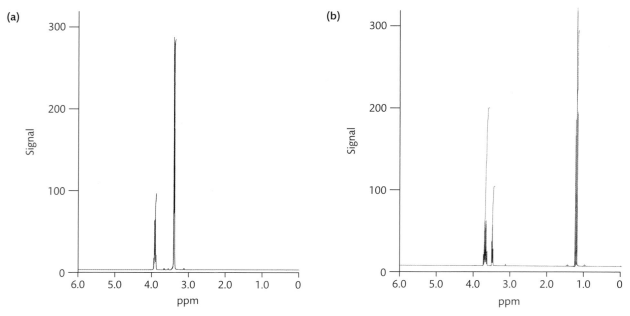

FIGURE 11.17
Nuclear magnetic spectra for (a) methanol and (b) ethanol.

words, the NMR signal is obscured by the presence of so many hydrogen atoms making it impossible to identify the signal from any one specific atom. Even using extremely strong magnets to increase the frequencies of the radio waves required to change the spin states of the nuclei, which increases the energy of the waves and, therefore, the resolving power of the technique, does not improve the situation. However, since the 1980s a variety of two-dimensional NMR spectroscopy techniques have been developed, which use a sequence of pulses separated by different time intervals, rather than the single burst associated with one-dimensional NMR. These techniques produce additional peaks that arise from the interactions of protons that are less than 0.5 nm apart in the molecules. This sort of data, together with other structural information, such as the lengths and angles of covalent bonds, allows the three-dimensional structures of complex molecules, such as those of relatively large proteins, to be predicted.

FIGURE 11.18
Typical values of chemical shifts for protons influenced by a single chemical functional group.

Clinical applications of nuclear magnetic resonance spectroscopy

Magnetic resonance spectroscopy (MRS) is a non-invasive technique used to investigate neurological disorders such as epilepsy, brain tumours, Alzheimer's disease, and strokes. This technique produces a spectrum or graph that identifies the types of chemicals in the brain (or other organs) and quantifies the amounts of each. The information provided is a useful indicator of the metabolic processes and biochemical changes occurring in diseased tissues. In contrast, nuclear magnetic resonance imaging (MRI, Box 11.2) provides a picture of the anatomical changes that occur in disease conditions.

11.9 **Mass spectrometry**

Mass spectrometry (MS) is a technique that allows the molecular masses of analytes to be accurately determined on the basis of their mass (m) to charge (z) ratios (m/z). Analytes must

BOX 11.2 *Nuclear magnetic resonance imaging*

Nuclear magnetic resonance imaging (MRI) is a medical imaging technique used to examine structures and functions of the body. The technique uses the nuclear magnetic resonance of protons in the water of healthy or diseased portions of the body to produce a signal that reflects the proton density. Since different tissues of the body contain different amounts of water, a computer can interpret the data in the signal to produce an image of the area being examined. The images are of higher resolution than those provided by other techniques such as X-ray photographs, computed tomography (CT), and positron emission tomography (PET) scans and the technique has the additional advantage of being non-invasive.

Magnetic resonance imaging is particularly applicable to viewing soft tissues areas, such as the spinal cord and the brain (Figure 11.19) and for cardiovascular, musculoskeletal, and oncological investigations. Nuclear magnetic resonance imaging can differentiate between pathological and normal, healthy tissues. For example, MRI is used to produce pictures of the blood vessel supplying the brain, neck, thoracic, and abdominal aorta, renal system, and legs. This information can be of use when investigating arteries for any abnormal narrowing or stenosis, or dilations of vessel walls or aneurysms. The technique also can be used to accurately identify the location, shape, and orientation of a tumour in a patient prior to radiation therapy.

A major advantage of MRI is that it is harmless to the patient because it uses strong magnetic fields and non-ionizing radiation. In contrast, the ionizing radiation utilized in traditional X-ray techniques may increase the likelihood of developing a malignancy particularly in foetuses. Furthermore, the resolution provided by MRI is as good as that provided by CT, but offers better contrast resolution, i.e. the ability to distinguish between two similar but not identical tissues.

FIGURE 11.19

Nuclear magnetic resonance image of a head. The MRI scan shows much more detail than an X-ray investigation. Courtesy of Dr G. Williams, Wolfson Brain Imaging Centre, Cambridge, UK.

be ionized to allow their masses and charges to be determined by passing them through electric or magnetic fields in a mass spectrometer. Molecular mass is usually expressed in daltons (Da); one Da is approximately equivalent to the mass of a hydrogen atom, or proton, or neutron. Thus, the mass in Da is numerically equal to the M_r. Mass spectrometry can be used to determine the M_r of compounds ranging in size from a few hundred, as would occur in pharmaceutical drugs, to several hundred thousand, such as those of macromolecular complexes. Mass spectrometry can also provide information regarding chemical structures and the primary structures of peptides and proteins. Both pure samples and complex mixtures from organic and inorganic sources can be subjected to mass spectrometric analyses. Thus, MS is a powerful technique, with a range of applications in the biomedical and biological sciences. Here, we can give only an overview of the general principles of MS and describe some of the techniques developed since the 1990s and their applications to biomedical science.

Mass spectrometers

Mass spectrometers generally consist of three parts: the ionizer or source region, the analyser, and the detector (Figure 11.20 (a) and (b)). Each part is held at high vacuum, of between 10^{-2} and

FIGURE 11.20
(a) Schematic showing the general organization of a mass spectrometer. (b) Overview of the operation of a mass spectrometer. An electric potential difference is used to accelerate ions through the system.

10^{-10} Pa, to limit collisions between analyte and residual molecules of air, which could reduce the accuracy of the analysis or even cause compounds to fragment. Typically subfemtomole ($<10^{-15}$) to low picomole (10^{-12}) amounts of sample are introduced into the mass spectrometer through a staged vacuum lock. Atoms or molecules in the sample need to be ionized by adding a H^+ to them to form a positive ion ($M + H^+$) or removing one to give a negative ion ($M - H^-$). Ionization occurs in the source region using one of a number of different techniques. Samples can be liquids, as with electrospray ionization (ESI), where they are dispersed into the gas phase using high voltage electric currents. Alternatively, solid samples can be mixed with a UV-absorbing matrix compound and irradiated using a UV laser. This forms sample ions in a gas phase, as is the case with matrix-assisted laser desorption ionization (MALDI). Less commonly used ionization techniques are electron ionization (EI), chemical ionization (CI), and fast atom bombardment (FAB). In EI, an electron beam is passed through the vaporized sample causing it to ionize. Samples can be chemically ionized by introducing a gas, such as methane, ammonia, or isobutene, in the source region and allowing it to collide with molecules of the analyte. In FAB, the analyte is mixed with a non-volatile matrix and then ionized by bombarding it with a beam of high-energy atoms, typically of an inert gas, for example argon or xenon. You can see the different uses and some applications of these ionization techniques in Table 11.3.

Whatever the method used, once ionized, the application of a potential difference is used to transfer the sample into the analyser as you can see in Figure 11.20 (b). The analyser uses static or dynamic electric or magnetic fields to separate the ions produced from the different components of the initial sample. Separation of the ions occurs because each has a characteristic, but different movement through the analyser, which depends upon its mass.

TABLE 11.3 Ionization techniques used in mass spectrometry

Ionization method	Analytes	Sample introduction method	Mass range/Da
Electron impact (EI)	Small molecules volatile	GC Solid Liquid	1–1000
Chemical ionization (CI)	Small molecules volatile	GC Solid Liquid	1–1000
Fast atom bombardment (FAB)	Carbohydrates Organometallics Peptides Non-volatile	Solid/liquid dissolved in glycerol	1–6000
Electrospray ionization (ESI)	Small molecules Proteins/peptides Nucleotides Carbohydrates	Liquid HPLC	1–2,000,000
Matrix-assisted laser desorption ionization (MALDI)	Small molecules Proteins/peptides Nucleotides Carbohydrates	Solid/liquid mixed with a solid or liquid matrix	1–5,000,000

TABLE 11.4 Common mass analysers used in mass spectroscopy

Analyser	Capability
Sector (magnetic/electrostatic)	High resolution, exact mass
Quadrupole (Q)	Unit mass resolution, fast scan, low cost
Time-of-flight (TOF)	Wide mass range, high throughput
Quadupole ion-trap (QIT)	MS/MS and MS^n capability
Ion cyclotron resonance (ICR)	Very high resolution, exact mass

The most common types of analysers used for biological applications are listed in Table 11.4; each type has advantages for different applications. A quadrupole (Q) MS consists of four (hence, quad) parallel circular rods. Oscillating electric fields are applied to opposing rod pairs, which causes the ions to separate in the quadrupole based on differences in the stabilities of their trajectories between the rods. In time-of-flight (TOF) analysers, electric fields are used to accelerate the ions to the same kinetic energy, so that the velocity of the ion over the flight path is dependent on its m/z ratio. Thus, the time to reach the detector also varies with the m/z ratio. Since all the particles travel over the same known distance to the detector, the times required for each particle to reach it can be measured allowing the M_r of the particle to be determined.

Quadrupole ion-trap (QIT) analysers are composed of three hyperbolic electrodes: a ring electrode and two end cap electrodes. The application of a potential difference to these electrodes, traps the ions in stable, oscillating trajectories within the cavity formed by the electrodes. These trajectories are governed by the applied voltage and the m/z ratio of the ions. Altering the voltage allows the ions to be ejected from the trap in order of their increasing m/z ratios, thus allowing the M_r to be determined. Once the ions are separated by the analyser, they are directed towards the detector again because of the applied electric field, which you can see in Figure 11.20 (b). The current produced by the ion impacting on, or passing through the detector surface, is recorded. This value is converted into an m/z value and a mass spectrum, typically a graph of m/z against ion signal intensity, produced as you can see in Figure 11.21.

FIGURE 11.21

Mass spectrum of the peptide with the sequence EGVNDNEEGFFSAR, where each letter stands for its usual amino acid residue, and of M_r 1569.65. The value for the ionized peptide $(M+H)^+$ was found to be 1670.68.

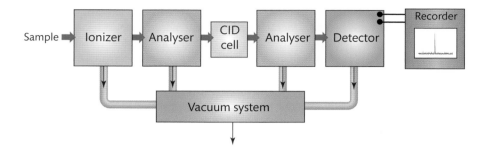

FIGURE 11.22
Schematic showing the
general organization of a
tandem mass spectrometer.

Mass spectrometers that have a single mass analyser can only perform simple M_r determinations, as described above. However, mass spectrometers can be designed with two mass analysers arranged in sequence. These instruments are referred to as tandem mass spectrometers. If you look at Figure 11.22, you can see the layout of a tandem MS, which produces spectra called MSMS.

The sample is introduced into the ionizer and the individual components of the mixture separated from one other in the first analyser and a selected ion is transmitted into the collisional dissociation (CID) cell. Here, the ion is fragmented when it collides with molecules of an inert gas, typically helium or argon, present in the CID cell. Fragments of ions from an individual component are analysed in the second analyser to give an MSMS spectrum that is representative of that compound. The fragmentation of the ions from any one individual component of the sample usually occurs in a predictable manner. Thus, the sizes of the fragments produced can be pieced together like a jigsaw to give information about the structure of the component.

In some specific cases, as with quarupole ion-trap instruments, several rounds of mass analysis can be performed in sequence within the same analyser. This analysis is performed by sequentially trapping and fragmenting the ions generated in the source. Specific fragments are trapped and fragmented in turn. This process can be repeated, giving up to 10 consecutive rounds of mass analysis. The resulting spectrum is commonly referred to as an MS^n spectrum, where n is equal to the number of consecutive mass analyses.

SELF-CHECK 11.9

Briefly describe the use of tandem mass spectrometry (MSMS).

The types of mass spectrometers most commonly used in biomedical sciences couple ESI or MALDI ionization methods with Q, QIT, or time-of-flight (TOF) analysers. These machines can determine the M_r of a wide range of materials. They have a high throughput: for example, one sample per minute for MALDI TOF mass spectrometers. Moreover, this speed of analysis is coupled to excellent accuracy, typically to 5–100 ppm of the real mass, and sensitivity. Furthermore, the results obtained using such techniques are reproducible, which enables the data obtained to be compared with that held in electronic databases.

Applications of mass spectrometry

Mass spectrometry can be applied to many different areas of the biomedical sciences. It is routinely used in hospital laboratories to screen newborn babies for inborn errors of metabolism, for example PKU, and the determination of illicit drugs in biological fluids, as we describe below.

Phenylketonuria is a genetic disorder caused by a deficiency of hepatic phenylalanine hydroxy-lase (PAH). This enzyme is necessary for catalysing the conversion of phenylalanine to tyrosine. Thus, in PKU, phenylalanine accumulates in the blood and if the condition is left untreated causes mental retardation. Phenylketonuria cannot be cured but can be treated by restricting the amount of dietary phenylalanine while ensuring an adequate intake of tyrosine. However, the mental retardation caused by PKU is irreversible; hence, its early detection in newborns is essential. Tandem mass spectrometry is routinely used to measure the molar ratio of phenyla-lanine to tyrosine in bloods spots collected from newborn babies. This method of investigat-ing PKU has been shown to be more effective than other types of screening. It is also used to screen for other treatable inborn errors of metabolism.

Mass spectrometry is also routinely used to test for illicit drugs in biological fluids. These test are normally performed on urine samples, as the concentrations of the drugs and their metab-olites in urine are relatively high compared with other fluids. Liquid chromatography or gas chromatography coupled to tandem mass spectrometry is a powerful technique in this area, providing selectivity, sensitivity, and reliability. It is also used for the study of drug metabolism, and for the investigation of the dynamics of endogenous biologically active substances, as well as contributing to the development of new therapeutics.

The research applications of MS include the analysis of peptide/protein (proteomics), metabo-lite studies (metabolomics), and carbohydrate analysis (glycomics). Emerging, although lesser used areas include mass spectrometric imaging and microorganism identification.

The proteome is all the proteins present in a given cell, tissue, or organism. Proteomics is the study of proteomes. However, the proteome changes in disease cells or tissues compared with healthy ones, both in composition and in the amounts of specific proteins present. Mass spec-trometry techniques involving ESI–MS and MALDI–MS can identify and quantify the proteins present in a sample, allowing tissues from normal (reference) and diseased states (patients) to be directly compared. For example, increases in the concentrations of proteins such as amy-loid-A protein are associated with the inflammatory response and can be detected in samples of plasma from patients with rheumatoid arthritis. Thus, proteins in a given tissue that change during the disease can be identified as a potential marker for that disease and/or as targets for drug therapy. This is often referred to as biomarker discovery.

Comparison of the proteins in clinical samples is generally done in an indirect manner called peptide mass fingerprinting (PMF). Samples of tissues are digested using a proteolytic enzyme, such as trypsin, to generate a set of peptides. The peptides are then analysed by MS and MSMS and the resulting mass spectrum is visually interpreted, as shown in Figure 11.23; alternatively it is matched against electronic protein databases containing similar data for known samples. The outcome of these types of analyses is a list of the proteins present in each sample, which can identify those proteins that may be a specific biomarker of the disease investigated.

In most diseases, however, it is not the presence or absence of a specific protein, but rather the *amounts* of a given protein, which increase or decrease compared with a healthy state. To identify such changes requires a quantitative technique to be used. In most cases a chemical labelling system is used where the proteins from the two states, diseased and healthy, are modified with tags of different masses to form heavy and lighter versions of molecules, which still have the same chemical formula. The heavy form is produced by incorporating a number of stable isotopes such as ^{2}H, ^{13}C, ^{15}N, or ^{18}O into the protein molecules thus forming a small, but easily measurable mass difference. The diseased and healthy states are labelled with the light and heavy versions, respectively. The two types are mixed, enzymatically digested, and the resulting peptides analysed by MS.

FIGURE 11.23

MSMS spectrum of a peptide. Note the difference in mass ($\Delta m = 97$ Da) between the consecutive peaks in the series is equivalent to the residue mass of the amino acid proline $M_r = 97$). Thus, a valid interpretation of the data gives the sequence, PPPPPPP.

Analysis of the resulting spectrum allows the quantitative ratio of a particular protein present in samples from patients and healthy individuals to be calculated. Analysis requires examination of the relative signal intensities of the ions from the heavy and light versions of both sets of peptides. A PMF analysis of the peptides allows the proteins in question to be identified. This, combined with the ion ratios allows the amounts or concentrations of any identified protein in the diseased and healthy samples to be calculated. Those proteins that change significantly may be potential biomarkers or indicators of disease.

In addition to the discovery of potential biomarkers, proteomics also includes the study of protein modifications, many of which have been implicated in disease pathways. These modifications occur post-translationally and may be catalysed by enzymes, for example phosphorylation and glycosylation, or occur non-enzymatically as is the case with glycation. Often proteins are not active until they are enzymatically modified in the cell. Analysis of protein modifications tends to exploit MSMS-based techniques, which are used to identify signature ions or differences in masses characteristic of a particular modification. The fragmentation pattern obtained by MSMS analysis can also identify the specific sequence of the modified peptide, thus identifying the site where the modification has occurred.

The *metabolome* is all the metabolites, which are molecules of small M_r, present in a biological system. These include, for example, many hormones, signalling molecules, as well as the chemical intermediates of metabolic pathways. It is believed that about 3000 metabolites are essential for normal human growth and development. Studying variations of these metabolites could be helpful in determining if a patient is fighting a disease, reacting badly to a drug therapy, or responding to another form of stress. This type of knowledge could have essential roles in optimizing therapy and management of a disease.

Cross reference

You can read about the various forms of chromatography in Chapter 13.

Mass spectrometry, along with NMR, is a major technique for analysing metabolomes. However, in this role MS is usually combined with a chromatographic separation method such as GLC or high performance liquid chromatography (HPLC), which are described in Chapter 13.

Gas liquid chromatography and HPLC greatly simplify the analyses by separating the components of complex clinical samples (urine, serum, cerebrospinal fluid for example) that are typically studied in metabolomics. This simplification means each component can be subjected to MS analysis in turn. This separation, coupled with the speed, high throughput, accuracy, and sensitivity of MS provides a powerful tool for the validation, evaluation, and monitoring of drug therapies, biomarker discovery, and clinical diagnosis.

Imaging by mass spectrometry was first used in 1999 and specifically uses MALDI MS. A section of tissue is placed directly in the source region of the MALDI mass spectrometer. The sample is moved in two dimensions as a laser is fired at it. Specific ions are formed at each point on the section by the action of the laser light. This procedure allows an image of the tissue section surface to be generated using a computer with appropriate software. This image can be manipulated to show the location of any given mass in the tissue section and therefore track, for example, where a drug and its metabolites are accumulating. This is illustrated in Figure 11.24, which shows mass spectrometric images of a mouse brain section highlighted for three different masses.

Mass spectrometry can be used to identify species of bacteria and fungi. Traditional methods for identifying these groups are time consuming and often relatively complicated. Using MS, identification can be carried out using standard proteomics-type techniques, as described above or, more elegantly, by an approach called intact-cell MALDI. The organisms are grown on agar plates until mature colonies are visible, then a single colony is transferred into the

Increasing signal intensity ➡   Increasing signal intensity ➡

m/z 14187   m/z 15035   m/z 6733

FIGURE 11.24

Mass spectrometric imaging of a section of tissue from a mouse brain. The upper panels show the relative distributions of compounds with m/z ratios of 14,187, 15,035 and 6,733, respectively. The corresponding lower panels are tissue sections to allow the sites of the compounds identified by the imaging to be mapped to specific areas of the brain.

MALDI source region and irradiated with a laser. As with proteomics, the analysis of these samples provides a pattern or fingerprint of the microorganism being investigated. This fingerprint is then compared with a database of known organisms and so identified provided it has previously been subjected to analysis. This technique has obvious uses in identifying pathogenic organisms in clinical samples.

SELF-CHECK 11.10

Which techniques are typically combined with mass spectrometry to simplify the analysis of complex mixtures?

 # SUMMARY

- Spectroscopy is concerned with the interactions of electromagnetic radiation and matter, particularly the absorption of radiation by matter and, in some cases, its emission and can be used to quantify molecules of interest in biomedical laboratories.

- Many molecules absorb light of characteristic wavelengths in the visible, ultraviolet, or infrared regions of the electromagnetic spectrum, and this is used as the basis for determining their concentrations using spectrophotometers and colorimeters.

- The absorbance of light of a specific wavelength by a solution of biological molecules depends on their concentration and the distance travelled by the light through the solution as described by the Beer–Lambert law.

- Infrared and Raman spectroscopy are specialized techniques in spectroscopy with applications in identifying biological molecules.

- Light scattering methods include turbidimetry and nephelometry. Turbidimetry can be used to estimate concentration by measuring the amount of light scattered by a suspension, whereas the more sensitive technique of nephelometry measures the amount of light scattered at an angle to the incident beam of light.

- Some molecules are fluorescent and are able to absorb light of a characteristic wavelength and then emit radiation of a longer wavelength. Fluorescence-based methods are highly sensitive, but are limited to molecules that fluoresce or which bind to one that is.

- Nuclear magnetic resonance (NMR) utilizes magnetic properties of nuclei within molecules and provides information on positions of atoms within their structures. Its value is in identifying molecules of biological interest and in determining their structures.

- Nuclear magnetic resonance imaging can distinguish between healthy and diseased tissues and is of value, for example, in locating the positions of tumours in patients.

- Mass spectrometry allows us to determine the masses of molecules based on their charge to mass ratios. The technique has wide applications in biomedical science, ranging from detecting phenylalanine when screening newborn babies for phenylketonuria and drugs of abuse in urine of suspected individuals, to identifying and sequencing proteins. This sophisticated technique is also the basis of locating the positions of molecules of interest in patients.

FURTHER READING

- Chary KVR, Govil G. *NMR in Biological Systems: From Molecules to Human*. Springer, Berlin, 2008.

- Gore MG. *Spectrophotometry and Spectrofluorimetry: a Practical Approach*. Oxford University Press, Oxford, 2000.

- Hoffman E, Stroobant V. *Mass Spectrometry: Principles and Applications*. John Wiley & Sons, Chichester, 2007.

- Honour JW. Benchtop mass spectroscopy in clinical biochemistry. *Annals of Clinical Biochemistry*, 2003; **40**: 628–638.

- Huettel SA, Song AW, McCarthy G. *Functional Magnetic Resonance Imaging*, 2nd edn. Sinauer Associates, Massachusetts, 2009.

- Moolenaar SH, Engelke UFH, Wevers RA. Proton nuclear magnetic resonance spectroscopy of body fluids in the field of inborn errors of metabolism. *Annals of Clinical Biochemistry*, 2003; **40**: 16–24.

- Riddle P. Principles of spectrophotometry. *Biomedical Scientist*, 2005; **49**: 135–137.

- Smith E, Dent G. *Modern Raman Spectroscopy: A Practical Approach*. John Wiley & Sons, Chichester, 2004.

- Stuart BH. *Infrared Spectroscopy: Fundamentals and Applications*. John Wiley & Sons, Chichester, 2004.

- Stuart B, George B, McIntyre P. *Modern Infrared Spectroscopy*. John Wiley & Sons, Chichester, 1998.

- Thomas MJK. *Ultraviolet and Visible Spectroscopy: Analytical Chemistry by Open Learning*. John Wiley & Sons, Chichester, 1996.

- Watson JT, Sparkman OD. *Introduction to Mass Spectrometry: Instrumentation, Applications, and Strategies for Data Interpretation*, 4th edn. John Wiley & Sons, Chichester, 2007.

QUESTIONS

1. Which of the following requirements must be satisfied to establish a colorimetric method to measure the concentrations of a specific analyte (X)?

    (a) Pure analyte (X) is needed for the production of a standard or calibration graph.

    (b) Analyte X should absorb light of a particular wavelength or it should be possible to convert it to one that can.

    (c) The wavelength of light at which maximum absorbance by X occurs needs to be determined.

(d) The relationship between concentration and absorbance of light for X must obey the Beer–Lambert law.

(e) All of the above criteria must be satisfied.

2. State which of the following are **TRUE** or **FALSE**?

(a) Spectrophotometers differ from colorimeters in only being able to analyse coloured compounds.

(b) Tungsten lamps in spectrophotometers are used to measure light in the ultraviolet region of the spectrum.

(c) Light of the desired wavelength can be selected using a monochromator.

(d) Plastic cuvettes are used when measuring light in the ultraviolet region of the spectrum.

(e) The relationship between concentration and absorbance is not linear when an analyte obeys the Beer–Lambert law.

3. A 1.0 mmol dm^{-3} solution of a compound gives an A_{340} of 0.6. Calculate (a) the molar extinction coefficient (ε) and (b) the absorptivity (a) of the compound.

4. What would be the advantage of using a standard curve formed from a range of concentrations of an analyte compared with estimating its value from the absorption of a single solution of known concentration?

5. An aqueous solution of yeast DNA has an absorptivity of 25 m^2 mol^{-1} at 260 nm. Calculate the percentage transmission of a solution of the DNA of concentration 2.0 mol m^{-3} if the absorbance is measured in a standard-sized cuvette.

6. What happens to light as it passes through a turbid suspension contained in a cuvette in a colorimeter?

7. In ^{1}H-NMR spectroscopy, the amino acid alanine [$^+NH_3CH(CH_3)COO^-$] produces two major chemical shifts at 1.5 and 3.8 ppm, respectively, using TMS as standard. Which hydrogens (protons) are responsible for each of these signals?

8. Describe the major components of a mass spectrometer and give an example of each.

9. Give three examples of how mass spectrometry is used in the biomedical sciences.

Answers to self-check and end of chapter questions are available in the Online Resource Centre accompanying this book.

Go to www.oxfordtextbooks.co.uk/orc/glencross

12

Centrifugation

Qiuyu Wang, Nessar Ahmed, and Chris Smith

Learning objectives

After studying this chapter, you should be able to:

- Describe the basic features of a centrifuge
- Explain the general principles of centrifugation
- Calculate the speeds and forces involved when separating mixtures by centrifugation
- Describe the major types of centrifuges, tubes, and rotors encountered in clinical laboratories
- Discuss the principal techniques used in centrifugal separations
- Outline the precautions to be taken in the safe use of a centrifuge
- List some of the uses of centrifuges in a clinical laboratory

Introduction

Given sufficient time, gravity will cause many particles, such as cells, suspended in a liquid to eventually settle to the bottom of their container. However, this is not usually a practical method of separating or isolating the particles as the time required for them to sediment is normally too long. Indeed, particles that are extremely small in size will not separate at all unless they are rotated at a sufficiently fast speed that the rotational or centrifugal force generated causes the particles to move radially away from the centre of rotation. **Centrifugation** is the term applied to the mechanical process of separating mixtures by applying centrifugal forces. It is one of the most widely used techniques in biochemistry, molecular and cellular biology, and in the biomedical sciences. Applications of centrifugation are many and may include harvesting viruses and cells from growth media, the separation of subcellular organelles, the collection of lipids, and the isolation of macromolecules, such as DNA, RNA, and proteins. For example, centrifugation can be used to separate blood into plasma and blood cells, or obtain cells from body liquids, such as the ascites fluid that accumulates in the abdomen in some pathological conditions. Centrifugation can be applied to a great many clinical and biological materials, so for the sake of convenience in this chapter we will generally refer to them all as *particles*.

Centrifugation is performed in a **centrifuge**. At its simplest, a centrifuge is a mechanical device that produces a centrifugal force by spinning a sample around a central axis. They consist essentially of two compartments. One contains a motor whose drive shaft enters the second compartment, called the rotor chamber, where it can be attached to a specialized container called the **rotor**. You can read about rotors in Section 12.3. The rotor contains liquid samples in suitable centrifugation tubes or bottles, although in some applications the samples are centrifuged in plastic bags. The centrifugal force generated by the motor-driven rotation of the rotor separates components of the mixture that differ in size and/or density. In general, the mixture is separated into two fractions. One fraction is the **pellet** at the bottom of the tube that contains the sedimented material; the other is the liquid **supernatant** that contains unsedimented material. Whatever the type, all centrifuges operate in essentially similar ways. However, different types are designed for a variety of uses, which you can read about in Section 12.4, and so differ in design details. **Centrifuges** vary from relatively simple bench top models to sophisticated floor standing instruments that must be sited with care (Figure 12.1). All types of centrifuges have a brake to rapidly slow and stop the spinning rotor following centrifugation and incorporate a number of safety features, such as locking lids that can be opened only when the rotor has stopped spinning.

12.1 **Basics of centrifugation theory**

The rate of sedimentation of a particle when spun in a centrifuge depends upon the **centrifugal field** or **acceleration** (G), which is directed outwards from the centre of rotation. The magnitude of this field is described by the equation:

$$G = \omega^2 r$$

where ω is the angular velocity of the rotor in radians per second and the **radius** of the particle from the centre of rotation is r in cm (Figure 12.2).

FIGURE 12.1
Outline structures of (a) bench and (b) floor standing centrifuges.

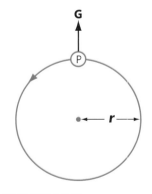

FIGURE 12.2
Diagrammatic representation of the centrifugal field (G) operating on a spinning particle (P), which is a distance of r cm from the centre of rotation.

One revolution equals 2π radians. However, the speeds of centrifuges are always given in revolutions per minute (rev min^{-1}), although often expressed as rpm. Hence, the angular velocity of a spinning rotor is given by:

$$\omega = 2\pi \text{ (speed in rev min}^{-1}\text{)}/60$$

So, if $G = \omega^2 r$

$$G = 4\pi^2 \text{ (speed in rev min}^{-1}\text{)}^2 \, r/60^2$$

It is usual to express G as a **relative centrifugal field** (RCF), that is in terms of the earth's gravitational field (g), which is an acceleration of 981 cm s^{-1} per s, giving:

$$\text{RCF (in } g) = 4\pi^2 \text{ (speed in rev min}^{-1}\text{)}^2 \, r/60^2 \times 981,$$

which simplifies to:

$$\text{RCF (in } g) = 1.12 \times 10^{-5} \text{ (speed in rev min}^{-1}\text{)}^2 \, r$$

Alternatively, the RCF can be determined using a **nomogram** (Box 12.1).

If it is to be repeated successfully, an initial report of an experiment or procedure that involved centrifugation must state the rotor dimensions, so that r can be determined, and the speed of the centrifugation, or the RCF and rotor radius plus, in each case, the duration of the centrifuge run.

Sedimentation velocity and sedimentation coefficient

The value of G (that is the RCF) influences the rate of sedimentation or sedimentation velocity (v) of a particle, i.e. how fast it is moving through the liquid when it is centrifuged. However, v also depends upon a number of other factors, including the size of particle, its density and **frictional coefficient**, and the density of the solvent, which are interrelated in the following manner:

$$v = M_r(1 - \bar{v} \, \rho) \, \omega^2 r/Nf$$

The mass of the particle measured in grams is m_0, but this is converted into its relative molecular mass (M_r) by the introduction of Avogadro's number (N, 6.02×10^{23}). The densities are accounted for by $\bar{v}$ and ρ, which are the **partial specific volume** of the particle (that is the change in volume that occurs when the particle is added to a large excess of solvent) and the density of the solvent, respectively. The sedimentation of a particle through the solvent is opposed by a frictional resistance that is equal to the product of its velocity and frictional coefficient (f), that is the vf of the particle. Thus as v and f become larger, the resistance also increases. The frictional coefficient depends upon the shape and size of the particle. It is larger

BOX 12.1 Nomograms

Nomograms consist of three scales. A straight line that connects a value on one scale to that of another on either of the other scales gives the corresponding value on the third scale. The use of nomograms relating the radius of rotation, speed of the rotor, and RCF can save a lot of arithmetic! Figure 12.3 shows how a nomogram can be used to estimate the speed needed to produce a specified RCF for a given radius of rotation.

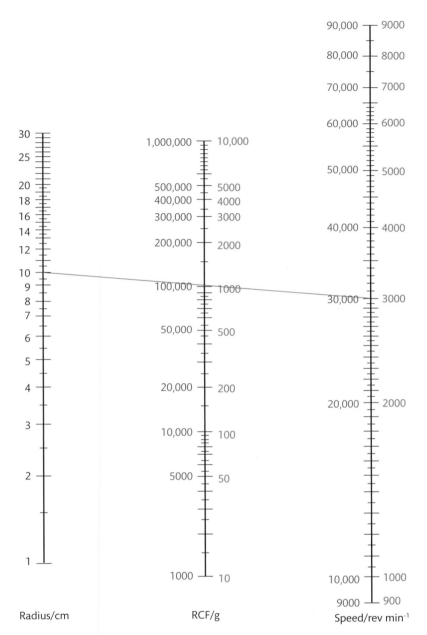

FIGURE 12.3
Nomogram to estimate the speed in rev min^{-1} needed to produce a specified RCF for a given radius of rotation. Note the line giving the RCF of 100,000 or 1000 g for a radius of 10 cm and specified speeds of 30,000 and 3000 rev min^{-1}, respectively.

SELF-CHECK 12.1

Use Figure 12.3 to estimate the speed required to produce an RCF of 50,000 g for a rotor of radius 7.67 cm.

---

for asymmetric molecules, for example nucleic acids and fibrous proteins, than for spherical ones, such as globular proteins, of similar M_r.

The rather complex expression given above that relates all these terms can be simplified by introducing a new term, the **sedimentation coefficient** (*s*) of the particle, which is the ratio of its velocity to the applied centrifugal field, that is

$$M_r(1 - \bar{v}\,\rho)/Nf = v\,/\omega^2 r = s$$

or, more simply:

$$s = v/\omega^2 r$$

The units of the sedimentation coefficient (*s*) are those of time. Many biological particles have values of *s* of 1×10^{-13} to several thousand $\times 10^{-13}$ second. For convenience, they are usually given as **Svedberg units** (S), where $1\ \text{S} = 1 \times 10^{-13}$ s (see Box 12.2). Try not to get too confused with all the forms of the letter s!

Bigger particles generally have larger sedimentation coefficients (whether given in s or S!), but the sedimentation coefficients of individual particles in an aggregate cannot simply be added together to give that of the complex, since the sedimentation velocity is also affected by the *shape* of the particle. For example, the two subunits of a bacterial ribosome have M_r of approximately 1.0×10^6 and 1.8×10^6 and sedimentation coefficients of 30 and 50 S, respectively. The subunits combine to form the complete ribosome with the expected M_r of 2.8×10^6. However, its sedimentation coefficient is only 70 S, a situation that arises because the value of *s* depends also on the *shape* of the particle. Temperature and the nature of the solvent also affect the value a of particle's sedimentation coefficient, so they are normally corrected to the defined conditions of 20°C in water, which are expressed as $s_{20,\,w}$. Figure 12.4 lists the sedimentation coefficients of some particles of biomedical interest.

BOX 12.2 *Theodor Svedberg*

The Svedberg (S) is a non-SI unit of time, equal to 1×10^{-13} s, used in characterizing the behaviour of particles during centrifugation. It was named after Theodor Svedberg (1884–1971), a Swedish physicist and chemist. Svedberg studied at the University of Uppsala, where he obtained his doctorate in 1907. He was elected Professor of Physical Chemistry at the University of Uppsala in 1912. Svedberg's work largely involved the chemistry of colloids and proteins whose macromolecules are too small to settle out of the surrounding water by gravity alone. To better study these particles, Svedberg and co-workers began developing the **ultracentrifuge**, for which he is best remembered, to produce the centrifugal forces necessary to sediment them. This involved combining mechanics and optics in the instrumentation and using mathematics to describe and analyse the behaviour of the particles. In 1924 they constructed an ultracentrifuge capable of generating a centrifugal force of up to 5000 *g*. Svedberg was able to perform experiments with the machine to calculate the M_r of the milk protein, casein, and eventually that of haemoglobin. Later centrifuges generated hundreds of thousands of times the acceleration of gravity. In 1933–34 RCFs of more than half a million *g* could be achieved. Studies based on the ultracentrifuge by Svedberg were instrumental in demonstrating that the molecules of certain pure proteins are all of a single size.

Svedberg was awarded the 1926 Nobel Prize in Chemistry for his development of the ultracentrifuge.

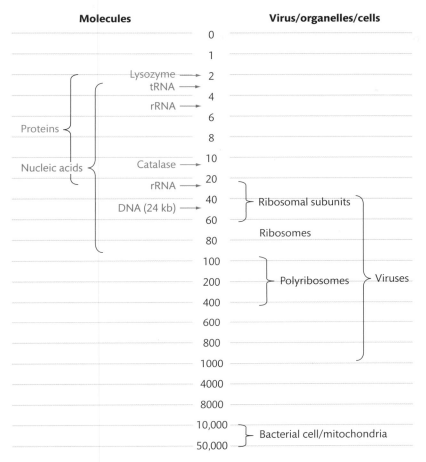

FIGURE 12.4
The sedimentation coefficients (s) of some structures (particles) of biological and biomedical interest.

SELF-CHECK 12.2

An enzyme has M_r 14,400 and a sedimentation coefficient (s) of 1.9 S. However, when bound to an inhibitor of only M_r 220, the sedimentation coefficient of the complex was only 1.6 S. How can you account for this change?

12.2 **Types of centrifuges**

The main types of centrifuge most likely to be encountered in biomedical or biological sciences laboratories are:

- Bench centrifuges
- Continuous flow centrifuges
- High speed refrigerated centrifuges
- Ultracentrifuges

Bench centrifuges (Figure 12.1(a)) are the commonest type and so the most likely type to be encountered. Small bench centrifuges generally have maximum speeds in the range of

4000 to 10,000 rev min^{-1} and generate RCFs of 3000–7000 g. They are mainly used to collect material that sediments rapidly, such as erythrocytes, yeasts, and others cells, and relatively granular precipitates of, for example, proteins. Larger bench centrifuges develop similar RCFs, but can accommodate a greater variety of tubes with sample volumes of 5–250 cm^3. They can also accommodate a number of adapters that allow, for example, the 96-well plates used in enzyme-linked immunosorbent assays you can read about in Chapter 15, to be centrifuged. Many high throughput biochemical-based assays that require the rapid yet efficient separation of rather coarse precipitates or cells depend upon bench centrifuges.

Microfuges are small bench top centrifuges capable of reaching speeds of 12,000–15,000 rev min^{-1} producing RCFs of about 12,000 g in only a few seconds. They are used to process samples of less than 1.5 cm^3 contained in Eppendorf tubes. Microfuges are indispensable tools in molecular and cell biology, being used to harvest small volumes of cells, isolate micro- and milligram quantities of proteins and nucleic acids, and even ensure that all the reagents used in some experiments, which may have volumes of only a few microlitres, are 'pelleted' to the bottom of the tube and so mix together thoroughly.

Continuous flow centrifuges are a form of high speed centrifuge but the sample material is added continuously throughout the separation process. The process is shown graphically in Figure 12.5. A vertical rotor is accelerated to the desired rotational speed and the suspension to be separated is added continuously through an input port and rotated to form a vertical liquid column. The solvent and any unwanted smaller particles are immediately washed to the surface and are discharged through an outlet pipe. The larger particles are, however, forced towards the rotor wall, where they accumulate and can be recovered when centrifugation is completed. Alternatively, they can be collected in an outflow, as shown in Figure 12.5. In some circumstances, continuous

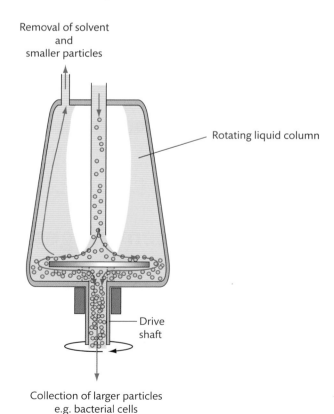

FIGURE 12.5
Outline of a continuous flow centrifuge. See text for details.

flow centrifuges can save up to 80% of the time required by discontinuous centrifugation and provide a rapid and reliable means of separating large volumes of suspension that have low solid matter content. This makes them ideal for harvesting cultures of bacterial and human or animal cells from fermentors and bioreactors in just a single centrifugation.

High speed refrigerated centrifuges maintain a predetermined fixed temperature, often set at 4°C, during the centrifugation. Refrigeration is necessary to counteract the heat produced by friction between the rapidly spinning rotor and air molecules in the chamber. Low temperatures help maintain the stability and activities of clinical and biological specimens. High speed refrigerated centrifuges can accept sample volumes of up to 1.5–2 dm^3 and are capable of speeds of about 25,000 to 28,000 rev min^{-1}, with corresponding maximum RCFs of up to approximately 100,000 g. High speed refrigerated centrifuges are used to collect most precipitates, harvest larger viruses and bacterial, animal, and plant cells, and in separating many cell organelles from homogenized tissues.

The distinction between bench and high speed centrifuges is becoming increasingly blurred, given some bench top types can achieve speeds of approximately 20,000 rev min^{-1}.

Ultracentrifuges (Figure 12.1(b)) are not only refrigerated, but the rotor chamber is also evacuated to prevent friction between air molecules and the rotor increasing the temperature of the rotor and sample during centrifugation. They are capable of speeds of about 100,000 to 150,000 rev min^{-1}, with corresponding maximum RCFs of up to approximately 800,000 g. Two types of ultracentrifuge are used: preparative and analytical types. Preparative ultracentrifuges are used, for example to collect membrane fractions and ribosomes, or to prepare samples of viruses and isolate macromolecules. As with high speed refrigerated centrifuges, preparative ultracentrifuges can use a variety of types of rotors and so a range of sample volumes can be processed depending upon the size and type of rotor accepted by the centrifuge.

Analytical ultracentrifuges incorporate some form of optical system to observe the sedimentation of the particles, which are often macromolecules or molecular aggregates. This allows biophysical analyses to be performed on the particles during centrifugation. The use of small, tabletop micro-ultracentrifuges has made ultracentrifugation particularly popular for investigating the hydrodynamic properties of biological particles, as outlined in Box 12.3.

BOX 12.3 Analytical ultracentrifugation

Estimating the purity of samples and determining the M_r of molecules is usually performed using chromatographic and electrophoretic methods, which are quick and convenient. You can read about these procedures in Chapters 13 and 14, respectively. However, the M_r obtained by these methods is estimated by comparison with standards of known size. An advantage of analytical ultracentrifugation is that the M_r value and other biophysical properties of macromolecules are given from first principles. Such analytical methods and the analyses of data were relatively time consuming. Also, while high speed refrigerated centrifuges and preparative ultracentrifuges are bulky floor standing instruments, large by any standards, analytical ultracentrifuges were even bigger! However, analytical ultracentrifugation can now be performed with microprocessor controlled tabletop micro-ultracentrifuges. These instruments accept small volume samples, up to about 4 cm^3, but can reach speeds of 100,000–150,000 rev min^{-1} and generate RCFs of over 800,000 g. This allows analytical experiments to be performed in a fraction of the time previously associated with analytical centrifugation. This, together with contemporary methods of data capture and analysis means that analytical ultracentrifugation is often the method of choice for determining a range of hydrodynamic properties of biological molecules. These properties may include, for example, determining the numbers of subunits that bind together in complexes, the value of the associated equilibrium constant, and characterizing the conformational changes undergone by a macromolecule on binding with a ligand.

12.3 **Tubes and rotors**

During centrifugation the samples are generally contained in specialized tubes. The tubes are made of a variety of materials, including glass, polyallomer, polycarbonate, and polypropylene, and range in volume from less than 1 cm^3 to over a litre. Safe working practice means choosing the most suitable type of tube, rotor, and centrifuge for the separation or experiment to be performed. Most centrifuges can be used with a variety of rotors that accept tubes which differ in volume, or by using a single rotor that has adapters for different tubes.

Rotors can be broadly classified into three common types:

- Swing out bucket rotors
- Fixed angle rotors
- Vertical tube rotors

You can see how the rotors differ from one another in Figure 12.6. In swing out bucket rotors, the tubes are placed in pivoted buckets that move from the vertical to horizontal

FIGURE 12.6
Schematics showing the basic structures of (a) swing out bucket (b) fixed angle and (c) vertical tube rotors. Note that in each case, the effective radius of rotation varies along the length of the tube from a minimum (r_m) to the maximum (r_{max}) value. Relative centrifugal fields are often quoted for the average radius (r_{av}).

as the rotor gathers speed. As their name implies, fixed angle rotors hold the tubes at a constant angle, which generally varies from 14° to 40°. Vertical tube rotors are a variant of the fixed angle type, where the tubes are at right angles to the horizontal. During centrifugation, rotors are subjected to considerable stress (see Section 12.5) and must be made of appropriately strong material. Those used in bench centrifuges are generally composed of steel. However, high speed and ultracentrifuges traditionally used rotors made of aluminium alloy or titanium.

Aluminium is easily corroded especially by alkalis. However, aluminium rotors are given an anodized (aluminium oxide) surface that resists corrosion. Unfortunately, it is thin and easily scratched by sharp metal objects. Titanium is resistant to corrosion and can withstand higher centrifugal forces than aluminium, but has the disadvantage that it is much more expensive. Metal rotors have the further drawback of being heavy; not only to handle, but also to accelerate in the centrifuge. The weight of a metal rotor also contributes to the potential hazards involved in using centrifugation (see Section 12.5), since the forces involved are the *product* of the weight of the fully loaded rotor and the RCF applied. If a rotor fails during a centrifugation, then the large forces involved mean nearby persons are in potential danger. Also, considerable damage can be done to the centrifuge and its surroundings.

Rotors made of carbon fibre composites are available, which are lighter than metal ones in weight. These can be accelerated and decelerated more rapidly than metal types resulting in quicker centrifugations and reduced maintenance costs.

SELF-CHECK 12.3

What is the force (apparent weight) of a rotor of weight 13.6 kg in (a) tonnes and (b) tons, when centrifuged at 150,000 *g*? Assume 1 kg is 2.2 lb.

12.4 **Separation methods using centrifuges**

If you examine Table 12.1, you should notice that many materials of biomedical interest differ in size and/or density. They can therefore be separated or studied by centrifugation. The two commonest methods used to separate particles in centrifugation are differential centrifugation and density gradients.

TABLE 12.1 **Sizes and densities of some materials of biological and biomedical interest**

Particle/structure	Size	Density/g cm^{-3}
Haemoglobin	~8 nm	~1.35
Nucleic acids	variable	~1.6
Viruses	20–50 nm	1.3–1.45
Ribosomes	~25 nm	~1.6
Mitochondria	0.5–2 μm	1.19
Lysosomes	0.05–1.0 μm	1.21
Peroxisomes	0.1–1.5 μm	1.23

Differential centrifugation

Differential centrifugation, as the name implies, separates particles mainly on *differences* in the sizes of the particles. It is the commonest centrifugation technique and is illustrated diagrammatically in Figure 12.7. However, to achieve an effective separation of two particles they must differ in size by an order of magnitude, that is one must have a diameter 10-times larger than the other.

The centrifuge tube is filled with the sample solution and then centrifuged. Depending upon its size and the conditions applied during the centrifugation, any one component of the initial mixture may end up in the supernatant, the pellet, or even be distributed between the two fractions. However, since all the components were initially distributed evenly throughout the sample, a pellet of larger particles will always be contaminated with smaller ones. It is, of course, possible to carefully decant off the supernatant. The pellet can then be resuspended and recentrifuged, which will improve its purity. The supernatant can also be recentrifuged but at a higher speed and possibly longer time to allow smaller particles to be obtained as a pellet.

Separation by differential centrifugation is commonly used in simple pelleting to isolate cells from blood and in obtaining partially pure preparations of subcellular organelles and macromolecules for example.

Density gradient centrifugations

Density gradient centrifugation is performed in a centrifuge tube that contains a column of liquid whose density increases towards the bottom of the tube. This separates particles in a mixture with greater resolution than discontinuous centrifugation because larger particles are not contaminated with smaller ones. The increase in density can be in discrete steps forming a discontinuous gradient or it can be continuous as shown in Figure 12.8. Discontinuous or step gradients are suitable for separating whole cells or subcellular organelles and or isolating some types of viruses. Continuous gradients are the more commonly used. Like step gradients they are used in isolating cell organelles and some viruses but also hormones, enzymes, and other proteins, as well as nucleic acids.

Whether continuous or discontinuous, the gradient is formed by dissolving a suitable solute in an appropriate solvent. Sucrose dissolved in water or buffer is common but, depending on the sample, other materials can be used. The sample to be separated is generally layered on top of the gradient and then centrifuged. Two methods of density gradient centrifugation are routinely used: rate zonal and isopycnic as you can see in Figure 12.9.

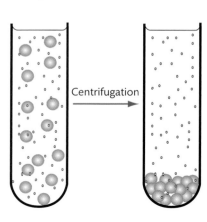

FIGURE 12.7
Outline of separation by differential centrifugation.

Rate zonal centrifugation (Figure 12.9(a)) separates particles because they sediment through the gradient at different velocities that depend upon their sedimentation coefficients. They thus separate into discrete zones, with each consisting of one type of particle. The zones of the separated particles are stabilized because the gradient slows diffusion and convection currents. The length of the gradient must be sufficient to allow the zones to separate. However, given that the densities of the particles in the sample must exceed that of the densest part of the gradient, the centrifuge run must be terminated before any or all of the zones reach the bottom of the tube. In isopycnic density gradient separation (Figure 12.9(b)), the density gradient is such that it exceeds the densities of all the particles in the sample. Thus, particles will sediment in the gradient to a position that equals their own density. Once all the particles have reached these positions, they will remain there and further centrifugation will not increase the migration of the particles.

A variety of materials are used to form gradients suitable for density gradient separations in biomedical and biological applications. For example, gradients of sucrose and CsCl are commonly used to separate subcellular fractions and nucleic acids, respectively. In many density gradient centrifugations, separation is really achieved by a combination of rate-zonal and isopycnic principles. Rate zonal centrifugation is often used to purify cellular organelles (Box 12.4) and to isolate proteins, such as immunoglobulins, since all classes of immunoglobulins have similar densities but differ in their masses. Rotors must be selected with care, since the choice affects the efficiency of the separation.

Choice of rotor

In fixed angle rotors, the particles move radially outwards during centrifugation as shown in Figure 12.10(a). Thus they travel only a short distance before hitting the tube wall. The particles then slide down the wall and form a pellet at the bottom of the tube. This makes fixed angle rotors effective for the differential collection of precipitated materials and harvesting cells. Swing out bucket rotors are prone to cause convection currents during deceleration and the particles also have long sedimentation paths. This makes them inefficient for pelleting particles by differential centrifugation but useful for rate-zonal (Figure 12.10(b)) and isopycnic techniques. Vertical tube rotors are popular for isopycnic centrifugation. During the centrifugation the sample and gradient reorientate as shown in Figure 12.10(c). This result is an effectively short sedimentation distance and rapid separation. However, the resolution of the particles is less effective than when a swing out bucket rotor is used.

(a) [Sucrose] /% (w/v)

5

10

15

20

(b)

FIGURE 12.8
(a) Continuous and (b) discontinuous density gradients.

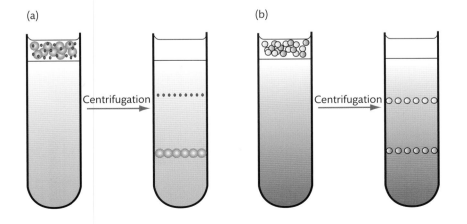

(a)   (b)

Centrifugation →   Centrifugation →

FIGURE 12.9
Outline of separations by (a) rate zonal and (b) isopycnic density gradient (vertical rotor), centrifugations.

FIGURE 12.10
The behaviour of particles during (a) differential in a fixed angle rotor, (b) rate-zonal (swing out rotor) and (c) isopycnic centrifugations.

BOX 12.4 *Subcellular fractionation*

If a cell homogenate is subjected to differential centrifugation, then generally larger particles, such as organelles in this case, will sediment faster than the smaller ones. This is the basis of obtaining fractions enriched in specific subcellular organelles. Thus, centrifugation of the homogenate at 1000 g for 10–20 minutes will result in a pellet consisting largely of unbroken cells and larger nuclei. The supernatant can be recentrifuged at 10,000 g for 20 minutes, which will mainly pellet mitochondria and the larger types of lysosomes and microbodies. Centrifuging the supernatant at 105,000 g for 60–90 minutes will sediment smaller subcellular structures, such as ribosomes and the artificial vesicles formed from fragments of the endomembrane system and the plasma membrane that are often called microsomes. The particles remaining in solution are largely macromolecules and are generally regarded as the soluble fraction.

Obtaining partially purified organelles by differential centrifugation is usually only a preliminary purification step; the fraction is normally purified further using some form of density gradient separation.

FIGURE 12.11
Surface view showing alternative arrangements for balancing four tubes in a rotor capable of accommodating eight. Filled tubes are indicated by the solid surfaces. Notice how the tubes are distributed symmetrically, with filled tubes always opposite one other.

Balancing rotors

Whatever the type of rotor or centrifugation procedure to be used, the rotor must be **balanced**; that is the weight of the tubes and their contents must be distributed evenly around the centre of rotation. If the rotor is imbalanced, it can damage the drive shaft of the centrifuge and even cause it to move. You can see an example of how tubes can be added to a rotor to balance it in Figure 12.11. It is necessary to attach *all* of the buckets when using a swing out rotor even if you are not using all of them for sample tubes. This is because the buckets and their lids are an integral part of the balancing of the rotor.

12.5 **Safety and centrifuges**

Centrifuges, especially high-speed and ultracentrifuges, are potentially dangerous pieces of equipment. Care and attention to detail must be applied at all times during their use. However, if you follow the procedures outlined below, then the associated risks will be reduced to a minimum. Note that staff in charge of any centrifuge must read standard operating procedures for their particular machine, taking note of any Medicines and Healthcare Products Regulatory Agency (MHRA) or Health and Safety Executive (HSE) safety alerts.

Cross reference
General aspects of health and safety are described in Chapter 4.

Qualified users

Centrifuges must only be operated by suitably trained persons or under their supervision. All staff/students must be trained by someone who is competent in their use before they use any centrifuge. If you notice a fault, report it promptly, do not attempt repairs yourself: only authorised and suitably trained persons may service or repair a centrifuge. Do not use the centrifuge until the fault has been repaired. A log book must be kept for ultracentrifuges and their rotors, as the number of hours used determines the life of the rotor. Excessive use can lead to metal fatigue, where microscopic changes in the metal structure can lead to small cracks, which eventually enlarge to the point of causing metal failure. Where necessary, the details of the centrifugation must be entered in the centrifuge log book. Before using a centrifuge, the rotor chamber should be checked to ensure that it is clean and free of dust or loose debris and condensation. Check also that the drive shaft is clean and free of scratches or burrs.

Rotor faults

Faults with a rotor can lead to its failure: the greater the speed used, the more likely this is to occur and the more extreme the results. Always examine a rotor for *cleanliness* and *damage*. Any extraneous material including dirt in the rotor cavities will affect its balance. Dirty rotors must be cleaned by an approved method using, for example, plastic coated brushes and mild detergents. Soaps and many detergents are alkaline and can attack the anodized surfaces of aluminium rotors. The body of a swing out bucket rotor should never be immersed since hanging mechanisms are difficult to dry and can corrode. Only the buckets should be washed. After cleaning, the rotor must be rinsed with deionized water and air dried with the buckets or cavities pointing downwards. Corrosion, cracks, or dents on any part of the rotor are potentially hazardous and even the most careful design will not protect against its misuse and abuse. When centrifuged, rotors are subjected to stress causing them to be stretched. You can see the effects of stress diagrammatically in Figure 12.12. The first stage of stretching constitutes the elastic limit and the rotor is able to return to its original size when the stress, that is the centrifugation, stops. However, above a certain level of stress, plastic strain occurs and the rotor will not regain its former dimensions. Plastic damage can lead to cracks in the fabric of the rotor and cause its eventual failure. Damaged rotors must *not* be used and must be reported to the appropriate supervisor.

Check that all **O rings**, which are gaskets consisting of a flat rubber or plastic ring used to seal a joint against high pressure, are in place and in good condition. Rotors must never be used if their O rings are missing or perished. Chill the rotor in a refrigerator to the desired temperature before placing it in the rotor chamber since this minimizes the chance of it seizing to the drive shaft. Close the rotor lid when pre-cooling it to prevent condensation in the chambers or buckets.

Tube checks

Use only correctly fitting tubes and check that they are made of a material that is compatible with the sample solvent. Balance the rotor to within its specified limits as described in Section 12.4. Do not operate the centrifuge without the appropriate rotor lid securely fitted and its seals in place. Never exceed the maximum stated speed for any rotor. Following centrifugation, do not remove the centrifuge tubes or bottles from the rotor with a metal object. Use only the tools provided by the centrifuge manufacturer for that purpose.

After checks

Check the rotor chamber and the rotor for any leaks and clean if necessary. Return the rotor to its normal storage place. Rotors must be stored inverted in a dry environment.

FIGURE 12.12
Schematic showing the effects of stress on a typical rotor alloy.

SELF-CHECK 12.4

You need to isolate viral particles from a cell culture lysate, when you notice that the aluminium rotor you were going to use has some fine hairline cracks in its surface. You do, however, have a pressing need to prepare the sample. What should you do?

12.6 **Examples of clinical centrifugation**

In addition to freestanding centrifuges, a number of automated instruments, for example DNA extractors (see Chapter 16), encountered in clinical laboratories include 'built-in' centrifuges to help process specimens. Specialist centrifuges for use in routine clinical laboratories are uncommon. One specific type, the **cytocentrifuge** or slide centrifuge is described more fully in the *Cytopathology* text of this series. A cytocentrifuge is essentially a low speed centrifuge with a modified rotor that accommodates microscopic slides as well as liquid samples. The centrifugal force separates the cells and sediments them as a monolayer onto the slides, while preserving their integrity. This simple technique can be used, for example, to isolate cells from samples of cerebrospinal fluid or fine needle aspirates directly for cytological examination. Haematocrits are a type of centrifuge that were used to determine the ratio of the relative volumes of erythrocytes and plasma in blood samples but they have been superseded largely by automated blood analysis systems (see Chapter 17), which can determine all haematological indices. The multibucket centrifuges used for serum separations are probably the commonest type found in biochemistry, immunology, and haematology pathology laboratories; although microfuges are used in a range of applications as discussed in Chapter 16, and refrigerated centrifuges find use in specialist complement work (Chapter 15) and some biochemistry assays. For example, adaptors can allow the 96-well plates used with enzyme-linked immunosorbent assays (ELISAs) to be directly centrifuged allowing a high throughput of samples (Figure 12.13).

Blood is the most frequently requested of clinical samples that require centrifugation. Once collected, it is added to a blood tube. Most blood tubes are made from a strong plastic like

Cross reference

You can read about immunosorbent assays in Chapter 15 'Immunological techniques'.

FIGURE 12.13

Centrifuge with a specialized rotor (R) for centrifuging enzyme linked immunoassays (ELISA) plates. An ELISA plate is indicated (P). Courtesy of Department of Clinical Biochemistry, Manchester Royal Infirmary, UK.

polyethylene terephthalate (PET), and so the blood sample can be directly centrifuged. Many blood tubes also have an inert gel barrier added to them during their manufacture, which aids in the centrifugal separation. The specific gravity of the gel is between those of the blood clot and the serum. During centrifugation the barrier gel moves upward and forms a distinct dividing barrier between the plasma and cells or the serum and clot, respectively (Figure 12.14). Thus, aliquots of the sample can be taken directly from the tube for analysis. Most of the analytes routinely investigated in clinical laboratories remain stable several days if the tubes are used according to manufacturer's instructions and if stored at 4°C. The introduction of separator gels has enhanced the use of *primary* samples for clinical analyses. A primary sample is one presented to the analyser in the same container in which it was originally collected from the patient. Aliquots may be taken directly from the serum/plasma and directly sampled by the analyser.

Cross reference

You can read more about obtaining blood samples in Chapter 7, and more about blood tubes in Chapter 17.

Manufacturers generally colour code blood tubes depending on whether they have a gel barrier, and whether they include an anticoagulant or not and, if so, what type of anticoagulant is present. For example, tubes with grey coloured caps contain oxalate and fluoride, and are used for samples to determine blood glucose measurements. The oxalate binds Ca^{2+} and prevents coagulation, while fluoride is an inhibitor of hexokinase, and so prevents glycolysis. The effect of the two additives is to maintain the concentration of blood for several days. Purple-capped tubes also contain an anticoagulant, in this case the metal chelator, ethylenediaminetetraacetic acid (EDTA). A green cap indicates the tube contains the anticoagulant lithium heparin. These tubes are available with and without gel barriers. The heparin allows the sample to be centrifuged immediately, so whole blood or plasma can be tested. Tubes with red and yellow caps both contain clot activators and are used in most routine chemistry tests on serum. Centrifugation of the former prevents analytes moving between the serum and the clot. The yellow capped tubes have a gel barrier, which allows for primary sampling.

(a)   (b)

FIGURE 12.14

(a) Blood tube containing a blood sample (B). Note the position of the separator gel (G). (b) Blood tube with a separator gel following centrifugation. Note how the gel (G) has moved up the tube to form a layer between the cells and clot (C) and serum (S). Courtesy of Department of Clinical Biochemistry, Manchester Royal Infirmary, UK. See also Figure 7.4.

The automated chemistry analysers used to perform many of the standard clinical tests on patient samples contain centrifuges, which may be stand alone devices or occur as discrete units on a tracked system. The tubes are centrifuged in specialized racks. Robots can perform all the operations associated with centrifugation of the samples. Thus, they transfer the sample tubes onto a balance to weigh them and then allocate each tube to an appropriate position on a centrifuge rack. This ensures the final load in each centrifuge rack is balanced before they are transferred to the centrifuge. Blood specimens are centrifuged at 1000 g or higher for eight to 12 minutes. Following centrifugation, the robot removes the racks from the centrifuge and may decap the tubes before transferring the racks onto a track to transfer to other units for clinical testing. Centrifugation is used extensively in blood banks for a number of procedures (Box 12.5).

Cross reference

The sampling of blood, the centrifugation of clinical specimens, and the use of primary samples are described more fully in Chapters 1 and 2 of *Clinical Biochemistry*.

BOX 12.5 Blood banks

For a number of clinical reasons, blood that contains leukocytes cannot be transfused legally in the UK. Hence, usually within a few hours of collection, leukocytes are removed from blood by filtering it in a process called leukodepletion. The filters used are leukocyte-specific and trap them but not the smaller erythrocytes or platelets. Leukodepletion does not involve centrifugation, but centrifugation is used extensively in blood banks for a number of procedures.

To ensure safety, the blood is tested to check that it is not contaminated with harmful microorganisms and to determine its blood group. Blood must be typed for the ABO and Rhesus (Rh) blood group systems. The ABO system depends on the presence of antibodies in the serum of individuals, which are complementary to the antigens on the erythrocytes. It is these antibodies that determine the identity of the blood group. Thus, individuals of group A have antibodies in their serum to blood group B antigens and those of blood group B possess antibodies to group A. Group AB individuals do not have either type of antibody; however, group O individuals possess both. ABO blood groups are identified for all donors and recipients to ensure that blood used for transfusion is of the same or a compatible type. If ABO incompatible blood is administered to a patient, the antibodies may mediate an immediate and fatal transfusion reaction.

The terms Rh positive and negative refer to the presence or absence of the Rh(D) antigen. An Rh incompatibility between mother and foetus is a common cause of haemolytic disease of the newborn in Caucasians.

The majority of transfusion laboratories now use gel technology for blood grouping and compatibility testing, which involves some centrifugation. Recent years have seen increasing use of the Diamed typing system to detect haemagglutination. This system uses typing monoclonal antibodies (mABs), which are distributed in a gel contained in individual tubes set in plastic 'cards'. Erythrocytes to be typed are added to the antibodies and the cards are centrifuged using a specialized rotor. Specific blood groups are recognized by the complementary mAB and agglutination occurs. In those tubes where agglutination occurs, the agglutinates remain on top of the gel during centrifugation, whereas nonagglutinated cells sediment through it to the bottom of the tube (Figure 12.15).

In addition, the *production* of blood products, such as erythrocytes, platelets, and plasma, all involve the large-scale centrifugation of blood using dedicated centrifuges.

(a)

(b)

FIGURE 12.15

(a) A centrifuge with a specialized rotor that allows gel cards to be centrifuged for blood group typing as outlined in the text. (b) A typical gel card used in blood group typing. Courtesy of the Manchester Blood Transfusion Service, UK.

SUMMARY

■ Centrifugation is a separation technique conducted in an instrument called a centrifuge. A centrifuge holds a spinning rotor containing samples in tubes that are rotated around a central axis by a motor. This spinning of the rotor generates a centrifugal force that separates the components in the mixture on the bases of differences in size and/or density.

■ Particles sediment when subjected to a radial acceleration or centrifugal field (G). This is normally expressed as a relative centrifugal force (RCF) as multiples of the earth's gravitational field (g).

■ A variety of centrifugation tubes, rotors, and centrifuges are available, which allows a range of different centrifugation methods to be used as appropriate.

■ The major centrifugation techniques used are differential and density gradient centrifugation. The former separates particles on differences in size. Density gradient separation can use rate zonal and isopycnic centrifugations. Particles that differ in their sedimentation coefficients can be separated by rate zonal centrifugation, while isopycnic methods rely on differences in the densities of the particles.

■ All appropriate safety considerations must be observed when using centrifuges. For example. only clean and undamaged rotors must be used.

■ Centrifugation has applications in biomedical science where it is commonly used, for example, in separating blood materials or isolating DNA. Clinical laboratories use specialist cytocentrifuges to isolate components onto slides prior to cytological analysis. In research laboratories, centrifugation is used in preparing fractions of cellular constituents such as macromolecules and organelles.

FURTHER READING

● Cole JL, Hansen JC. Analytical Ultracentrifugation as a contemporary biomolecular research tool. *Journal of Biomolecular Techniques*. 1999; **10**: 163–176. Laue T. Biophysical studies by ultracentrifugation. *Current Opinion in Structural Biology*. 2001; **115**: 579–583.

These two journal articles provide a valuable synopsis of the uses of analytical ultracentrifugation in characterizing the behaviour of macromolecules.

● Goodman, T. Centrifuge safety and security. *American Laboratory*, 2007; **39**: 20–21.

An interesting article detailing design features relating to safety and security that are available on a number of centrifuges.

● Goodman T. Centrifuge rotor selection and maintenance. *American Laboratory*, 2007; **39**: 12–14.

An interesting short read that describe the different types of centrifuge rotors, their applications, and also outlines procedures for their safe handling and maintaining their life.

- **Goodman T. Choosing the right centrifuge for your application. *American Laboratory*, 2008; 40: 28–29.**

  A short, useful article, whose title says it all.

- **Graham JM. *Biological Centrifugation*. Bios Scientific Publishers, Oxford, 2007.**

  Excellent coverage of the basics of centrifugation. Also includes a number of useful centrifugation protocols.

- **Rickwood D. (ed.) *Preparative Centrifugation – a Practical Approach*. Oxford University Press, Oxford, 1992.**

  A very helpful reference text.

- **Rickwood D, Ford T, Steensgaard, J. *Centrifugation: Essential Data*. John Wiley & Sons, Chichester, UK, 1994.**

  A pocket-sized (128 pp) volume that provides core data in tabulated formats for performing centrifugal separations of macromolecules and cells.

QUESTIONS

1. Calculate the RCF to the nearest 100 g, experienced by a particle 8.0 cm from the centre of rotation when it is centrifuged at 3500 rev min^{-1}.

2. Which of the following centrifuges:

    - Microfuge
    - Ultracentrifuge
    - Bench centrifuge
    - High speed refrigerated centrifuge

    would be your first choice to:

    (a) Isolate bacterial cells from liquid medium

    (b) Collect a fine precipitate of DNA molecules

    (c) Isolate mitochondria from a tissue homogenate

    (d) Harvest the relatively small poliomyelitis viral particles (diameter 28 nm)

3. Which of the following would not affect the sedimentation velocity of a particle during centrifugation?

    (a) Solvent density

    (b) Atmospheric pressure

    (c) Particle density

    (d) Distance of the particle from the centre of rotation

    (e) Angular velocity

    (f) M_r of the particle

4. A viral particle takes 10 minutes to move 1 cm when centrifuged at 20,000 rev min^{-1}. How long would it take to move the same distance if the speed was doubled?

5. Which would have the larger frictional ratios (f) and sedimentation coefficients (s): enzyme molecules or DNA fragments of comparable M_r?

6. Non-filamentous influenza viral particles have a sedimentation coefficient (s) of 750 S. How far would a virus particle travel in 10 minutes, when centrifuged at 30,000 rev min^{-1} at the r_{av} position of 6.12 cm?

7. Serum albumin has a sedimentation coefficient (s) of 4.6 S. Determine its M_r, given its partial specific volume ($\bar{v}$) and frictional coefficient (f) are 0.74 cm^3 g^{-1} and 6.63×10^{-8} s^{-1}, respectively. The density of water at 20°C is 0.998 g cm^{-3}.

8. You are going to perform a centrifugation experiment using CsCl (a corrosive salt) when you spill a *little* of the solution into a rotor bucket. You do, however, have to complete the experiment quickly, since the results are needed in a hurry. Should you ignore this small spillage and proceed with the experiment?

Answers to self-check and end of chapter questions are available in the Online Resource Centre accompanying this book.

 Go to www.oxfordtextbooks.co.uk/orc/glencross

13

Chromatography

Qiuyu Wang, Nessar Ahmed, and Chris Smith

Learning objectives

After studying this chapter, you should be able to:

- Define the term chromatography
- Discuss the general basis of chromatographic separations
- List some of the major chromatographic techniques
- Describe the theoretical bases of specific chromatographic techniques
- Discuss the relative efficiencies of some chromatographic methods
- List some of the uses of chromatography in the clinical and biological sciences

Introduction

Chromatography was first used by the Russian botanist Mikhial Tsvet, who used it to separate plant chlorophyll and carotenoid pigments. Tsvet first gave a written description of the method in 1903, although he did not coin the term **chromatography**, which he adopted from the Greek words chroma, colour in reference to the plant pigments, and graphein, meaning to write, until 1906. Chromatography is now the collective term for a family of analytical techniques used to separate the components of mixtures of molecules for their identification and possible estimation of their concentrations in the original mixture. A major difficulty that is often encountered when investigating clinical and biological samples is the presence of substances that can interfere or be confused with the substance or **analyte** under investigation. This means it is often necessary to separate and isolate a particular analyte from other contaminants in the mixture to allow its concentration to be determined. The principal uses of chromatographic methods are to remove contaminating materials from the analyte of interest, to aid in the identification of unknown analytes, and to estimate the concentration of analyte in the sample.

All chromatographic techniques are based on differences in the relative affinities of the molecules in a mixture of two different and **immiscible** phases, one of which is **mobile** and the other **stationary**. Mobile phases may be liquids or gases that flow over or through the stationary phase, which can be a solid or an immobilized liquid. Different combinations of phases can be used. For example, in column chromatography, particles of a hydrated gel are used as

FIGURE 13.1
Schematic illustrating the separation of two substance using differences in their relative affinities for two phases in chromatography.

the stationary phase and a liquid moving through the gel is the mobile phase. In many cases these have been developed into extremely high resolution techniques by forcing the liquid through the column under high pressure. These latter methods are usually referred to as high performance liquid chromatography or HPLC. Whatever the specific type of chromatography, different substances in a mixture can partition differentially between the two phases and so are separated from one another. You can see this schematically in Figure 13.1. Even complex mixtures of closely related molecules can be separated if an appropriate combination of phases is selected.

Most types of chromatography are used in biomedical science laboratories to assist in analysing and purifying analytes from a variety of clinical samples. You will be introduced to a number of these applications in the chapter.

13.1 **Partition or distribution coefficient**

The basis of all forms of chromatography is the **partition** or **distribution coefficient**, (K_d), which describes the way in which a substance distributes at equilibrium between two immiscible phases. For example, the K_d of a substance X, which distributes between the two immiscible phases A and B is given by:

$$K_d = \text{The concentration of X in A/The concentration of X in B}$$

given that the distribution of X between A and B is at equilibrium. You can see an illustration of this in Figure 13.2.

Partition coefficients are normally determined at 20 or 25°C. Often one of the two phases is water and the other a hydrophobic solvent such as octanol. This gives an octanol–water partition coefficient, K_{OW}. Note that the K_{OW} is not the same as the ratio of the solubility of the substance in octanol to that in water because the two phases are not fully immiscible! At equilibrium, octanol contains 2.3 mol dm^{-3} of water and the water phase 4.5×10^{-8} mol dm^{-3} of octanol. The measured values of K_{OW} for organic compounds vary from 10^{-3} to 10^{7}. This is an enormous range. Hence they are often quoted as their logarithmic value (log K_{OW}), which condenses the range from −3 to 7. The K_{OW} value describes the relative hydrophobicity/hydrophilicity of the compound in question: the lower the value, the more hydrophilic the compound; the larger it is, the more hydrophobic.

$$k_d = \frac{[x]_A}{[x]_B}$$

FIGURE 13.2
The definition of the partition coefficient for a solute (X) following its equilibrium partitioning between the two immiscible phases, A and B.

SELF-CHECK 13.1

What are the units of K_d, K_{OW}, and log K_{OW}?

In chromatography, the mobile and stationary phases must be chosen such that the molecules to be separated differ in their partition coefficients. This, in turn, depends upon them differing in some particular physical property. If you examine Table 13.1 you will see some of the properties upon which different types of chromatographies rely.

Polar molecules have an unequal distribution of electrons due to differences in the electronegativity of the atoms from which the molecule is composed. For example, water is a polar molecule, as oxygen has a greater electronegativity than hydrogen. This means the electrons are more attracted to the oxygen leaving less of an electron density around the hydrogens. As a result, the oxygen acquires a slightly negative charge and the hydrogens have an equal but positive one (Figure 13.3).

The polarity of molecules influences their solubility and **adsorption** on to surfaces. Thus, in chromatographic methods based on differences in solubility and polarities, the particles partition to different extents between two immiscible liquids, and so separate. This is often

Adsorption is the accumulation of molecules from a solution on the surface of an adjoining solid without penetrating its interior.

TABLE 13.1 A summary of the differences in physical properties of molecules in the mixture that can be used as a basis for their separation by various forms of chromatography

Property	Chromatographic techniques
Solubility and polarity	Paper chromatography Thin layer chromatography (TLC)
Size (mass)	Gel filtration chromatography
Charge	Ion-exchange chromatography
Hydrophobicity	Hydrophobic chromatography Gas–liquid chromatography (GLC)
Ligand binding	Affinity chromatography Metal chelation chromatography

FIGURE 13.3
Structure of a water molecule.

described as 'differential partitioning'. We will describe how differences between the solubilities and adsorption properties of molecules are exploited in paper and thin layer chromatographies in Section 13.2. Differences in the *sizes* of molecules can also be exploited in gel filtration chromatography. Molecules may become charged under certain conditions. These charges can be used as a basis for their separation, as in ion-exchange chromatography. Hydrophobic and gas–liquid chromatographies exploit differences in the partitioning of hydrophobic molecules between two phases. Finally, differences in how molecules bind to ligands are utilized as the basis for their separation in affinity and metal chelation chromatographies (Sections 13.2–13.5).

The major types of chromatography likely to be encountered in biomedical science laboratories are those grouped together as:

- Planar techniques
- Column chromatographies
- Gas–liquid chromatography

Each of these and some of their uses will be discussed in turn.

13.2 Planar chromatography

Planar or open bed chromatography uses a stationary phase consisting of a thin, essentially two-dimensional, sheet. The two major types you are likely to encounter are paper and thin layer chromatography. In paper chromatography, a sheet or a narrow strip of paper binds the stationary phase. In thin layer chromatography (TLC), a film of solid particles, bound together for mechanical strength with a binder such as calcium sulphate, is coated on a supporting plate and also binds a stationary liquid phase. Both are examples of partition chromatography because molecules distribute themselves between bound and mobile liquid phases, although adsorptive effects to the particles in TLC also contribute to the separation.

Paper chromatography

In paper chromatography, the stationary phase consists of a layer of water or other polar solvent molecules that are bound to the cellulose fibres of the paper, even though it may appear dry. Materials to be separated are applied to the paper, generally as a series of discrete spots. One end of the paper is then dipped into the mobile liquid phase, which moves through the paper by capillary action. The molecules to be separated are differentially partitioned between the stationary and mobile phases and so move along the paper at different rates and separate. Paper chromatography is a useful method for assessing the purity of substances but is now seldom used in clinical laboratories having been superseded by TLC.

Thin layer chromatography

Thin layer chromatography relies on the differential partitioning of substances between stationary and moving liquid phases. In this respect, it resembles paper chromatography. A solvent is used to suspend an adsorbent material, which is then bound to an inert glass, plastic, or aluminium plate as a thin layer and allowed to dry. The thin layer of solvent remaining in the bound adsorbent forms the stationary phase (hence the name). A range of materials are

TABLE 13.2 Suitable adsorbents for TLC and their applications

TLC adsorbent	Used in separating
Aluminium oxide	Steroids
Calcium phosphate	Proteins
Cellulose	Proteins
Magnesium carbonate	Porphyrins
Silica gel	Steroids, amino acids, lipids

suitable for forming the adsorbent layer and, indeed, mixtures of materials can be used. Some of them are listed in Table 13.2; the choice depends on the material to be analysed.

Again, a mobile liquid phase migrates through the thin layer by capillary action. A wide range of solvents may be used as mobile phase, including water, ethanoic acid, ether, hexane, methanol, and benzene, which can be mixed together in different combinations and proportions as appropriate. A general procedure for TLC is outlined in the Method Box 'Generalized procedure for TLC' and illustrated in Figure 13.4.

METHOD Generalized procedure for TLC

■ The adsorbent material is prepared by mixing it with a binding agent such as calcium sulphate or plaster of Paris (10%) and water to form a slurry. The binding agent helps hold the adsorbent to the plate.

■ The slurry is applied to the plates as a thin uniform layer using a plate former and allowed to set.

■ The plates are activated by heating in an oven at 100–120°C to remove free liquid, with the remainder forming a thin layer bound to the adsorbent particles. The final thickness of the adsorbent layer is between 0.25 and 2 mm.

■ A 1–20 μdm^3 (μL) sample is applied to the plate as a concentrated spot or band using a micropipette or microsyringe and allowed to dry.

■ The TLC plate is placed in a tank which contains the mobile phase and developed for 30–90 minutes, as shown in Figure 13.4. The atmosphere within the tank must be saturated with solvent vapour in equilibrium with the mobile phase to minimize evaporation from the plate surface and prevent the composition of mobile phase changing due to losses of its volatile components. The mobile phase travels up the plate by capillary action. Development is stopped when the moving edge of the mobile phase (usually referred to as the *solvent front*) approaches the top of the plate (Figure 13.4).

■ Following their separation on the TLC plate, the positions of the separated components of the sample can be determined and the presence of the analyte of interest confirmed and its concentration in the sample possibly estimated.

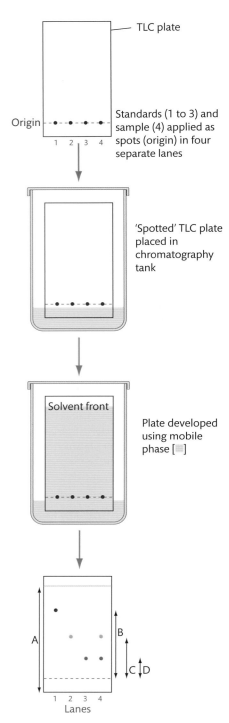

FIGURE 13.4

General procedure for thin layer chromatography. See text and Method box 'Generalized procedure for TLC'. In this case, lanes 1, 2, and 3 have been used to 'run' substances of known identity ('standards'). Lane 4 was used for an unknown sample. On the basis of its migration, the unknown would appear to contain standards 2 and 3.

As with all chromatographic techniques, the substances to be separated partition differentially between two phases. Initially, the stationary phase consists of the liquid retained by the adsorbent layer during its drying. Separation is achieved because as the mobile phase carries sample and standards up the thin layer, the more polar molecules in them will associate with the thin layer. In contrast, the relatively hydrophobic substances will be more soluble in the mobile phase and be carried further along the plate. However, the mobile phase may be a complex mixture and contain polar liquid molecules. As these move through the thin layer by capillary action, they may bind to the adsorbent material and will form what is essentially a new stationary phase, which may assist in the separation. Thin layer chromatography plates have a large surface area. Hence, significant adsorption of sample molecules to the particles can also take place. This ability is often due to the presence of hydroxyl groups (–OH) on their surface, which can form hydrogen bonds with molecules in the sample. The more polar the molecules in the sample, the greater the strength of the bonds and the more adsorption occurs. Adsorption to the particles can increase the effectiveness of separation. It is apparent that the factors that affect separation in TLC are complex!

Following its development, it is necessary to detect the separated components on the plate. A number of methods are available to achieve this, including the use of:

- Ultraviolet (UV) light
- Specific colouring reagents
- Autoradiography
- Sulphuric acid

Examining the TLC plate under ultraviolet (UV) light will show the positions of compounds that absorb or fluoresce in the presence of UV light. For example, nucleotides and nucleic acids absorb UV light, and so appear as dark spots. Spraying of plates with specific colouring reagents can be used to stain certain compounds. For example, ninhydrin will react with amino acids to form purple-coloured complexes, which will indicate their location on the plate. Alternatively, the TLC plates can be sprayed with sulphuric acid and then heated to 110°C. Most organic compounds are charred by this procedure and show up as brown-black spots or bands. Radioactively labelled compounds will present as dark spots or bands on X-ray film after the TLC plates have been subjected to autoradiography (Box 13.1).

Cross reference
You can also read about autoradiography in Chapter 10 'Radioactivity and radiation'.

It is usual to calculate the **R_f** (relative to front) value, which is the distance travelled by any component relative to that moved by the solvent using the expression:

$$R_f = \text{migration distance of analyte/migration distance of solvent front}$$

BOX 13.1 *Autoradiography*

Autoradiography is a simple and sensitive method of producing a photographic image of a specimen containing radioactively labelled material(s). The image is formed by placing an X-ray film directly in contact with a chromatogram, histological preparation, or electrophoresis gel. These are then wrapped in a material which prevents external light fogging the film. However, radiation from the labelled materials impinges on the film causing the deposition of silver grains in its emulsion. The length of exposure varies depending upon the intensity of the radiation source: it can be as little as a few minutes or up to several days. The package is often stored in a freezer at −70°C during longer exposures to reduce the diffusion of the molecules in the sample. Once the film is developed, it forms a radiograph, whose patterns of deposited silver record the location(s) of radioactive substance(s) in the specimen. The intensity of the developed spots or bands on the film is proportional to the dose of radiation and, thus, the concentration of the labelled material.

Cross reference
See the *Histopathology* book in the same series and Chapter 14 'Electrophoresis'.

In theory, R_f values are identical for any given substance analysed under the same chromatographic conditions. However, given experimental conditions vary, it is more usual to also 'spot' known reference compounds, in addition to the samples to be analysed, and subject them all to TLC. Component(s) in a mixture can then be tentatively identified by comparing the R_f value(s) to those of the standards. You can see this diagrammatically in Figure 13.4, where the unknown compounds in the sample (Lane 4) have migrated the same distances (C,D) relative to the front (A) as the standards in Lanes 2 and 3.

SELF-CHECK 13.2

Is it possible to have an R_f value of 1.2 for a particular analyte? What effect would changes in temperature have on the TLC? What difference would an increase in room temperature make to R_f values?

Thin layer chromatography has a number of advantages over some other chromatographic techniques, particularly over paper chromatography. One particular advantage is its versatility. If you examine Table 13.2 you will see it can be applied to the separation of a number of differing types of substances. Moreover, separation is quick, often requiring as little as 30–90 minutes, it is simple to perform and is inexpensive. Hence, it is often the first method of choice, particularly when the components of the mixture are known to be structurally similar. Its major disadvantage is that it is not especially suitable for quantitative measurements, for example for determining the amounts of each substance in a mixture. However, TLC is used successfully to identify a number of substances of clinical interest, including drugs, lipids, amino acids and vitamins. Figure 13.5 illustrates its use in separating a number of drugs of abuse.

Once separated using TLC, analytes can be located on the plate using UV light, and any of potential interest can be extracted from the plate and identified using gas chromatography-mass spectrometry.

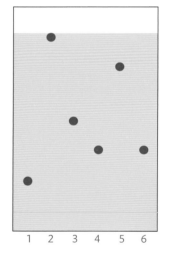

1   2   3   4   5   6

FIGURE 13.5

Thin layer chromatographic separation of morphine (Lane 1), cocaine (2), nicotine (3), methadone (4), heroin (5), and a clinical sample (6) using silica gel as the adsorbent material and a mobile phase consisting of ethanol: dioxin: benzene: ammonium hydroxide (4: 40: 50: 5).

CASE STUDY 13.1 Investigating galactosaemia

Joe, a child of 7 months age, was suspected of having galactosaemia. His urine sample and three known sugars, glucose, fructose, and galactose, were subjected to thin layer chromatography. The following R_f values were obtained:

Glucose	0.42
Galactose	0.35
Fructose	0.39
Urine sample	0.35

QUESTIONS

(a) Which sugar is most likely to be present in Joe's urine sample?

(b) Is the diagnosis supported?

Calculate the R_f value of the drugs and the unknown analyte used in Figure 13.5. What is the most likely identity of the unknown analyte?

13.3 Column chromatography

In column chromatography, the stationary phase is organized into a column by packing it into a glass or steel cylinder (Figure 13.6), which is in equilibrium with a liquid mobile phase. Equilibrium is achieved by using a gravity-based system or motorized pump to move the

FIGURE 13.6

Outline of a general column chromatography system (see text for details).

mobile phase through the column. The sample is added to the top of the column as a band, which is then flushed through the column by the mobile phase. Individual components in the sample are separated within the column and eluted from it at different times as the mobile phase leaves the column. The liquid leaving the column and carrying the (hopefully) separated molecules is often called the eluant. It is normally collected in discrete fractions in a series of tubes, either manually or automatically using a fraction collector. The different fractions can be analysed and an elution profile or chromatogram constructed by plotting a graph of the amounts of substances eluted against time, eluant volume, or fraction number, as shown in Figure 13.6. Hopefully, different fractions will contain different components of the original mixture and so it will at least have been partially separated.

Whatever the form of column chromatography, it is necessary to be able to detect the analytes of interest as they leave the column in the elution buffer. If the analytes are coloured then this is not a problem, but many materials of biomedical interest are colourless. However, most proteins and nucleic acids are colourless, but absorb light at about 280 and 260 nm, respectively This is also the case for many other compounds of clinical interest. Thus they can all be detected by monitoring the column using UV light, which will indicate when they are being eluted.

It is also possible to exploit biological activities of some substances to monitor their separation. For example, in the cases of enzymes the fractions of eluant can be analysed for the reaction they catalyse. Carbohydrates are often detected by monitoring the refractive index of the eluant, which changes when a sugar is present.

The movement of any one analyte through a column is described by its individual retention time (t) or elution volume (V_e). The retention time for an analyte is the period between addition of this substance to the column and its maximum concentration in the eluant. The term V_e is the volume of mobile phase required to elute a particular analyte from the column. It is used particularly when referring to separations by gel filtration (see below). The time taken for an analyte to pass through the column without any interaction with the stationary phase is the dead time (t_o) and the volume of mobile phase required for its elution is the void volume (V_0).

Following column chromatography, the resulting chromatogram will show discrete peaks for each of the separated components, which are symmetrical in nature (Figure 13.7 (a)). In some cases, asymmetrical peaks are produced where there is a slow rise at the beginning of the peak and a sharp fall after the peak. This is called **fronting** (Figure 13.7 (b)) and is due to overloading of the column with the sample. Occasionally peaks occur where there is a normal rise before the peak but a slow fall after the peak and this is referred to as **tailing** (Figure 13.7 (c)). It is usually due to retention of analytes on certain binding sites, usually hydroxyl groups, on the stationary phase. Tailing can be eliminated by removing these hydroxyl groups by treating the gel with a silanizing agent, such as hexamethyldisilazine.

The ability of a column to distinguish and separate two similar analytes a and b is given by its **selectivity** (α), which is essentially the difference in retention times between two analytes (a and b). It is calculated using the expression:

$$\text{Selectivity } (\alpha) = (t_b - t_0)/t_a - t_0$$

where t_a and t_b are retention times for a and b, respectively. The greater the value of α, the more selective the column.

The **resolution** (R) of any separation technique is its ability to separate (or resolve) completely one analyte (a) from another, similar one (b). In column chromatography, it can be determined using the expression:

$$R = 2(t_{Rb} - t_{Ra})/W_a + W_b$$

Cross reference

See Chapter 11 'Spectroscopy' and 15 'Immunological techniques'.

(a)

(b)

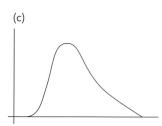

(c)

FIGURE 13.7

(a) Shows the symmetrical peaks following an ideal separation of two components by column chromatography. Components isolated by column chromatography showing (b) fronting and (c) tailing.

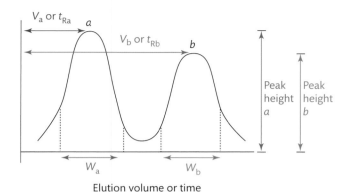

FIGURE 13.8
Definitions of elution volumes or times, peak heights, and peak widths.

where W_a and W_b are the base widths of the peaks for a and b, respectively, as shown in Figure 13.8, and t_{Ra} and t_{Rb} represent their respective elution volumes. However, in the case of high performance liquid chromatography, which we will discuss in Section 13.4, they correspond to the retention times. Values of 1 or more for R are generally acceptable since they correspond to a separation of 98% for two consecutive symmetrical peaks.

A number of different forms of column chromatography are routinely used in biomedical science laboratories. These include:

- Gel filtration chromatography
- Ion-exchange chromatography
- Affinity chromatography
- Metal-chelation chromatography
- Hydrophobic chromatography

We will now consider each of these in turn.

Gel filtration

Gel filtration, also called permeation chromatography or 'molecular sieving', is a form of column chromatography that separates molecules according to differences in their sizes and to some extent shapes. The column is composed of polymeric materials hydrated in the form of microscopic porous beads. These are in equilibrium with a buffered solution suitable for use with the molecules to be separated, which usually has a constant composition and pH. This is an **isocratic elution**.

Molecules are washed out of the column in order of *decreasing* size. You can see this diagrammatically in Figure 13.9. Each bead is permeated by a three-dimensional network of pores that can only be penetrated by molecules of a specific size range. Molecules bigger than the pore dimensions are completely excluded from the gel particles and will pass through the spaces between the beads as they are washed through the column. The volume of liquid required to elute such molecules from the column is its void volume (V_0). Smaller molecules that can enter the pores of the beads completely, however, will be distributed between the solvent that is both inside and outside the beads and will therefore be eluted by a volume equal to the total volume of the column (V_t). Molecules between these two extreme sizes will be eluted by intermediate volumes. Hence, if a mixture of different sized molecules is applied to the

FIGURE 13.9
Schematic outline of separation by gel filtration. The separation of two differently sized particles in a mixture is illustrated. See text for details.

Movement of tubes

column the largest ones will pass through the column faster than the smaller ones. Thus the larger molecule will require a lesser volume of eluant than the smaller ones to move it through the column.

A variety of commercial materials are available for gel filtration. Sephadex is a commonly used material and consists of particles of dextran that have been modified to give varying degrees of cross-linking and, therefore, different pore sizes. Other substances used include agarose and polyacrylamide. They are available in the form of dry 'beads' that swell in aqueous buffers to form gels with pores of differing sizes. This allows various ranges of molecules to be separated (Table 13.3).

SELF-CHECK 13.4

Which of the Sephadex materials listed in Table 13.3 would be most suitable to separate a mixture of the proteins haemoglobin (M_r 64,500) and insulin (M_r 11,470)?

TABLE 13.3 Some commonly used gel filtration materials

Trade name	Polymer		M_r fractionation range
Sephadex	Dextran	G10	<700
		G25	1×10^3 to 5×10^3
		G50	1.5×10^3 to 3×10^4
		G100	4×10^3 to 1.5×10^5
		G200	5×10^3 to 6×10^5
Sephacryl	Dextran	S200	5×10^3 to 2.5×10^5
		S300	1×10^4 to 1.5×10^6
		S400	2×10^4 to 8×10^8
Sepharose	Agarose	2B	1×10^4 to 4×10^6
		4B	6×10^4 to 2×10^7
		6B	7×10^4 to 4×10^7
Bio-Gel	Polyacrylamide	P2	1×10^2 to 1.8×10^3
		P6	1×10^3 to 6×10^3
		P30	2.5×10^3 to 4×10^4
		P100	5×10^3 to 1×10^5
		P300	6×10^4 to 4×10^5

Column performance

The commonest types of partition chromatography are paper and TLC chromatographies. However, gel filtration is also a partition method, in which separations are based on differences in size. Thus, molecules to be separated by gel filtration must have different distribution coefficient (K_d) values for their successful separation. In gel filtration, the K_d calculated for a given type of molecule represents that fraction of the stationary phase available to it. In practice, K_d is difficult to determine in relation to gel filtration, and it is usually replaced by a K_{av} value since there is a constant relationship between K_{av} and K_d. A value of K_{av} for a solute can be determined using the equation:

$$K_{av} = (V_e - V_o)/(V_t - V_o)$$

where V_e is the elution volume of the solute. Molecules must have values of K_{av} that differ significantly if separation is to be effective.

While components of a mixture are moving through a gel filtration column, their separation is hindered by the diffusion of their molecules, which tends to mix them together. This can restrict the resolution and lead to an incomplete separation, with the individual components overlapping with one another as they leave the column. The net result is a less efficient separation, which you can see diagrammatically in Figure 13.10.

The efficiency of a column is a measure of the diffusion of the sample component during separation. A column can be thought of as consisting of a number of adjacent zones called

FIGURE 13.10
Overlapping peaks resulting from the incomplete separation of two components in a mixture during column chromatography.

theoretical plates. The length in the column (*L*) is the height of a theoretical plate (*H*) multiplied by their number (*N*). The solute reaches equilibrium between the mobile and stationary phases within each plate; thus, *N* is a measure of the efficiency of the column. It can be calculated using the formulae:

$$N = 16 \ (t/W)^2 \text{ or } 5.54 \ (t/W_h)^2$$

where *W* is the peak base width and W_h the peak width at half the peak height as shown in Figure 13.8, and *t* is the retention time. If *W* and W_h are expressed in units of time, then *N* is unitless.

The ability of a column to separate (resolve) different substances increases with the length of the column, *L*, since this increases *N*. Calculating the values of *N* for different columns can be used to compare their relative efficiencies. Indeed, the simplest way to increase the efficiency of a gel filtration step in purification would be to use a column of increased length. There are obvious practical limitations to the maximum size of *L* that can be used!

The number of theoretical plates is related to the surface area of the particles forming the stationary phase. Smaller particles have a relatively greater surface area than larger ones, thus the smaller the size of the particles of the stationary phase, the more effective the resolution of the column. If the size of *N* is calculated for a column of known length, the size or height equivalent to a theoretical plate (HETP) can be calculated using the formula:

$$HETP = L/N$$

and is normally expressed in μm. The smaller the value of HETP, the better the chromatographic performance. Thus, the HETP value can be used to assess changes in the efficiency of the same column under different chromatographic conditions.

SELF-CHECK 13.5

Calculate the HETP of the column of length 64 cm that eluted a protein with a V_e of 90 cm^3 using a suitable buffer at a flow rate of 45 cm^3 h^{-1}. The base width of the eluted protein peak was equivalent to 30 cm^3.

Applications of gel filtration

Gel filtration chromatography has a number of applications including:

- The purification of analytes
- Desalting samples
- Determining the M_r of biological molecules (analytical gel filtration)

Gel filtration has been used to purify molecules of biomedical interest by separating them from larger or smaller contaminating molecules. For example, it has been applied to the purification of a large range of proteins, including enzymes, hormones, and antibodies, as well as nucleic acids, polysaccharides, and viruses.

Desalting is the process of removing unwanted small molecules, particularly ions, from preparations of macromolecules. An example is the removal of phenol and ammonium sulphate from partially purified nucleic acids and proteins, respectively. A column of Sephadex G-25 is often used because the macromolecules elute in the void volume, while the smaller contaminant molecules are retarded on the column. This process is faster and more effective than the alternative desalting process of dialysis.

The M_r of molecules can be estimated using analytical gel filtration to determine the K_{av} values of molecules of known M_r and similar type to that being investigated. A calibration curve can be constructed by plotting the K_{av} values on the y axis against the logarithm values of M_r on the x axis. The M_r of the unknown molecule can then be estimated from the graph once its K_{av} has been determined. This method allows different columns to be compared. If this comparison is not required, then a simple graph of elution volumes against the logarithm values of M_r will suffice to estimate that of the unknown analyte.

Ion-exchange chromatography

Ion-exchange chromatography uses a stationary resin in the form of a charged gel that is able to bind ions of the opposite charge, and a mobile liquid phase that is able to selectively release the bound ions, as we explain below. Ion-exchange resins are insoluble porous cross-linked polymers that contain chemical groups of a particular charge. The extent of cross-linking of the resin influences the size of its pores. Polymers commonly used include those of cellulose, polystyrene, and agaroses modified by the attachment of charged functional groups, such as sulphonate $(-SO_3^-)$ and ammonium derivatives $(^+NR_3-)$. These are regarded as strong exchangers because they are ionized over a wide range of pH values. In contrast, carboxylate $(-COO^-)$ and diethylammonium $[-H^+N(CH_2CH_3)_2]$ are weak exchangers because they are ionized only within a narrow range of pH.

Positively charged resins are *anion* exchange resins, while those with a negative charge are *cation* exchangers. In both cases, ions of the opposite charge can bind to the column. Ion exchange occurs rapidly and is an equilibrium process, which you can see in Figure 13.11 (a). The strength

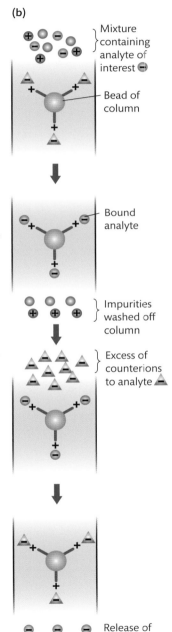

(a) Column matrix

Cation exchanger

$$-SO_3^- \, Na^+ + {^+NH_3}-R \; \longleftrightarrow \; -SO_3^- \; {^+NH_3}-R \; + \; Na^+$$

Anion exchanger

$$\begin{array}{c} CH_3 \\ | \\ CH_2 \\ | \\ -N^+HCl^- \; + \; {^-OOC}-R \; \longleftrightarrow \; -N^+H \; {^-OOC}-R \; + \; Cl^- \\ | \\ CH_2 \\ | \\ CH_3 \end{array}$$

FIGURE 13.11
(a) Schematic illustrating the binding and release of ions from gels used in ion-exchange chromatography. (b) Binding of an anionic analyte from a mixture of materials and its release from an anion exchange resin.

of binding depends on the extent of ionization of the analytes in the mobile phase. The more highly charged an ion, the tighter it will bind to oppositely charged groups on the column. The total capacity of a column, which is the maximum number of exchangeable ions it can bind, must not be exceeded. If it is, the 'excess' charged analyte molecules will simply pass through the column without binding. Manufacturers normally give total capacities of ion-exchange resins as milliequivalents of exchangeable groups per mg of dry resin.

The ion-exchange column must initially be equilibrated in a buffer whose composition and pH will (a) stabilize the analyte of interest and (b) allow it to bind to the column. Once bound, the components of the mixture need to be released (eluted) sequentially (Figure 13.11 (b)). This may use an isocratic buffer system, which differs in composition and pH from the one used to equilibrate the column. More often, a buffer of gradually changing pH or one whose concentration of salt progressively increases (Figure 13.12) is used and this is referred to as **gradient elution**. The former changes the charge of the bound molecules so they are released from the column in to the mobile phase. The effect of increasing the concentration of salt is that its ions displace the bound components from the column.

SELF-CHECK 13.6

Give the most probable order of elution of the following amino acids alanine [$^+$NH–CH(CH$_3$)COO$^-$], glutamate ($^+$NH$_3$CH(CH$_2$CH$_2$COO$^-$)COO$^-$) and lysine ($^+$NH$_3$CH(CH$_2$CH$_2$CH$_2$CH$_2$$^+NH_3$)COO$^-$) from a cation exchanger gel column at pH 7.

Affinity chromatography

Affinity chromatography relies on a specific binding between the analyte of interest and some type of ligand, which you can see diagrammatically in Figure 13.13. The ligand is immobilized by covalently attaching it to the column matrix (Box 13.2). Dextrans, agarose, and polyacrylamide have all been used as matrices in affinity chromatography. The ligand is often attached to the matrix using a chemical with an extended structure. This forms a spacer arm that separates the ligand from the bulk of the column preventing the matrix interfering with the specific binding of ligand and analyte.

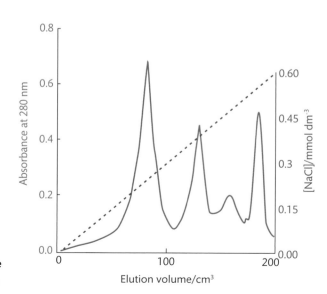

FIGURE 13.12

Sequential release of four proteins from an ion-exchange column using a gradient of increasing salt concentration.

When the sample is added to the column, only those molecules that can bind to the ligand that is those of the analyte, will be attached to it. The column is then washed with buffer to remove the nonbinding impurities. It is then possible to release the analyte of interest and elute it from the column in a pure form. The release of the bound molecule requires that the affinity between them and the ligands be reduced or that molecules be added to the eluting buffer that have a greater affinity than the bound ligand and so can displace the analyte from the ligand. For example, bound enzymes can be eluted from the column with solutions of substrate, coenzyme, or an inhibitor that competes for the enzyme's binding sites. If binding requires the presence of a metal ion, ethylenediaminetetraacetic acid (EDTA), which chelates and removes free metal ions, can be added. Often changing the pH of the mobile phase can alter the conformation of bound proteins allowing their release.

Key Points

Chelation is the binding and therefore removal from solution of metal ions by substances such as some amino acids and EDTA.

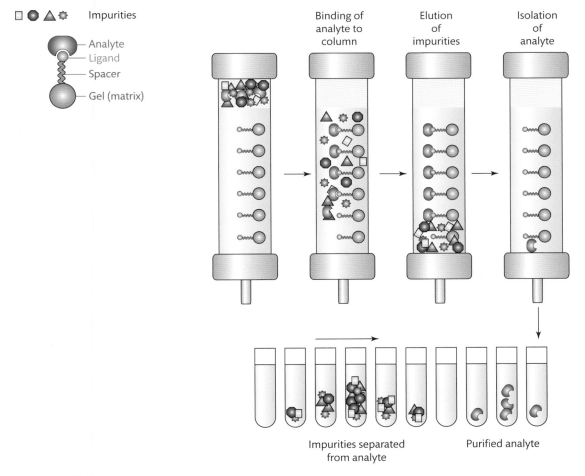

FIGURE 13.13
Outline of the use of affinity chromatography in purifying an analyte.
See text for details.

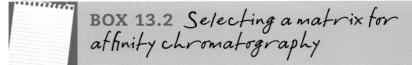

BOX 13.2 *Selecting a matrix for affinity chromatography*

An ideal matrix for affinity chromatography will have at least some of the following characteristics. It must:

- Form a stable link between the ligand to the matrix
- Be stable during binding of the macromolecule and its subsequent elution
- Not interact with the macromolecules generally to avoid non-specific adsorption
- Allow the mobile phases to flow freely through the matrix readily
- Be a hydrophilic polymer with pore sizes large enough to allow free passage of molecules.

Affinity chromatography has been used to isolate:

- Enzymes
- Antibodies/antigen
- Lectins
- Receptor proteins
- Nucleic acids

Enzymes can be isolated and purified from complex mixtures by using their substrates, coenzymes, or competitive inhibitors as ligands attached to affinity columns. Antibodies and/or antigens can be purified by immobilizing either the complementary antigen or antibody, respectively, on the column matrix. Lectins are glycoproteins that can bind other glycoproteins by recognizing all or part of their carbohydrate component. Affinity chromatography columns that use immobilized lectins have been used to isolate glycoproteins, particularly those solubilized from cell membranes. The receptors of a number of hormones have been isolated by using immobilized forms of the appropriate hormone as an affinity ligand. Once bound to the column, the receptor is released in ways that resemble the purification of enzymes with, for example, eluting buffers containing the hormone in free solution.

Nucleic acid strands are able to bind other strands that possess a complementary sequence of nucleotides, as explained in Chapter 16. Individual nucleotides or short single stranded oligonucleotides can be immobilized for use as affinity ligands. These have been used to extract nucleic acids with the complementary nucleotide sequence or to isolate nucleic acid binding proteins that are capable of recognizing and binding the nucleotide or oligonucleotide.

Metal-chelation chromatography

Metal-chelation chromatography can be regarded as a form of affinity chromatography which exploits differences in the strengths of binding of different amino acids to the immobilized ions of some heavy metals. Metal ions, such as Cu^{2+}, Zn^{2+}, Hg^{2+}, and Ni^{2+}, are immobilized by chelation with aminodiacetic acid, which, in turn, is covalently attached to suspended agarose gel. The suspension of agarose is then poured into a tube to form a column. The side chains of a number of amino acid residues found in proteins, including imidazole groups of histidine, thiol groups of cysteine, and indole groups of tryptophan, can bind to the metal ions. When mixtures of proteins are washed through the column, some will bind to the metal ions and be retained. The bound proteins are eluted sequentially by, for example, adding free amino acids to the mobile phase, which can displace the bound proteins. Thus, mixtures of proteins are at least partially separated.

Hydrophobic interaction chromatography

Hydrophobic interaction chromatography can be regarded as a form of affinity chromatography that is based on differences in the binding of hydrophobic sites on the surfaces of protein molecules to immobilized hydrophobic ligands attached to a solid support matrix in the form of a column. Hydrophobic interaction chromatography is generally used to separate mixtures of proteins. Proteins often have hydrophobic sites on the surfaces of their molecules due to the presence of certain amino acid residues, such as those of valine, leucine, isoleucine, and phenylalanine. However, the hydrophobic sites of different proteins vary in their nature and area, so the strengths of binding of different proteins to the column vary.

The bound proteins are eluted from the column by reducing the hydrophobic interactions between sample molecules and the stationary phase by, for example, eluting the column with a solution of a non-ionic detergent, such as Triton X–100, or by changing the pH or temperature of the eluting mobile phase.

13.4 **High performance liquid chromatography**

Increasing the length of a column and its number of theoretical plates improves its ability to resolve mixtures into their components (Section 13.3). However, long and densely packed columns of small particles have a greater resistance to the flow of the mobile phase. To overcome this resistance, and gain the advantages of faster and increased resolution, columns have been developed that can withstand the relatively high pressures produced when high pressure pumps are used to force the mobile phase through the column. The pumps used should give continuous flows of at least 10 cm^3 per minute. This group of techniques is called **high performance liquid chromatography** (HPLC). A typical HPLC system is shown in Figure 13.14. The columns are contained in tubes made of stainless steel that can withstand pressures up to 5.5×10^7 Pa. Columns typically have lengths of 15–50 cm and are normally 1–4 mm in diameter. The stationary phase consists of tightly packed small particles of material giving a correspondingly large surface area.

Three types of material are used in forming these particles:

- Microporous
- Pellicular
- Bonded

Microporous particles have diameters of 5–10 μm and contain pores with diameters of less than 2 nm. Pellicular particles are porous and coated onto an inert solid core, such as a glass bead of about 40 μm diameter. Bonded particles have a stationary phase that is chemically joined to an inert support, such as silica. Columns packed with pellicular material offer efficiencies, but only accept samples of small volumes compared with those with microporous

FIGURE 13.14

General components of a high performance liquid chromatography (HPLC) system.

material. Hence, the latter are generally preferred. Bonded phases give excellent chromatographic separations, and are chemically and mechanically stable, giving long-lasting columns.

Partition chromatography (see below) usually uses a liquid stationary phase coated onto an inert microporous or pellicular support. A problem with such supports is the tendency of the stationary liquid phase to be washed away by the mobile phase. However, bonded phases have been developed where the liquid stationary phase is chemically joined to the silica to overcome this problem.

The choice of mobile phase depends on the nature of the analytes to be separated. The polarity of the solvent chosen should be such that effective partitioning of components in the sample occurs between the two phases. All the solvents used must be of high purity because contaminants can affect the column and also interfere with the detection system (see below). Solvents also require degassing, not only because gas in the solvent can adversely affect the separation of the sample components, but also because it can interfere with the continuous monitoring of the eluant by, for example, absorbing UV light. Unfortunately, some solvents, for example, acetone, also absorb at these wavelengths. The mobile phase should not react with the stationary phase nor break the chemical bonds linking it to its supporting material, since this will, of course, decrease the effectiveness of purification and/or interfere with the detection of the components leaving the column.

1. Oxytocin
2. Angiotensin II
3. Angiotensin I
4. Insulin

FIGURE 13.15

Use of reverse phase high performance liquid chromatography to separate a mixture of proteins and protein hormones. The hormones separated are 1, oxytocin; 2, angiotensin II; 3, angiotensin I, and 4, insulin. Chromatogram courtesy of Varian, Inc.

The correct application of samples onto HPLC columns is necessary to obtain the best resolutions. The sample is injected onto the column using a microsyringe. There are two ways of applying the sample: stop flow and loop injection systems. In the former, the pump is switched off so that the pressure inside the column falls. The sample is then applied and the pump then switched back on. Loop injection systems allow the sample to be added while there is still a high pressure in the column. Repeated additions of impure samples can cause the column to lose its resolving power, which can be prevented by using a guard column. This is a short column of material similar to that of the main one, sited between the sample injector system and the analytical column. It is able to retain contaminating material, so protecting the main column and extending its lifespan.

High performance liquid chromatography methods generally involve adsorption and partition effects. Partition chromatography exploits differences in the solubilities and partition coefficients of polar substances as the bases for separating polar substances between stationary and mobile phases. The two usual techniques of partition chromatography are normal phase and reverse phase liquid chromatography. In normal phase, the stationary phase is a polar substance whereas the mobile phase is a non-polar solvent. These are reversed in reverse phase liquid chromatography (Figure 13.15), hence the name! Here, the stationary phase is a non-polar substance and the mobile phase is a polar solvent. This is the commonest type of HPLC. In reverse phase liquid chromatography, non-polar molecules dissolve in the non-polar stationary phase to varying extents, whereas polar molecules pass rapidly through column within the mobile phase.

However, HPLC can also be used with ion-exchange (Figure 13.16), exclusion and affinity chromatographies.

Detector systems used in HPLC

Whatever the form of HPLC, a sensitive method is required to detect the separated components, including the analytes of interest, as they are eluted from the column. Often the eluant passes through a spectrophotometric cuvette or flow cell, which are described in Chapter 11, and its absorbance in the ultraviolet region of the spectrum is continuously monitored to detect proteins and nucleic acids, which absorb maximally at about 280 and 260 nm, respectively. However, other detection systems are deployed. These include fluorescence and electrochemical detectors. Fluorescence detectors have greater sensitivity than simple ones based on absorbance. Their applications are, however, limited due to the relatively small number of compounds of interest that fluoresce. Electrochemical detectors are increasing in popularity.

1 = ovalbumin
2 = cytochrome *c*
3 = lysozyme

FIGURE 13.16

Separation of a mixture of proteins using a high performance liquid chromatography system with an ion-exchange column. The proteins concerned are 1, ovalbumin; 2, cytochrome *c*, and 3, lysozyme.

Cross reference

See Chapter 9 'Electrochemistry'.

Most are amperometric: they measure a current produced by oxidation–reduction reactions of the analyte at an electrode.

Whatever the method of detection, calibration of the system is necessary to compensate for differences in the volumes of samples, variations in their injection into the system, evaporation of solvent, and other irregularities. This is usually achieved using an internal standard (IS) that is similar in structure to the analyte. The same quantity of IS is added to the sample and to each solution of standard. You can see this in Figure 13.17. The peak height ratio (Figure 13.17 (a)) is measured for each eluted analyte or standard concentration using the expression:

Peak height ratio = height of analyte peak/height of internal standard peak

A graph of peak height ratio on the y axis against the corresponding concentrations of standards on the x axis will give a straight line graph as you can see in Figure 13.17 (b).

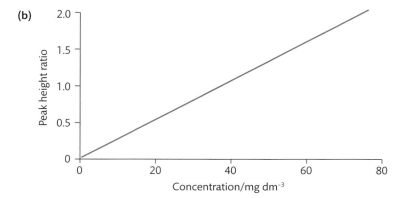

FIGURE 13.17

(a) Schematic showing the separation of an analyte X (or a standard of fixed concentration) and an internal standard (IS). The ratio of the heights of their peaks give a peak height ratio for that concentration. Calculating peak height ratios for a number of concentrations of the standard gives a calibration graph as shown in (b), which can be used to determine the concentration of an analyte in a sample.

A method to determine the concentration of drug X in clinical samples was established using HPLC. This method was calibrated using a range of standard concentrations of X and an internal standard, IS. A serum specimen from a patient was analysed and the chromatogram produced is shown in Figure 13.17 (a). Use this chromatogram and the accompanying calibration curve (Figure 13.17 (b)) to determine the concentration of X in the serum of the patient.

Applications of high performance liquid chromatography

The major advantages of HPLC are its speed, wide applicability, and sensitivity. The versatility of HPLC ensures it has many applications in biomedical science. It has been used to separate and analyse a wide range of materials of interest including amino acids, proteins, nucleotides, nucleic acids, drugs, and vitamins. For example, ion-exchange resins are used in HPLC to determine the proportion of glycated haemoglobin (HbA_{1c}) to the total haemoglobin in the blood of diabetic patients (Figure 13.18). In erythrocytes, glucose binds to haemoglobin non-enzymatically and the extent of this glycation depends on the degree and duration of hyperglycaemia. Measurements of HbA_{1c} give an estimate of average glycaemia over the preceding two months. In non-diabetics, the proportion of HbA_{1c} is less than 6% whereas that in diabetic patients is often in excess of 9%. A high relative proportion of HbA_{1c} reflects a poor control of blood glucose concentration. It is well established that such patients are more susceptible to long-term micro- and macrovascular complications.

Cross reference
You can read more about diabetes in Chapter 13 'Diabetes mellitus and hypoglycaemia', in the companion text *Clinical Biochemistry*.

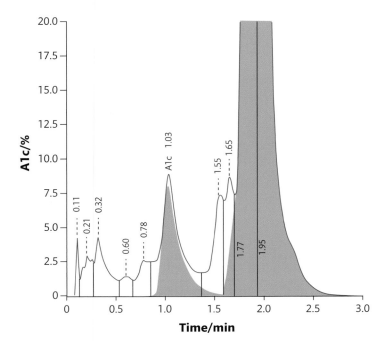

FIGURE 13.18
Use of high performance liquid chromatography to separate and estimate the proportion of glycated haemoglobin (HbA_{1c}) to total haemoglobin.
Courtesy of Department of Clinical Biochemistry, Manchester Royal Infirmary, UK.

13.5 **Gas–liquid chromatography**

Gas–liquid chromatography (GLC) partitions molecules between a stationary phase consisting of a microscopic layer of liquid or polymeric material contained within a glass or metal tubing, called a column, and a mobile gas phase. The mobile phase is forced continuously through the column and carries the components of the mixture being analysed. Samples for analysis are injected into the system where they are vaporized and introduced into the gas phase. The gas is heated to comparatively high temperatures to keep the components of the sample in the gas phase, allowing them to be carried along in the mobile phase. Those components that have a greater affinity for the stationary liquid phase have their movement through the column retarded. Those more soluble in the gas are carried through the column faster. The partitioning of the components of the mixture (sample) between the stationary and mobile phases causes different compounds to elute from the column at different retention times.

A typical GLC system is shown in Figure 13.19. The columns have capillary diameters between 0.03 and 1 mm and may be as long as 100 m. Two types of capillary columns are used: wall-coated open tubular (WCOT) and support-coated open tubular (SCOT). In the former, a liquid stationary phase is coated directly onto the wall of the capillary column, whereas the latter uses a liquid stationary phase coated onto a support material which is then bonded to the walls of the column. Stationary phases must be non-volatile and not enter the gaseous phase at the temperatures used. They are usually high boiling point organic compounds such as silicone greases.

FIGURE 13.19
Typical outline of a gas-liquid chromatography (GLC) system.

The mobile phase is usually an inert gas such as helium or argon or a nonreactive one, such as nitrogen. The gas is forced through the column at a rate of 40–80 cm^3 per minute. The column is contained within an oven that maintains the column at higher than ambient temperature during separation. This is necessary because, in general, the higher the temperature, the greater the volatility of molecules. Temperature therefore influences the separation because higher temperatures speed up the movement of more volatile molecules through the column. Temperatures may be set at constant values between 50 to 250°C to give isothermal conditions. However, for some applications it may be necessary for it to be pre-programmed to increase by, for example, 10°C every minute.

The sample for analysis is dissolved in a solvent, such as acetone or methanol, and injected into the GLC system using a microsyringe. The injection region of the column is usually at a higher temperature than the column itself. This ensures the rapid and complete vaporization of the sample. Wall-coated open tubular columns have a relatively small amount of stationary phase and so only small volumes of samples can be analysed. The sample injection port has a splitter system, which allows a small fraction of injected sample to reach the column and the remainder is vented as waste. The greater capacity of SCOT columns means there is no need for a splitter system and the sample is injected directly into the column.

In some ways, GLC resembles column chromatography. However, it differs from other forms of chromatography in that the molecules must be *volatile*, that is have low boiling points, if they are to move through the column and be separated. Gas–liquid chromatography is therefore particularly suitable for separating hydrophobic molecules or ones of low polarity, which therefore have low boiling points.

Molecules that are non-volatile can, however, often be converted to volatile derivatives prior to their separation. For example, long chain fatty acids can be converted to their methylesters and monosaccharides to trimethylsilyl compounds. Both of these types of derivatives have much lower boiling points than their parent compounds and can be analysed by GLC (Figure 13.20).

Cross reference
You can read about mass spectrometry in Chapter 11 'Spectroscopy'.

Like thin-layer chromatography (Section 13.2), GLC can be combined with other analytical techniques. For example, clinical samples can be subjected to GLC and the separated components identified by tandem mass spectrometry. The peroxisomal diseases adrenoleukodystrophy, Zellweger syndrome, and Refsum's disease, although rare, are extremely debilitating and are associated with high concentrations of very long chain fatty acids (VLCFAs) in the blood of patients. Thus, GLC–tandem mass spectrometry is used, for example, to investigate the VLCFA composition of samples of plasma from patients suspected of having these diseases.

Gas–liquid chromatography is thus a versatile technique and can be used to analyse a wide range of substances including lipids, alcohols, phenols, and a number of drugs. The major advantages offered by GLC in separating such materials are:

- Excellent resolution of a wide range of materials
- Reproducibility of separations
- Excellent sensitivity of its analyte detection systems.

Detector systems used in GLC

The most commonly used detectors in GLC are flame ionization and electron capture detectors (flame ionisation detector (FID) and electron capture detector (ECD), respectively). In the former, the gas from the column is passed through a hydrogen flame. Most organic

FIGURE 13.20
Separation of a number of triacylglycerols (TAGs) by gas–liquid chromatography (GLC).
The lengths of the fatty acids in each TAG are given as the number of carbon atoms
present. Courtesy of Varian, Inc.

compounds produce ions and electrons as they move through creating a small current, which is amplified and used as a detection signal. The main advantages of FID are that it is extremely sensitive, capable of detecting as little as 0.1 pg, which is 1.0×10^{-13} g of an analyte, and has a linear response over a wide range of concentrations. Its principal disadvantage is that the sample is destroyed during analysis.

In electron capture detectors the gas from the column is ionized by a radioactive source to release electrons. These produce a current across two electrodes. However, when an electron-absorbing compound is eluted from the column, it captures some of the electrons produced from the gas. The drop in current is recorded and so signals the elution of the compound from the column. Like FID, electron capture detectors are sensitive but their obvious disadvantage is that they can only be used with substances capable of capturing electrons, for example polychlorinated compounds such as dichloro-diphenyl-trichloroethane (DDT).

In both types of systems, it is customary to incorporate a known amount of an analyte into the test sample as an internal standard. This allows variations between samples, such as changes in volume, to be accounted for. However detected, the amount of a compound eluted from the column is recorded against its retention time. Retention times have a characteristic value for a given compound under set conditions as shown in Figure 13.20, for example. Comparison of the retention times for different compounds is equivalent to the analytical power of GLC, in a manner similar to HPLC (Section 13.4).

SELF-CHECK 13.8

Predict the order of elution of the methylesters of the following fatty acids:

Decanoic acid ($CH_3 (CH_2)_8 COOH$)

Eicosanoic acid ($CH_3 (CH_2)_{18} COOH$)

Hexadecanoic acid ($CH_3 (CH_2)_{14} COOH$)

Octadecanoic acid ($CH_3 (CH_2)_{16} COOH$)

Tetradecanoic acid ($CH_3 (CH_2)_{12} COOH$)

 SUMMARY

- Chromatography is a set of techniques that separate mixtures whose components differ in their distribution or partition coefficients

- Chromatographic techniques can also be useful in identifying individual components of the mixture and in providing quantitative data in some instances

- Planar chromatographies include paper and thin layer chromatographies. Both separate mixtures using a mobile liquid phase moving through a bound stationary liquid. Thin layer chromatography is a simple, quick, and inexpensive technique, suitable for qualitative but not quantitative analyses. It is used to separate a wide range of components, which can often be identified by their R_f values using those of known standards

- Column chromatographies comprise a range of techniques in which an immobilized solid forms a cylinder of material, through which a mobile liquid phase is allowed to percolate. Molecules in a mixture are applied to the column and separate due to differences in their sizes as in gel filtration chromatography, their charges in ion-exchange chromatography, hydrophobicities in hydrophobic chromatography, and ligand binding in affinity and metal chelation chromatographies. Whatever the type of column chromatography, components of the mixture differentially distribute between the stationary and mobile phases. Thus, individual components leave the column and can be identified by their elution volumes or retention times. An effective separation of the components of a mixture is indicated if they appear as discrete peaks on the chromatogram that do not overlap or show tailing or fronting

■ Specialized types of column chromatography materials are used to increase the surface area of the stationary phase, allowing the mobile liquid phase to be forced through the column at high pressure, as is the case in high performance liquid chromatography. This gives an improvement in resolution over low pressure systems

■ Gas–liquid chromatography separates mixtures of volatile compounds using a stationary liquid phase and a mobile gaseous phase. Partitioning of the individual components of the mixture between the two phases, effects their separation. Individual components leaving the column are detected by their absorbance, fluorescence, or electrochemical properties, and identified by their retention times

 # FURTHER READING

● Bollag DM. **Ion exchange chromatography.** *Methods in Molecular Biology*, 1994; **36**: 11–22.

● Cutler P. **Size-exclusion chromatography.** In: P. Cutler (ed.) *Protein Purification Protocols*. Humana Press, Totowa, 2004.

● Dong M. *Modern HPLC for Practising Scientists*, John Wiley & Sons, Harlow, 2006.

● Ho WF, Stuart B. *High performance liquid chromatography* (Coordinating author Prichard, E). Royal Society of Chemistry, Cambridge, 2003.

● Lunn G. *HPLC methods for pharmaceutical analysis*, Vols 2–4. Wiley, New York, 2004.

● Stuart B. *Gas Chromatography*. (Coordinating author Prichard, E.). Royal Society of Chemistry, Cambridge, 2003.

● Wall PE. *Thin Layer Chromatography: A Modern Practical Approach*. Royal Society of Chemistry, Cambridge, 2005.

● Zachariou M. *Affinity Chromatography: Methods and Protocols*, Humana Press, Totowa, 2008.

Three Useful Journals

■ *Biomedical Chromatography*

■ *Chromatography Today*

■ *Journal of Chromatography B: Biomedical Sciences and Applications*

 # QUESTIONS

1. Chloramphenicol, fluorouracil, and paracetamol are three therapeutic drugs. Their log K_{OW} values are 1.14, –0.85, and 0.48, respectively. List them in order of their *increasing* water solubility.

2. In chromatography, the separation of a mixture of molecules is determined by their differential distributions between two different phases. Which ONE of the following terms best describes this relationship?

   (a) Retardation factor
   (b) Partition coefficient
   (c) Elution volume
   (d) Void volume
   (e) Electrophoretic front

3. Which of the following statements concerning thin layer chromatography (TLC) is **INCORRECT**?

   (a) The adsorbent material often used is silica gel
   (b) The mobile phase moves up the thin layer by capillary action
   (c) The Rf value is always less than 1
   (d) TLC is a simple, relatively inexpensive technique
   (e) The separation of molecules is based on differences in their sizes

4. Predict the probable order of elution from a column of Sephadex G-100 of the following proteins: carbonic anhydrase (M_r 30,000), catalase (M_r 250,000), chymotrypsinogen (M_r 23,250), fibrinogen (M_r 330,000), haemoglobin (M_r 64,500), and insulin (M_r 11,470). You may find it helpful to consult Table 13.3 when attempting this question.

5. Gel exclusion chromatography was used to determine the M_r of catalase. The elution volumes (V_e) of five proteins of known M_r from a column of Sephacryl were determined, as was that of catalase. Chromatography was carried out at 4°C using a dilute buffer of pH 7.4. Under these conditions, the structures of all the molecules concerned are preserved. Use the data in the following table to determine M_r of catalase.

Protein	M_r	V_e/cm^3
Hexokinase	102,000	24.7
Aldolase	158,000	22.7
α-amylase	199,000	21.6
Fibrinogen	330,000	19.4
RNA polymerase	450,000	18.0
Catalase	?	20.7

6. Amino acids are polar molecules and mixtures of them cannot be resolved by gas–liquid chromatography. Explain a general strategy for overcoming this limitation.

7. Which ONE of the following properties of molecules generally forms the basis for their separation by affinity chromatography?

   (a) Size
   (b) Volatility
   (c) Charge
   (d) Shape
   (e) Polarity

8. Galactosyl transferase catalyses the reaction:

    $$\text{UDP-galactose} + \text{acceptor} \rightarrow \text{UDP} + \text{galactose-acceptor}$$

    The enzyme binds to a UDP-agarose affinity column in the presence of Mn^{2+}. Suggest a plausible method of eluting it from the column.

9. Two molecules, X and Y were separated using HPLC. They had retention times of 3 minutes 30 seconds with a base width of 38 seconds, and 5 minutes and 30 seconds and base width 42 seconds, respectively. A molecule Z, which was completely excluded from the column, had a retention time of 1 minute and 30 seconds. Determine (a) the selectivity of the column and (b) the resolution of X and Y. Comment on each value.

Answers to self-check and end of chapter questions are available in the Online Resource Centre accompanying this book.

 Go to www.oxfordtextbooks.co.uk/orc/glencross

Electrophoresis

Qiuyu Wang, Nessar Ahmed, and Chris Smith

After studying this chapter, you should be able to:

- Explain the underlying principles of electrophoresis
- Describe the uses of the major types of electrophoretic separation techniques
- Describe how the positions of separated components may be determined following electrophoresis
- Explain the underlying principles of isoelectric focusing
- Discuss the uses of pulsed field and two-dimensional gel electrophoresis

Introduction

Electrophoresis is the movement of charged molecules or **ions** in an **electric field** (E). Ions that differ in their charge and/or mass will move at different rates such that, if they have a common starting point they will separate over time as some will move faster than others. As a result, the components of a mixture will be separated. Thus, electrophoresis is a separation technique. However, it can also be adapted to provide analytical data, such as the size (M_r) of molecules. Many biological molecules, such as amino acids, peptides, proteins, nucleotides (nt), and nucleic acids, possess groups that can ionize, and give the molecules an overall negative or positive charge. Hence, electrophoresis has many applications in the biomedical sciences: in separating proteins and nucleic acids, in assessing purity of biological samples, and in estimating the sizes of molecules.

14.1 **Principles of electrophoresis**

If a potential difference or voltage (V) is applied between two electrodes, it produces an electric field or **potential gradient** (E). The value of E can be calculated by dividing the applied voltage by the distance (d) between the electrodes:

$$E = V/d$$

Anode   Cathode

FIGURE 14.1
Schematic showing the movements of anions and cations in an electric field.

Values of E are normally quoted in units of V cm^{-1}. When an E is applied to molecules carrying a charge of q, the force produced is equal to Eq; this force drives the molecules towards the electrode of opposite charge. For example, the positive electrode or **anode** will attract negatively charged molecules or **anions**; whereas the negative electrode or **cathode** attracts positively charged ones or **cations** (Figure 14.1).

SELF-CHECK 14.1

Calculate the value of the electric field (E) when a potential difference of 200 mV is applied between two electrodes 20 cm apart.

During electrophoresis, the applied potential difference, usually expressed in volts (V), is normally kept at a constant value. However, in some cases, electrophoresis is performed using a constant current, and the voltage will vary with the electrical resistance of the system. Occasionally, electrophoresis is carried out under conditions of constant power, which is the *product* of the current and voltage.

The strength of E can be increased by simply increasing the voltage applied between the electrodes. An increase in voltage increases the force that the molecules experience, making them move more quickly. As a result, an increase in voltage shortens the time needed to complete electrophoresis experiments. However, it is not always practical to use high voltages because of their heating effects. Heating can increase the diffusion of molecules in the sample and may also cause convection currents, both of which decrease the efficiency of separation. Higher temperatures can also result in the thermal denaturation of some types of biological molecules, such as proteins and nucleic acids.

Electrophoretic mobility

Cross reference
You met frictional coefficients in Chapter 12 'Centrifugation'.

The movement of ions subjected to an E is opposed by a frictional resistance, which depends, in turn, upon the frictional coefficient (f) of the ions.

The velocity (v) of the ion is given by:

$$v = Eq/f$$

However, the movement of an ion is often described by its **electrophoretic mobility** (μ) that is its velocity per unit of electric field. Electrophoretic mobility can be obtained by dividing v by E:

$$\mu = v/E$$

Combining these equations, gives the expression:

$$\mu = q/f$$

Thus, in theory it should be possible to obtain the value for f of a molecule from its mobility in an electric field if its net charge is known, which should make it possible to obtain information about its shape and size. In practice, this has not proved possible, because the simple equations given above do not adequately describe electrophoretic processes. For example, interactions between migrating molecules and the medium in which electrophoresis is performed all complicate the system. However, what is clear is that during electrophoresis, molecules that differ in their values of μ will separate from one another.

Initially, electrophoresis experiments were carried out in free solution, that is, a buffer, but this poses problems associated with diffusion of sample molecules, which opposes their movement during electrophoresis. Subsequently, porous mechanical support media were introduced, through which both buffer ions and sample molecules migrate during electrophoresis. This type of technique greatly reduces the diffusion of sample molecules, allowing the different substances in the sample to migrate as discrete zones or bands during electrophoresis. Indeed, the separations can be made extremely efficient by performing the electrophoresis in support media that offer greater frictional resistance to larger molecules than to smaller ones (Section 14.2).

14.2 Factors affecting electrophoretic separations

The equipment used for electrophoresis typically consists of an electrophoresis unit connected to a power pack, as you can see in outline in Figure 14.2. The electrophoresis units may be arranged to hold the supporting material in either a horizontal or a vertical arrangement. The power pack provides a flow of direct current to the electrodes in each compartment of the unit. Buffer solution in the compartments and in the supporting medium that links them then carries electricity between the two electrodes to generate a flow of electricity through the system.

The electric field generated during electrophoresis is associated with the current that is carried by the buffer connecting the electrodes, although a relatively small proportion of it is also conducted by the ions of the sample. If the ionic strength of the buffer is increased, the proportion of current carried by it will also increase; whereas that carried by the sample ions will decrease. Consequently, their velocities will slow. If the ionic strength of the buffer is decreased, the converse happens; the proportion of the current carried by the ions of the sample will increase and their rate of migration will speed up. However, although a low ionic buffer decreases the overall current and causes less heat production, there is greater diffusion of the sample due to an increase in separation time and, therefore, some loss of resolution or separation may occur. Thus, the E chosen is always a compromise between the time required to separate the molecules and the resolution obtained.

During electrophoresis, all ions are subjected to the same E and separation occurs because different substances have different electrophoretic mobilities. Individually, the mobility of charged ions is influenced by a number of factors including:

- Their sizes and shapes
- The buffer in which electrophoresis is performed
- The supporting medium

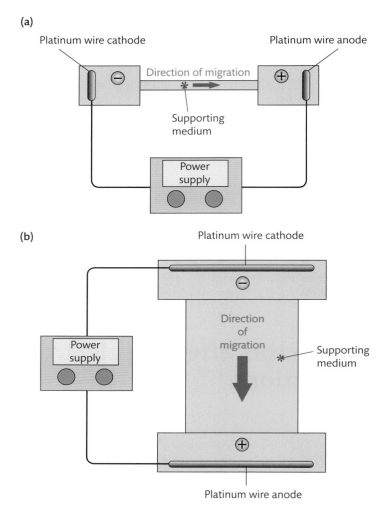

FIGURE 14.2
Schematics showing (a) horizontal and (b) vertical electrophoresis systems. Note how in each system, the electrodes are positioned in their own buffer compartments and the buffer completes the electric circuit between them. The electrodes in both systems are made from a platinum wire.

The sign of the charge (+ or −) of an ion will determine whether it moves towards the anode or cathode, as described in Section 14.1. However, the mobility of an ion during electrophoresis at constant voltage depends solely on the ratio q/f irrespective of the direction in which it is moving. For mixtures of molecules with *similar conformations*, for example, globular proteins or linear DNA molecules (Chapter 16), then the variables affecting their mobility are restricted to the individual charges and f values. The value of f depends to a great extent upon the mass of the molecule. Thus, each type of molecule in the mixture will migrate in a given electric field at a velocity proportional to its mass-to-charge ratio. The shape of molecules also influences their mobility. For example, globular proteins migrate differently to fibrous proteins of the same mass-to-charge ratio.

During electrophoresis, a buffer is used to saturate the supporting medium. The choice of buffer can affect the migration of particles in a number of ways. Its pH influences the ionization of some chemical groups within the sample molecules, in particular that of amino and carboxyl groups of polypeptides, and so determines the overall charge of the ions. pH is temperature dependent, which is a major reason for performing electrophoresis under conditions in which the temperature remains constant.

The ionic composition of the buffer is also a factor to be considered. Ions in the buffer, such as Na^+, K^+, Mg^{2+}, Cl^-, and those of phosphate, can all interact with charged groups on molecules of the sample. This results in them becoming surrounded with ions of the opposite charge, and decreases both their overall mass-to-charge ratio and mobilities.

In general, electrophoresis is carried out in buffers of low ionic strength that are compatible with the solubilities of biological macromolecules. Such compatibility reduces the types of interaction described above to minimal levels. In some cases, however, the binding of buffer constituents to sample molecules can be exploited. For example, borate buffers are used when separating carbohydrates as their components bind to carbohydrates. This converts neutral carbohydrates to charged complexes, which can then be subjected to electrophoresis.

The support medium used in electrophoresis is usually inert, although it can lead to **molecular sieving**, **adsorption** and **electro-osmosis**, each of which can influence the rate of migration of sample ions. Some support media consist of gels that contain pores (Sections 14.6 to 14.8) that restrict the flow of ions, particularly those of large M_n, during electrophoresis. Smaller ions experience a lower frictional resistance as they migrate through the gel and so travel faster and therefore further than larger ones for a given time. This effect is referred to as molecular sieving and, as we shall see in Sections 14.6 to 14.8, allows separation of molecules based on their size alone.

Adsorption is the retention of sample molecules onto the *surface* of the support medium. Adsorption slows the migration of sample molecules and causes tailing of the sample so that it moves in the shape of a comet, that is, in an extended area, rather than as a distinct zone. Thus, the individual components of a mixture do not separate into distinct bands and the resolution of separation is impaired.

Cross reference

We discuss adsorption in the context of chromatography in Chapter 13 'Chromatography'.

Electro-osmosis occurs in support media that possess negatively charged groups on their surface. These groups attract cations present in the buffer. However, the cations, like all ions in solution, are hydrated and surrounded by a layer of water molecules. When an electric field is applied, the cations and associated water are attracted towards the cathode and move towards it in a flow that opposes the migration of anions towards the anode. The force generated by movement of the water is called electro-osmosis and may be so large that weakly charged anions can be carried towards the cathode! Electro-osmosis is the basis of capillary electrophoresis (CE), which we will discuss in Section 14.9.

14.3 **Detection of separated components**

Following electrophoresis, the positions of the separated components on the support medium need to be detected, other than in the case of capillary electrophoresis (Section 14.9). The two major groups of compounds we will consider are proteins and nucleic acids.

If the substances are coloured, such as the proteins haemoglobin, myoglobin, or the cytochromes, then their positions are apparent because they are directly visible on the gel. However, most proteins are colourless and, in general, are detected by **staining** them with coloured or fluorescent dyes or by reacting them with reagents that convert them to coloured complexes. These treatments allow them to be seen as coloured bands in the gel.

Normally, staining is performed by immersing the gel in a solution of the dye. Following staining, the gel is soaked in a solvent to remove excess stain and clear the background: a procedure usually called **destaining**. Table 14.1 lists a number of dyes and staining procedures that are available.

TABLE 14.1 Common stains used in electrophoresis

Stain	Substance(s) detected	Comment
Acridine orange (Section 14.8)	DNA single stranded RNA	Use with denatured RNA
Alcian blue (Section 14.7)	Glycoproteins	Stains only the carbohydrate portion
Amido black	Proteins	Lacks sensitivity, gives high background stain with paper and cellulose acetate, shrinks polyacrylamide gels, generally replaced by Coomassie dyes
Coomassie blue dyes (Section 14.7)	Proteins	Probably the most widely used dye to detect proteins on native and SDS polyacrylamide gels
Ethidium bromide (Section 14.8)	DNA RNA	Sensitive for double-stranded nucleic acids, used with agarose and polyacrylamide gel electrophoresis
Enzyme activities (Section 14.7)	Enzymes	Sensitive, specific for enzyme activity, can be time consuming
Ninhydrin (Section 14.4)	Amino acids, peptides, proteins	Sensitive reagent, mainly used with paper electrophoresis
Ponceau S (Section 14.5)	Proteins	Used with cellulose acetate and starch gels, stains rapidly
p-anisidine/ethanol and sodium periodate/acetone (Section 14.4)	Sugars	Two reagent technique for detecting sugars following paper electrophoresis
Periodic acid-Schiff (PAS) reaction (Section 14.7)	Glycoproteins	Stains only the carbohydrate portion, lacks sensitivity
Pro-Q emerald 300 (Section 14.7)	Glycoproteins	Sensitive fluorescence stain for detecting sugar portions of glycoproteins
Silver staining (Section 14.7)	Proteins	More sensitive than Coomassie dyes but more expensive and staining can be variable
Sypro orange and red (Section 14.7)	Proteins	Sensitive fluorescence dyes for detecting proteins on polyacrylamide gels

Following staining, scanning densitometry may be used to estimate the amounts of the separated proteins or nucleic acids present. The densitometer passes a narrow beam of light from a laser through the gel and measures the proportion of light transmitted by each band of protein. The central processor of the densitometer or a computer to which it is connected can then calculate the amount of each protein present from the amount of light it absorbs: the smaller the amount of light transmitted by the gel, the greater the amount of sample present. These data are often presented as graphs of absorbance against the distance each protein migrated during electrophoresis, as you can see in Figure 14.3. However, the estimated amounts of each protein must be treated with caution. Absorbance is only linear up to a particular concentration of protein, in accordance with the Beer–Lambert law described in Chapter 11. Furthermore, equal amounts of different proteins do not stain to the same extent. Thus, estimates of the amounts of proteins present are, at best, semi-quantitative.

Cross reference

You can read about the Beer–Lambert law in Chapter 11 'Spectroscopy'.

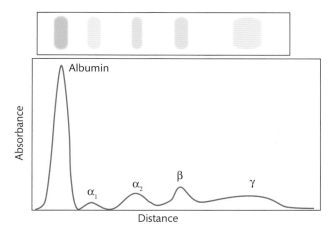

FIGURE 14.3
Densitometric scan of proteins, in this case serum proteins, separated by electrophoresis.

Gel documentation systems are now replacing densitometers. These consist of a computer linked to a video imaging system enclosed in a small dark compartment containing a white or ultraviolet (UV) light transilluminator. Transilluminators are instruments that project a narrow beam of visible or UV light through a translucent sample, such as an electrophoresis gel, to allow the stained bands of the separated components to be observed and photographed. Images of the gel can be stored on the computer and the intensities of the protein bands analysed.

SELF-CHECK 14.2

Examine Figure 14.3 and deduce from the densitometric scan which protein is present in serum in the greatest concentration.

How the separated components are detected and analysed depends upon their nature and the type of electrophoresis performed. Although the principles and points outlined in Sections 14.1 and 14.2 underpin all types of electrophoresis, there are major differences between the various techniques. These include the types of support media used, which in turn depends upon the sizes and nature of the molecules to be separated, and whether the equipment has a vertical or horizontal geometry.

The support media used in electrophoresis include:

- Paper (cellulose)
- Cellulose acetate
- Starch
- Polyacrylamide
- Agarose

Horizontal electrophoresis systems (Figure 14.2 (a)) are often used in paper, cellulose acetate, or agarose gel electrophoresis. In contrast, polyacrylamide electrophoresis generally employs a vertical system (Figure 14.2 (b)).

We will now review the different types of support media used in electrophoresis in more detail, starting with paper electrophoresis.

FIGURE 14.4
Structure of cellulose.

14.4 **Paper or cellulose electrophoresis**

Paper is composed of the polymer of β-glucose called cellulose (Figure 14.4). It was first used as a support medium in the 1950s in both vertical and horizontal geometries. Paper electrophoresis is largely used to separate small molecules, many of which, for example, amino acids and sugars, are of biomedical importance. However, the use of paper in electrophoresis has a number of flaws. These include its fragility, particularly when soaked in buffer, making it difficult to handle. Furthermore, a number of biological molecules, for example those of proteins, interact with the hydroxyl groups of the cellulose retarding their mobility and resulting in tailing (Section 14.2).

Following electrophoresis, the dye ninhydrin (Table 14.1) can be used to detect amino acids because it reacts with their amine groups to give a blue–purple coloured complex (Figure 14.5). Imino acids, proline, and hydroxyproline, are secondary amino acids and react to yield yellow–orange coloured products.

A number of reagents are available to detect sugars following their electrophoresis. For example, the electrophoresis paper can be immersed in a solution of *p*-anisidine/ethanol followed by one of sodium periodate/acetone, which converts the sugars to coloured derivatives. This method can detect as little as 20 µg of a variety of sugar types, including polyols, pentoses, hexoses, sugar acids, amino sugars, and their acetyl derivatives.

Cellulose contains a relatively small number of carboxyl groups, but these can ionize in many buffers and result in electro-osmosis. Both tailing and electro-osmosis hinder resolution and the separation of mixtures. For these reasons, other support media have largely replaced paper.

FIGURE 14.5
The reaction of ninhydrin with an amino acid.

14.5 **Cellulose acetate electrophoresis**

The use of cellulose acetate in the late 1950s overcame some of the disadvantages of paper described in Section 14.4. Cellulose acetate is made by acetylating the hydroxyl groups in cellulose (Figure 14.6). The absence of hydroxyl groups reduces interactions with sample ions and produces a structure that is less hydrophilic and, therefore, holds less water, both of which increase resolution. The pores inside cellulose acetate are uniform and of sufficient size to allow passage of all except the largest molecules.

Single strips of cellulose acetate, typically 2.5×12 cm, are used for single samples although wider ones are available to allow the analysis of multiple samples simultaneously. The strips are first fully wetted in electrophoresis buffer, and placed on a glass or plastic sheet in a horizontal geometry. Buffer-soaked filter papers are used to connect the strip to the electrode compartments (Figure 14.7). Care must be exercised to ensure the strips never dry out. The pH of the buffer used (8.6) ensures proteins have a negative charge and move towards the anode. A small, 1 or 2 µL volume of sample is applied as a 1 cm wide band to the cathode end of the strip.

Electrophoresis is carried out for about 45 minutes and since different proteins have different mass-to-charge ratios, they migrate at different rates and separate. Following electrophoresis, proteins are fixed in their positions in the cellulose acetate by immersing the strip in a solution of trichloroacetic acid (TCA), which denatures proteins and reduces their diffusion. The strips are stained to detect the proteins often with the dye, Ponceau S (Table 14.1 and Figure 14.8).

FIGURE 14.6
Structure of cellulose acetate.

FIGURE 14.7
Typical apparatus for carrying out cellulose acetate electrophoresis.

FIGURE 14.8
Structure of Ponceau S.

The negatively charged dye binds to positively charged amino groups of the protein but it also binds through hydrophobic interactions with non-polar parts of the protein.

Alternatively, the strips can be rendered transparent by immersing them in a mineral oil whose refractive index equals that of cellulose acetate. The positions of the proteins can then be determined spectrophotometrically.

Cross reference

You can read about spectrophotometry in Chapter 11 'Spectroscopy'.

Cellulose acetate electrophoresis lacks the resolution associated with gel-like materials, such as starch, polyacrylamide, and agar. However, it has the advantages that it is rapid, simple, and relatively inexpensive, and is used in pathology laboratories to analyse serum proteins, for example, in the investigation of multiple myeloma (MM).

Multiple myeloma and paraproteins

Multiple myeloma is the cancerous proliferation of immunoglobulin-producing plasma cells located in the bone marrow. It has an incidence of 4 in 100,000 worldwide. The condition is more frequent after the age of 50 years, but can occur before the age of 30.

B lymphocytes arise in the bone marrow and would normally leave it and circulate to lymphoid tissues. Here, they can develop into plasma cells when appropriately stimulated. Plasma cells from a single clone all synthesize and secrete an identical immunoglobulin (Ig) or monoclonal antibody and as these cells multiply, large amounts of a particular Ig are produced. However, in MM, plasma cells in the *bone marrow* produce excessive amounts of their specific Ig, which is released into the circulation. The large amounts of Ig are usually IgG molecules, but can be IgA or IgD. Following electrophoresis of serum from the patient, the Ig produced is detected as a discrete band, often called a **paraprotein**. Thus, the presence of paraproteins aids in the diagnosis of MM. The large amount of the particular Ig produced suppresses the production of other antibodies and reduces their concentrations in the blood of the patient. Immunoglobulins are large molecules, and the paraproteins cannot be removed from the blood by the kidneys and so increase the viscosity of the blood. However, in some patients, a defect in the plasma cells results in the production of only light chains of the Ig being produced, which can be filtered by the kidneys and lost in the urine. These can be detected and referred to as Bence-Jones protein.

Cross reference

You can read more about Bence-Jones protein in Chapter 18 'Specific protein markers' in the book, *Clinical Biochemistry*.

Patients suffering from MM present with a variety of symptoms that include bone pain, coma, and retinopathy, the latter two a consequence of blood hyperviscosity. They have an increased susceptibility to infections due to the decline of other immunmoglobulins, present with anaemia due to a decline in other bone marrow functions, such as erythrocyte, white blood cell and platelet syntheses, and suffer kidney damage due to deposition of

FIGURE 14.9

Separation of serum and plasma proteins by cellulose acetate electrophoresis.
Lane 1 shows the separation of a sample of serum from a healthy individual.
Lane 2 is serum from a patient with multiple myeloma. The position of the
separated paraprotein is indicated. Lane 3 shows the separation of plasma
from a healthy individual. The position of fibrinogen is indicated.

Ig molecules. About one-third of patients with MM die of renal failure and the average survival time is 2–3 years.

A diagnosis of MM is assisted by demonstration of a paraprotein in the serum following electrophoresis. It is essential that serum and not plasma is used for paraprotein analysis as the
latter contains fibrinogen, which can present as a discrete band and be mistaken for a paraprotein (Figure 14.9). Other clinical tests include demonstrating the presence of abnormal
malignant cells in the bone marrow following a biopsy. X-rays of bones can also be used to
aid in the diagnosis as characteristic holes are seen in affected areas of skull, pelvis, ribs, and
vertebrae in patients suffering from MM. Treatment of patients involves use of radiotherapy
and cytotoxic drugs.

14.6 **Starch electrophoresis**

Starch was first used in the late 1950s as a molecular sieve in gel electrophoresis to separate
proteins. The usual type of apparatus is shown in Figure 14.10. As you can see, it has a horizontal geometry, which holds a slab of starch connected to the buffer compartments with wicks,
rather like the system encountered in cellulose acetate electrophoresis (Section 14.5).

Starch gels are normally prepared using hydrolysed potato starch. The starch is suspended
in the electrophoresis buffer and heated until a transparent solution forms. The solution is
then poured into a suitable mould and solidifies to form a gel as it cools. Slots are cut into
the gel into which the samples for electrophoresis are added. Following electrophoresis,
the gel can be sliced into a number of layers using a fine wire, which provides several replicate gels, each of which can be analysed using different detection methods. For example,
different slices of gels can be examined for different enzyme activities (see Section 14.7),
particularly to demonstrate the presence of isoenzymes, such as those of serum lactate
dehydrogenase.

Cross reference

You can read about isoenzymes
in Chapter 7 'Clinical
enzymology and biomarkers'
in the accompanying textbook,
Clinical Biochemistry.

FIGURE 14.10
Schematic showing typical apparatus
for starch gel electrophoresis.

The major drawbacks of using starch as a support medium in electrophoresis are that, like cellulose, starch contains some negatively charged groups that can interact with sample cations, and also lead to electro-osmosis. Also, the pore sizes of starch gels vary with the concentration used. This is a major disadvantage to its use as a sieving medium. If the concentrations of starch are too extreme, the gel is too soft to handle or excessively stiff to use, which restricts the range of sizes of molecules that can be analysed.

14.7 **Polyacrylamide gel electrophoresis**

Polyacrylamide gel electrophoresis (PAGE) separates mixtures of molecules of proteins or nucleic acids through a gel composed of polyacrylamide. The gel is formed *in situ* between two glass plates that are clamped to form a sandwich, with the plates being held apart by plastic spacers that also retain the reaction mixture used to form the gel.

Polyacrylamide gels are formed by polymerizing acrylamide units into long chains. The polymerization is initiated by free radicals generated in the mixture by a reaction between ammonium persulphate and N,N,N′,N′-tetramethylethylenediamine (TEMED) as shown in Figure 14.11 (a) and (b). In Figure 14.11 (a) the free radical is represented as X·, and the acrylamide unit as A. The free radical X· reacts with A to form X–A·, which is still reactive as it has an unpaired electron. Thus, it will react with another molecule of A to form XAA· and so polymers of acrylamide units are built up in a stepwise fashion. As the chains grow, they are cross-linked by bisacrylamide (N,N′-methylene bisacrylamide), which consists of two acrylamide units linked by a methylene group. The cross-linking converts the solution to a solid gel, whose structure is shown in Figure 14.11 (b), and consists of a three-dimensional network of pores.

Acrylamide gels can also be formed in a photopolymerization reaction, where the ammonium persulphate and TEMED are replaced by riboflavin. Exposure of the gel solution to bright light for 2–3 hours leads to the photodecomposition of riboflavin, which generates free radicals that catalyse polymerization of the acrylamide units.

Separation during PAGE is based not only on differences in the electrophoretic mobilities of the different biological molecules, but also on the molecular sieving effects of the pore size. The sizes of the pores in the gel can be varied by altering the concentrations of acrylamide and bisacrylamide. Typically, gels consist of 3–30% concentrations of acrylamide. Those with

(a)

A Acrylamide unit
X• Free radical

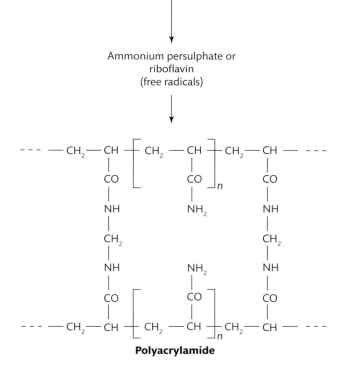

FIGURE 14.11
(a) Schematic illustrating the polymerization of acrylamide. See text for details. (b) Structure of polyacrylamide.

a low percentage of acrylamide have larger pores and are more suitable for separating larger proteins or nucleic acids as described in Chapter 16. Smaller molecules move through the gel easier than larger ones and so the relative rates of migration are the smallest ones fastest and the largest slowest.

There are a number of elementary safety precautions which should be observed for electrophoresis (see Health and Safety Box 'Some elementary precautions for electrophoresis').

Some elementary precautions for electrophoresis

- Mouth pipetting must not be used in preparing any of the solutions used for gel formation, staining, blotting, or detection.

- All acrylamide solutions should be prepared in a fume cupboard. Unpolymerized acrylamide and bisacrylamide are a skin irritant and a potential neurotoxin, respectively. Thus safety gloves must be worn throughout all handling procedures and a face mask when weighing the dry powder. Mercaptoethanol has a powerful and unpleasant odour, and should be handled only in a fume cupboard.

- Acrylamide begins to polymerize when tetramethylethylenediamine (TEMED) and ammonium persulphate are added to it. The ammonium persulphate solution should be freshly prepared and it and the TEMED added to the acrylamide solution *immediately* before it is poured into the gel apparatus.

- Do not touch the electrophoresis chamber or wires while electrophoresis is in progress. Voltages may be routinely as high as 200–300 V and shocks may be fatal.

- Ethidium bromide is a mutagen. Always wear gloves when handling solutions or agarose gels containing the compound. Dispose of the final electrophoresis buffer containing ethidium bromide according to directions from your instructor and local health and safety procedures.

- Avoid looking into ultraviolet (UV) light during the detection of DNA fragments on the agarose gel. Wear UV protective goggles.

SELF-CHECK 14.3

What basic safety precautions must you take when you prepare and handle a polyacrylamide gel?

Polyacrylamide gel electrophoresis occurs in a vertical gel (Figure 14.2 (b)). A photograph of a typical commercial vertical apparatus for carrying out vertical gel electrophoresis is shown in Figure 14.12 (a), which is shown schematically in (b). The typical dimensions of a polyacrylamide gel are 12 × 14 cm with a thickness of 1–2 mm, although gel dimensions can vary enormously.

Whichever method is used to polymerize the acrylamide, a plastic comb or well former is inserted between the plates into the polymerization mixture before it sets. The spacers and comb are removed following polymerization. The indents in the gel formed by the comb form wells to which samples are subsequently added.

The buffer provides the electrical contact between the electrodes and the gel. After assembly, the lower electrophoresis tank buffer surrounds the gel plates helping to keep them cool. Polyacrylamide gel electrophoresis can use **continuous** or **discontinuous** buffer systems.

Continuous buffer systems

In a continuous system, the same type of buffer is used to dissolve the sample as that present in the gel and in the compartments of the electrophoresis equipment. Sucrose or glycerol are added to the samples, which are loaded directly into the wells. This increases their densities ensuring they settle quickly to the bottom of the wells. The samples also contain a so-called

(a)

(b)

FIGURE 14.12
(a) Photograph of a
typical commercial gel
electrophoresis system for
performing polyacrylamide
gel electrophoresis.
(b) Schematic showing typical
apparatus for performing
polyacrylamide gel
electrophoresis.

tracking dye, usually bromophenol blue, which is a small molecule and passes through the pores in the gel to act as a marker for the electrophoretic front.

Electrophoresis is performed for a defined period ranging from an hour to overnight, usually until the tracking dye has neared the bottom of the gel. The current is then switched off.

SELF-CHECK 14.4

If 6 cm³ of a 30% combined acrylamide and bisacrylamide solution was made up to 18 cm³ with the other gel forming reagents, what is the concentration of acrylamide in the final gel?

Following electrophoresis, the gel is removed from the glass plates. The separated proteins are fixed in position usually using a mixture of TCA and methanol; the gel is then stained to reveal the positions of the separated proteins.

Discontinuous buffer systems

In a discontinuous buffer system, the buffers used to dissolve the samples, and those present in the gel and reservoir compartments of the equipment differ in terms of their composition and pH. In addition, the gel is formed in two parts. The main separating or resolving gel is made as described above. However, it occupies only part of the space between the glass plates and does not contain wells. Once this has polymerized, a much smaller stacking gel is formed above it, which generally has a buffer of lower pH to the resolving gel. The wells for samples are formed in the stacking gel, which typically consists of only 4% acrylamide or even less. The large size of its pores allows the sample molecules to move through it freely during electrophoresis.

Samples are added to the wells of the stacking gel generally in the same type of buffer used to form the resolving gel. When an electric field is applied to the system, the molecules of proteins or nucleic acids become sandwiched between various ionic species of the buffer, which concentrates them as a thin band within the stacking gel. Although the pH is higher in the resolving gel, which would tend to increase the mobilities of the ions, its pore size is smaller and separation of the various molecular species occurs as described above for continuous systems. However, because the proteins or nucleic acids have entered the resolving gel as narrow concentrated bands discontinuous systems have *enhanced resolution* over continuous systems.

Polyacrylamide gel electrophoresis can be performed under non-denaturing conditions as described above, in which case the technique is referred to as native PAGE. Native PAGE, although useful for many applications, is not generally applicable for determining the M_r of molecules or of resolving molecules that have similar sizes, because mobility under these conditions is influenced by both charge and size. However, when it is unnecessary to preserve the structures and functions of biological molecules, PAGE can also occur in denaturing conditions, such as in the presence of the detergent **sodium dodecyl sulphate** (SDS).

Sodium dodecyl sulphate electrophoresis

Cross reference

The sequencing of nucleic acids is outlined in Chapter 16 'Molecular biology techniques'.

The most widely used PAGE technique is known as sodium dodecyl sulphate polyacrylamide gel electrophoresis (SDS-PAGE). Sodium dodecyl sulphate-PAGE is particularly useful in analysing proteins to assess their purity or to determine their M_r, and was extensively used in sequencing DNA molecules, as described in Chapter 16, before being replaced by capillary electrophoresis (Section 14.9).

Sodium dodecyl sulphate-PAGE, as the name implies, uses the anionic detergent SDS (Figure 14.13), which is added to samples and also incorporated in the acrylamide gel and added to the electrophoresis buffers. Samples are boiled for several minutes or kept at 37°C overnight in a buffer containing SDS. The SDS strongly binds to molecules and disrupts the hydrophobic interactions that are largely responsible for maintaining their structures. In the case of protein samples, β-mercaptoethanol is often added to reduce the disulphide bonds that help stabilize the protein structure and which hold the monomers of some quaternary proteins together. Urea is added to nucleic acid samples to denature them to single stranded forms, suitable for analysis by SDS-PAGE as, for example, in the diagnosis of some genetic diseases such as cystic fibrosis (CF), which is discussed in Box 14.1.

FIGURE 14.13
Structure of sodium dodecyl sulphate (SDS).

$$CH_3 - CH_2 - CH_2 - CH_2 - CH_2 - CH_2 - CH_2 - CH_2 - CH_2 - CH_2 - CH_2 - CH_2 - SO_4^- Na^+$$

In general, proteins bind SDS in quantitative amounts with one SDS molecule binding to every two amino acid residues; furthermore, the amount of SDS bound overwhelms the native charges carried by the proteins. Since the negative charges of the sulphate groups of SDS repel each other, the protein molecules adopt rod-like structures (Figure 14.14). This means that all proteins in the presence of SDS and β-mercaptoethanol have a similar rod-shaped confor-mation and possess similar mass-to-charge ratios. Thus, the negatively charged proteins will move towards the positive electrode. Furthermore, all the protein molecules will now have the same charge per unit length and, therefore, travel with the same mobility in a given electric field. However, the gel through which they migrate offers frictional resistance and, therefore, smaller molecules migrate faster than larger ones through the pores in the polyacrylamide gel. Thus, molecules separate from each other according to differences in their *sizes*.

Gels are prepared with different pore sizes depending on the range of sizes of molecules to be separated. For example, 15% polyacrylamide gels are typically used to separate molecules with M_r ranging from 10,000 to 100,000. Gels with a lower concentration of acrylamide and, therefore, larger pore sizes are used for larger molecules. For example, 5% acrylamide gels separate molecules with M_r in the range of 60,000 to 350,000.

Visualizing proteins following separation

Perhaps the most commonly used dyes to stain proteins in polyacrylamide gels are the Coomassie types (Table 14.1 and Box 14.2). Coomassie dyes bind non-specifically to vir-tually all proteins through hydrophobic and van der Waals interactions, and because the negative charges of their sulphonate groups make ionic interactions with arginine, lysine, and histidine residues of the proteins. Gels are stained by immersing them for 2–4 hours in a solution of the Coomassie brilliant blue dye in a mixture of methanol and TCA, until the gel is a uniformly blue colour. The solvents used precipitate the proteins and so fix their positions in the gel. Following this immersion, the gels require destaining by gently shaking them in a mixture of methanol and acetic acid until the background is clear, which usually takes four to 24 hours. The proteins will then appear as darkish blue bands against a light background (Figure 14.17). The inclusion of positively charged resins speeds up the destain-ing time considerably.

FIGURE 14.14
Schematic showing the denaturation of proteins by β-mercaptoethanol and sodium dodecyl sulphate.

BOX 14.1 Investigating cystic fibrosis

Cystic fibrosis (CF) is the commonest fatal, inherited homozygous recessive disorder of Caucasian populations. It affects approximately 1 in 2000 people in the UK and about 1 in 20 carry a defective gene. The condition results from mutations in the gene *CFTR* that encodes the CF transmembrane conductance regulator protein. Thus, there is reduced permeability to Cl⁻ in the apical membranes of epithelial tissues that line the lungs and other organs, and viscous mucus accumulates in the lungs, digestive tract, and associated organs and epididymis. Patients suffer chronic respiratory disease, malabsorption, cirrhosis, and electrolyte disturbances.

Possible cases of CF are investigated by applying pilocarpine to the skin of patients to stimulate a flow of sweat, which is then sampled. Concentrations of Cl⁻ in the sweat greater than 70 mmol dm⁻³ are indicative of CF. Unfortunately, the sweat test is not always reliable in the first 6 weeks of life or in adulthood, and some patients with CF test normally. Several screening tests for CF based on its effects on pancreatic function are available. One measures immunoreactive trypsin (IRT) in dried spots of blood. The amount of IRT increases significantly in infants with CF compared with healthy ones. However, after the first few weeks of life pancreatic insufficiency develops and IRT decreases, and the test cannot be used.

Identifying mutations in *CFTR* of CF patients can not only confirm diagnosis, but also assist with assessing the carrier status of their relatives. It also provides an accurate and rapid prenatal diagnosis of CF once the parental genotypes are confirmed. Unfortunately, over 1300 mutated forms of *CFTR* have been reported, which makes their identification in individual cases difficult. The commonest CF mutation is called ΔF508 and is the deletion of the codon in *CFTR* that specifies a phenylalanine residue at position 508 in the amino acid sequence of the encoded protein. Molecular techniques offer relatively simple ways to detect this mutation. The gene is hydrolysed and the relevant portion amplified to increase its concentration using specific enzymes. The commonest technique produces a DNA fragment from healthy individuals that is 98 base pairs (bp) in size. However, if the ΔF508 mutation is present, the product is only 95 bp long. Sodium dodecyl sulphate-PAGE can separate the appropriate gene fragments as shown in Figure 14.15 and detect the deletion. The test is not totally specific for ΔF508 given that other 3 bp deletions in this region of the gene would also result in the same difference in size. Mutations other than ΔF508, can also be identified because they produce differing banding patterns following SDS-PAGE.

Cross reference

See Chapter 16 'Molecular biology techniques'.

98bp
95bp

1   2   3

FIGURE 14.15

Use of electrophoresis in detecting the ΔF508 mutation in cystic fibrosis. Lane 1 shows the position of the 95 bp fragment associated with the mutation. Lane 2 has the two bands that occur in the heterozygote condition, the normal 98 bp fragment and the 95 bp one. Lane 3 is the normal condition where only a single 98 bp fragment is observable.

BOX 14.2 Coomassie dyes

Coomassie dyes (Table 14.1), which are sometimes called Coomassie brilliant dyes, were introduced in the 1960s as stains to detect proteins following electrophoresis in the 1960s. They are particularly used following native PAGE and SDS-PAGE. The Coomassie dyes were original developed as acid woollen dyes (they stain wool in acid conditions). They were named to commemorate the 1896 British occupation of Kumasi, which was formerly called Coomassie, in Ghana, West Africa.

Coomassies dyes constitute a group of chemically related compounds. The first to be developed was Coomassie Blue R-250; the R stands for reddish tint of the dye, while the 250 is an indicator of the strength of the dye on a scale used by the original manufacturer, Imperial Chemical Industries. Coommassie Blue G-250, G for greenish, and Coomassie Violet R-150 later followed. The structures of R-250 and G-250, which are the most commonly used for staining proteins in electrophoretic gels, are shown in Figure 14.16 (a) and (b). Coommassie R-250 generally gives better resolution and is more often used than G-250, even though the latter has greater sensitivity.

Coomassie dyes can be added to native (non-denatured) proteins, rather than SDS prior to electrophoresis. The name of this technique is Blue Native- or BN-PAGE. Like SDS, Coomassie dyes impart a net negative charge to the proteins, allowing them to separate during PAGE. Only negatively charged proteins would migrate without the presence of the dye (or SDS) during native PAGE.

The Bradford method is one of the commonest used to determine the concentrations of proteins in solution. Coomassie dyes are one of the components of the Bradford reagent. Coomassie Brilliant Blue G-250 can bind to proteins in acid solution and, when it does so, its wavelength of maximum light absorbance shifts from 465 to 595 nm. Equal amounts of Bradford reagent are added to solutions containing known concentrations of proteins, as well as to the samples whose concentrations are unknown. The absorbances of the resulting coloured solutions are then measured in a spectrophotometer. The data from the known solutions are then used to construct a calibration graph from which the concentration of protein in the samples can be determined.

FIGURE 14.16
Structures of Coomassie blue (a) R-250 and (b) G-250.

SELF-CHECK 14.5

What is the structural difference between Coommassie R-250 and G-250?

Cross reference

You can read more about spectrophotometry and standard graphs in Chapter 11 'Spectroscopy'.

Coomassie Blue R-250 has a detection limit of about 0.3–1 µg of protein per band on the gel. If greater sensitivity is required to detect even smaller amounts of proteins, then Coomassie blue in a *colloidal* form may be used. A solution of colloidal and molecularly dispersed Coomassie blue in equilibrium is prepared by adding methanol and ammonium sulphate to a solution of the dye in dilute phosphoric acid. During staining, only the molecular form enters the gel

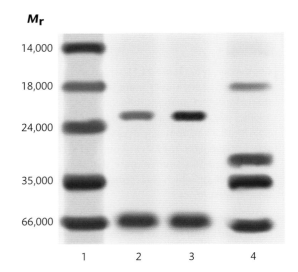

M_r

14,000

18,000

24,000

35,000

66,000

1   2   3   4

FIGURE 14.17

Photograph of a polyacrylamide gel showing the separated proteins stained with Coomassie blue R-250. Lanes 1 shows the positions of proteins of known M_r. Lane 2–4 show the positions of a number of proteins of differing M_r from three separate samples. Courtesy of Dr M. S. Ahmad, School of Healthcare Science, Manchester Metropolitan University, UK.

and stains the proteins preferentially, while the colloidal form is unable to enter the gel and so background staining does not occur.

The fluorescent Sypro orange and red dyes (Table 14.1) have comparable sensitivities to colloidal Coomassie dye and can detect about 1–2 ng of protein. Stained proteins can be seen on the gel by illuminating them with light of approximately 300 nm. Although both fluorescent dyes have similar staining properties, Sypro orange is the more sensitive, but the red form gives less background interference.

Silver staining is the most sensitive procedure and has a detection limit about 300 times less than that of conventional Coomassie Blue R-250. Proteins must be fixed in the gel using TCA prior to silver staining. The gel is stained by immersing it in a solution of silver nitrate. Silver ions react with thiol and basic groups in the proteins, and are then reduced using formaldehyde, which causes the deposition of grains of metallic silver on the protein. Thus, the protein bands can be visualized as they are stained brown or black against a light amber stained gel.

Despite being the most sensitive method of detecting protein following electrophoresis, problems can occur with silver staining. Different proteins can respond differentially to the stain and, indeed, a number of proteins do not stain at all. Hence, the observed sensitivity varies between proteins.

SELF-CHECK 14.6

Calculate the approximate lowest sensitivity in ng of silver staining to detect the presence of proteins following PAGE and SDS-PAGE.

Cross reference

Radioisotopes are described in Chapter 10 'Radioactivity and radiation'.

Proteins can also be detected by labelling them with radioisotopes such as ^{125}I and their positions on the gel are detected by autoradiography (Section 10.6 and Box 13.1 Autoradiography).

Detection of glycoproteins

Glycoproteins can be detected on the gels using Alcian blue stain or the periodic acid-Schiff (PAS) reaction (Table 14.1). The former is now little used. You can see an outline of the PAS procedure in Figure 14.18. The periodic acid oxidizes sugars that contain hydroxyl groups on

Sugar in glycoprotein

FIGURE 14.18
Schematic showing the periodic acid
Schiff reaction of sugars.

Pink to red coloured product

adjacent carbon atoms in the carbohydrate part of the glycoprotein to a dialdehyde product. The aldehydes subsequently react with Schiff reagent to form a coloured product. However, the method lacks sensitivity, being only able to detect bands with at least 2.5 ng glycoprotein. Thus, the weakly coloured pink or light red bands are often difficult to see on the gel.

The Pro-Q emerald 300 dye (Table 14.1) also reacts with periodate-oxidized carbohydrates but gives a bright green fluorescent signal when excited with light of wavelength 300 nm. This sensitive fluorescence stain can detect as little as 0.5 ng of glycoprotein. However, rather than using these methods to detect glycoproteins, blotting techniques using lectins as sensitive and specific probes of the carbohydrate portion of the molecule are often employed.

Cross reference
You can read about the blotting of gels and the use of probes in Chapter 16 'Molecular biology techniques'.

If the proteins are enzymes, then their position in the gels can be detected by exploiting their catalytic activities. This method of detection is particularly applicable to native PAGE and starch gel electrophoresis (Section 14.6) because these support media are permeable to the substrates, cofactors, and dyes required to detect enzyme activities. For example, hydrolases can be detected using p-nitrophenylphosphate. This artificial substrate is hydrolysed by the enzyme to phosphate and p-nitrophenol:

$$\text{p-nitrophenylphosphate} + H_2O \xrightarrow{\text{Hydrolase activity}} \text{p-nitrophenol} + \text{phosphate}$$

The substrate is colourless, but p-nitrophenol is yellow coloured. Hence, the positions of hydrolases on the gel show as bright yellow bands.

SELF-CHECK 14.7

Why is it not possible to detect the positions of enzymes following SDS-PAGE by monitoring their catalytic activities?

Whatever staining procedure is used, it may cause hydration or dehydration of the gel, and so alter its size. Such changes must be taken into account when measuring the mobilities of the separated components. Thus, the relative mobility of, for example, a protein can be determined using the expression:

$$\text{Relative mobility} = (\text{migration distance of protein/migration distance of tracking dye})$$
$$\times (\text{length of gel } before \text{ staining/length of gel } after \text{ staining})$$

If the M_r of an unknown protein was being estimated by SDS-PAGE, then proteins of known size would be included in the electrophoresis as shown in Figure 14.17. A calibration graph would then be drawn of the logarithm of size (M_r) on the y axis against relative mobility on the x axis. The M_r of the unknown protein may be estimated by determining its relative mobility using the calibration graph.

14.8 **Agarose gel electrophoresis**

Agarose is one of the components of agar, which, in turn, is a mixture of polysaccharides isolated from some seaweeds. It is a linear polysaccharide consisting of units of agarobiose, that, in turn, consist of alternating galactose and 3,6-anhydrogalactose residues linked together by alternating α1-4 and β1-3 glycosidic bonds (Figure 14.19).

Agarose is able to gel because hydrogen bonding within and between the agarose polymers forms a three-dimensional lattice containing a continuous system of pores. The size of the gel pores is determined by the concentration of agarose. A low concentration of agarose gives a gel with larger pores and *vice versa*. The use of an appropriate concentration allows mixtures of different sized molecules to be resolved during electrophoresis because the smaller molecules are able to move faster through the gel than the larger ones. Like all supporting media, the gel also limits the diffusion of the molecules, which improves separation.

Molecules that differ in their M_r by as little as 1% can be resolved on agarose gels. In addition to its ability to separate molecules with a wide range of M_r with high resolution, other advantages of using agarose include its lack of toxicity and relative ease of use. Agarose is generally used in electrophoresis to separate nucleic acids as we discuss below, but is also used in isoelectric focusing (IEF) to separate proteins (Section 14.10).

FIGURE 14.19
Structure of agarose.

$$- 3) - \beta - \text{Gal} - (1{-}4) - 3,6 - \text{anhydro} - \alpha - \text{Gal} - (1-$$

Agarose gel electrophoresis is often used to separate and analyse mixtures of nucleic acid molecules, particularly fragments of DNA. These are usually of large size and their separation requires gels with larger pores. The size of double stranded DNA is often expressed as the number of base pairs (bp) or kilobase pairs (Kbp). However, even a fragment of DNA as small as 1 Kbp has a M_r of 620,000. In the case of RNA and single-stranded DNA, size is usually given as the number of nucleotides. You can read more about the structures of nucleic acids in Chapter 16.

Cross reference

Chapter 16 'Molecular biology techniques'.

Every nucleotide in DNA and RNA includes a negatively charged phosphate group. Thus, the mass-to-charge ratio for all nucleic acid molecules is identical. In this respect they resemble polypeptides in the presence of SDS (Section 14.7 and Figure 14.14). This means that DNA and RNA molecules will move towards the anode with identical mobilities when subjected to an electric field. Hence, in theory, they should not separate during electrophoresis. However, in agarose electrophoresis, the gel matrix resists the movement of DNA and RNA molecules: larger fragments have the most difficulty in getting through the pores, while the smaller fragments move faster because they experience the least resistance. Indeed, the mobility of very large nucleic acids may be blocked completely. Therefore, separation of DNA and RNA molecules by agarose gel electrophoresis depends solely on differences in their sizes, like proteins in SDS-PAGE.

The use of appropriate concentrations of agarose with different pore sizes allows nucleic acids of varying ranges of size to be resolved. Those most frequently used include 0.3% agarose gels, which are used to separate fragments of DNA 5–60 Kbp in size, 0.8% for 0.5–10 Kbp fragments, and 1–2% agarose gels for 0.1–3 Kbp fragments. Using an appropriate concentration of agarose means the distance moved by a limited range of nucleic acid fragments is inversely proportional to the logarithm of their sizes, as was the case for SDS-PAGE described in Section 14.7. Whatever the concentration of agarose to be used, the gel is formed by dissolving dry agarose in buffer and boiling the mixture until a clear solution forms. The solution is poured into the gel former and a plastic comb inserted in it before it sets. Once the solution cools, it sets to form a rigid gel. Removal of the comb leaves wells into which samples can be added. The gel is placed in the electrophoresis tank and covered by the buffer (Figure 14.20 (a) and (b)).

Samples for electrophoresis are, as in PAGE (Section 14.7), prepared by dissolving them in a buffer that contains sucrose or glycerol to increase their densities above that of the electrophoresis buffer. The samples are loaded directly into the wells using a micropipette and settle quickly to the bottom of each one. Also, as in PAGE, a dye, most commonly the small molecule bromophenol blue, is normally included in the sample buffer. The dye moves rapidly through the gel in advance of all but the smallest DNA fragments. It therefore acts as a marker for the electrophoretic front allowing electrophoresis to be stopped before the samples migrate so far that they run out of the end of the gel!

Staining nucleic acids in agarose gels

Ethidium bromide (Table 14.1 and Figure 14.21) is the stain of choice for double stranded DNA molecules that have been separated by agarose gel electrophoresis. It is normally added to the electrophoresis buffer and so is present throughout the electrophoresis run. Since ethidium bromide consists of relatively small, planar molecules, it binds to all the DNA molecules present by intercalating between their stacked base pairs (Figure 14.22 (a)). In this position, the ethidium bromide enhances the normally weak fluorescence of the purine and pyrimidine bases by

(a)

(b)

FIGURE 14.20

(a) Photograph of a typical commercial gel electrophoresis system for performing agarose gel electrophoresis. (b) Schematic showing side and top views of typical apparatus for performing agarose gel electrophoresis.

FIGURE 14.21
Structure of ethidium bromide.

(a)

Ethidium bromide (EtBr)

(b)

Size/bp

9416
6557
4631

2322
2027

564

FIGURE 14.22

(a) Intercalation of ethidium bromide into double stranded DNA. (b) A photograph of an agarose gel showing the separated DNA molecules stained with ethidium bromide. Lanes 1, 2, and 3 show DNA fragments from samples stained with ethidium bromide. Lane 4 shows stained fragments of known size. See also Figures 16.17 and 16.30.

about 25 times. Hence, the separated DNA molecules fluoresce with an intense orange colour when the gel is irradiated with UV light. Fluorescence stains are sensitive means of detecting the fragments: as little as 1 ng of DNA can be detected (Figure 14.22 (b)).

Following staining, it is possible to determine the relative mobility of each band from the formula:

$$\text{Relative mobility} = \text{distance moved by band from origin} / \text{distance moved by electrophoretic front from origin}$$

Since the ethidium bromide is usually added to the electrophoresis buffer prior to electrophoresis, it is not necessary to correct for changes in the size of the gel during staining. It is possible to determine the sizes or M_r of unknown DNA fragments by including fragments of DNA of known size in the electrophoresis (Figure 14.22 (b)). A calibration graph of the logarithm of size (in Kbp or M_r) on the y axis against relative mobility on the x axis can then be drawn. The values of the unknown nucleic acids can then be determined by using their relative mobilities and the calibration graph. This is essentially the same method as used to estimate the M_r of proteins using SDS-PAGE (Figure 14.17 and its associated text).

FIGURE 14.23
Structure of acridine orange.

Agarose gel electrophoresis can be used to separate molecules of RNA, as well as DNA. Single-stranded molecules of RNA are able to double back on themselves. Consequently, *intrachain* double stranded regions can form that will bind ethidium bromide, allowing their detection. However, determining the sizes of RNA molecules requires the electrophoresis be performed in the presence of denaturants, such as formaldehyde, glyoxal, or methylmercuric hydroxide, which are compatible with agarose. These denaturants bind to bases in the RNA and so prevent double stranded regions forming.

The dye, acridine orange (Table 14.1 and Figure 14.23) can detect the single stranded RNA molecules following electrophoresis. Its binding involves a complex mixture of electrostatic interactions and the intercalation of dye molecules between bases of the single-stranded polymer. When excited by blue light, RNA-acridine orange complexes fluoresce orange but DNA-acridine orange emits yellow-green light. However, in many experiments, RNA is radio-labelled and so can be detected by autoradiography.

Cross reference

See Section 10.6 and Box 13.1 Autoradiography, and the *Histopathology* text of this series.

SELF-CHECK 14.8

Explain why ethidium bromide is not a suitable stain for single stranded RNA molecules.

Determining the order of bases of a DNA molecule (sequencing) also involves separating single stranded nucleic acids, in the case of DNA. Chapter 16 gives a description of sequencing, which involves separating fragments of *single* stranded DNA by electrophoresis. The methods initially developed used SDS-PAGE gels containing urea but contemporary techniques use capillary electrophoresis (Section 14.10). Urea, like SDS, is a denaturant and together they ensure the DNA chains remain single stranded during electrophoresis. Following their separation, the different lengths of DNA can be interpreted to give the equivalent sequences of bases in the DNA sample. You can see examples of sequencing DNA using gel electrophoresis and capillary electrophoresis in Figures 16.19 and 16.20 respectively.

Pulse field gel electrophoresis

Agarose gel electrophoresis is unable to resolve molecules of DNA larger than 15 Kbp as molecules of this size and above tend to migrate as a group during electrophoresis. Pulsed field gel electrophoresis (PFGE) is an electrophoretic technique able to separate such large DNA molecules. It does this by periodically altering the magnitude of the applied electric field across the agarose gel. Thus, during PFGE, two different electric fields are applied alternately at different angles to the gel for a defined period of time, for example, 60 seconds (Figure 14.24).

Each field causes the DNA molecules to become stretched and to move in one direction through the gel. The application of the alternate field then causes the DNA molecules to change direction. Even though all the DNA molecules are relatively large, the smaller ones are able to realign faster when the direction of field changes. This enables the smaller molecules to move faster through the gel than larger ones, leading to separation of the mixture.

Pulsed field gel electrophoresis is used in genotyping or genetic fingerprinting. It is also used in epidemiological studies of pathogenic organisms, for example in identifying subtype strains of *Listeria monocytogenes* in clinical samples.

Cross reference
Chapter 13 'Chromatography'.

14.9 **Capillary electrophoresis**

Capillary electrophoresis (CE), as the name implies, separates the components of samples in narrow bore tubes, whose internal diameters are only 25–100 μm, but may be 40–100 cm in length. Like all electrophoresis, charged molecules move in an electric field. However, because capillary tubes have a high surface area to volume ratio, the high voltages used and resistances encountered allow the considerable heat generated to dissipate rapidly. Hence, convection currents and diffusion of sample components are minimized, and separation can be carried out in solution without the need for the support media necessary in other forms of electrophoresis.

Capillary electrophoresis combines the high resolution of electrophoresis with the speed of high performance chromatography, which you can read about in Chapter 13. For this reason, CE is also called high performance capillary electrophoresis (HPCE).

Figure 14.25 shows a typical arrangement for CE. The capillary tubes are made of fused silica and coated externally with a polymer to provide mechanical strength. Samples are injected into the capillary by one of two methods:

- The inlet of the capillary tube is removed from the anode reservoir and immersed in the solution of sample. A high voltage is then applied across the tube, which attracts sample ions into the tube

- The sample can be forced into the capillary tube by pressurizing the sample container with an inert gas

Whatever the method, a major advantage of the technique is that sample volumes as small as 5–10 nL can be analysed.

Once the sample has entered the tube, a high voltage is applied across the capillary tube and the constituent molecules in the sample migrate along its length at different rates and so separate. Typical voltages used range from 7 to 10 kV, but can be as high as 50 kV depending upon the samples to be separated.

The basis of separation in CE is rather different to that encountered in other types of electrophoresis. Normally, charged particles would move towards the electrode of the opposite charge. However, in CE, electro-osmosis causes all the charged molecules to be carried towards the cathode because the electro-osmotic flow is greater than their electrophoretic mobility. Electro-osmosis occurs in the capillary tubes due to a net negative charge on the fused silica surface. This attracts cations, which migrate to the inner wall forming an electrical double layer. The cations, in turn, are surrounded by water molecules. When a potential difference is applied to the column, the cations migrate towards the cathode. Since these cations are solvated and clustered at the walls of the capillary tube, they drag the rest of the solution with them including the anions (Figure 14.26).

Although all components of the sample, both charged and neutral, migrate towards the cathode, cations reach it first because they move by a combination of electro-osmotic flow

FIGURE 14.24
Schematic showing the operation of pulsed field electrophoresis. See text for details.

DNA sample in well

Agarose gel

Direction of field

Direction of field

Direction of field

Separated and stained DNA fragments

FIGURE 14.25
Schematic showing typical apparatus for performing capillary electrophoresis.

Cross reference

Chapter 11 'Spectroscopy' and Chapter 9 'Electrochemistry'.

and electrophoretic migration. Cations are followed by neutral molecules and then anions. Thus, the components of the sample are all separated in typically 10–20 minutes. As they flow towards the cathode, the individual components are detected by UV or visible spectrophotometry, fluorescence, or using electrochemical detection methods.

Amino acids, peptides, proteins, DNA fragments, drugs, and even metal ions can all be separated by CE. A major use is in sequencing DNA molecules and assessing plasmid homogeneity (Box 14.3). The main application of CE in biomedical laboratories is to analyse serum proteins and to monitor disease. For example, CE is used in the detection of paraproteins in the sera of patients as their presence indicates MM, a malignant proliferation of B cells in the bone marrow (Section 14.5). Haematology laboratories use CE to separate haemoglobins for the qualitative analysis of haemoglobin S in diagnosing sickle cell anaemia.

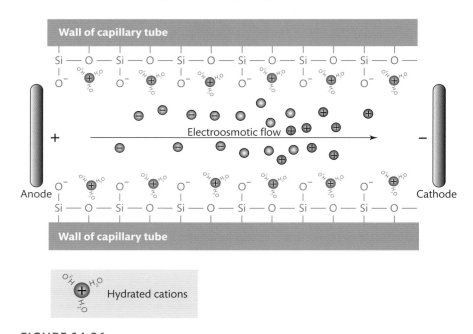

FIGURE 14.26
The role of electro-osmosis in capillary electrophoresis. See text for details.

BOX 14.3 *Separating plasmids using electrophoresis*

Plasmids are extrachromosomal, double-stranded DNA molecules found in bacteria, yeasts, and some other eukaryotic cells. A specific type of plasmid can occur in several forms that differ from one another in conformation and size. One conformation is the supercoiled covalently closed circular (ccc) structure. In this most compact form, the circular DNA double helix is interwoven, rather like a twisted rubber band. Loss of this compact structure happens if one of the DNA strands breaks, which allows the circular molecule to relax with a loss of coiling producing the open circular (oc) or nicked conformation. If both strands of the DNA are hydrolysed, then the closed circular structure is lost and a linear plasmid formed. Further structural heterogeneity results from the ability of plasmids to dimerize.

Plasmids are used as vectors in the production of recombinant DNA molecules (Chapter 16 'Molecular biology

techniques'). Thus, both purity and heterogeneity of preparations of plasmids must be assessed before their use as a vector. Pharmaceutical grade plasmid DNA is of adequate homogeneity if more than 90% of the preparation is in the form of ccc molecules.

Agarose gel electrophoresis (AGE) is the traditional method for assessing the homogeneity of plasmid DNA. However, AGE suffers some drawbacks: it cannot be automated and is only semi-quantitative. Furthermore, the assignment of bands to the different plasmid conformations is difficult, since their mobility changes with electrophoretic conditions. However, capillary electrophoresis (CE) is a more useful technique for the routine analysis of plasmid preparations (Figure 14.27). It has high resolution, sensitivity, and reproducibility, in addition to being readily automated.

1. ccc monomer of pUC19 plasmid
2. ccc dimer of pUC19 plasmid
3. Linear monomer of pUC19 plasmid
4. Linear dimer of pUC19 plasmid
5. Marker of 5.7 Kbp size
6. oc monomer of pUC19 plasmid
7. oc dimer of pUC19 plasmid

FIGURE 14.27
Separation of forms of the plasmid pUC19 by capillary electrophoresis. Courtesy of Dr. Martin Schleef, PlasmidFactory, Bielefeld, Germany. See also Figure 16.28.

14.10 **Isoelectric focusing**

Isoelectric focusing (IEF) is used to separate molecules usually peptides and proteins, according to differences in their isoelectric points (*pI*). The *pI* of a molecule is the pH at which it lacks a net charge. The charge on a protein (or amino acid or peptide) is pH dependent. If the pH of the solvent is below the *pI* of the protein, then the protein will be positively charged and will move towards the cathode. Conversely, if the pH of the buffer is above the *pI*, then the protein will be negatively charged and will move towards the anode. If the pH, however, is equal to that of the *pI* of the protein it will not migrate because it carries no charge. Thus, separation occurs in a gradient of pH values formed using a mixture of **ampholytes**. Ampholytes are synthetic low M_r polyamino-polycarboxylic or polysulphonic acids that differ from one another in the value of their *pI* (Figure 14.28).

In most IEF systems, focusing is carried out in horizontal, large pore gels (for example, 4% acrylamide, or agarose gels of low concentration) on glass or plastic plates or in horizontal glass tubes. The gel stabilizes the pH gradient and the large pores reduce any molecular sieving. Thus, for example, gels can be prepared as thin layers between two glass plates, using a solution containing riboflavin and acrylamide as described in Section 14.7. The solution also contains a range of ampholytes. Different ranges of pH, for example, 3–10, 7–8 or 2–12 are available. Naturally, a range is chosen that encompasses the *pI* values of the molecules to be separated. Once polymerization of the gel has occurred, one of the glass plates is removed and thick strips of wetted filter paper are used as wicks to connect the gel to the electrode compartments. The anode compartment contains phosphoric acid and the cathode chamber sodium hydroxide. At this stage, the pH is constant throughout the gel. However, when a potential difference is applied, the ampholytes migrate and form an appropriate pH gradient within the gel. The current is then switched off and the samples applied to the gel as a concentrated, salt-free layer on its surface. Alternatively, the sample can be added directly to the gel preparation solution. When the current is switched back on, the proteins are subjected to an electric field and migrate either towards the anode or the cathode depending upon their charge and position in the pH gradient (Figure 14.29). Once they reach a position where the pH of the gradient is equal to their *pI*, they will no longer have a net charge and will cease moving (Figure 14.29). Given that different proteins differ in the value of their *pI*, they will move to different parts of the pH gradient and so separate.

Following focusing, the gels are treated with TCA, which removes the ampholytes and fixes the proteins in the gel. The positions of the proteins on the gel can then be found by staining them as described in Sections 14.3 and 14.7.

FIGURE 14.28
General structure of an ampholyte.

R = H or $- (CH_2)_n - COOH$

n = 2 or 3

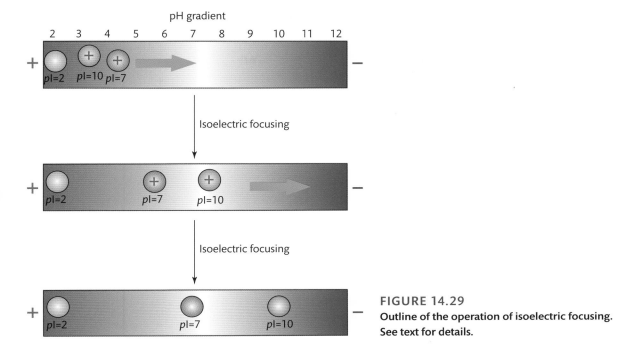

FIGURE 14.29
Outline of the operation of isoelectric focusing.
See text for details.

Isoelectric focusing can resolve proteins that differ in their pI by a pH value of as little as 0.01: it is, therefore, a high resolution technique. It can also be used to determine the pI of a protein if a mixture of proteins of known isoelectric points are incorporated onto the same gel as the test protein. After staining, the distance of each protein from one electrode is measured and a calibration graph constructed of the distance moved by each protein against its pI. The pI of the test protein is obtained as the value corresponding to its distance on the graph. Box 14.4 illustrates the use of IEF in investigating clinical conditions.

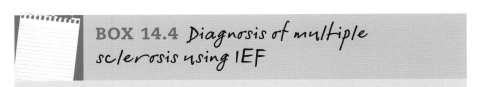

BOX 14.4 *Diagnosis of multiple sclerosis using IEF*

Multiple sclerosis (MS) is an autoimmune condition where antibodies or immunoglobulins (Ig) attack nerves within the central nervous system (CNS) causing their demyelination. These antibodies are produced within the CNS and are not found in blood. The disease is commoner in females than males and its onset is typically during young adulthood. Multiple sclerosis affects between 2 and 150 per 100,000 people.

In the majority of patients, MS can be diagnosed using IEF. Samples of serum and cerebrospinal fluid (CSF) are obtained from patients, the latter after performing a lumbar puncture (see Figure 7.8), and subjected to IEF. Up to 79–90% of samples from patients with MS show 2–5 so-called oligoclonal bands following IEF. Oligoclonal bands are molecules of IgG or their fragments secreted by plasma cells. The presence of oligoclonal bands in CSF together with their absence in serum indicates that they are produced in the CNS as

Cross reference
You can learn more about immunoglobulins in Chapter 15 'Immunological techniques'.

you can see in Figure 14.30. It is therefore normal to subtract bands in serum from bands in CSF when investigating CNS diseases.

FIGURE 14.30
Use of isoelectric focusing in diagnosing multiple sclerosis (MS). Oligoclonal bands are absent in cerobrospinal fluid (CSF, lane 1) and serum (lane 2), indicating the person concerned does not suffer MS. Oligoclonal bands are present in CSF (lane 3), but absent in serum (lane 4), which indicates the individual suffered from MS.

14.11 **Two-dimensional electrophoresis**

Two-dimensional electrophoresis is a high resolution technique that combines IEF and SDS-PAGE. In the first dimension, the samples are subjected to IEF, which separates proteins on the basis of differences in pI (charge). The IEF is carried out in an acrylamide gel containing ampholytes for making an appropriate pH gradient and 8 mol dm^{-3} urea and a non-ionic detergent. These conditions denature the protein molecules, which separate according to their pI values.

Following IEF, the gel is equilibrated in a buffer containing the detergent SDS, which gives the denatured protein molecules a uniform mass-to-charge ratio. The gel is then transferred to the top of an SDS-polyacrylamide gel. The proteins are now subjected to SDS-PAGE, which forms the second dimension by separating the proteins according to differences in their sizes. The gels are usually silver stained to show the separated components as described in Section 14.7. Large (20 × 20 cm) 10% acrylamide two-dimensional gels are often used in this technique and have been reported to resolve up to 10,000 proteins in samples isolated from cells or in tissue extracts (Figure 14.31).

Two-dimensional electrophoresis can be used to detect and compare changes in gene expression. For example, in two different tissues where one is healthy and the other is diseased. An extract from each of these tissues is subjected to two-dimensional electrophoresis and the gel patterns of the two can be compared. Such a comparison may reveal the presence

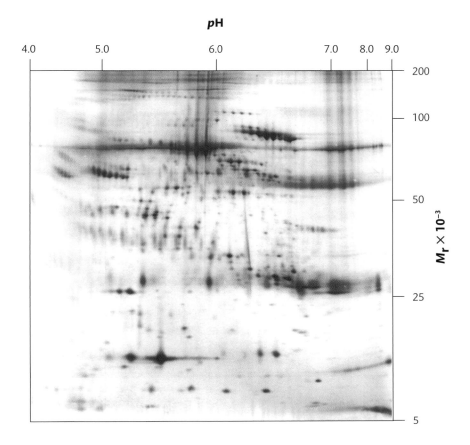

pH

FIGURE 14.31
Photograph of a two-dimensional electrophoretic separation of proteins in a sample of cerebrospinal fluid. Courtesy of P.R. Burkhard, Faculty of Medicine, University of Geneva, Switzerland.

or absence of a particular protein in the diseased state, which may ultimately be identified and its gene expression investigated. Thus, this technique can increase understanding of the disease.

Proteomics is that branch of science that studies the complete complement of proteins encoded by a genome. It includes identifying and quantifying all the proteins produced (expressed) by a cell, tissue, or organism during the development of a disease. Two-dimensional electrophoresis aids proteomic studies in a number of ways, particularly when combined with other analytical techniques. For example, separate two-dimensional 'maps' can be produced using extracts from a cell or tissue from a healthy individual and from a patient, respectively. The gels are difficult to compare because of the large amounts of data (the number of spots). However, there is commercially available software that can compare the two different 2-D maps, and analyse these data both quantitatively and qualitatively. Any proteins that differ between the two maps are of clinical interest and require further analysis. Thus, they may be subjected to peptide mass-fingerprinting. The spots of interest are cut out of the gel and digested with trypsin, which hydrolyses the polypeptide at the carboxyl terminal of each arginine and lysine residue to produce a number of peptides. Matrix-assisted laser desorption ionization-mass spectrometry (MALDI-MS) is then used to give quantitative measurements of the peptides. The set of peptides produced is highly specific for any particular protein; hence, the pattern of peptides produced from the protein of interest can be compared with those in an electronic database and so identified. In a similar way, it is possible to explore the effects of exposure to a particular drug or treatment regime, or toxin.

Cross reference
You can read more about mass spectrometry in Chapter 11 'Spectroscopy'.

SUMMARY

- Electrophoresis is the movement of charged molecules or ions in an electric field. Separation of the ions is influenced by their nature, the support media used and the potential difference used.

- Electrophoresis is performed using different types of support media, some that provide only physical support such as paper and cellulose acetate, while others are gels, for example, starch, polyacrylamide, and agarose. The latter offer improved resolution because of their molecular sieving effect.

- Sodium dodecyl sulphate-PAGE and agarose gel electrophoresis are the methods of choice for separating proteins and nucleic acids, respectively. In both procedures, mixtures of molecules have uniform negative charges and are separated on the basis of their sizes by the molecular sieving effect of the gel.

- Pulsed field gel electrophoresis is used to separate larger molecules of DNA by periodically altering the size and direction of the applied electric field.

- Capillary electrophoresis combines the high resolution of electrophoresis with the speed of high performance liquid chromatgraphy, and is used to separate analytes such as amino acids, proteins, DNA, and drugs.

- Isoelectric focusing is a specialized technique where molecules are separated on the basis of differences in their isoelectric points.

- Electrophoretic techniques have applications in biomedical science, including investigation of diseases such as MM, CF, and multiple sclerosis.

FURTHER READING

- Hames BD. *Gel Electrophoresis of Proteins: a Practical Approach*. Oxford University Press, Oxford, 1998.

- Harris LR, Churchward MA, Butt RH, Coorssen JR. Assessing detection methods for gel-based proteomic analyses. *Journal of Proteome*, 2007; **6**: 1418–1425.

- Perrett D. Capillary electrophoresis in clinical chemistry. *Annals of Clinical Biochemistry*, 1999; **36**: 133–150.

- Righetti PG. Electrophoresis: the march of pennies, the march of dimes. *Journal of Chromatography A*. 2005; **1079**: 24–40.

- Sasse J, Gallagher SR. Staining proteins in gels. *Current Protocols in Molecular Biology*, 2009; **10**: 10.6.1–10.6.27.

- Westermeier R. *Electrophoresis in Practice*. Wiley-VCH, Weinheim, 2006.

QUESTIONS

1. You normally separate fragments of DNA by electrophoresis using a potential difference of 100 mV. The distance between the electrodes in your apparatus is 150 mm. However, this apparatus is unavailable and the larger system available does not have a protocol or manufacturer's instructions. The distance between its electrodes is 200 mm. What potential difference is necessary to produce the electric field you would normally use?

2. Why are glycerol and the dye bromophenol blue added to gel loading buffers?

3. A protein was subjected to native PAGE and centrifugation to determine its size. Native PAGE indicated that the molecules had a M_r of 64,000, which was confirmed by the centrifugation studies (Chapter 12). However, its M_r was found to be only 16,000 when subjected to SDS-PAGE. Are these data compatible? If so, explain your answer.

4. When DNA molecules are separated by agarose gel electrophoresis, which of the following statements best describes their migration?

    (a) Shorter molecules move faster than longer ones because they have a greater density

    (b) Shorter molecules move faster than longer ones because they can move through the pores in the gel more easily

    (c) Longer molecules move faster than short ones because they are subjected to more frictional resistance

    (d) Longer molecules move faster than shorter ones because of their larger mass

    (e) Longer molecules move faster than short ones because they have a greater mass-to-charge ratio

5. Which of the following are used to stain proteins and which to stain nucleic acids?

    (a) Acridine orange

    (b) Amido black

    (c) Coomassie blue R-250

    (d) Ethidium bromide

    (e) Silver stain

6. A number of standard proteins of known M_r were subjected to SDS-PAGE. The bromophenol blue tracking dye moved a distance of 57.0 mm. The distances moved by the proteins were measured. The distances and their M_r are given in the accompanying table. A protein of unknown M_r was also included in the experiment and the distance it migrated is also shown in the table. Staining and destaining did not affect the dimensions of the gel.

$M_r \times 10^{-3}$	Migration distance/mm
14.3	50.0
18.4	43.0
24.0	30.0
34.7	20.0
66.0	10.0
?	38.0

Calculate the relative mobilities of each of the proteins and plot a suitable graph of these data. Use this graph to determine the M_r of the unknown protein.

7.   A number of standard proteins of known pI were subjected to isoelectric focusing. Following staining, the distances of the proteins from the anode edge of the gel were measured. The distances and their pI values are given in the accompanying table. Two test proteins, A and B, of unknown pI were also included in the experiment and their distances are also shown in the table.

Standard proteins	pI	Distance/mm
	3.50	4.5
	4.55	17.0
	5.20	33.5
	5.85	53.0
	6.55	69.0
	6.85	79.5
	7.35	90.0
	8.15	103.0
	8.45	116.5
	8.65	125.0
	9.30	140.0
Test proteins		
A	?	85.5
B	?	31.5

Plot a suitable graph of these data and determine the pI values of A and B.

Answers to self-check and end of chapter questions are available in the Online Resource Centre accompanying this book.

 Go to www.oxfordtextbooks.co.uk/orc/glencross

15

Immunological techniques

Christine Yates

Learning objectives

After studying this chapter, you should be able to:

- Outline the components and functions of the immune system
- Describe the *in vivo* production of antibodies
- Explain the principles of antigen–antibody reactions
- Discuss the uses of immunological techniques in the clinical laboratory
- Discuss the uses of monoclonal antibodies in the clinical laboratory.

Introduction

A grasp of the underlying principles of the immune system and the fundamentals of immunoassay are essential to the understanding of routine immunological diagnostic assays. Immunological assays are techniques that depend upon the visualization of interactions between antibodies and antigens to detect and measure the concentrations of a wide range of substances in health and disease. They are relative newcomers to the clinical laboratory having developed from basic agglutination tests (Box 15.1). The applications were revolutionized in the mid 1970s with the development of monoclonal antibodies. The rapid growth in our understanding of the immune system and the utility of antibodies as diagnostic markers and reagents has led to the development of a wide range of applications that assist in diagnosis and treatment.

Immunological techniques largely depend upon the visualization of antibody–antigen reactions, detecting the formation of immune complexes or the attachment of labelled antibodies to cells or an artificial surface by a variety of methods, which we describe in this chapter. A range of immunological techniques have been developed to detect and measure the concentrations of cellular and humoral elements of the immune system in health and disease and, in some cases, assess their biological activities. They have also been used to raise

BOX 15.1 *Beginnings of research into the immune system*

The discovery of antigen–antibody reactions took place at the beginning of the 20th century. Michael Heidelberger measured antibodies using a precipitation reaction in the 1930s producing a globulin fraction of 70% purity and established the existence of complement (Box 15.2). In the decades that followed, electrophoretic and ultracentrifugation studies increased our knowledge of immunoglobulins and antibody activities in the immune response. In 1972, Rodney Porter and Gerald Edelman received the Nobel Prize in Medicine for their work on the structure of immunoglobulin molecules. These studies were the beginning of our understanding of the immune response and led the way for the techniques that we take for granted in clinical laboratories of the present day.

antibodies to other analytes, such as hormones and tumour markers increasing the range of immunoassays available to biomedical scientists.

15.1 Outline of the immune system

The immune system comprises a number of physical barriers, lymphoid organs, cells, and humoral components that work in harmony to protect the body from infection. Its primary function is to destroy or disable microorganisms, for example, viruses, bacteria, fungi, and protozoa, and parasites, such as helminth worms, which are present in the environment.

Lymphoid organs are divided into two types: primary and secondary organs. Primary organs consist of the bone marrow and thymus, where the cells of the immune system are generated and undergo maturation. Secondary lymphoid organs consist of the spleen, lymph nodes, mucosa-associated and gut-associated lymphoid tissues (MALT and GALT respectively). These are essential for the development of the immune response and cells circulate between the tissues and lymphoid organs in the blood and lymph and maintain the process known as immune surveillance. The cells and humoral elements of the immune system are listed in Table 15.1.

Innate immunity

Innate immunity is that present at birth. It is non-specific, does not develop an immunological memory, and is most effective against pyogenic bacteria, fungi, and multicellular parasites. Physical barriers, such as the skin that present an impermeable surface and secretions that wash mucosal tissues, are the first line of defence of the innate systems. Humoral components such as complement (Box 15.2), mannan-binding lectin, and proteolytic enzymes, which are present in blood, mucous, saliva, and other body fluids are also part of the innate response. The cells of the innate system are neutrophils, eosinophils, mast cells, and natural killer cells. All of these factors combine to disable and dispose of potentially pathogenic organisms.

Acquired immunity

Acquired immunity is also known as the adaptive response. Unlike innate, acquired immunity is not present at birth, but develops throughout life. The response to microorganisms

TABLE 15.1 **Cells and humoral elements of the immune system**

Cells	Myeloid:
	• Neutrophils
	• Eosinophils
	• Macrophages
	• Mast cells
	• Dendritic cells
	Lymphoid cells:
	• Lymphocytes
	• Plasma cells
	• Natural killer cells
Humoral elements	Antibodies
	Complement (Box 15.2)
	Mannan binding lectin
	Proteolytic enzymes, cytokines
	Adhesion molecules

BOX 15.2 *Complement*

The complement system is part of the innate immune system and, as such, it is not adaptable and does not change over the course of an individual's lifetime. Plasma contains over 40 complement proteins. They are often abbreviated as C and given a number, for example, C1q, C2, C3; although others, such as mannose-binding lectin (MBL), factors B, D, and so on, obviously do not follow this rule. Complement proteins work in a sequential cascade fashion; individual proteins in the sequence activating several molecules of the next component. The effect of such a cascade is to greatly magnify the final effect. The cascade may initially follow one of three pathways:

- classical pathway, which is activated by an antibody
- the pathway activated by MBL
- the alternative pathway, which is activated by direct contact with bacteria

However, all three pathways converge in a common final pathway that destroys the cells of the pathogen. For example, complement kills bacteria by piercing their cell surface membrane. A further function of the complement system is the disposal of immune complexes that are produced naturally as the immune system combats infection.

An inappropriate activation of such a powerful process could lead to many potential problems, hence control of the complement pathway is strict, and includes a number of inhibitory proteins, for example, C1-inhibitor, factor I, and factor H.

is specific. Acquired immunity has the ability to develop an immunological memory so the responses increase in intensity with repeated exposure to a microorganism.

The active components of the adaptive immunity are lymphocytes derived from the bone marrow. These produce antibodies in response to protein antigens on the surfaces of viruses and bacteria. The antibodies bind specifically to the organisms carrying their complementary antigen and interact with complement and receptors on granulocytes and mononuclear phagocytic cells leading to the death of the target organism.

Cytokines

In both innate and acquired immunities, chemical messengers called **cytokines** are produced by cells. These allow the cells to communicate with each other, and co-ordinate and control their growth, differentiation, and functions. **Chemokines** and **adhesion molecules** are also instrumental in the movement of cells between the different lymphoid organs.

15.2 **Synthesis of antibodies *in vivo***

Antibodies are glycoproteins that are able to bind to antigens, usually proteins that are normally bound to cell surfaces. They belong to a group of proteins called immunoglobulins, which are so called because they have a role in acquired immunity and appear in the globulin fraction when serum proteins are separated by electrophoresis (see, for example, Figures 14.3 and 14.9). Immunoglobulin molecules are formed of two heavy and two light chain polypeptide chains joined together by disulphide bonds. The structure of their molecules can be thought of as Y-shaped. It consists mainly of a relatively conserved constant region. However, there is a region at the end of each of the two arms of the Y called the variable region, which forms the antigen binding site and which varies in different types of antibodies (Figure 15.1).

Immunoglobulins may be divided into one of five classes depending largely upon their structures and which of the five main types of heavy chain are present. There are two types of immunoglobulin light chain that are named lambda and kappa. The different light chains have no particular function except to contribute to the diversity of the variable region. Immunoglobulins that exist as multiples of immunoglobulin G (IgG) molecules are joined together by a protein called the J chain.

The five classes of immunoglobulins are:

- Immunoglobulin G
- Immunoglobulin A

Cross reference

You can read about electrophoresis in Chapter 14 'Electrophoresis'.

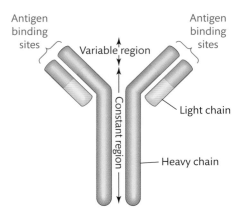

FIGURE 15.1
Structure of an immunoglobulin G (IgG) molecule.

- Immunoglobulin M
- Immunoglobulin E
- Immunoglobulin D.

Immunoglobulin G (IgG) is the most abundant immunoglobulin in samples of serum. It may be subdivided into four subclasses. Immunoglobulin G is most effective when it associates with its antigen. It can then bind to complement proteins, which results in the recruitment and activation of phagocytes. An interesting aspect of IgG activities is its ability to cross the placenta and give the newborn baby some protection against infection during the first few months of life.

Immunoglobulin A (IgA) it found in relatively low concentrations in serum in a form that resembles two IgG molecules joined together. It is secreted generously in saliva, tears, and other fluids that bathe the mucosal surfaces limiting microbial entry to the body through these vulnerable routes.

Immunoglobulin M (IgM) is the first antibody produced in the primary immune response to infection. It is formed of the familiar four polypeptide chains but in a structure where they are repeated five times. This gives it multiple antigen-binding sites. Immunoglobulin M is a potent activator of complement and phagocyte recruitment.

Immunoglobulin E (IgE) occurs at low concentrations in serum samples from healthy individuals but this concentration increases in patients with parasitic infections or allergies. Immunoglobulin E principally interacts with mast cells that produce substances capable of deactivating parasites and which contribute to the development of allergic reactions.

Immunoglobulin D (IgD) is found on the surface of immature B lymphocytes (Section 15.2). It is normally present in very low levels in serum and its function is not fully understood.

SELF-CHECK 15.1

(a) List the primary lymphoid organs. (b) Define *innate immunity*. (c) State the major activity of an antibody.

Polyclonal and monoclonal responses

Antibodies are produced by plasma cells that are activated through receptors on the surface of T and B lymphocytes. These cells recognize specific parts of antigens called antigenic determinants or **epitopes**. The recognition sites of antibodies interact with the complementary epitope surface rather like pieces of a jigsaw puzzle. Plasma cells which produce and secrete antibodies develop from B lymphocytes. Each type of plasma cell is a clone and uniquely produces a single, specific type or monoclonal antibody. However, antigens are complex molecules that potentially have multiple epitopes on their surfaces. They, therefore have the potential to stimulate the production of many different types of antibodies. Most infections would elicit the production of multiple antibodies, which is known as a polyclonal response. Measurable monoclonal responses are usually confined to clinical cases where a plasma cell tumour leads to the over-production of a single clone of cells, or when plasma cells are manipulated under laboratory conditions.

When an antigen is first encountered, the immune system mounts a primary response within 5–10 days, usually producing an IgM antibody, whose concentration increases then falls to a low level. However, if the same antigen (epitope) is encountered again, *memory* plasma cells respond by producing a secondary response. This is usually in the form of an IgG antibody that is produced more rapidly than the initial IgM and is of a higher intensity and specificity.

Cross reference
You can read more about immunoglobulins in the relevant chapter of the *Immunology* book of this series.

Associated disease states

Diseases of the immune system are varied, but are generally categorized by an inability to mount an effective immune response or by an inappropriate activation of the immune system. Individuals with immunodeficiency disorders are unable to raise an appropriate immune response. This may be due to an abnormality in the development of the cells of their immune system, a defect in communication between the immune cells, human immunodeficiency virus (HIV) infection, or a fault in another essential process, such as the complement pathway.

Autoimmune disorders occur when immune responses are directed against an individual's own tissues or specific organs. This is normally avoided by the development of tolerance mechanisms during the maturation of T and B cells. Individuals may be genetically predisposed to developing autoimmune disease, but it usually takes the added trigger of an environmental stimulus, such as a viral infection, to start the chain of events that lead to the development of disease. In some cases, an autoantibody can be identified as a diagnostic marker, which may or may not have direct pathological activity. The autoimmune disorders are divided into organ specific where a particular organ is targeted, for example the pancreas in Type I diabetes, or non organ specific, where antibodies are directed at cell components such as nuclear proteins as in systemic lupus erythematosus (SLE).

In multiple myeloma (see Figure 14.9 and associated text), a clone of malignant plasma cells produces excessive amounts of a single immunoglobulin, which may or may not behave as an antibody. This may inhibit the production of other immunoglobulins effectively creating an immunodeficient state and making the individual particularly vulnerable to infections.

Cross reference
You can read about diseases of the immune system in the *Immunology* volume of this series.

Allergic symptoms develop in susceptible individuals when an antibody reaction is elicited by a normally harmless environmental substance or food. The presence of the allergen, which may be inhaled pollen or an ingested food substance, stimulates the production of IgE. The molecules of IgE cross-link receptors on mast cells, leading to the release of histamine, and other factors that produce the allergic reaction.

SELF-CHECK 15.2

(a) Describe the structure of an immunoglobulin. (b) What is the main function of a plasma cell? (c) Name the two factors that may combine to predispose an individual to develop an autoimmune disease.

15.3 Production of polyclonal and monoclonal antibodies *in vitro*

Polyclonal antibodies are heterogeneous pools of antibodies that arise from a secondary immune response to an antigen; that is, they are directed against multiple epitopes. They are produced by injecting subjects with antigenic material that may be from the same species or a different species: a procedure known as immunization. A variety of animals are used to raise antibodies such as the antihuman globulins used in immunological assay systems (Section 15.5). Antigens are injected into the subject accompanied by a carrier called an **adjuvant**. It is unclear how adjuvants work, but it is thought they slow down the release of the antigen in the subject resulting in a series of small 'hits' to the immune system. Examples of adjuvants include Freund's adjuvant, Syntex adjuvant, Gerbu adjuvant, saponin, and aluminium hydroxide, which is one of the few licensed for use in humans. Protocols vary depending on the response that is required and the routes of administration, but the procedures take into account the size of the subject, the form of the antigen, whether it is particulate, cellular, or soluble material, and the amount of antigen available.

Polyclonal antibodies may be raised in human volunteers but the range is limited by the choice of acceptable antigens and other ethical considerations. An example of the human antibody production is given in Box 15.3. Animal sources (Figure 15.2) are also controlled by stringent legal requirements, but offer a wider range of possible antigens and anti-human options in antibody production. Polyclonal antibodies are versatile and easy to produce; they may recognize a range of epitopes on a single antigen, which is often an advantage in immunoassay. However, drawbacks to their use include limited reproducibility, cross-reactivity, low quantity, and the high cost of maintenance of laboratory animals.

Monoclonal antibodies are homogenous pools of antibodies that are raised from one clone of B cells, which produce antibodies directed against one specific epitope. The technology was developed in the 1970s by Georges Köhler and César Milstein, and is considered one of the most significant contributions to the biological sciences. B lymphocytes from the spleen of mice immunized with antigen are fused with myeloma cells. The resulting hybridoma forms a single clone of cells that produces a monoclonal (pure) antibody. You can see this procedure in outline in Figure 15.3. The hybridoma cells may be grown in culture for many years raising large quantities of antibody to almost any antigen.

Monoclonal antibodies are specific and may be produced in unlimited supply maintaining high quality. They are invaluable reagents for research and diagnostic purposes, and have been developed to treat some immune disorders and cancers. However, high levels of investment are required for their production as sophisticated methods of selection, characterization, and purification are necessary; especially if the final product is a therapeutic agent (see Box 16.4).

BOX 15.3 The production of polyclonal anti-D antibodies in human volunteers

Approximately 15% of the UK population have the rhesus negative blood group, which is determined by the absence of the rhesus D antigen on their erythrocytes. When a rhesus negative woman conceives a rhesus positive baby there is a risk that the mother may produce anti-D antibodies to her unborn child. This may have implications when she has a second child, as the secondary immune response will produce IgG antibodies that can cross the placenta and enter her baby's bloodstream. The antibodies will react with the foetal blood cells, damaging them and causing anaemia, jaundice, and an enlarged liver and spleen. This is known as haemolytic disease of the newborn, and potentially could affect 70,000 babies per year in the UK. However, this life-threatening condition can be averted by immunizing mothers with antibodies to the D antigen during their pregnancy. Thus, the anti-D antibodies can attach to D positive cells, destroying them before they can illicit a reaction from the mother's immune system.

The anti-D antibodies are obtained by raising a polyclonal response in male human volunteers who donate plasma regularly. The volunteers are injected with rhesus positive blood cells, which stimulate their immune systems to produce anti-D antibodies. Research into producing a monoclonal source of the antibody is continuing.

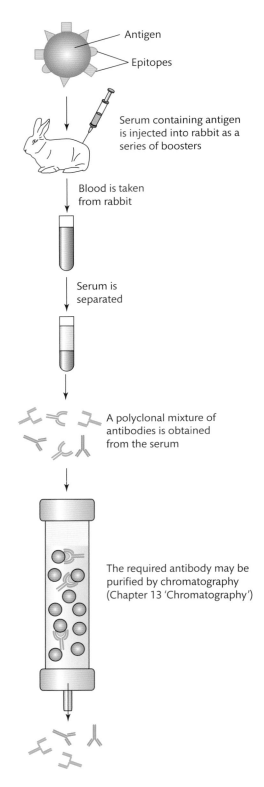

FIGURE 15.2
Schematic showing the production of polyclonal
antibodies in a rabbit. See main text for details.

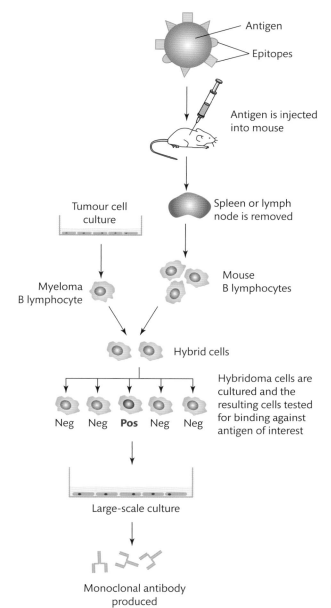

FIGURE 15.3

Schematic showing the production of monoclonal antibodies in a mouse. See main text for details.

How are adjuvants thought to work?

15.4 **Antigen–antibody interactions**

To appreciate the reaction between an antigen and an antibody, two points are worth empha-
sizing. First, not all antigen–antibody reactions occur with the same strength; that is antibodies

have different affinities for antigens. Secondly, antigen-antibody reactions are reversible, and differ in that they may not have the same capacity to remain bound to each other.

In the primary response to an infection, many antibodies will be produced to the antigenic stimuli presented on the surface of the pathogen. Some of the antibodies fit better to epitopes on the antigen than will others, that is their binding is stronger or of higher affinity. As the infection progresses, further antibody production will favour those antibodies with the best fit. If the infection is encountered again the memory B cells will be those with the highest affinity. Primary antibody responses are generally IgM types with relatively low affinities but with many binding sites due to the five-fold IgG structure of IgM. Secondary responses are largely IgG in nature with much higher affinity due to the fine tuning of immunoglobulin production.

Antibodies vary in their number of binding sites depending on the immunoglobulin type. Thus, IgG has two binding sites, IgM has 10. If one binding site of an IgG antibody binds to an antigen, then the other is also likely to bind. However, if the antigen-antibody immune complex disassociates, the two binding sites must release at the same time. In contrast, if the antibody is an IgM, there is potential for all ten binding sites to be bound and the likelihood of simultaneous release becomes much less. In other words IgM molecules may have lower affinity but they will usually have much higher avidity.

Affinity is an indication of the strength of antigen-antibody binding. **Avidity** is a measure of the ability of an antibody to bind multiple binding sites.

Determining the affinity and avidity of an antibody is a complex procedure. It is generally confined to research or commercial laboratories, where antibodies are produced as research tools or commercial products.

Methods of detection based on antigen-antibody interactions

Cross reference
You can read more about black boxes and automation in Chapter 17 'Laboratory automation'.

High quality, fast turnaround assays are performed in today's clinical laboratories, largely in 'black boxes' where the sample goes in one end and the answer appears on a VDU screen. However, to understand the limitations and capabilities of the analysis systems at our disposal, we need to understand the basic principles on which they are based.

As knowledge of the components of the immune system expanded, the need to demonstrate their presence and activities in controlled laboratory conditions increased. The increasing precision and sophistication of research has led to the development of assays that could be reliably validated for diagnostic uses. Early precipitation experiments led to the purification of immunoglobulins (Box 15.1) and the ability to measure their concentrations in serum reasonably accurately.

The quantitative precipitation curve

The quantitative precipitation curve was devised by Michael Heidelberger and Forrest Kendall in the 1950s, and is the basis for all quantitative studies of antigen-antibody reactions. In essence, *increasing* aliquots of antigen are added to a series of tubes containing a *constant* volume of a complementary antibody. Antigen-antibody immune complexes form and precipitate in each tube. The precipitates may be separated by centrifugation and weighed or the particles of precipitate dispersed and the absorbance of the resulting suspension may be measured. The amount of antigen-antibody formed can then be plotted against that of antigen added, which gives the type of graph you can see in Figure 15.4. The Heidelberger curve

FIGURE 15.4
Precipitation curve for human serum albumin (HSA). An increasing amount of HSA is added to a fixed concentration of anti-HSA. Free antibody is present in solution prior to the point of equivalence. However, after the point of equivalence, antigen will be in excess. See text and Box 15.4 for details.

varies according to the species from which the antibody or antigen is derived. In the initial tubes, the amount or concentration of antibody is in excess of that of antigen. However, in the later tubes, the antigen is in excess and so the amount of precipitation will decrease because of the formation of soluble complexes. These arise because, although the binding sites of the antibody molecule are blocked with antigen, they are unable to form cross-links, as explained in Box 15.4.

This type of experiment illustrates the basic principles of all immunological test systems that depend upon the formation of immune complexes. An equivalence of antigen and antibody is reached just before maximum precipitation occurs (Box 15.4). There are points of *equal* precipitation on both sides of the curve shown in Figures 15.4 and 15.5: calculations based on readouts from this type of curve must take this into account. The method can be adapted to quantify the amounts of antibody in samples of serum, to determine antigenic valency and is, indeed, the basis of all other precipitation reactions. Light scattering methods, such as nephelometry and turbidimetry (Section 15.5) have, however, largely replaced precipitation methods, making the quantitation of immune complexes more accurate and more sensitive.

Other immunological techniques visualize immune complexes in media such as gels or observe the agglutination of cells or particles with antigens attached to plastic surfaces, or are based on the attachment of antibodies to thin sections of solid organs or sheets of tissue cultured cells.

SELF-CHECK 15.4

(a) Define the meanings of avidity and affinity in relation to antigen–antibody binding.
(b) At what point is equivalence reached on the quantitative precipitation curve?

BOX 15.4 *To precipitate or not*

Antigen–antibody or immune complexes can precipitate from solution or in a gel or other formats, which are described in this chapter. Antibodies have a number of sites per molecule that can bind to the corresponding antigens. Molecules of IgG antibodies have two binding sites, in the case of IgM there can be ten. Thus, the immune complexes potentially consist of large aggregates or lattices of combined molecules that form precipitates. In experiments when a comparatively small amount of antigen is added to a constant amount of antibody, all of the antigen molecules will be bound up in the precipitate. This is called the state of antibody excess. When more antigen is added, the amount of precipitate formed will increase until the ratio of antigen to antibody is equal. This is the point of equivalence. Adding more antigen at this point will cause precipitation to decrease and a state of antigen excess is reached.

Sometimes the antigen or antibody that is being measured may appear to be absent or a negative result is obtained on analysis. However, when the sample is diluted a positive result is obtained. This is known as the prozone phenomenon, when relatively high levels of the antigen or antibody prevent complete precipitation by blocking binding sites, without completing the formation of a lattice of immune complexes. The diagram in Figure 15.5 illustrates this situation. At points A and B the readout for the assay is the same but the quantity of antigen present is very different. This is the basis for the prozone phenemenon.

Y Antibody

◇ Antigen

FIGURE 15.5

Prozone phenomenon. The basis of the prozone phenomenon is that at points A and B, the results for the assay are the same even though the quantities of antigen present are different. See text for details.

15.5 Immunological techniques used in clinical laboratories

Immunological techniques depend in some way upon visualizing and quantifying interactions between antigens and antibodies to detect and measure the concentrations of analytes in clinical samples. We will discuss them under the following headings:

- Immunodiffusion techniques
- Agglutination techniques
- Immunoelectrophoresis techniques
- Immunoassays
- Nephelometry and turbidimetry techniques
- Enzyme-linked immunosorbent and enzyme linked immunospot assays
- Flow cytometry and fluorescence activated cell sorting
- Multiplex techniques
- Microarrays
- Complement assays
- Indirect and direct immunofluorescence techniques
- Immunocytochemistry

Immunodiffusion techniques

Immunodiffusion techniques depend upon the diffusion of antigens and antibodies from holes or wells cut into an agarose gel. The antibodies employed may be monoclonal or polyclonal, depending on the design of the testing protocol. Where antibody meets antigen in the gel, recognition, and binding occur forming a white line of precipitated immune complexes.

The gel may be inert or may contain defined quantities of known antibody or antigen. Polyethylene glycol may also be added to enhance precipitation or Coomassie blue may be used to stain and clarify the lines of precipitation. The incubation time and temperature employed depends on the type of the immunoglobulin employed in the method. Incubation times range from 24–72 hours, although with some IgG antibodies, reactions can be observed after only four hours if incubated at 37°C.

The use of gel diffusion was a significant development in the transfer of immunology methods from the research area to the clinical laboratory. The techniques require smaller volumes of reagent and have diverse applications in diagnosis.

Single radial immunodiffusion

In single radial immunodiffusion (SRID), antigen of unknown concentration is added to the wells in a gel that contains uniformly distributed known antibody or antiserum, which is serum that contains known antibodies and so may be used as an analytical reagent. The gel is usually incubated at room temperature for 24–48 hours. The antigen diffuses through the gel and on binding with its complementary antibody forms a precipitate in the shape of a ring that surrounds the well and moves outwards from it. Movement of the ring stops at a diameter where the point of equivalence is reached, that is the antigen and antibody are present in equal quantities. The final diameter of the ring is directly proportional to the concentration of the

FIGURE 15.6
Photograph of a single radial immunodiffusion experiment, showing precipitation rings of varying diameter relative to the quantity of antibody in the sample. Courtesy of D. Raine, Immunology Department, Hull Royal Infirmary, UK.

antigen present (Figure 15.6). This is measured using a calibrated eyepiece that magnifies the ring of precipitation or an electronic reader. A standard curve can be constructed using known concentrations of antigen in a number of wells to allow the semi-quantitation of unknown antigen samples. The method can be reversed by incorporating antigen in the gel to quantify the amount of antibody in a sample.

Single radial immunodiffusion has limited applications in present day clinical laboratories as the determinations are not sufficiently accurate and are operator dependent, being easily affected by the delivery of sample into the wells and inattention to the test protocol. The method is also time consuming and inappropriate for the analyses of large numbers of samples. This method is, however, routinely used to measure the components and inhibitors of the complement system in rare immunodeficiency disorders. In these cases, the number of samples is likely to be low and few other analytical options are available.

Cross reference

The *Immunology* volume of this series has further details about complement assays and the practical applications of gel diffusion methods.

Double diffusion and Ouchterlony double diffusion techniques

Double diffusion techniques involve placing antigen and antibody in opposing wells of gel and observing the precipitation of bands where immune complexes form. The commonest application is in the Ouchterlony format, where a number of antigens or antibodies can be compared around a single well of known antigen or antiserum. In the Ouchterlony double diffusion method, a number of wells are cut in the gel such that one is placed centrally and the rest are equidistant from it. Figure 15.7 outlines the Ouchterlony double diffusion technique.

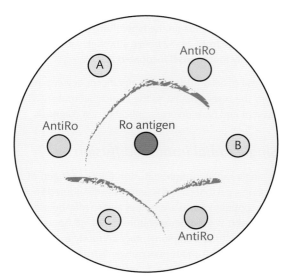

FIGURE 15.7
Schematic showing the results of an Ouchterlony double diffusion assay. Wells in the gel plate were loaded with three test sera, A, B, and C. The central well contained a purified nuclear antigen (Ro, and alternate surrounding wells contained known anti-Ro. Following incubation, the line of precipitate formed to give the results shown. Thus, sample A is positive for anti-Ro as it shows a line of full identity with the Ro antiserum. Sample B is negative as there is no line of precipitation and sample C shows some reactivity with the Ro antigen but cannot be fully identified as anti-Ro. This latter type of reaction can be caused by impurities in the antigen preparation.

If a known antibody is placed in the central well and a variety of unknown antigens placed in the surrounding ones, then precipitation lines will appear between them that may be one of three characteristic patterns:

1. Line of full identity
2. Line of non-identity
3. Line of partial identity

A line of full identity is a continuous one and indicates that the antigens reacting with the antibody are all identical. In contrast, a line of non-identity is cross-shaped, thus the antigens reacting with the antibody are different. Lastly, a line of partial identity appears as a single spur on a line indicating that the antigens reacting to the antibody have some activity in common.

The Ouchterlony double diffusion technique may be used to differentiate the different anti-nuclear antibodies in autoimmune disease. However, it is not a quantitative method and its use in clinical laboratories is becoming less frequent as automated methods have replaced such time-consuming manual techniques.

SELF-CHECK 15.5

(a) What happens when an antigen meets its corresponding antibody in an agarose gel?
(b) What term is used to describe the situation when high concentrations of analyte prevent the formation of an antigen–antibody precipitate?

Agglutination techniques

Visible agglutination, that is the binding together of particles, may be observed in a liquid matrix on mixing antibodies complementary to an antigen carried on the surfaces of cells or particles, as you can see in Figure 15.8. Erythrocytes and latex particles can be used to give a semi-quantitative assay for estimating the concentrations of antibodies. This is achieved by preparing a range of serum dilutions from high to low concentration. Equal quantities of each dilution and appropriate cells or particles are then mixed on slides or in microwell plates. The strength of the antibody is expressed as a titre which is the dilution at which agglutination is still, but only just, visible; as you can see in Figure 15.9.

In blood group serology, the agglutination of erythrocytes with known antisera is used to identify ABO and Rhesus blood groups. When donor blood is selected for transfusion into an individual, a cross-match procedure is performed to determine the compatibility of the donated blood with that of the recipient. Incompatibility is indicated if agglutination occurs

FIGURE 15.8
Schematic showing the agglutination of erythrocytes and latex particles on a slide. On the left erythrocytes (pink) and latex particles (blue) are clumped or agglutinated, giving a grainy appearance. On the right the suspension is smooth in appearance because the cells and particles are not agglutinated.

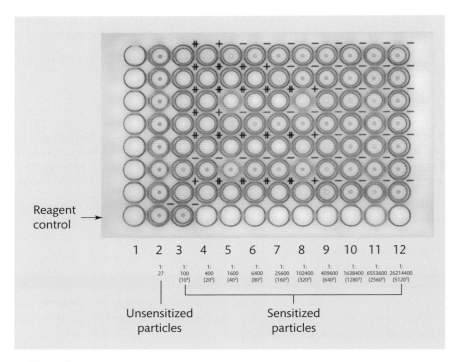

FIGURE 15.9

Agglutination of particles in multiwell plate. Particles coated with thyroid antigen have been mixed with varying dilutions of test sera. Non-agglutinated cells settle in a small pellet at the bottom of the well. Agglutinated particles form a smooth layer on the well and partially agglutinated or weak reactions settle as a ring in the well.

when the donor's blood is mixed with the recipient's plasma. The ABO blood group are antibodies, which are usually of the IgM type. They will agglutinate erythrocytes of an opposing ABO group within a few seconds at room temperature. However, IgG antibodies have fewer binding sites and their agglutination reactions are more temperature dependent. They may attach to cells and bind complement but not achieve visual agglutination. In the case of IgG Rhesus antibodies, tests are performed at 37 °C and antihuman globulin antibodies are used that attach to the heavy chains of molecules of IgG and complete the agglutination reaction. Antihuman globulin reagents are polyclonal antibodies raised in laboratory animals to specific immunoglobulin isotypes and complement components.

In the past, animal erythrocytes were used fresh or preserved and coated with antigen or antibody in a range of assays, for example, the Rose Waaler test for rheumatoid factor; the detection of thyroid antibodies and the Paul Bunnell test for infectious mononucleosis. In some cases sera containing antibody required heat inactivation at 56 °C to destroy complement activity that could potentially haemolyse the cells during incubation. Most of these tests have now been replaced by automated analysis, nephelometry, or other forms of immunoassay. The use of erythrocytes has largely been replaced by polystyrene or latex particles. Commercial kits (see Box 16.1) have reduced the use of reagents obtained from animals and increased the shelf life of products. Simple rapid slide tests may be used to detect rheumatoid factor, C reactive protein, and a wide range of other immunological assays that are varied in quality and utility. However, these kits are not suitable for high volume work in contemporary clinical laboratories.

Cross reference

You can read more about blood transfusion in the accompanying text of the series, *Transfusion and Transplantation Science.*

Immunoelectrophoretic techniques

Diffusion methods are confined to systems that use few reactants and where the precipitation lines can be easily observed. However, the resolution of *multiple* antigens and antibodies can be achieved by combining immunodiffusion with electrophoresis.

Proteins in serum and urine, including immunoglobulins can be separated, by electrophoresis and visualized by staining them with one of a number of different dyes (Table 14.1). The quantities of the separated proteins can be determined by densitometry. This is the method of choice when detecting and monitoring monoclonal paraprotein, which is a measure of tumour load and a feature of myeloma and other B cell malignancies.

The simplest application of antibodies and electrophoresis is immunofixation. Antisera to the range of immunoglobulin isotypes present in the sample is added to proteins separated by electrophoresis, which identifies the light chain type of the immunoglobulin paraprotein. However, the early methods were unsatisfactory. Adding a range of antisera to immunoglobulin isotypes following their electrophoretic separation formed lines of precipitation in the agarose gel, where each antibody reacted with a different immunoglobulin.

> **Cross reference**
>
> Electrophoresis is described in Chapter 14.

Counter immunoelectrophoresis

Since immunoglobulins migrate towards the cathode during electrophoresis and most other proteins migrate to the anode, a combination of double immunodiffusion and electrophoresis can be used to speed up the reaction. This is particularly useful in the preparation of antisera and antigens for use as reagents as their specificities can be assessed quickly.

Electrophoresis in gels containing antibodies

Single radial immunodiffusion, which we described above, can be combined with electrophoresis. The precipitate forms in rocket-shaped zones as the antigen moves through the gel and binds with the antisera. The area under the rocket is proportional to the concentration of antigen.

Crossed or two-dimensional immunoelectrophoresis

In crossed or two-dimensional immunoelectrophoresis, antigens may be separated once by electrophoresis then again at right angles to the first separation, but this second time, through a gel that contains antibody. This technique has been used historically to separate and analyse components of the complement system (Box 15.2). You can see a photograph showing the results of this technique in Figure 15.10.

Immunoassays

The term, *immunoassay* is broadly applied to methods of analysis that use antibodies to detect and determine the concentration of any substance present in serum and urine in health and disease. This is made possible by attaching a detectable label, which is usually an enzyme, radioisotope, or a fluorescent tag to an antibody or antigen. Immunoassays are flexible and this has led to the development of a number of manual and automated techniques, which are used in clinical biochemistry, virology, histology, cytology, and immunology laboratories. Many immunoassays are now available for diagnostic purposes using a wide range of tests on

FIGURE 15.10
Photograph showing the results of immunoelectrophoresis. Samples have been placed in wells (W) and antibody in a trough (T). Following electrophoresis, arcs of precipitation have formed and been stained with Coomassie blue dye. See text for details. Courtesy of D. Raine, Immunology Department, Hull Royal Infirmary, UK.

large high throughput analytical platforms. Clinical tests are available to measure the concentrations of, for example, hormones, tumour markers, autoantibodies, allergens, and for helping to diagnose many infectious diseases. Biomedical scientists are often challenged in validating selected assays for clinical use, bearing in mind the wide variation in analytical and clinical sensitivity and specificity, and the quality of technical performance.

Since it makes no difference in principle whether the concentrations of an antibody or antigen are being measured by the assay, the unknown substance is usually referred to as the *analyte* and the labelled substance, *reagent*.

Early immunoassays were based on the principle of saturation analysis where excess labelled antigen (the analyte in this case) is reacted with a limiting amount of antibody (reagent). When known aliquots of unlabelled antibody are added they will inhibit the binding of labelled antigen; determining the amount of bound labelled antigen indicates the amount of unlabelled antigen that has been added. In this way an inhibition standard curve can be drawn to estimate the amount of analyte by the level of inhibition it induces. Ideally, the antibody concentration should be relatively low, however, this can intensify the effects of other factors, including the accuracy of the method for detecting the labelled reagent, the speed of reaction, and the concentrations, avidities, and affinities of the reactants. The development of immunoassays was initially limited by the cross-reactivity of polyclonal antibodies raised in animals, and by the difficulty in producing large quantities of reliable antisera. However, with the advent of commercially available monoclonal antibodies (see Box 16.4) such limitations have been largely overcome producing high affinity, specific, reagents of good quality that give accurate and reproducible results. A schematic that outlines the use of radioactively labels is given in Figure 15.11.

Immunoassays may be classified using a number of factors:

- Choice of label, whether radioisotope, enzyme, or fluorochrome
- Method of signal detection
- Competitive or non-competitive assay
- Whether the assay requires the separation of bound and unbound reactants

Radioisotope labels

Cross reference
You can read more about health and safety in Chapter 4.

Radiolabelled immunoassays with radioisotopes, such as ^{125}I, ^{3}H, and ^{32}P, are rarely performed in clinical laboratories because of the stringent health and safety restrictions that apply to their use.

Such safety regulations define the location where an experiment may be performed, the monitoring of staff, the storing of reagents and disposal of waste, as well as requiring a trained radioisotope safety officer to oversee all such activities. These limit the practicality of the technology in routine clinical laboratory tests.

A number of technical aspects to the use of radioisotopes also pose problems; especially when measuring the relatively low concentrations of analytes in serum and urine. Measuring the activities of radioisotopes requires low background levels and isotopes with a radioactive half-life that is short enough for the assay protocol, but long enough to give a reasonable shelf life. The radioactive signal can be counted using a γ scintillation counter or liquid scintillation counter, which is sensitive and accurate. However, the signal of an isotope is determined by its rate of radioactive decay; to obtain useful results the isotope must bind to millions of reagent molecules to give a count that is statistically significantly above the background (see KP below).

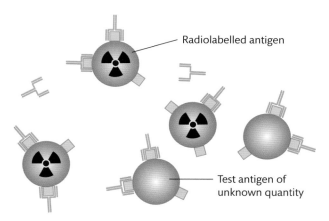

Radiolabelled antigen

Test antigen of
unknown quantity

Antigens compete to bind antibody

Separation stage

Scintillation counter

The number of radiolabelled antigens
bound to antibody can be used to
calculate the concentration of test antigen

FIGURE 15.11
Schematic showing a competitive radioisotope labelled
immunoassay. See text for details.

Key Points

Theoretically a minimum of 600 molecules of labelled reactant (equivalent to 10^{-21}
mole) must be detected in an analysis to obtain a statistically relevant result.

Enzyme labels

Attaching an enzyme to a reactant provides a convenient way of generating a sensitive signal.
In many cases, a chromogenic substrate is used which the enzyme can convert into a coloured
product. Since a single enzyme molecule is capable of converting many molecules of substrate to product, a detectable colour signal is readily available. Unfortunately, the photom-
eters or spectrophotometers used to measure the intensity of the colour are limited by the
optical quality of the cuvettes and microtitre trays used in analyses. The commonest enzymes
used in immunoassays are alkaline phosphatase (AP) and horse radish peroxidase (HRP),

together with chromogenic substrates such as tetramethylbenzidine (TMB) and azino-*bis*-ethylbenzthiazoline sulfonic acid (ABTS).

Some immunoassays combine the advantages of both radioactive and enzyme labels by using an enzyme-labelled antibody or antigen with a radioisotope-labelled substrate. Thus, the action of the enzyme releases the radioactive label, which can then be isolated and accurately measured.

Fluorescent labels

Cross reference

Fluorescence is described in Chapter 11 'Spectroscopy'.

Fluorescent molecules absorb radiant energy, but then emit energy as light of a longer wavelength because energy is lost in the process. Using fluorescent labels in immunoassays presents a number of problems, such as high background fluorescence and quenching of the emitted light by other components in the assay system. The advantage of using fluorescent labels, for example, fluorescein and rhodamine β-isothiocyanate in immunoassays is the sensitivity they give to the system.

Whatever type of label is used, radiolabelled, enzyme, fluorescent, all can be combined with the biotin-avidin system outlined in Box 15.5.

Competitive and non-competitive assays

Immunoassays using labelled reagents may be subclassified into homogenous and heterogeneous types. Homogenous assays are quicker and easier to perform. Heterogeneous assays require an extra step in the procedure to remove unbound antigen or antibody from the test system. They may be competitive or non-competitive.

Competitive immunoassay is when the unknown analyte competes with a similar labelled antigen to bind with an antibody. The quantity of detectable labelled reagent is therefore inversely proportional to the concentration of analyte because the greater the amount of labelled reagent that is able to bind means there are fewer binding sites available for the analyte. In non-competitive or sandwich immunoassays, the unknown analyte is bound to an antibody that is usually attached to a solid phase. Thus, when a labelled antibody is attached to the *bound* analyte it forms a type of molecular 'sandwich'. In this case, measurements of the signal from the labelled antibody are directly proportional to the quantity of the analyte.

BOX 15.5 Biotin and avidin

Biotin is a coenzyme that is synthesized by many plants and microorganisms, particularly those present in the flora of the gastrointestinal tract. Avidin is found in egg white and can also be isolated as streptavidin from *Streptomyces avidinii*.

Biotin can bind to antibodies, enzymes, and many other proteins without significantly affecting their biological activities. Avidin can bind to antibodies and be linked to a wide variety of enzymes, fluorochromes, and radioisotopes. Avidin also has an extremely high affinity for biotin; indeed, when combined, they may be considered for practical purposes to be linked by the equivalent of a covalent bond. Avidin has four binding sites for biotin. If a range of proteins are biotinylated, and all four sites on the avidin molecule are used to recognize and bind to them, this gives many possible combinations of complexes. Many variations of biotin–avidin systems are available commercially for use in immunoassays.

Separation of bound and unbound reactants

Incomplete separation of the labelled component from the other components of the immunoassay mixture is a potential error in the test result. Separation may be achieved in a number of ways:

- Adsorption onto activated charcoal
- Precipitating the antibody with 50% saturated solution of ammonium sulphate
- Precipitation with polyethylene glycol
- Using a secondary binding antibody such as antihuman globulin (AHG)
- Attaching a component to a solid phase, such as plastic tubing or a microtitre plate so that unbound components can be washed away
- Attaching a component to a solid phase in the form of beads or particles so that the bound reactant can be separated by simple centrifugation

Adsorption and precipitation methods have largely been replaced in clinical laboratories by the effective use of antibodies in solid phase assays that can be adapted for use on automated analyzers with the capacity for high volumes of tests.

> **Cross reference**
>
> Automation is described in Chapter 17 'Laboratory automation'.

Nephelometry and turbidimetry

Nephelometry and turbidimetry techniques are based on the absorbance or deflection of monochromatic light by immune complexes in suspension. They differ in that turbidimetry measures the amount of light *absorbed* by the complex, while nephelometry measures the amount of light scattered at an angle to the original light path.

> **Cross reference**
>
> You can read some more about nephelometry and turbidimetry in Chapter 11 'Spectroscopy'.

The most advanced instruments are fully automated, incorporate laser light sources and can detect quantities of analyte. Rate nephelometry is a frequent method of choice in clinical immunology laboratories; this measures the amount of immune complex production against time. As explained earlier (Section 15.4), the principles of the quantitative precipitation curve must be taken into account. This is especially true regarding the phenomenon of antigen excess, where the antibody reagent may be completely consumed by relatively high amounts of antigen in the sample, leading to an inaccurate result.

The concentrations of immunoglobulins in samples of serum are normally measured by nephelometry or turbidimetry in the clinical laboratory. An antihuman globulin reagent is employed and the heavy chain of the immunoglobulin that is being measured becomes the antigenic target. A wide range of analytes, such as complement, rheumatoid factor, $\beta,2$-microglobulin, and $\alpha,1$-antitrypsin may be measured by these methods when investigating B cell malignancies, autoimmune and liver diseases, and immunodeficiency.

SELF-CHECK 15.6

(a) List three ways in which immunoassays may be classified. (b) What is the main practical disadvantage of using radioisotope-based immunoassays? (c) Name three separation techniques used in immunoassays.

Enzyme-linked immunosorbent assays

Enzyme-linked immunosorbent assays (ELISAs), also known as enzyme immunoassays (EIAs), are one of the commonest immunoassay techniques used in clinical laboratories. The general technique has many applications in immunology, virology, and other analytical areas.

It is easy to perform manually, but there are many automated ELISA platforms. This is a highly commercially driven area that has expanded rapidly. Clinical laboratories have a wide choice of readily available commercial kits for a multitude of different analyses. Enzyme-linked immunosorbent assays are commonly used in the diagnosis of autoimmune and infectious diseases, such as when investigating possible human immunodeficiency virus (HIV), hepatitis B viral infections, and a range of autoantibodies, complement components, and immunoglobulin subtypes.

The basis of ELISA technology is that unknown antigens are attached to a solid phase in the form of a plastic multiwell plate. Areas of the plate that might non-specifically bind antibody are prevented from doing so by blocking them with an inert protein. Enzyme-labelled antisera that can detect the antigen are added to the wells. After a period of incubation the unbound antibody is washed away. Removing the excess antibody or reagent by repeated washings at each stage of the procedure is essential to the technique since residual reagent or analyte in the well can potentially interfere with the next step of the process. Washings are usually performed using an automated plate washer or as part of a fully automated procedure. Following washing, a chromogenic substrate is then added. The enzyme catalyses the release of a coloured product. Colour development is stopped after a defined period of incubation using substances that prevent enzyme activity, usually by changing the pH of the solution. Colour intensity is measured as the absorbance or optical density at an appropriate wavelength on a specially adapted spectrophotometer designed to monitor ELISA plates. The intensity of colour development is directly proportional to the quantity of antigen bound to the plate (Figure 15.12).

Some ELISAs use a standard curve to quantify the level of antibody in patient sera. This employs a range of solutions containing known concentrations of antibody, which are measured at the same time as the unknown serum. The absorbances obtained on measuring these solutions

Cross reference

More on ELISAs is given in Chapter 16 'Molecular biology techniques'.

FIGURE 15.12

Photograph showing the final stage of an ELISA with different intensities of colour development in the wells. The colour developed in an assay using a biotinylated antigen with an avidin-linked peroxidase and ABTS substrate. Courtesy of D. Raine, Immunology Department, Hull Royal Infirmary, UK.

are plotted against their known antibody concentrations producing a standard curve. In many cases, there is no recognized international standard for the antibodies measured by ELISA, so often results are expressed in arbitrary ELISA units. If a recognized standard does exist, the reagents in commercial kits should be calibrated to it and the results may be expressed as international units.

Fluorescent labels and other detectable probes that are used in similar ways with multi-well plates are usually referred to as ELISAs because they use similar protocols, even though enzymes are not involved.

Modifications to the basic ELISA protocol include:

- Sandwich ELISA
- Competitive binding ELISA
- Automated procedures

Sandwich ELISA is a sensitive and robust variation to the basic ELISA protocol, which is outlined in Figure 15.13. A known antibody is bound to the wells of the microtitre plate at a predetermined concentration. Diluted sera containing unknown concentrations of the analyte is then added to the wells. The bound reagent will capture the analyte within the wells. Following washings, a second, antibody is added. The second antibody, which has an enzyme label attached to it is the detection antibody, and binds to the immobilized analyte. The rest of the procedure is as described above for the basic technique. Competitive binding ELISA is commonly used for detecting small molecules with limited numbers of epitopes. The sera containing unknown quantities of analyte are added to wells that have been coated with a capture antibody, along with an enzyme-labelled reagent of similar antigenic material. If the colour change of the chromogenic substrate is less than that obtained from a well containing only labelled reagent, then the difference is proportional to the quantity of the analyte.

Commercially developed automated techniques based on the principle of the sandwich ELISA use individual plastic wells or nitrocellulose sponge discs as the solid phase. However, when the enzyme comes into contact with its substrate, a light-emitting fluorescent product is formed. The fluorescence is measured using an ultraviolet light source and its value can be compared to a standard curve to determine the amount of analyte in the sample of serum. This type of analysis is the basis for determining IgE values of serum and a wide range of allergens and autoantibodies. Other adaptations, using a range of absorbent materials, attach antibodies to the solid phase such that the colour change produced forms a line or dot if the result is positive. This technique underlies home pregnancy kits and a range of rapid techniques that are used in laboratories, including autoantibody typing and the diagnosis of malaria.

Key Points

ELISA techniques depend upon the purity and quality of the immunological reagents and robust binding to the solid phase.

Enzyme-linked immunospot assays

Enzyme-linked immuno*spot* (ELISPOT) assays are an adaptation of the ELISA technique, which were developed to measure antibodies produced by plasma cells. The cells are isolated from immunized mice and added to antigen-coated microtitre tray wells where they bind to antigen

Well

Test sample is added to antibody-coated well

Antigen binds to antibody and excess is washed away

Wash step

Detection of antibody using a secondary antibody with attached enzyme (⊙)

Wash step

Substrate (▲) is added and colour intensity (absorbance of product) (■) is measured

FIGURE 15.13
Outline of the sandwich ELISA. See text for details. See also Figure 16.25 and its associated text.

and are captured on the solid phase. The cells are then removed by washing and enzyme-labelled antihuman globulin (AHG) is added to the wells, followed by a substrate suspended in agarose gel. The appearance of coloured spots in the gel shows where the plasma cells are located in the wells.

Modifications to the original ELISPOT method have been made to investigate the production of cytokines by T lymphocytes. Stimulated T cells are isolated and added to wells containing a monoclonal antibody to the cytokine of interest. The procedure is the same as the original ELISPOT, but uses a labelled antibody directed against the cytokine. The sites of colour development indicate the locations of cells secreting the cytokine as you can seen in Figure 15.14. If the spots are counted, it is possible to estimate the proportion of T-cells secreting the cytokine present, based on the number of cells originally added to the plate.

Commercial variations on the ELISPOT have been developed that have increased the reproducibility of the assays and made them more suitable for use in clinical laboratories. They are used for the detection of interferon γ secretion by T lymphocytes that have been sensitized to *Mycobacterium tuberculosis*. This is of particular importance in the diagnosis of immunodeficient individuals who may have latent or subclinical tuberculosis.

SELF-CHECK 15.7

(a) Why are washing steps essential in ELISAs? (b) Which components of the immune system are detected by ELISPOTs?

Flow cytometry and fluorescence-activated cell sorters

Flow cytometry is a method of counting and evaluating cells or particles by passing them in a thin stream through a laser beam. A fluorescence-activated cell sorter (FACS) is able to separate the cells as they are identified. Figure 15.15 outlines their operation.

Flow cytometers are incorporated into analysers that count and identify leukocytes in routine haematology tests. More complex flow cytometers with multiple lasers are used to differentiate subpopulations of a range of cells. This is an important tool in the immunophenotyping of cells when classifying leukaemias and in the classification of immunodeficiency diseases.

FIGURE 15.14

(a) Positive control for an ELISPOT assay. (b) Sample of a patient showing a positive ELISPOT result. Courtesy of D. Spradbery, Immunology Department, Hull Royal Infirmary, UK.

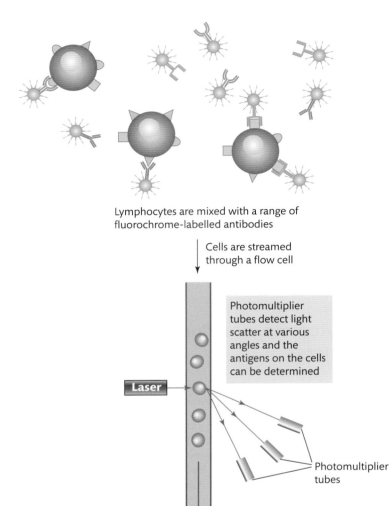

Lymphocytes are mixed with a range of
fluorochrome-labelled antibodies

Cells are streamed
through a flow cell

Photomultiplier
tubes detect light
scatter at various
angles and the
antigens on the cells
can be determined

Laser

Photomultiplier
tubes

FIGURE 15.15
Basic principles of flow cytometry.
See text for details.

Different types of leukocytes, which may superficially appear similar when stained and examined by light microscopy (see Figure 8.1 (a)), have, however, many differences in their cytoplasm and on their surfaces allowing them to be distinguished and identified. Some surface markers are termed *clusters of differentiation* (CD). Many CD markers have been identified and numbered. For example, we refer to CD4 and CD8 lymphocytes when monitoring cell counts in acquired immunodeficiency syndrome (AIDS) patients.

Cells are incubated with monoclonal antibodies directed against a predetermined panel of CD markers or intracellular components. The antibodies are labelled with fluorescent dyes called fluorochromes that will vary in their response to lasers of differing wavelengths. As the dilute suspension of cells is forced through a fine tube they pass through the laser beam. Sensitive photomultiplier tubes detect the extent of light *scattering*, which will indicate the size and granularity of the cells. They also measure the fluorescence emission: this indicates which of the labelled antibodies is attached to the cells. The data are electronically stored for analysis and can be displayed in the form of a graph or histogram if only one antibody is

FIGURE 15.16

A dot plot showing the separation of CD4 and CD8 T lymphocytes in a sample from a patient with HIV. Courtesy of D. Spradbery, Immunology Department, Hull Royal Infirmary, UK.

used and as a two-dimensional dot plots if two or more antibodies have been used as you can see in Figure 15.16.

The use of complex analysers with multiple lasers, and the wide variety of fluorochromes and monoclonal antibodies have increased the use of this versatile technology. Flow cytometers are able to process large numbers of cells and calculate the number of cells carrying a particular marker. This is necessary in the routine testing of patients who are being treated for HIV infection. Along with FACS analysis, flow cytometry has added enormously to our knowledge of the cells of the immune system.

Cross reference

Immunology in this series gives more information on flow cytometry.

Multiplex technology

Multiplex technology is an advanced development in the clinical laboratory, which makes it possible to measure multiple analytes in one assay. There are a number of variations on the basic principle, but all allow complex panels of antibodies and antigens to be evaluated simultaneously.

Microscopic polystyrene beads or microspheres of approximately 5 µm diameter with differing ratios of fluorescent dyes incorporated into them form the solid phase of the assay. The combination of the dyes can be varied to produce up to a hundred different coloured beads; each bead can then have a different antigen or antibody attached to it. The beads are added to a sample of serum that potentially contains antibodies or antigens that will attach to an appropriate bead. A secondary fluorochrome-labelled antibody, which recognizes and binds to the primary one, is then added. The beads are then analysed using flow cytometry, which

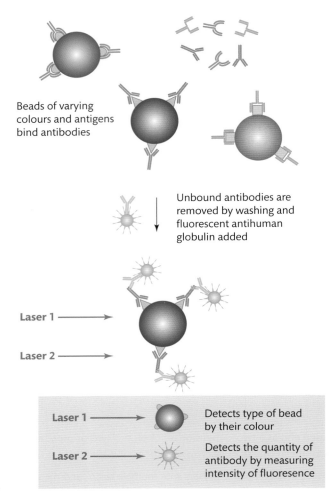

FIGURE 15.17
Basic principles of multiplex analysis. See text for details.

we described above. As the beads pass through a flow cell, they pass through two laser beams: one detects the bead and the other excites the fluorochrome. The data obtained will identify the bead and determine the type and quantity of antibody or antigen bound to the surface of the bead (Figure 15.17).

The potential of this technology to assist clinical laboratories and research facilities is impressive, providing extensive data on a wide range of analytes.

Microarrays

Typically, microarrays are used for DNA analysis where they are known as 'DNA chips'. However, their use has widened with the introduction of microarrays to measure cytokines. Glass slides with minute wells etched into their surface are loaded with a range of monoclonal antibodies that are used to bind ('capture') the cytokine of interest. When the serum to be analysed is added to the microarray, the capture antibodies will bind to some of the proteins in the serum. The analysis is performed very much like an ELISA, a further antibody that has an attached fluorescent label is added. The slides are read using an array reader with compatible image-analysis software. The presence of a fluorescence signal indicates the positions of the bound cytokines.

Cross reference

For more information on microarrays see Chapter 16 'Molecular biology techniques'.

SELF-CHECK 15.8

(a) What name is given to the cell surface markers that are detected by flow cytometry?
(b) What is the main advantage of multiplex technology over other antibody-based methods of analysis?

Complement assays

Severe or chronic infections will cause a continuous activation of complement (Box 15.2) and increased concentrations of circulating immune complexes, which are a feature of some autoimmune diseases. Both types of condition increase consumption of complement proteins and therefore may lower their concentrations in the blood. Complement proteins are also labile and care must be taken to prevent their activation in the sample tube prior to testing. This means samples should be tested promptly or frozen immediately until required for testing. Some procedures may require the use of the anticoagulant ethylenediamintetra-acetic acid (EDTA) in samples or the addition of a protease inhibitor.

Complement components C3 and C4 are present in low, but relatively stable and measurable quantities in normal sera. Both are routinely analysed by nephelometry and other automated immunoassay systems as a first line complement screen. If a complement immunodeficiency is suspected, many of the other proteins can be determined. A number of complement proteins can be measured by single radial immunodiffusion (SRID).

Haemolytic assays are sometimes used when investigating complement deficiencies to determine which activation pathway is affected. Erythrocytes are used to demonstrate the presence of complement by creating an environment in which the complement system pierces the erythrocyte membrane and releases haemoglobin into solution. The procedure was referred to as the *total haemolytic complement test* or CH100. However, it is difficult to determine the final end point of haemolysis, and therefore the working end point is usually accepted as occurring when 50% of the erythrocytes are lysed (CH50).

The classical pathway is investigated using sheep cells that have IgG anti-sheep antibodies attached to their surfaces. They are incubated at 37°C and observed for haemolysis at regular intervals until 50% haemolysis is reached. You can see an example of this in Figure 15.18. This test may be performed in test tubes, multiwell plates or on SRID gels containing erythrocytes. The time and detection systems vary with method of choice. A number of in-house protocols and commercial kits are available.

If 50% lysis is achieved under comparable conditions to a 'normal' serum control, the complement activity of that pathway is normal.

The alternative pathway is tested in the same way as the classical, but antibody is not required to activate complement. Rabbit, guinea pig, or chicken cells that are susceptible to direct haemolysis by human complement may all be used.

In virology laboratories, complement fixation tests are used to detect the presence of microorganisms such as mycoplasma, those causing Q fever and psittacosis, and a range of influenza viruses. The antibody to be detected is incubated with an appropriate antigen in the presence of complement. Sheep erythrocytes that have been coated with anti-sheep IgG are

FIGURE 15.18
The CH50 test (to determine where 50% of the erythrocytes are lysed) for haemolytic complement activity. The photograph shows erythrocytes in the gel. Circles of haemolysis can be observed where complement activity is present. Courtesy of D. Raine, Immunology Department, Hull Royal Infirmary, UK.

then added and after a further incubation, the suspension is centrifuged. If there is still free complement in the mixture the erythrocytes will be haemolysed, that is the complement will have interacted with the IgG and pierced the cell surface membrane. This means that the test is negative as the complement has not been bound by an antigen–antibody reaction during the first incubation.

Indirect immunofluorescence and immunocytochemistry

Indirect immunofluorescence and immunocytochemistry techniques differ from those already described in that the solid phases necessary to detect the analytes are sections of solid organs, or cells grown in tissue culture that have been fixed to a glass or plastic slide.

Indirect immunofluorescence

Indirect immunofluorescence (IIF) is a commonly used technique in clinical immunology laboratories to detect and identify antibodies in the serum of individuals with autoimmune diseases. Many autoantibodies may be detected using this technique by employing a wide range of substrates that may be sections of animal tissues or cultured cell lines. Table 15.2 shows you examples of antibodies and tissues used in investigating autoimmune diseases. The AHG used is normally directed at IgG, however, anti-IgA is used for some cases.

The principle of the technique is outlined in Figure 15.19. A range of animal specimens are prepared by cutting sections of frozen tissue and fixing them in an alcohol–acetone mixture.

TABLE 15.2 Tissue substrates used in the detection of autoantibodies by indirect immunofluorescence

Cell/tissue	Source	Target antigen	Disease state
Hep2 cells	Human epithelial cells	Nuclear components, e.g. nucleolus, centromere	SLE
Liver	Rat, mouse	Mitochondria, nuclear components	PBC
Oesophagus	Primate	Tissue transglutaminase	Coeliac disease
Neutrophils	Human	ANCA	Vasculitis

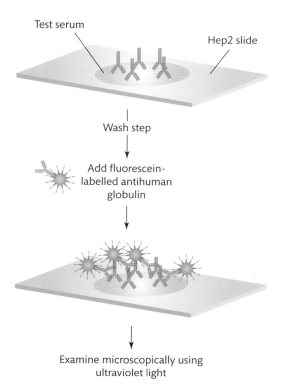

FIGURE 15.19
Outline of indirect immunofluorescence. See text for details.

Cultured cell lines used include Hep2, which are human epithelial-type cells derived from a carcinoma, that have large nuclei and a rapid rate of division. This allows observation of many nuclear antibody patterns. The tissue or cultured cell substrate is fixed to a glass slide, which is then flooded with dilute serum, which may contain antibody. Following a period of incubation, the serum is washed away leaving antibodies attached to their target antigens. Fluorescently-labelled polyclonal AHG is then added to the slide and will bind to the antibodies attached to the antigens. After further incubation, the excess labelled antibody is removed by washing. As with ELISAs (see above) the washing steps are essential to ensure excess unbound components are removed (Figure 15.20). The sections or cells are then immersed in mounting medium and a cover slip added.

FIGURE 15.20
(a) Photograph showing commercial indirect immunofluorescence slides that have tissue sections attached to them and diluted serum placed in the wells prior to incubation.
(b) Photograph showing the careful washing of the indirect immunofluorescence slides.
Courtesy of D. Raine, Immunology Department, Hull Royal Infirmary, UK.

Cross reference

Chapter 8 'Microscopy' has further information on fluorescence microscopy.

The slides are then examined using fluorescent microscopy and the patterns formed by the location of antibody in the tissues or cells recorded. Figure 15.21 shows examples of the types of staining patterns. Depending on the results obtained in the IIF panel of tests, immunology laboratories will follow up positive results with more specific assays such as ELISA for subtypes of antinuclear antibodies, liver antibodies, and vasculitis-related antibodies.

The amount of antibody present can be estimated by testing a range of dilutions of serum on the slide and expressing the result as a titre, that is the dilution at which fluorescence can still be observed in the cells.

The commonest fluorochromes employed in IIF are fluorescein, rhodamine, and phycoerythrin.

Many antibodies can be identified by indirect immunofluorescence. Antinuclear antibodies include a number of antigenic targets and create a variety of patterns on the nuclei of

(a)

(b)

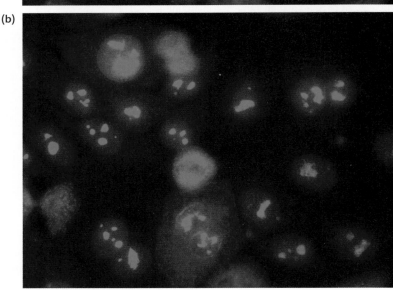

FIGURE 15.21

Photographs of indirect immunofluorescence patterns. (a) Nucleoli of Hep2 cells are indicated by the presence of an anti-nucleolar antibody. (b) Diffuse, granular staining of kidney tubules indicates the presence of mitochondrial antibodies. Courtesy of D. Raine, Immunology Department, Hull Royal Infirmary, UK.

Hep2 cells, for example smooth homogenous staining of the nucleus is seen in systemic lupus erythamatosis (SLE). Rat or mouse kidney and liver sections will reveal the presence of mitochondrial antibodies as seen in primary biliary cirrhosis (PBC). Pancreatic islet cells and sections of adrenal gland, ovary, testes, and oesophagus are commonly used when investigating for autoantibodies such as those for anti-tissue transglutaminase, which occur in coeliac disease. Human neutrophils are used to demonstrate the antineutrophil cytoplasmic antibodies (ANCA) that are a feature of some forms of vasculitis.

The drawback of IIF in routine clinical laboratory testing is that it is extremely operator dependent and time consuming. The method relies on the expertise of individuals in slide preparation and microscopy. Higher than normally expected numbers of staff are also required to achieve an appropriate throughput of samples. This has been alleviated to some extent by using automated slide processors, but eventually many of the high volume antibody tests are likely to be performed by multiplex technology-based assays (see above).

Cross reference

The *Immunology* volume of this series covers an extensive range of autoimmune disorders and their associated antibodies.

Direct immunofluorescence

Direct immunofluorescence is used to detect substances that may be deposited in tissues and organs during some disease states, for example, some skin disorders can be diagnosed when antibodies and complement can be demonstrated in all layers of the epidermis. A biopsy of the patient's tissue is removed from a site adjacent to a skin lesion and fixed on a glass slide. Antihuman globulin with a fluorescent marker is added to the slide, which will bind to antibodies present in the tissue sample. The procedure is the same as indirect immunofluorescence in that excess antihuman globulin is removed by washing, and then the tissue is examined for the fluorescence that will demonstrate the presence of autoantibodies.

Immunocytochemistry

Immunocytochemistry is similar to the fluorescence methods described above, in that direct and indirect methods can be used. The techniques were developed to identify tumours by accurately targeting tumour cell surface markers. This technology has been automated and is a growing area in laboratories that specialize in histological and cytological investigations. Although fluorescent labels can be used, the methods are often based on colour development using enzyme-labelled antibodies and substrates similar to those employed in ELISAs.

Cross references

You can read more about cytological and histological investigations in Chapter 7 'Samples and sample collection' and Chapter 8 'Microscopy'.

You can learn more about direct immunofluorescence and immunocytochemistry in the *Histopathology* book of this series.

SELF-CHECK 15.9

(a) What is the difference between direct and indirect immunofluorescence? (b) Name two examples of fluorochromes that may be used when detecting antibodies by indirect immunofluorescence.

SUMMARY

- The immune system consists of the interaction of specialized organs, cells and humoral components that protect the body from infection.

- Innate immunity is present at birth, is immediate acting, but non-specific. Acquired immunity is slower to develop, more specific and has immunological memory.

- Immunoglobulins, in the form of antibodies, are essential components of the immune system, which have been adapted as diagnostic tools.

- The formation of antibody–antigen immune complexes and their behaviour form the bases of immunoassays.

- The use of radioisotopes, enzymes, and fluorochromes to label antibodies and antigens was an important advance in the development of immunoassay.

- The use of monoclonal antibodies in immunoassays has revolutionized laboratory medicine.

- Advances in areas, such as flow cytometry and multiplex technology, will have a lasting impact on clinical laboratories that were previously dominated by manual methodology; small-scale manual immunoassays are being overtaken by automated analytical methods in all specialities in clinical laboratories.

 FURTHER READING

- **Cruse JM, Lewis RE.** *Historical Atlas of Immunology*. **Informa Healthcare, 2005.**

  An interesting and well illustrated account of the development of our knowledge of immunology.

- **Hay FC, Westwood OMR.** *Practical Immunology*, **4th edn. Blackwell Science, Oxford, 2002.**

  Practical Immunology is a useful bench textbook for research scientists and gives insight into the origins of immunological techniques and increases the understanding of bio-medical scientists.

- **Murphy KM, Travers P, Walport M.** *Janeway's Immunobiology*, **7th edn. Garland Science, 2007.**

  A comprehensive text that gives a detailed, well written account of the immune system. Excellent diagrams and a CD-ROM with animations and videos. Although a little advanced for undergraduates, it is an invaluable resource for studying all aspects of immunology.

- **Peakman M, Vergani D.** *Basic and Clinical Immunology*, **2nd edn. Churchill Livingstone, London, 2009.**

  This excellent book describes the immune system and associated disease states. It is well laid out and relevant to clinical laboratory practice.

- **Price CP, Newman DJ.** *Principles and Practice of Immunoassay*, **2nd edn. Macmillan, Oxford, 1997.**

  Comprehensive account of the development and practice of immunoassay in the clinical laboratory. Covers all aspects of methodology, standardization, and quality.

- **Sompayrac LM.** *How the Immune System Works*, **3rd edn. Wiley-Blackwell, Harlow, 2008.**

  Brilliant, readable account of how our immune system works; recommended to anyone who wants to learn about immunology. Well written, humorous, and explained in simple terms; difficult concepts are presented in a highly imaginative way making them easier to remember.

QUESTIONS

1. Select three characteristics of a high affinity antibody from:

   (a) An antibody with more than two binding sites

   (b) An antibody produced in the secondary immune response

   (c) Is most likely to be IgG

   (d) Forms immune complexes that do not easily dissociate

   (e) Produced by memory B cells that have refined 'best fit' antibodies

2. Which of the following statement(s) is/are **TRUE**.

   (a) Acquired immunity is present at birth

   (b) Complement can kill pathogens without the involvement of antibodies

   (c) Antibodies are produced by T lymphocytes

   (d) Genetic predisposition alone commonly causes autoimmune diseases

   (e) The precipitation curve is not affected by the source of the antibody or antigen

3. Discuss the difference between a polyclonal antibody response and a monoclonal response.

4. What are the drawbacks of using polyclonal antibodies?

5. Give a short description of the three types of antibody labels that may be used in immunoassays with examples of assays that use them.

Answers to self-check and end of chapter questions are available in the Online Resource Centre accompanying this book.

 Go to www.oxfordtextbooks.co.uk/orc/glencross

16

Molecular biology techniques

Qiuyu Wang, Nessar Ahmed, and Chris Smith

Learning objectives

After studying this chapter, you should be able to:

■ Outline the structures of nucleic acids and proteins

■ Describe the general functions of biological macromolecules

■ Explain the basic molecular biology encountered in clinical laboratories

■ Discuss the principal techniques of molecular biology used in clinical tests

■ Outline the uses of molecular biology techniques in clinical laboratories

Introduction

The term, **molecular biology** is thought to have been first used by Warren Weaver in a 1938 report to the Rockefeller Foundation. Molecular biology is the study of biological phenomena in terms of the physical and chemical properties of the molecules of organisms. It can be described generally as the study of the structures, functions, and interactions of biological macromolecules. Two major types of macromolecules common to all cells are the nucleic acids, deoxyribonucleic acid (DNA) and ribonucleic acid (RNA), and proteins. DNA and RNA carry biological information. Deoxyribonucleic acid contains the genetic information of all cells and some viruses, that is to say it is the material of genes. Ribonucleic acid only holds the genetic information of some viruses. However, it is present in all cells in a number of forms, including messenger RNA (mRNA), transfer RNA (tRNA) and ribosomal RNA (rRNA) molecules that are necessary to allow the genes to be expressed and form proteins. Proteins perform most of the activities in cells. They catalyse metabolic reactions, perform mechanical work in muscles, cilia, and flagella, act as sensors, perform regulatory, and control functions, and also constitute some of the major structural features of the cell.

The definition of molecular biology given above is rather limited in that it fails to describe how the subject overlaps with other areas of biology, chemistry, microbiology, and computational biology, in particular biochemistry, genetics, immunology, genomics, proteomics, and bioinformatics. The techniques of molecular biology have reached sophisticated levels, and are used extensively in the biological and biomedical sciences. Many of these techniques have been automated and are performed by robots, which makes them particularly suitable in the diagnosis of clinical conditions and monitoring treatments. This chapter will provide an overview of the scope of molecular biology and introduce you to the structures, properties, and functions of some of the molecules involved. It will describe some of the experimental techniques of molecular biology and outline a number of their applications in the biomedical sciences.

Cross reference
You can read about automation in Chapter 17 'Laboratory automation'.

16.1 **Structure and properties of macromolecules**

Macromolecules are ones that are large, that is have M_r in excess of approximately 5,000. Biological macromolecules are polymers, formed by covalently linking much smaller units together in condensation reactions (Figure 16.1). Within the polymer, such units are often referred to as *residues*. Linking two units gives a dimer, three a trimer, and so on to eventually give a polymer, which may contain hundreds or thousands or millions of residues.

FIGURE 16.1
A general condensation reaction.

Nucleic acids

The complete hydrolysis (see Section 16.4) of any nucleic acid gives:

- Pentose (five carbon) sugars
- Pyrimidine and purine bases
- Phosphate groups

The sugars are ribose (Rib) in the case of RNA and deoxyribose (dRib) in DNA. Deoxyribose nucleic acid contains the purine bases adenine (A) and guanine (G), and the pyrimidines, cytosine (C) and thymine (T). Ribonucleic acid also contains A, G, and C, but the primidine uracil (U), rather than T (Figure 16.2). Minor amounts of other, chemically related, bases are also released. The phosphate groups are negatively charged at physiological values of pH.

The combination of a base and sugar forms a **nucleoside** (Figure 16.3 (a)). Note that the atomic positions in sugars and bases need to be distinguished, and this is achieved by adding a prime (') to the sugar positions as you can see in Figure 16.3 (a). Bases are attached to the 1' carbon atoms of the sugars. The addition of a phosphate(s) to a nucleoside gives a **nucleotide** (Figure 16.3 (b)). Phosphates may be bonded to the 5' or 3' positions giving nucleoside 5' phosphate or nucleoside 3' phosphate respectively. Table 16.1 summarizes the nomenclature of the major bases, nucleosides, and nucleotides.

Both DNA and RNA are polymers of nucleotides; hence, their alternative name of **polynucleotides** and both are similar in structure. Phosphodiester bonds link the nucleotides together. If you examine Figure 16.4, you will see these bonds link the 3' and 5' carbons of adjacent sugar residues giving the link directionality: the convention when writing or drawing them is to place the 5' to the left and the 3' to the right (as in Figure 16.5).

This means the *sequence* of bases also runs in a 5' to 3' direction forming the primary structure of the nucleic acid, which differs in nucleic acids from different sources. Primary structures of nucleic acids can be described by one of several conventions, which you can see in Figure 16.5 (a) and (b).

Adenine (A)

Guanine (G)

Thymine (T)

Cytosine (C)

Uracil (U)

FIGURE 16.2
Structures of the sugars and bases
of nucleic acids.

(b)

(a)

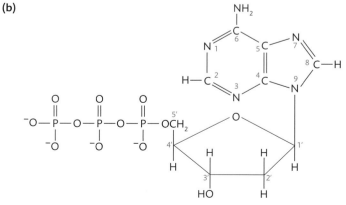

FIGURE 16.3
Structures of (a) the ribonucleoside, uridine, and (b) the nucleotide, deoxyadenosine
5′-triphosphate (dATP). See also Table 16.1.

TABLE 16.1 Standard nomenclature of the major nucleic acid bases, nucleosides, and nucleotides

Bases	Ribonucleoside	Ribonucleotide*†
Adenine (A)	adenosine	adenosine 5′-monophosphate (AMP)
Guanine (G)	guanosine	guanosine 5′-monophosphate (GMP)
Uracil (U)	uridine	uridine 5′-monophosphate (UMP)
Cytosine (C)	cytidine	cytidine 5′-monophosphate (CTP)
Bases	**Deoxyribonucleoside**	**Deoxyribonucleotide (as 5′-monophosphate)**
Adenine (A)	deoxyadenosine	deoxyadenosine 5′-monophosphate (dAMP)
Guanine (G)	deoxyguanosine	deoxyguanosine 5′-monophosphate (dGMP)
Thymine (T)	deoxythymidine	deoxythymidine 5′-monophosphate (dTMP)
Cytosine (C)	deoxycytidine	deoxycytidine 5′-monophosphate (dCTP)

* Where the attachment for the phosphate is not specified, the 5′ can be assumed by default. Other sites, for example C-3′, must be denoted.

† Other nucleotides with two (di), three (tri), and, rarely four (tetra) phosphates are known. For example, adenosine 5′-triphosphate (ATP), deoxythymidine 5′-diphosphate (dTDP).

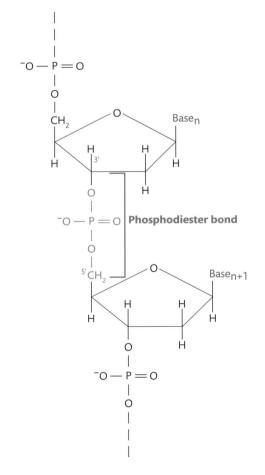

FIGURE 16.4
Structure of a phosphodiester link.

(a)

... pCpApGpU ...

... pC – A – G – U ...

... pC A G U ...

(b)

... pdCpdApdGpdT ...

... pdC – dA – dG – dT ...

... pdCdAdGdT ...

FIGURE 16.5
Various conventions used in showing the primary structures of (a) RNA and (b) DNA.

Native DNA exists as two complementary strands held together by hydrogen bonds. The strands form the famous double helical or duplex structure first described by Crick and Watson, and constitute the secondary structure of DNA. You can see this in Figure 16.6. The strands run in opposite directions, an arrangement described as antiparallel. The bases occur as complementary pairs, with A hydrogen bonded to T and G with C. In each case, a purine base is complementary to a pyrimidine meaning the lengths of each base pair (bp) are approximately the same and the strands can line up as a double helix of regular diameter. Two hydrogen bonds hold each A-T pair together unlike the G–C combination that has three, making it the more stable of the two pairs. The sequence of bases of one strand automatically gives that of the other. Ribonucleic acid is single stranded. However, it can form double helical regions, when the strand double backs on itself. Its base pairing rules are similar to those of DNA, but A pairs with U and G with C. It is also possible to form heteroduplexes that consist of complementary strands of DNA and RNA.

Complementary base pairing is the basis for replicating genes and for allowing their expression. During replication, each strand of DNA acts as a separate template for the synthesis of a new complementary strand. **Deoxyribonucleic acid dependent DNA polymerases** (usually abbreviated to DNA polymerase or DNA pol), assisted by a variety of other proteins, copy each template to give a new complementary strand. Thus, the two new DNA molecules each consist of one old (parental) and one newly synthesized strand, hence, the process is called semi-conservative replication. Molecules of RNA are formed when DNA dependent RNA polymerases use a single strand of DNA as a template to catalyse the formation of a complementary RNA copy. This process is called **transcription**.

Cross reference

You can learn more about replication and transcription of DNA in Elliott & Elliott (2009) and Craig et al. (2010), which are listed in the Further reading section.

SELF-CHECK 16.1

A strand of DNA has the sequence ATGCCGTTAGGAGTA. Give the complementary sequence to it, but written in the 5′ to 3′ direction.

FIGURE 16.6
Computer generated model of double-stranded DNA.

10 bp = 3.4 nm

Diameter
2 nm

All chromosomes contain a single molecule of DNA. In prokaryotic cells, the genetic material usually occurs as a single chromosome containing a single circular DNA molecule. However, mutual repulsion between the negative charges of the phosphate groups opposes the bending of the molecule. In addition, bacterial chromosomes have M_r of at least 2×10^9, and are over 1 mm in circumference and need to be folded to fit into bacterial cells that are only approximately 1 μm diameter. The charges on the phosphate groups are neutralized by binding of basic, that is positively charged, proteins to the DNA. This makes the bending and elaborate folding of DNA possible. Bacterial, and some yeasts, which are eukaryotes, also contain extrachromosomal DNA structures called **plasmids**. Like the bacterial chromosome, plasmid DNA is circular and double stranded. However, it is much smaller with M_r of 2×10^6 to 20×10^6, which corresponds to 3000–30,000 bps. Bacterial plasmids normally contain genetic information for proteins that confer a specialized and sometimes protective phenotype on the organism often, for example, imparting antibiotic resistance. Plasmids are self-replicating and their DNA is duplicated before every cell division and copies of the plasmid DNA segregate to each daughter cell, assuring continued propagation of the plasmid through successive generations of the host cell.

The paired chromosomes of eukaryotes are much larger than bacterial ones. For example, a human haploid cell contains DNA molecules of total length approximately 3×10^9 nucleotides divided between 23 chromosomes (22 autosomes, and a sex chromosome, X or Y). Each eukaryotic chromosome contains a linear DNA molecule many millimetres long. These can only fit into a nucleus by being even more elaborately folded than bacterial types, following their combination with a variety of chromosomal proteins.

SELF-CHECK 16.2

Calculate the total length of DNA in a diploid human cell. You may find it helpful to consult Figure 16.6.

The sequences of bases in the nucleotides of a gene that encodes a protein are transcribed to eventually form a mRNA molecule. The mRNA is then used in the synthesis of the corresponding polypeptide in a process called translation; the sequence of bases in the mRNA encodes the sequence of amino acid residues in the polypeptide. Translation of mRNA occurs on ribosomes and requires the activities of many other nucleic acids and proteins. Each of the groups of three bases called codons in the mRNA specify each of the 20 types of amino acids used in translation. Hence, the sequence of bases in the gene (DNA) determines the sequence of amino acid residues in the polypeptide.

Ultraviolet absorption by nucleic acids

The presence of the purine and pyrimidine bases (A, T, G, C, and U) means nucleic acids absorb significant amounts of light in the ultraviolet (UV) region of the spectrum (see Chapter 11), with maximal absorbance occurring at about 260 nm (Figure 16.7). Hence, solutions of pure double and single stranded DNA and RNA of concentrations 50 and 40 μg cm⁻³, respectively, have absorbances of 1.0 at 260 nm. Proteins, the major contaminants in isolated DNA samples, also absorb in the UV region, but maximally at 280 nm (Figure 16.7). Thus, the ratio of the absorbances, A_{260}/A_{280}, is often used to assess the purity of isolated samples of nucleic acids. The typical A_{260}/A_{280} for isolated DNA with little contaminating protein is about 1.8. Lower ratios indicate increased contamination with proteins. The A_{260}/A_{280} ratio of extracted RNA is generally 1.8–2.0.

Cross reference

Chapter 11 'Spectroscopy', gives much more information on measuring absorbances.

SELF-CHECK 16.3

A 10 μL sample of a purified DNA solution was diluted to a final volume of 1 cm³. The A_{260} of the diluted solution was 0.55. What is the concentration of DNA in mg cm⁻³?

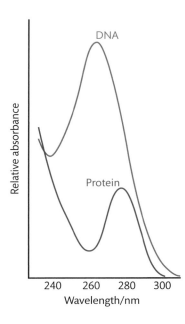

FIGURE 16.7
Ultraviolet absorption maxima of DNA and proteins.

Separating the double helical structure of purified DNA with denaturing agents, such as heat, alkali, organic solvents, to form two single strands causes its absorption at 260 nm to increase markedly. This hyperchromic effect arises because the transition from an ordered double-helical structure to unpaired single stranded DNA means base–base interactions are now at a minimum. The resulting change in the properties of the rings of the bases means they can absorb more UV light. If the strands are allowed to reassociate, or **reanneal**, to give the original double helical structure, the UV absorption returns to its initial value. Thus, the processes of DNA disassociation and reannealing under various experimental conditions can be followed by monitoring the absorbance at 260 nm.

SELF-CHECK 16.4

Would you expect RNA molecules to show a hyperchromic effect?

Proteins

Proteins are polymers of amino acid residues, which all have a common structure, but differ from one another in the nature of their side chains, which you can see in Figure 16.8.

Individually, proteins have more complex structures than nucleic acids because they are formed from 20 different types of amino acids, which are linked together by peptide bonds (Figure 16.9).

The structures of most proteins are subdivided arbitrarily into hierarchical levels. The four main levels are the primary, secondary, tertiary, and quaternary structures, which you can see illustrated in Figure 16.10, although other structural levels have also been described. Note, like any classification, this is an artificial division. The primary structure of a protein is the sequence of its amino acid residues or side chains. The secondary structure is the regions of local folding of the polypeptide chain. The famous α helix and β sheet are examples of secondary structures. Tertiary structure specifies the positions of all the atoms in the polypeptide. Quaternary structures occur only in proteins composed of more than one polypeptide chain. For example,

$$H_3^+N - \underset{\underset{R}{|}}{\overset{\overset{COO^-}{|}}{C}} - H$$

R =

Glycine — H
Alanine — CH_3
Serine — CH_2OH
Aspartic acid — CH_2COO^-
Lysine — $CH_2 - CH_2 - CH_2 - CH_2 - NH_3^+$
Leucine — $CH_2 - CH \begin{smallmatrix} \diagup CH_3 \\ \diagdown CH_3 \end{smallmatrix}$

Phenylalanine — CH_2–⬡

Cysteine — CH_2SH

Glutamine — $CH_2 - CH_2 - C \begin{smallmatrix} \diagup\!\!\!= O \\ \diagdown NH_2 \end{smallmatrix}$

Histidine — CH_2 ⎤

$\overset{+}{HN}$ NH

FIGURE 16.8
Structures of some α amino acids.

the quaternary structure of adult haemoglobin is $\alpha_2\beta_2$, meaning its molecules are composed of two so-called α polypeptides and two β polypeptides. Individual polypeptides of a quaternary structure are usually called subunits.

Protein molecules are only marginally stable. The energy of a folded (native) protein is only relatively slightly below that of an unfolded or denatured protein; hence, samples containing

FIGURE 16.9
Formation of a peptide bond by condensing two amino acids.

(a) G-I-V-E-E-C-C-A-S-V-C-S-L-Y-E-L-E-D-Y-C-D

FIGURE 16.10
Examples of the four main structural levels of proteins. (a) Primary structure of the A chain of insulin denoted using the standard one-letter abbreviations for each of the amino acid residues. (b) Secondary structure (largely α-helical) of a myoglobin molecule. The red portion is a haem group. (c) Tertiary structure of myoglobin. (d) Quaternary structure ($\alpha_2\beta_2$) of haemoglobin.

proteins have to be treated with care. The structure of any protein is determined largely by its primary structure that is the sequence of its side chains. In an aqueous environment, the polypeptide folds so that side chains that cannot interact favourably with water molecules segregate to the interior of the molecule. This is referred to as the **hydrophobic effect**. The hydrophobic effect ensures each type of protein has a unique conformation that determines the properties of that protein and the types of molecules with which it can interact.

Proteins generally function by recognizing and specifically interacting with other molecules called ligands. Ligands are normally small, that is, they are not macromolecules, but proteins can also bind to relatively small areas of other proteins and nucleic acids. Protein molecules have discrete sites on their surface that are complementary to the ligand, which allows the protein to recognize and specifically bind to it in a reversible manner. The protein–ligand complex can dissociate back to the free molecules or it can induce a change in the conformation of the protein, which induces some biological event to occur. For example, in the case of enzymes, the ligand would be a substrate and the event would be the catalysis of a reaction that converts it to product. If the protein were a receptor and the ligand a hormone, then changes would be stimulated in target tissues.

16.2 Molecular biology-based techniques

The complementary binding of pairs of bases in nucleic acids offers a convenient way of recognizing and isolating specific base sequences within a fragment of a DNA or RNA molecule or of isolated nucleic acid molecules. Similarly, proteins specifically recognize and bind to a variety of different types of molecules. The abilities of macromolecules to bind complementary substances is the basis of many routine clinical techniques, for example fluorescence *in situ* hybridization (FISH) and enzyme-linked immunosorbent assays (ELISAs), as we will describe later (Section 16.8).

Throughout this chapter you will encounter many molecular biology techniques where the specific binding properties of macromolecules has been exploited in recognizing, isolating, and analysing molecules in biological and clinical samples. In all cases, the specificity of recognition is such, that the techniques are extremely sensitive. They therefore require relatively small amounts of sample compared with more traditional chemical techniques and can be performed on volumes as small as a few microlitres, hence minimal amounts of reagents are required saving the health services' money. Experimental molecular biology kits are available (Box 16.1).

16.3 Isolation of nucleic acids

The study of nucleic acids and their analysis or use in clinically related procedures, for example, preparing the **probes** (Box 16.2) used in fluorescence *in situ* hybridization (FISH) described in Section 16.8, often involves their isolation from cells and tissue samples. The nucleic acid purified for clinical purposes is usually DNA, rather than RNA. Chromosomal DNA can be purified from bacterial or eukaryotic cells and this is generally described as *total* DNA. However, plasmids (Section 16.1) can also be isolated from bacterial and some types of eukaryotic cells, such as yeasts. In general, the separation of plasmids follows the same basic protocols as those used to obtain chromosomal DNA, although at some stage, procedures must be used to separate the plasmid from the bulk of the cellular DNA. This section will discuss general methods for isolation and purification of chromosomal DNA from bacteria and eukaryotic cells; the isolation of plasmids is described in Section 16.9.

Chromosomal DNA is difficult to isolate in an intact and undamaged form because of the large size and fragile nature. However, several isolation procedures have been developed to purify it from cells. Indeed, a number of them are commercially available in kit form. Used correctly, these procedures produce preparations of DNA that are stable and of large M_r, and relatively

BOX 16.1 Experimental molecular biology kits

Many of the techniques of molecular biology are well established and can be performed with experimental kits that can be purchased from commercial suppliers. Indeed, experimental kits are central to research in molecular biology and a number of the protocols described in this chapter make use of them. For example, you can buy kits to extract DNA or RNA from tissues (Section 16.3) and synthesize complementary DNA (cDNA) from the isolated mRNA (Section 16.9). Such kits provide all the reagents and other materials needed for the particular experiment, descriptions of the underlying method, and easy to read and follow protocols. Reasonably 'good' results can usually be obtained using a kit. Indeed, in many cases, experimental results are easier to obtain than with the traditional method of purchasing the materials separately and preparing individual solutions of the reagents. This saves time, but kits are generally the more expensive option. They are also inflexible in that their protocols are difficult to modify for different circumstances. Furthermore, their extensive use carries the danger of lowering one's general laboratory skills. You will never fully understand an experimental method if your depth of knowledge is that 3 μL of the reagent in the red tube must be added to 5 μL of sample and then heated to 30°C for 5 minutes! A thorough understanding of any experimental procedure will enable you to use it to its full effect, know what to do when it goes wrong, as it will on some occasion, and appreciate when a result is aberrant or outside the experimental limits, artefactual, or simply incorrect. The best advice is to learn and understand the technique before using the corresponding kit.

free of contaminating RNA and proteins. Whatever, its source, a number of points must be considered in the isolation of DNA. These include:

- pH
- Temperature
- Ionic strength
- Nuclease activities
- Mechanical stress

The isolation of DNA requires it be released into a medium of appropriate pH. This is necessary to preserve the interactions and chemical bonds that stabilize the molecules. For example, hydrogen bonding between the complementary strands is stable at pH 4–10, while the phosphodiester linkages become unstable below pH 3 and above 12. Furthermore, the N-glycosidic bonds joining purine bases (adenine and guanine) to the deoxyribose sugars are hydrolysed at pH values of 3 or less. Phosphodiester linkages and N-glycosidic bonds are stable at temperatures up to 100°C. However, temperatures of 80–90°C destabilize the hydrogen bonds holding the double helix intact causing it to unwind. Care must be taken in choosing an appropriate ionic strength for solutions used in DNA extractions. Deoxyribonucleic acid is most stable and soluble in salt solutions. However, concentrations less than 0.05 mol dm^{-3} weakens the hydrogen bonding between complementary strands. Enzymes, such as deoxyribonucleases or DNases (see Section 16.4) are present in all cells and can degrade DNA during its purification if their action is not inhibited.

BOX 16.2 Probes

Probes are labelled molecules that are complementary to the molecules one is studying. For example, if a solution of DNA is heated at 100°C or exposed to pH values above 13 the complementary base pairs are disrupted and the double helix dissociates into its two single strands. Complementary single strands of DNA, however, readily reform double helices by a process called annealing, hybridization, or DNA renaturation, if they are allowed sufficient time at temperature at or below 65°C. Hybridization reactions can also occur between any two single-stranded complementary polynucleotides forming DNA–DNA, RNA–RNA homoduplexes, or RNA–DNA heteroduplexes. These specific hybridization reactions are the basis of sensitive methods widely used to detect and characterize specific nucleotide sequences in both RNA and DNA in samples. In such cases, the probe could be a single-stranded DNA molecule that has been labelled in some way to identify and quantify the resulting **hybrid** (sample nucleic acid – DNA probe).

Native DNA molecules are highly asymmetric, being relatively long (mm to cm), but only 2 nm in diameter. The extraction of DNA involves removing the stabilizing basic proteins, which leaves the DNA prone to breakage by shearing forces generated by routine laboratory procedures, such as grinding, shaking, stirring, and pipetting. Hence, all extraction procedures must be performed with care. It is not, however, possible to isolate chromosomal DNA without fragmenting the molecules: isolation procedures always result in preparations containing DNA molecules whose lengths are much shorter than those found in chromosomes. However, damage to the secondary (double helical) structure does not usually occur.

The isolation of DNA and purification from cells and tissues are essential for many other molecular biological techniques. A number of methods have been developed for extracting and purifying DNA from different cells, tissues, and organisms, which are often modified by different laboratories to accommodate different types of samples. Here, we will give simplified protocols used in isolating DNA from bacteria and animals samples.

DNA extraction from bacteria

Deoxyribonucleic acids can be extracted from bacteria following the generalized procedure outlined below. Its isolation from specific species of microorganisms may require modifications to the procedures.

The cell wall and cell surface membrane of the bacterial cells must be disrupted, often using lysozyme and sodium dodecyl sulphate (SDS), to release the DNA. Lysozyme catalyses the hydrolysis of glycosidic bonds of cell wall peptidoglycans and lyses the cell wall, while the SDS disrupts the hydrophobic interactions that stabilize the membrane. Hence, DNA, together with other cellular components, is released into a medium in which it is soluble and protected from degradation. The medium of choice is usually a saline solution, buffered to pH 8 that contains ethylenediaminetetraacetic acid (EDTA). The DNA is soluble in the salt solution. The EDTA chelates divalent metal ions, such as Ca^{2+}, Mg^{2+}, Mn^{2+}, removing them from free solution so they cannot form salts with the phosphate groups of DNA and so inhibits the DNases that require their presence for activity. The relatively high pH also reduces nuclease activities. A secondary action of SDS is to act as a denaturant of DNases and other proteins.

Basic proteins bound to native DNA must be dissociated from it during the extraction. The presence of detergent, the high concentration of salt, or the addition of sodium perchlorate to the medium, and its mildly alkaline (pH 8) nature all reduce electrostatic interaction between the two. The DNA can now be isolated from other soluble cellular components by removing the proteins and then precipitating the DNA in a relatively pure form. The solution is deproteinized by adding a mixture of chloroform–isoamyl alcohol followed by centrifugation. Following centrifugation, three layers are produced as shown in Figure 16.11: an upper aqueous phase that contains the DNA, a lower organic layer, and a compact band of denatured protein at the interface between the aqueous and organic phases. Chloroform causes surface denaturation of proteins. Isoamyl alcohol is used because it reduces foaming and stabilizes the interface between the aqueous phase and the organic phase where the protein collects. The upper aqueous phase containing nucleic acids is removed from the centrifugation tube. The ionic nature of DNA is exploited in its purification because the addition of an organic solvent to the aqueous medium reduces its polarity so it is no longer a solvent for nucleic acids. Thus, the addition of ethanol precipitates the DNA as threadlike material that can easily be collected from the medium. Any remaining proteins contaminating the isolated DNA are removed by dissolving it in a saline medium and repeating the chloroform–isoamyl alcohol treatment until denatured protein no longer collects at the interface. While RNA does not normally coprecipitate with DNA during these procedures, it could still be present as a minor contaminant. Ribonucleases (RNases) may be added to the preparations during the procedure after the first or second deproteinization step to digest any RNA present. Its removal sometimes makes it possible to remove additional proteins using chloroform–isoamyl alcohol.

Aqueus phase containing nucleic acid

White precipitate of protein at interface

Organic phase

FIGURE 16.11
Final centrifugation stage in the extraction of DNA.

If highly purified DNA is required, several deproteinization and alcohol precipitation steps may be necessary. In general, up to 50% of cellular DNA can be recovered by this procedure, with an average yield of 1–2 mg DNA per gram of wet packed cells.

Phenol and chloroform are potentially dangerous materials and their use in extracting DNA from samples is often time consuming. Extraction procedures based on the reversable binding of DNA to columns of silica overcome these disadvantages because phenol and chloroform are not used. Thus, these methods are safer than the organic solvent extraction methods described above. They are also easier to perform because there is less handling and manipulation of the sample. Such methods rely on the property of DNA to bind to silica in the presence of high concentrations of chaotropic salts, such as sodium iodide, guanidine hydrochloride, or sodium perchlorate, but to be released when their concentrations are reduced. The procedure is outlined in Figure 16.12. The column is contained within a small tube that can be centrifuged in a microfuge to drive the various solutions through it, speeding up the purification considerably. High concentrations of salts dehydrate the DNA and favour the formation of hydrogen bonds between it and the silica. Thus, if the DNA is extracted in a medium of high salt concentration, the sample can be added directly to the column. The column is then extensively washed with alcohol-based solutions to remove the impurities. The DNA can then be recovered in a pure form by washing the column with a buffer containing a low concentration of salt, for example a 2-amino-2-hydroxymethylpropane-1,3-diol-(Tris)-EDTA buffer or even nuclease-free water.

Cross reference
Centrifuges, including microfuges, are described in Chapter 12 'Centrifugation'.

DNA extraction from animal cells

The major steps for isolating DNA from animal cells, such as clinical samples, resemble those used for bacteria described above. The tissue must be disrupted, the cellular components solubilized and the DNA separated from the other released cellular components. Care in the initial collection of animal tissue samples greatly reduces the problems encountered during DNA isolation. Ideally, samples should be immediately preserved in a freezer at –80°C or in

FIGURE 16.12
Column-based purification of DNA. See text for details.

(Figure labels, top to bottom:)
Sample containing DNA

Binding

Column

Centrifuge

DNA now bound to column

Discard

Washing buffer

Centrifuge — Two washes to remove impurities

Discard

Elution of DNA

Centrifuge

Purified DNA

liquid nitrogen. Excellent samples of DNA can be prepared from blood, heart, liver, kidney, stomach, intestine, cerebrospinal fluid, and skeletal muscle. Clinical samples are often blood and its cells should be separated from the plasma prior to freezing.

The simplest means of disrupting cells to isolate DNA is to boil them in water or to use an alkaline extraction medium. High temperatures (100°C) or pH values above 9 disrupt cell membranes, releasing the DNA into solution. However, the DNA obtained by these methods is denatured, but active degradation in low ionic strength solutions is stimulated by metal ions also released from the cells. Chelating resins have been used to remove metal ions from solutions, thus making feasible the rapid isolation of DNA. It should also be noted that because of the high alkalinity (pH 10–11) of the chelating resin, samples prepared with it may degrade much more rapidly than those prepared in other ways such as isolation using proteinase K with detergent SDS or with formamide. However, denatured DNA is amenable to some molecular biology techniques, for example polymerase chain reaction, but not others such as **recombinant DNA** cloning (Section 16.9), Southern **blotting**, or restriction fragment analysis (Section 16.7).

Extraction buffers used to isolate animal DNA are usually based on Tris and contain EDTA to chelate divalent cations such as Ca^{2+} and Mg^{2+} and thus inhibit nuclease activity, and a Na^+ or K^+ salt to stabilize the nucleic acids in an isotonic medium. A protease, usually proteinase K, is often added to digest cellular proteins. An anionic detergent, usually SDS, is normally included to solubilize cellular membranes and denature proteins, as is formamide, an ionizing solvent that dissociates DNA–protein complexes and denatures the released proteins. However, its presence significantly reduces the activity of proteinase K. Both the pH of the isolation buffer and the specific protective agents or detergents included may need to be modified to give optimal conditions for specific samples. Once released into solution, the DNA can then be purified by the methods outlined above for bacterial samples.

Robotic workstations are available that automatically purify genomic DNA from tissue samples. Figure 16.13 shows a typical instrument that can, for example, simultaneously process 8 or 16 of 5–10 cm^3 samples of whole blood with yields of up to 350–500 µg of DNA per sample. The DNA obtained is sufficiently pure for its immediate use in many of the techniques described in the remainder of this chapter.

Isolation and purification of RNA

Typical mammalian cells contain approximately 10^{-5} µg of RNA, of which 80–85% is rRNA, the remaining 15–20% is largely low M_r tRNA and small nuclear RNA molecules, which can be isolated by density gradient centrifugation, anion-exchange or high-performance liquid chromatography, or gel electrophoresis. Messenger RNA constitutes 1–5% of the total cellular RNA and is heterogeneous in both size, from several hundred bases to many kilobases in length, and sequence. However, most eukaryotic mRNAs have a tract of several 100 adenylate residues at their 3′ termini, the so-called poly(A) tail, that is exploited to allow their purification by affinity chromatography with a column of oligo(dT) cellulose. The yield of total RNA depends on the tissue or cell source.

Ribonucleic acids are chemically more reactive than DNA because they have hydroxyl groups at both the 2′ and 3′ positions. They are, therefore, more susceptible to hydrolysis by contaminating nucleases. Ribonucleases are released from cells during purification procedures and are also present on the skin. Unlike many DNases, RNases do not require divalent cations for activity and are not inactivated by EDTA or other chelators in buffer solutions, and are often resistant to prolonged boiling and mild denaturants. Thus, glassware and material are often autoclaved to denature any RNases present. Disposable gloves must always be worn.

FIGURE 16.13
Robotic workstation capable of extracting DNA from numerous samples simultaneously. Courtesy of S. Smith, Regional Genetics Services, St Mary's Hospital, Manchester, UK.

To avoid contamination with RNases during work with RNA, instruments, such as automatic pipettes should be reserved for that specific use and all procedures carried out in restricted areas or fume cupboards.

When isolating RNA from tissues such as pancreas or the gastrointestinal tract, that are rich in digestive enzymes, it is best to cut the dissected tissue into small pieces and add them quickly to liquid nitrogen. These fragments are stored in a freezer at about −70°C or used immediately for RNA extraction. Extraction requires the cells or tissues to be disrupted in denaturing solutions that contain phenol, and guanidinium thiocyanate or guanidinium isothiocyanate, that help disrupt the cells while solubilizing their components, and denature endogenous RNases simultaneously. The addition of chloroform followed by centrifugation separates the sample into a colourless upper aqueous phase, an interphase, and a lower organic phase. RNA remains exclusively in the aqueous phase and can be recovered by precipitating it with alcohol and then washing the precipitate with 75% ethanol before redissolving it in RNase-free water or 0.5% SDS solution.

The Health & Safety Box 'Precautions when isolating DNA' highlights some of the safety procedures necessary when using the harsh chemicals used in extracting DNA. Naturally, similar precautions are essential when isolating RNA from samples because for example guanidinium thiocyanate and isopropanol are harmful if ingested or absorbed through the skin or if the alcoholic fumes are inhaled.

> **Cross reference**
> You can read about centrifugation, chromatography, and gel electrophoresis in chapters 12, 13, and 14, respectively.

HEALTH & SAFETY

Precautions when isolating DNA

- Experimental work with bacterial cells presents a potential biohazard. Some bacteria recommended for use, for example strains of *Salmonella*, *Haemophilus*, are pathogenic and can cause disease. Eukaryotic tissues carry the danger of harbouring pathogens.

- Benches and other work surfaces must be washed with a 10% solution of bleach before and after DNA extractions. Do not pipette any solutions by mouth. Always wear a suitable laboratory coat, protective goggles, and disposable gloves, and wash hands well with hot water and soap when leaving the laboratory, and before eating or drinking.

- Chloroform is a volatile liquid that irritates the skin, eyes, mucous membranes, and respiratory tract. It is also carcinogenic and can cause liver and kidney damage. Work involving chloroform must always be performed in a fume cupboard.

- Ethanol is volatile and flammable. There must be no naked flames in the laboratory during its use.

- Isoamyl alcohol can cause tissue damage following inhalation, ingestion, or topical absorption.

- Lysozyme is an hydrolytic enzyme that can damage mucous membranes.

- Sodium dodecyl sulphate (SDS) is an irritating toxic detergent that can severely damage eyes. It can also cause tissue damage following inhalation of its dry powder form or ingestion of the powder or solution.

16.4 Hydrolysis of nucleic acids and restriction endonucleases

If you re-examine the structures of Rib and dRib in Figure 16.2 you will see that the only difference in their structures is the presence of the 2′-OH on the former but only a 2′-H on dRib. The name, in fact, says this: *deoxy* means *without oxygen*. The presence of this additional hydroxyl group means RNA is much more susceptible to alkaline hydrolysis than DNA. Mild acid hydrolysis releases purine bases from DNA.

The alkaline or acid hydrolyses of nucleic acids degrades susceptible bonds in a random fashion. However, enzymes are available that can catalyse the hydrolysis of nucleic acids in a more specific manner. These enzymes are called nucleases or phosphodiesterases and they catalyse the hydrolysis of phophodiester bonds, although they differ considerably in their specificities. Exonucleases catalyse the sequential removal of terminal nucleotides; endonucleases hydrolyse internal phosphodiester bonds. Some nucleases can only use DNA or RNA, respectively, as substrates, while others can degrade both. Some nucleases are only active against single stranded nucleic acids, but others can catalyse the hydrolysis of double stranded structures. Look again at the structure of the phosphodiester link shown in Figure 16.4. Note that it can be cut on one of two sides. If the hydrolysis occurs on the 3′ or *a* side then the product is a 5′-phosphate derivative; hydrolysis on the 5′ or *b* side gives a 3′ prime product. Figure 16.14 summarizes both hydrolyses. Again, different nucleases differ in *a* or *b* specificity. Some nucleases can only catalyse the hydrolysis of phosphodiester bonds associated with nucleotides that contain only purine or only pyrimidine bases. **Restriction endonucleases** (REs) are even more specific and only catalyse the hydrolysis of specific phosphodiester bonds within or near to short specific base sequences in double-stranded DNA. Restriction endonucleases are classified into three different groups: I, II, and III, type II REs are the most widely used in molecular biology.

Restriction enzymes

Type II REs are endonucleases produced by bacteria that typically recognize specific 4–8 bp sequences, called restriction sites, and then hydrolyse both strands of the DNA at this site at the same position. For example, the RE, *Eco*RI recognizes the sequence GAATTC and hydrolyses it at the positions indicated in Figure 16.15. If you examine Table 16.2, which lists a number of REs, notice how the sites of hydrolysis occur at corresponding bonds within

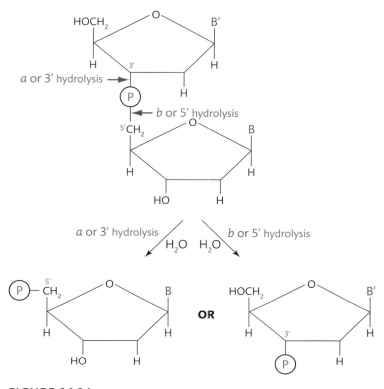

FIGURE 16.14

Definition of *a* and *b* hydrolyses of a nucleic acid. Note how each type of hydrolysis form a different product.

identical sequences of bases in both of the strands when they are read in the 5′ → 3′ direction. Restriction sites are therefore inverted repeats and are called **palindromes** because the sequence of the site is the same on each DNA strand. Different REs recognize and hydrolyse different palindromes. Different species of bacteria synthesize different REs, which protect them from viruses by degrading viral DNA. The bacteria DNA is not degraded because each RE has a corresponding enzyme that recognizes the same palindrome and modifies it, so it is no longer recognized by the RE (Box 16.3).

A **palindrome** is a word or phrase that has the same order of letters when read in the reverse direction, for example madam, tit.

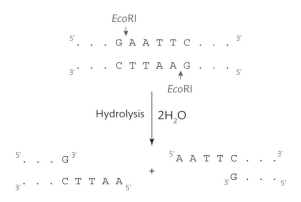

FIGURE 16.15

Palindrome recognized and hydrolysed by the restriction enzyme, *Eco*RI.

BOX 16.3 *Restriction endonuclease: sources and nomenclature*

Restriction endonucleases are produced by bacterial cells. They function in protecting the bacterium from attack by viruses by degrading the viral DNA. The host bacterial DNA is protected from hydrolysis because each RE has a corresponding methyl transferase with an identical palindrome specificity. Thus, the bacterial palindromes, but not the viral, have methyl (-CH₃) groups attached to bases near to the palindrome that prevents their recognition by the REs. Several hundred REs have been isolated from different bacterial species and characterized. The nomenclature adopted for them consists of a three-letter abbreviation that identifies the bacterial source, followed by a letter representing the strain of the species (if necessary) and, lastly, a Roman numeral designating the order of discovery of that enzyme when more than one RE has been isolated from that strain. Thus *Eco*RI was the first RE to be isolated and characterized from the R strain of *Escherichia coli*.

Many restriction enzymes make staggered cuts in the two DNA strands at their recognition site and produce products that have a single stranded portion at each end, as is the case with *Eco*RI. Look carefully at the products of the reaction. Note how they have overlapping strands that are often called 'sticky ends'. Such REs are used extensively in recombinant DNA technology, which we will describe in Section 16.9. However, some REs have actions that form products with blunt cut ends, as you can see is the case with *Eco*RV (Figure 16.16).

SELF-CHECK 16.5

Which REs in Table 16.2, besides *Eco*RI, produce products with sticky ends?

Uses of restriction endonucleases

The abilities of REs to recognize palindromes in DNA and restrict hydrolytic events to those sites means they can be used to degrade molecules of DNA into a specific number of fragments in a reproducible manner. Thus, REs are invaluable tools in molecular biology and used extensively to:

- Specifically hydrolyse large DNA molecules
- Produce physical maps
- Convert circular plasmids in to linear DNA molecules
- Produce recombinant DNA molecules
- Investigate restriction fragment length polymorphisms (RFLPs)

Restriction endonucleases can be used to hydrolyse large DNA molecules into smaller fragments that are more amenable to analysis. The fragments are readily separated and their size estimated by agarose gel electrophoresis as you can see in Figure 16.17 (Section 16.5 and Chapter 14, you may also wish to look at Figure 14.22 (b)). It is unlikely that the sets of fragments produced from any two different DNA molecules using the same RE will be the same. Hence, the pattern of DNA products is likely to be unique and so can be considered a 'fingerprint' of the DNA substrate.

Restriction endonucleases can be used to produce physical maps of DNA molecules. An understanding of the genetics, metabolism, and regulation of an organism requires knowledge

TABLE 16.2 Examples of restriction endonucleases and their palindrome specificities.
The red arrows indicate the sites of hydrolysis

Restriction endonuclease (bacterial species)	Palindrome recognized and sites of hydrolysis
*Alu*I (*Arthrobacter luteus*)	↓ . . . AGCT. . . . . . TCGA. . . ↑
*Bam*HI (*Bacillus amyloliquefaciens* H)	↓ . . . GGATCC. . . . . . CCTAGG. . . ↑
*Eco*RI (*Escherichia coli* RY 13)	↓ . . . GAATTC. . . . . . CTTAAG. . . ↑
*Eco*RV (*Escherichia coli* J62 pLG74)	↓ . . . GATATC. . . . . . CTATAG. . .
*Hae*III (*Haemophilus aegptius*)	↓ . . . GGCC. . . . . . CCGG. . . ↑
*Hind*III (*Haemophilus influenzae* Rd)	↓ . . . AAGCTT. . . . . . TTCGAA. . . ↑
*Kpn*I (*Klebsiella pneumonia*)	↓ . . . GGTACC. . . . . . CCATGG. . . ↑
*Msp*I (*Moraxella* species)	↓ . . . CCGG. . . . . . GGCC. . . ↑
*Pst*I (*Providencia stuartii* 164)	↓ . . . CTGCAG. . . . . . GACGTC. . . ↑
*Pvu*II (*Proteus vulgaris*)	↓ . . . CAGCTG. . . . . . GTCGAC. . . ↑
*Taq*I (*Thermus aquaticus*)	↓ . . . TCGA. . . . . . AGCT. . . ↑

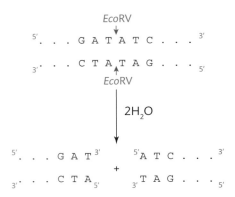

FIGURE 16.16
Reaction catalysed by *Eco*RV.

of the precise arrangement of its genetic material. The construction of a RE map for a DNA molecule provides some of this information since it will show the sites that can be cut by different REs and the number of fragments obtained after each digestion. Two characteristics for every DNA fragment produced by an RE digest are known: the nature of the fragment ends (from the known specificity of the individual RE) and its approximate M_r (from electrophoresis separation).

FIGURE 16.17
Separation of DNA fragments by agarose gel electrophoresis. Lane 1 shows a stained sample of DNA isolated from λ bacteriophage. Lanes 2, 3, and 4 show the fragments obtained by digesting the bacteriophage DNA with the restriction endonucleases *Hind*III, *Taq*I, and *Eco*RV, respectively. Lane 5 shows the positions of fragments of DNA of known sizes ('ladder'). Note how each of the enzymes has produced a different set of products by hydrolysing the DNA at different palindrome sites. Courtesy of N. Shaheen, School of Healthcare Science, Manchester Metropolitan University, UK. See also Figure 14.22 (b).

Other information about the fragments, for example the sequence of their bases (Section 16.6) can be obtained using other experimental methods. (See the Method Box 'Experimental procedure for performing a restriction endonuclease digest').

The circular DNA molecules present in many plasmids can be converted into linear molecules by the action of some REs. Note that it is necessary that the plasmid have only a single recognition site for the RE being used, otherwise digestion will fragment it! Restriction endonucleases are valuable tools in the construction of recombinant DNA molecules. Thus, once a circular plasmid has been 'linearized', it is capable of binding to other fragments of DNA produced by the actions of the same RE on different samples irrespective of the source of the DNA. This gives molecular biologists a way to cut and splice different pieces of DNA together to produce recombinant DNA molecules, which can be transferred into bacterial cells and cloned as described in Section 16.9. The use of REs not only opens plasmids for insertion of a DNA fragment, they can also destroy a phenotype. For example, if they cleave at a site that specifies resistance to a particular antibiotic, then when the recombinant plasmid is added back to a suspension of bacteria, those that are transformed by taking it up can be identified by their susceptibility to the antibiotic.

The digestion of DNA using REs and the separation of the digest fragments by electrophoresis can be used for clinical purposes, such as the diagnosis of inherited disorders for example sickle cell anaemia using RFLPs, which we describe in Section 16.7.

METHOD Experimental procedure for performing a restriction endonuclease digest

- Commercially available REs are usually dissolved in an appropriate buffer and posted in vials surrounded by dry ice. Disposable gloves must be worn when removing the enzyme container. All REs are heat-labile and should quickly be stored at −20°C in a buffer containing 50% glycerol. The solution of enzyme should be removed from the freezer just before use and stored on ice.

- A typical RE digest is performed in a microcentrifuge tube (Chapter 12). The mixture normally contains approximately 1 µg or less of DNA and about 1 unit of RE activity in the appropriate incubation buffer.

- Restriction nucleases have optimal reaction conditions for activity, especially regarding the composition and pH of the reaction buffer, and incubation temperature. The information for commercially available enzymes is readily available from the supplier or from published literature. The total reaction volume is usually between 20 and 50 µL. The contents of the tube are mixed by gently tapping it or briefly centrifuging the mixture in a microcentrifuge.

- Incubations are most often carried out at the recommended temperature, usually 37°C for about 1 hour.

- The reaction is then stopped by adding EDTA, which removes the divalent metal ions essential for nuclease activity from solution.

- The products of digestion are often analysed by agarose gel electrophoresis (Section 16.5 and Chapter 14). They can usually be temporarily stored at 0–4°C until subjected to electrophoresis.

16.5 **Gel electrophoresis**

Electrophoresis is the movement of ions, for example nucleic acids and proteins, in an electric field. You can read more about electrophoresis in Chapter 14. Here, we will briefly highlight the main points. Particles that differ in their mass-to-charge ratios will move at different speeds in the field and thus separate from one another. However, because phosphate groups are regularly spaced along DNA and RNA molecules, they all have the same mass-to-charge ratio and will all move at the same speed for a given electric field and so would not separate. However, performing electrophoresis in agarose or polyacrylamide gels means plasmids and smaller DNA and RNA molecules, and proteins can all be separated. The concentration of agarose or polyacrylamide in the gel determines the sizes of its pores. Using appropriate concentrations therefore allows mixtures of different sized molecules to be resolved during electrophoresis because the smaller molecules are able to move faster through the gel than larger ones.

Cross reference

Chapter 14 'Electrophoresis'.

In many types of gels, the distance moved by a limited range of nucleic acid fragments is inversely proportional to the logarithm of their sizes. The size of DNA fragments is usually expressed in bp or kilobase pairs (kbp), but can be given in M_r. In general, the lower the concentration of agarose in the gel, the larger the DNA molecules that can be separated. However, gels with concentrations of agarose below 0.3% are too fragile for ordinary use and this therefore forms a practical lower limit. Gels of this concentration allow the analysis of linear double-stranded DNA fragments of 5–60 kbp (M_r up to 150×10^6). Gels of 0.8% agarose can separate DNA in the range of 0.5–10 kbp; and 2% agarose is used to separate smaller DNA fragments of 0.1–3 kbp. In addition to the DNA molecules of analytical interest, fragments of DNA of known size are also subjected to electrophoresis, forming a so-called 'DNA ladder'. The sizes of unknown fragments can then be estimated by comparing their mobilities with those of known size in the ladder.

Once stained, separated fragments of DNA in the gel are usually detected using ethidium bromide as a stain (Figure 14.21). Ethidium bromide consists of relatively small, planar molecules, which can bind to DNA by intercalating between the stacked base pairs, which enhances the normally weak fluorescence of the purine and pyrimidine bases by about 25 times (Figure 14.22 (a)). Thus, when irradiated with UV light the separated fragments fluoresce with an intense orange colour giving a sensitive means of detecting the fragments: as little as 1 ng of DNA can be seen (Figure 16.17).

DNA fragments containing 1000 or fewer bp can be separated by sodium dodecyl sulphate polyacrylamide gel electrophoresis (SDS-PAGE). Such gels are also widely used to separate mixtures of proteins, as you can see in Figure 14.17. Normally, proteins are separated following their denaturation using β-mercaptoethanol and SDS. β-mercaptoethanol reduces the disulfide bonds that help stabilize protein structure and which hold the subunits of some quaternary proteins together, while the negative charges of the SDS bound to the proteins repel each other. In these conditions, protein molecules adopt rod-like structures. Proteins bind SDS in a size-dependent manner and since SDS overwhelms the native charges of the proteins, all proteins in the presence of SDS have similar mass-to-charge ratios. Polyacrylamide, like agarose, forms gels whose pore size is determined by the concentration used. Hence, if the electrophoresis is carried out in a gel also saturated with SDS (hence SDS-PAGE), smaller proteins or very small DNA fragments will migrate faster through it than larger ones and separate from one another. The higher the concentration of acrylamide, the greater the resolution of proteins of lower M_r; the lower the concentration the better the resolution of larger ones. Following separation, proteins can be detected using one of a variety of different staining procedures (Table 14.1).

As with agarose electrophoresis of nucleic acids, it is normal to treat and separate a mixture of proteins of known M_r to allow the sizes of unknown proteins to be estimated.

16.6 DNA sequencing

Sequencing of DNA, that is determining the order of its bases, is mainly performed using the chain termination or 'dideoxy' method first devised by Sanger in the 1970s. Sequencing methods are largely automated and performed by robotic machines whose results are analysed by computers. However, the basic procedure is relatively straightforward. The DNA must be pure and hydrolysed to give a series of suitably sized overlapping pieces. Each fragment is amplified (Section 16.9) before being independently sequenced. Since the fragments have overlapping regions of bases, it is possible to assemble them in the appropriate order to give the complete sequence.

The dideoxy method has two key features. One is that it exploits the specificity of DNA polymerase, the enzyme that copies DNA strands to produce a new complementary strand. This enzyme catalyses the extension of a polynucleotide chain by forming new 3'-5' phosphodiester bonds. Bonds are formed when the enzyme adds a new nucleotide to the terminal 3'-OH. The base sequence of the new strand is determined by that of the template. The second feature exploited in the method is the use of dideoxynucleoside triphosphates (ddNTPs). These lack a 3'-OH and, so once a ddNTP is added to the end of a newly forming chain, its growth ceases because the absence of a 3'-OH prevents the addition of a further nucleotide. You can see this diagrammatically in Figure 16.18.

Dideoxy sequencing occurs in a reaction mixture that contains a sample of the single stranded DNA to be sequenced (the template), an oligonucleotide primer that is complementary to part of the template, DNA polymerase, the four deoxynucleotides, dATP, dTTP, dGTP, and dCTP,

FIGURE 16.18
Outline of the dideoxy method for sequencing DNA. See text for details.

and relatively small amounts of each of the four ddNTPs. The DNA polymerase then catalyses the addition of nucleotides and extends the length of the DNA primer. The newly formed extension is complementary to the template. However, during extension of the primer the enzyme will sometimes use a ddNTP, rather than the normal dNTP. When this happens, extension of the growing chain will cease to occur. If the mixture contains an appropriate ratio of the respective ddNTPs to dNTPs, then when the reaction is completed, the mixture will contain a set of new but incomplete DNA chains complementary to the template. Extension of each will have started at the same point, but because the ddNTP was selected in a random fashion by the enzyme, they will differ in length by a single base.

The newly formed fragments in the reaction mixture can then be separated on a SDS-PAGE gel containing urea (Section 16.5) or by capillary electrophoresis. Each technique can separate DNA molecules that differ by only a single base in length. The denaturants SDS and urea ensure the DNA chains remain separate during electrophoresis. Following separation, the different lengths of DNA form a densely packed 'ladder' of bands on the gel, as you can see in Figure 16.19. These ladders can be interpreted to give the equivalent sequences of bases in the template. Each of the four ddNTPs is tagged with a different dye. A laser light passes along the gel and each dye fluoresces with a different coloured light that identifies the particular base used in terminating that length, for example, yellow for G(uanine), red for T(hymine). The length of the DNA fragments and their colours, which identify each terminal base, are analysed by a computer to give the base sequence of the template.

Sodium dodecyl sulphate-PAGE gels cannot resolve more than 500–800 bases in a single electrophoresis run, which limits sequencing experiments. However, modern DNA sequencers separate the fragments using capillary electrophoresis (Chapter 14) and automatically detect dye fluorescence and record the data output as a fluorescent peak trace (Figure 16.20). The use of such automation means many thousands to millions of bases can be sequenced in an hour.

Cross reference

Chapter 14 'Electrophoresis'.

FIGURE 16.19

Portion of a sequencing gel. The colours identifying each base are green for adenine (A), blue, cytosine (C); yellow, guanine (G), and red, thymine (T). See text for details. Courtesy of Wellcome Trust Sanger Institute, Cambridge, UK.

FIGURE 16.20
Trace showing the sequencing output (capillary electrophoresis) from a Beckman Coulter CEQ 8000 automated sequencer. Part of the corresponding base sequence is shown. Courtesy of Dr Patricia Linton, School of Healthcare Science, Manchester Metropolitan University, UK.

SELF-CHECK 16.6

Complete the sequence of bases shown in Figure 16.20.

The application of automated methods means that DNA sequencing is largely a routine and rapid procedure. Indeed, the genomes of many of the model organisms studied routinely in the laboratory have been sequenced. These include a number of different yeasts, the nematode *Caenorhabditis elegans*, the Drosophila fruit fly, the mouse, dog, and chimpanzee and, finally, yet of great clinical importance, that of humans. The complete sequences of many 100s of parasitic viruses, for example, the smallpox and Epstein–Barr viruses, large numbers of pathogenic bacteria, such as those responsible for cholera, tuberculosis, syphilis, gonorrhoea, Lyme disease, and stomach ulcers, as well as parasitic protozoa like *Plasmodium falciparum*, have also been fully determined (Table 16.3). Knowing these sequences should help produce more rapid and effective diagnostic tools, such as in designing specific probes (Box 16.2) for use in FISH (Section 16.8) and PCR (Section 16.9) methods. It is anticipated that analysis of their genomes will also provide useful information about their virulence and indicate more effective treatments for the clinical conditions they cause.

Sequencing specific genes can also be used to determine if a person is a carrier of a genetic disease or to confirm a provisional diagnosis, particularly in those conditions where the inheritance does not follow simple Mendelian rules, for example, Fragile X syndrome. This is

TABLE 16.3 Examples of pathogens whose genomes have been completely sequenced

Group	Virus or organism	disease
Viruses	Enterovirus 100,	Meningitis
	Immunodeficiency viruses 1 and 2	Acquired immunodeficiency syndrome (AIDS)
	Papillomavirus16	Cervical cancers
	Herpes virus	Herpes
	Influenza A	Influenza
	Poliovirus	Polio
Bacteria	*Bacillus anthracis*	Anthrax
	Bordetella pertussis	Whooping cough
	Campylobacter jejuni	Gastroenteritis
	Clostridium botulinum	Botulism
	Clostridium tetani	Tetanus
	Corynebacterium diphtheriae	Diphtheria
	Escherichia coli	Food poisoning
	Helicobacter pylori	Stomach ulcers
	Klebsiella pneumoniae	Pneumonia
	Legionella pneumophila	Legionnaires disease
	Mycobacterium leprae	Leprosy
	Mycobacterium tuberculosis	Tuberculosis
	Neisseria gonorrhoeae	Gonorrhoea
	Neisseria meningitidis	Meningitis
	Salmonella typhimurium	Gastroenteritis
	Shigella dysenteriae	Dysentery
	Staphylococcus aureus	Skin infections
	Streptococcus mutans	Dental plaque
	Treponema pallidum	Syphilis
	Vibrio cholerae	Cholera
	Yersinia pestis	Bubonic plague
Protozoa	*Plasmodium falciparum*	Malaria
	Trypanosoma brucei	Sleeping sickness

the commonest cause of mental retardation in the UK, affecting one in about 4100 males, but is less common in females, at approximately one in 8000, and who are generally less severely affected. The condition is associated with repeated copies of a trinucleotide, CGG, which occurs in the *FMR1* gene. Most people have a stable copy of *FMR1* containing about 30 repeats. Those with 45 to 55 copies have an increased chance of passing on an even larger number to their children. Individuals with 55–200 are said to possess a *premutation* because their children may have more than 200 CGG repeats, which is the full mutation associated with Fragile X syndrome. Sequencing *FMR1* or relevant portions of it will directly determine the number of repeats present and confirm the status of the patient.

16.7 **Blotting techniques**

Electrophoresis is a convenient way of separating nucleic acid and protein molecules. Unfortunately, it is difficult to identify specific molecules in the gel because probes (Box 16.2) do not readily diffuse into gels. Also, gel matrix materials, like agarose or polymerized polyacrylamide, are fragile and can be difficult to handle. These difficulties can be overcome by the common laboratory procedure of **blotting**, in which nucleic acids or proteins in the gel matrix are transferred from the gel to, for example, a piece of nitrocellulose paper or a nylon membrane. The positions of the molecules of interest on the paper/membrane can be identified and their analysis facilitated because they are bound to the surface of the paper or membrane.

Three major types of blotting techniques are used, namely:

- Southern blotting
- Northern blotting
- Western blotting.

The first blotting technique devised was Southern blotting by E. M. Southern, hence its name. The nomenclature of Northern and Western blotting were given with reference to it. These techniques allow the detection of specific DNA fragments, RNA molecules, and individual proteins, respectively. Thus, while genes or fragments of DNA can be detected by Southern blotting, the expression of specific genes is usually monitored by Northern or Western blotting, which detects the specific mRNA or protein molecules produced in the cell. In general, Southern and Northern blotting methods are insufficiently sensitive to detect the amount of nucleic acids obtainable from a single cell. However, the polymerase chain reaction or PCR, which we describe in Section 16.9, can be used to amplify the nucleic acid in the sample to amounts that are detectable.

Southern blotting

Southern blotting follows gel electrophoresis of double stranded fragments of DNA, which are separated on the basis of differences in their sizes. The DNA fragments present in the gel are then denatured with alkali and the resulting single stranded molecules transferred using a buffer onto a nitrocellulose filter or nylon membrane by blotting, as shown in Figure 16.21.

This procedure preserves the pattern of distribution of the DNA fragments in the gel, by forming a replica of the gel on the filter paper or membrane. The filter is then incubated under hybridization conditions with a specific radiolabelled DNA probe. The DNA restriction fragment that is complementary to the probe hybridizes with it, and its location on the filter can then be revealed, often by autoradiography (Box 13.1 'Autoradiography').

FIGURE 16.21
Outline of Southern blotting, which is explained in the general text.

However, autoradiography suffers from a number of disadvantages, including possible health hazards, the inconvenience of handling a radioisotope and the relatively short half-life of ^{32}P. The digoxygenin (DIG) system provides a sensitive alternative to autoradiography for analysing blots and is simple to use. Analysis involves adding a probe labelled with DIG to the nitrocellulose filter or membrane. An antiDIG-antibody (see Western blotting below) tagged with an enzyme, such as horseradish peroxidase (HRP) or alkaline phosphatase (ALP) is added to the reaction mixture. The addition of substrates that are broken down by the enzyme to give a chemiluminescence product means that even the binding of a single enzyme molecule is amplified because catalysis produces many molecules of product. The chemiluminescence can be detected using a high sensitivity detection device, such as a cooled charge-coupled device (CCD) camera system.

Digoxygenin-labelled DNA or RNA probes can be used to analyse the products of all types of hybridization reactions, including *in situ* hybridization and ELISAs (Section 16.8), microarrays (Section 16.10), and Northern blots (see below).

Restriction fragment length polymorphisms

Southern blotting is sufficiently sensitive to detect a single specific product in a complex mixture of DNA fragments as is formed when the entire human genome is hydrolysed with a restriction endonuclease. Southern blotting can detect and identify the genomes of specific viruses and microorganisms and the presence of mutations, or **polymorphisms**, in samples of DNA from patients produced for example by the deletion or the insertion of short DNA sequences; most single-base changes, however, cannot be detected in this way unless they form or break a palindrome that can be recognized by a RE. Sickle cell anaemia is a common disorder among some populations inhabiting or originating from malarial infested regions. Patients typically present with severe joint pain and a microscopic examination of their blood will show the presence of sickle-shaped erythrocytes. The disorder arises from a single point mutation in the gene for β globin, one of the two types of polypeptide that make up haemoglobin. A thymine substitutes for an adenine base resulting in a valine residue being incorporated at position 6 in the globin molecule rather than a glutamate, forming haemoglobin S. β globin genes are inherited from each parent in a Mendelian fashion. Hence, individuals can be homozygous for adult haemoglobin (Hb^A/Hb^A) or for haemoglobin S (Hb^S/Hb^S), or heterozygous (Hb^S/Hb^A) and form a mixture of adult and S haemoglobins. The condition can be diagnosed using foetal material obtained by amniocentesis using the restriction endonuclease (Section 16.4), *Mst*II. The normal globin gene has three restriction sites that can be hydrolysed by *Mst*II. However, the sickle mutation destroys one of these sites, resulting in a product that is larger in size. Thus, electrophoresis of the globin gene fragments followed by blotting produces different patterns for the hetero- and homozygotes as shown in Figure 16.22.

Northern blotting

The expression of a particular gene can be followed by detecting its corresponding mRNA by Northern blotting. This procedure is similar to Southern blotting, but electrophoresis is performed on a sample of mRNA. The RNA sample is often the total RNA purified from cells or tissues. It is denatured by treatment with agents, for example, formaldehyde, that disrupt the hydrogen bonds between any intrastrand base pairs, which ensures that all the RNA molecules have an unfolded, linear conformation. The sample is then subjected to gel electrophoresis and RNA molecules separated on the basis of differences in their sizes. As in Southern blotting, the denatured RNA molecules are transferred to a sheet of nitrocellulose or nylon membrane

FIGURE 16.22
Application of restriction fragment length polymorphism (RFLP) analysis using the restriction endonuclease *Mst*II and Southern blotting to diagnose sickle cell anaemia. See text for explanation.

which is then probed with a labelled DNA that is complementary to the RNA of interest. Molecules of RNA that hybridize to the probe on the paper/membrane are located by detecting the bound probe by autoradiography or by the DIG system described above. The sizes of the hybridized RNA molecules can be estimated by reference to RNA standards of known size that are subjected to electrophoresis side by side with the experimental sample.

Northern blotting is widely used to compare the amounts of a particular mRNA molecule in different cells grown in varying conditions. For example, it can be used to show if a mutation in a gene increases or reduces the amount of normal-sized mRNA produced. It can also be adapted to demonstrate if the mutation has resulted in the transcription of abnormally short mRNA molecules by testing the blotted paper/membrane with a series of shorter DNA probes, each complementary to short portions of the mRNA; this would also demonstrate which part of the normal RNA molecule is missing.

Western blotting

Western blotting (immunoblotting) combines several techniques to detect specific proteins in a sample. Like other blotting techniques, Western blotting gives information about the size of the protein or the relative amounts of protein produced in different cells, or the same cell in different conditions.

You can see an outline of this procedure in Figure 16.23. As with Southern and Northern blots, gel electrophoresis is used to separate native or denatured proteins, which are then transferred to a polyvinylidene difluoride (PVDF) or nitrocellulose membrane. Nylon membranes are not used in Western blotting because they are cationic and strongly bind acidic proteins.

SDS-polyacrylamide gel

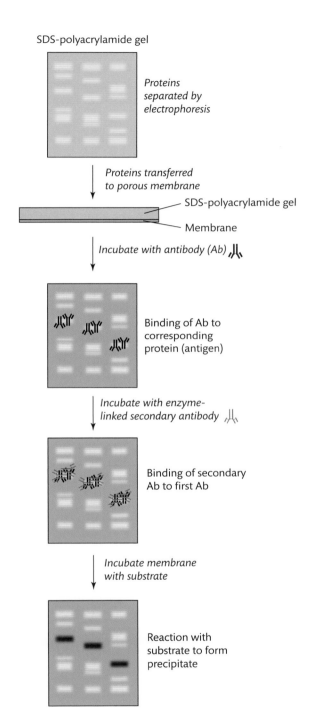

Proteins separated by electrophoresis

Proteins transferred to porous membrane

SDS-polyacrylamide gel

Membrane

Incubate with antibody (Ab)

Binding of Ab to corresponding protein (antigen)

Incubate with enzyme-linked secondary antibody

Binding of secondary Ab to first Ab

Incubate membrane with substrate

Reaction with substrate to form precipitate

FIGURE 16.23
Outline of Western blotting.
See text for details.

Thus, under many conditions they bind so much protein that detection procedures colour the membrane strongly and it is difficult to identify the proteins of interest. Polyvinylidene difluoride also binds proteins strongly, but gives light background staining after analysis. However, nitrocellulose membranes are the commonest choice for general use. They are the cheapest alternative and bind adequate amounts of protein, largely due to hydrophobic interactions. However, they are relatively fragile and do not stand up well to repeated probings and do not bind proteins of M_r less than 14,000 strongly. Following protein transfer, the uniformity and overall effectiveness of transfer of protein from the gel to the membrane can be checked by staining the membrane with a general protein stain such as Coomassie blue or Ponceau S dyes (Table 14.1).

Cross reference

Chapter 15 'Immunological techniques'.

In Western blotting, the membrane is probed with antibodies, which we describe in Chapter 15. To briefly recap, antibodies are glycoproteins synthesized when a foreign, that is non-self, agent such as an infectious microorganism, stimulates the immune system. Each type of antibody is produced in response to the presence of a macromolecular antigen, for example a protein or lipopolysaccharide. Individual antibodies recognize specific sites on antigens called epitopes. Thus, antibodies can be used in Western blotting because they can specifically bind to the target protein(s) and so locate their positions on the membrane. However, given that the membrane also binds proteins, but in a non-specific fashion, steps must be taken to prevent any interactions between it and the antibody, otherwise detection of the target protein would be hindered. This is achieved simply by adding an inexpensive protein, for example bovine serum albumin (BSA) or non-fat dry milk in a dilute solution of detergent such as Tween 20, to the membrane. These proteins attach to the membrane and block nonspecific binding sites. Hence, when the antibody is added, it can only bind to epitopes of the target protein. The excess BSA or dried milk is then removed by washing with a suitable buffer. The bound protein–antibody complex must now be detected on the membrane to pinpoint its position. Traditional detection uses a two-step process, in which the antibody used to detect the protein is called the *primary* antibody. The membrane is rinsed to remove unbound primary antibody, and a *secondary* antibody, which is species-specific to a portion of the primary antibody is added. This secondary antibody is usually linked to biotin or to a reporter enzyme such as ALP or HRP whose activities produce a coloured signal allowing the position of the bound protein to be identified (Figure 16.24). Commonly, a HRP-linked secondary antibody is used in conjunction with a chemiluminescent agent. The reaction product produces a luminescent signal, whose intensity is proportional to the amount of protein. The light may be

FIGURE 16.24

(a) Picture of a Western blot showing the presence of the protein, *p27* in five different samples. In each case, the product of the housekeeping gene, tubulin has also been analysed as a control. Since each sample contains the same quantity of tubulin, any experimental variation in the treatments of the samples can be eliminated.
(b) Densitometry to show the relative expression of *p27* in each of the samples.

detected on photographic film or by using a CDD camera that captures a digital image of the Western blot.

Images of Western blots can be analysed by densitometry, which is explained in Chapter 14. This allows us to compare the amounts of a protein present in different samples, if known quantities of a so-called house keeping protein are analysed at the same time (Figure 16.24). The M_r of the protein of interest can also be estimated if proteins of known sizes are also included in the analysis. In some applications, detection can be achieved by a one-step process, where the antibody probe, in addition to recognizing the target protein, also carries the detectable label (Box 16.4). When using primary antibodies or the one-step approach, incubation times vary from 30 minutes to overnight depending on the temperature. Temperatures can be varied between approximately 4–25°C. Higher temperatures are associated with enhanced specific binding of the antibody to the target protein, which increases the signal from the target on the membrane. Unfortunately, it can also produce more non-specific binding, which increases the background staining.

Cross reference
Chapter 14 'Electrophoresis'.

Western blotting is used in biochemical and clinical investigations because it is one of the most specific means of identifying the presence of individual proteins in complex biological samples, such as blood, serum, saliva, urine, cells, and tissues; it is thus the basis of many diagnostic assays. For example, the serodiagnosis of human immunodeficiency virus type 1 (HIV-1) infections relies on detecting antibodies to the virus in the sera of potential patients. Serum proteins from potential patients are isolated and blotted onto a membrane. Antihuman immunoglobulin G conjugated to an enzyme that produces a coloured product with an appropriate substrate is used to detect the presence of antiviral antibodies. As with all clinical tests, positive and negative control sera must be analysed simultaneously with that of the test serum. Western blotting also provides improved serological diagnoses for conditions such as Lyme disease, some forms of spongiform encephalopathy, congenital toxoplasmosis, syphilis, and in patients with rheumatic disease, and is the basis for many home diagnostic tests, for example for pregnancy testing kits.

BOX 16.4 Antibodies for Western blotting

The results of Western blotting depend on the quality of antibody probe used, in particular how specific it is for the target protein.

Both monoclonal and polyclonal antibodies (see Box 15.4) against a large number of target proteins are readily obtainable from commercial sources. If, however, the protein is a novel one, the antibody must be produced in-house, although commercial companies are available that will also do this for a fee! In either case, small amounts of the target protein must be purified to stimulate antibody production. A large number of species, such as rats, mice, rabbits, guinea pigs, chickens, goats, sheep, and donkeys, are available in which to raise antibodies. Generally, monoclonal antibodies take longer to develop than polyclonal ones and are therefore more expensive (Figures 15.2 and 15.3). Monoclonal antibodies are, however, usually more specific to their targets and bind more strongly to their epitope than polyclonal types. This means that fewer nonspecific bands are detected on the membrane and the background staining is of a lower intensity.

SELF-CHECK 16.7

List the principal uses of Southern, Northern, and Western blottings? State the major use of each technique.

16.8 Enzyme-linked immunosorbent assays and fluorescence *in situ* hybridization

Enzyme-linked immunosorbent assays (ELISAs) and fluorescence *in situ* hybridization (FISH) are two sensitive groups of techniques that rely upon the respective abilities of proteins (antibodies) and nucleic acids to specifically recognize and bind to complementary molecules, as we described in Section 16.1. Each of the techniques forms the basis of many clinical assays.

Enzyme-linked immunosorbent assays

Like Western blotting, enzyme-linked immunosorbent assays, or ELISAs, rely upon the ability of antibodies to recognize and bind to their complementary antigens. This makes them powerful reagents for detecting and quantifying analytes of clinical interest. The use of antibodies in this respect is called **immunoassay**. Immunoassays are used in all branches of the biomedical sciences. Thus, ELISAs use an antibody to which an enzyme has been attached to measure the amount of a specific antigen in a sample. The enzyme used is one that will convert a colourless substrate into, for example, a coloured soluble product that can be measured spectrophotometrically. Horseradish peroxidase and ALP are the most frequently used enzymes. The simplest ELISA format (Figure 16.25 (a) and (b)) is to allow a protein antigen to absorb onto the wells of a plastic microtitre plate. Aliquots of known concentrations of the antigen are also applied to other wells of the plate. An enzyme-labelled antibody that binds to the antigen is then added to the wells. Wells containing a large amount of antigen will bind more antibody and therefore more enzyme. The substrate for the enzyme is then added and, after a limited period, the reaction is stopped and the amount of product formed determined. The amount of product formed with the known concentrations of antigen can be used to form a standard curve from which the concentration of the unknown sample can be determined. Given each enzyme molecule forms many thousands of molecules of product, even relatively tiny amounts of antigen can be detected. Other labels that can be used in immunoassays include fluorescent labels and bio- and chemiluminescent labels.

Cross reference

See Chapter 15 'Immunological techniques'.

The development of ELISAs in 1971 by Engvall and Perlmann revolutionized medicine by replacing many awkward to perform and time-consuming diagnostic tests. Since then, the basic protocol has been subjected to many different adaptions that allow non-protein antigens to be measured or which have increased sensitivity. The specificity of antibodies allows biomedical scientists to measure the concentration of an analyte, such as a steroid hormone or autoantigens in patients with autoimmune diseases, in biological fluids such as plasma, which contain hundreds of other biomolecules, some of which are similar in structure to the analyte being measured. Enzyme-linked immunosorbent assays are widely used, for example, in diagnosing cases of malaria in Africa to home pregnancy testing kits in the developed world.

Cross reference

You can read about mutations in Chapter 20 'Inherited metabolic disorders and newborn screening' in the companion volume, *Clinical Biochemistry*.

FIGURE 16.25
(a) Schematic outline of an enzyme-linked immunosorbent assay (ELISA). See text for details.
(b) The outcome of a typical ELISA assay. See also Figure 15.13 and associated text.

Fluorescence *in situ* hybridization

In FISH, single-stranded DNA molecules chemically linked to a fluorochrome are used as probes to detect DNA samples that have sequences complementary to them; you can see this diagrammatically in Figure 16.26 (a). Fluorochromes are molecules or parts of molecules

Cross reference

Fluorescence is described more fully in Chapter 11 'Spectroscopy'.

that absorb light of one wavelength, often in the UV range and re-emit most of the energy, but at a longer wavelength. Thus, the fluorochrome appears bright against a dark background when viewed with a fluorescence microscope, giving a technique that can be sufficiently sensitive to detect single DNA molecules in microscopic sections of karyotypes, cells, or tissues.

Fluorescence *in situ* hybridization techniques are used for diagnosis. For example, the analysis of metaphase chromosomes is a major tool in cancer cytogenetics. Thin sections of the tumour obtained from biopsies are treated to separate the DNA strands, which are then hybridized *in situ* with fluorochrome labelled probes (Box 16.2) for known mutated forms of cancer-associated genes. When the slides are viewed using a fluorescence microscope, any fluorescent areas in chromosomes from cells in the metaphase indicate the presence of a mutation. For example, patients with DiGeorge anomaly (DGA) exhibit a range of signs including cardiac malformations, hypoparathyroidism, and hypocalcaemia. The majority of patients have a deletion or partial monosomy of chromosome 22. These genetic lesions are detectable *in utero* if cells obtained by amniocentesis are analysed using FISH with an appropriate probe (Figure 16.26 (b)).

The Quadruple Test for Down's syndrome calculates the potential risk of a pregnant woman carrying a Down's syndrome foetus. When the test indicates a higher risk of Down's syndrome, FISH, or polymerase chain reaction (PCR, Section 16.9) tests are available from many laboratories in the UK to help clarify the situation.

(a)

Chromosome analysed by FISH, showing presence of bound fluorescent probe

Excitation light

Fluorescent light

DNA probe with fluorescent label

Chromosomal DNA

(b)

FIGURE 16.26

(a) Schematic to outline the binding of a fluorescently labelled DNA probe to target specific chromosomal DNA sequences. (b) Picture of a fluorescence *in situ* hybridization (FISH) analysis confirming DiGeorge anomaly (DGA). The TUPLE probe fluorescing red shows the 22q11 region that is deleted in DGA. The ARSA green signal is diagnostic for a region at the end of the q arms of chromosome 22, confirming the deletion to this chromosome.
Courtesy of Paul Virgo, Department of Immunology, Southmead Hospital, Bristol, UK.

16.9 **DNA cloning**

Gene or **DNA cloning** is the preparation of many identical copies of a DNA molecule, that is the amplification of a particular piece of DNA. Cloning can be accomplished by preparing recombinant DNA in a cloning vector or by using the polymerase chain reaction.

Recombinant DNA

DNA recombination is the combining of a piece of the DNA one is interested in into another type of DNA to form a recombinant DNA molecule. A recipient cell is then induced to assimilate and replicate the recombinant DNA. Many copies of the recombinant DNA may be produced by the progeny of the recipient cell; thus, the DNA of interest is cloned. If the inserted fragment is a functional gene that encodes a specific protein then the protein could be produced in the host cell when the gene is transcribed and translated. This process is used for the large-scale production of proteins valuable in the clinical sciences, for example insulin, somatostatin, and other hormones, but which are difficult or expensive to prepare by other methods. You can see an overview of the steps involved in cloning DNA by recombination in Figure 16.27.

The carriers of the DNA to be cloned, often called vehicles or **vectors**, are usually plasmids (Section 16.1) or bacteriophage DNA (Boxes 16.5 and 16.6). The recombinant DNA molecule is formed by hydrolysing the vector with a RE to break both its strand at a specific point (as you can see in Figure 16.27). The DNA of interest is prepared as a fragment using the same RE as used to cleave the vector. The fragment and the vector are allowed to anneal at their complementary overlapping ends. DNA ligase is then used to catalyse the covalent joining of the ends of the restriction fragment and vector DNA together to form the recombinant DNA. Ligation is most efficient when the ends overlap, but the DNA ligase from bacteriophage T4 can join fragments and vectors that have been produced using REs that produce blunt ended DNA, although this is a less efficient means of ligating the DNA together.

A major breakthrough in recombinant DNA technology was the isolation of mutant strains of *Escherichia coli* that are not able to digest foreign DNA. These strains are suitable as host organisms for replicating recombinant DNA. The recombinant DNA must now be taken up by such a host cell in a process called **transformation**. For example, *E. coli* cells can be made transiently permeable to DNA and then mixed with recombinant vector DNA. Under certain conditions, a small fraction, about one in about 10,000, of the cells will take up the recombinant plasmid. The transformed cells can be distinguished from the others by using, for instance, a plasmid containing a gene that confers resistance to the antibiotic ampicillin. Thus, if the cells are grown in an ampicillin-containing medium, then only the transformed cells will be viable and can be selected. Replication of the plasmid during the growth of the transformed cells ensures that the inserted DNA is cloned.

It is sometimes necessary to isolate plasmid DNA from cells to allow it to be characterized and manipulated. We introduced general methods for purifying DNA earlier (Section 16.3). However, several methods for isolating plasmid DNA have been developed to separate plasmid from chromosomal DNA once the cells have been disrupted. Plasmid DNA differs from chromosomal DNA in having a tight circular conformation, whereas the chromosomal DNA usually consists of linear fragments formed when the chromosome is sheared. The plasmid DNA is still

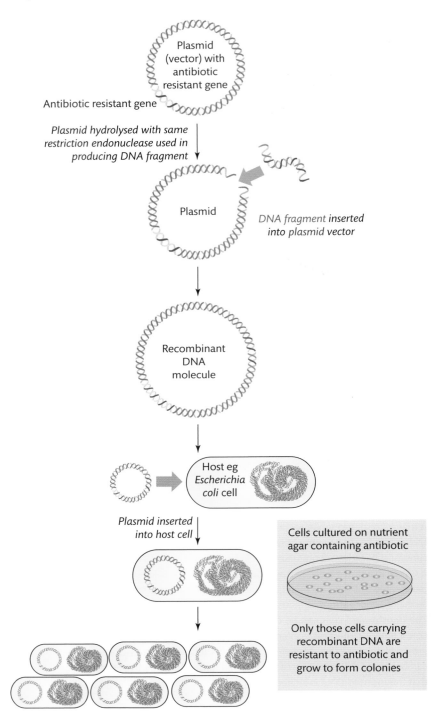

FIGURE 16.27

Overview of the steps involved in a typical recombinant DNA cloning experiment. A plasmid is used in this case as a vector. The plasmid and the piece of DNA of interest are spliced to form a recombinant DNA molecule. Bacterial cells are then transformed when they take up the recombinant DNA. Only transformed cells can grow in an antibioticrich medium because of the antibiotic resistance gene on the plasmid. Growth of the bacteria amplifies the DNA of interest.

much smaller than these fragments. These structural differences can be exploited to separate the two types of DNA. For example, plasmid DNA can be specifically adsorbed on to nitrocellulose microfilters and hydroxyapatite columns or the chromosomal DNA can be selectively precipitated by using buffers of extreme pH, high temperature or other denaturing conditions.

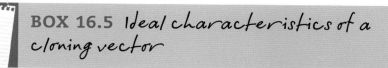

BOX 16.5 *Ideal characteristics of a cloning vector*

The ideal cloning vector should have a number of properties, including the following:

■ The vector should possess only a single site for a specific RE.

■ The vector should contain identifiable markers so that it is possible to screen transformed cells for uptake. At least two selective markers are desirable; one to confirm the insertion of foreign DNA into the plasmid, the second to allow the presence of the plasmid in the cell to be confirmed.

■ The vector should generally be small relative to the fragments of chromosomal DNA, making its isolation and purification easier and reducing the number of potential sites for attack by REs.

■ The vector should be capable of being replicated rapidly by transformed cells to ensure an efficient cloning of the recombinant DNA.

Finally, centrifugation can be applied to separate them because their size difference means they differentially sediment.

Plasmid vectors based on the naturally occurring F plasmid of *E. coli* are used to clone DNA fragments of 300,000 to 1 million nucleotide pairs. Unlike smaller bacterial plasmids, the F plasmid and its derivative, the bacterial artificial chromosome (BAC, Box 16.6) is present in only one or two copies per *E. coli* cell. The fact that BACs are kept in such low numbers in bacterial cells may contribute to their ability to maintain large cloned DNA sequences stably. Because of their stability, ability to accept large DNA inserts, and ease of handling, BACs are now the preferred vector for building DNA libraries of complex organisms, including those representing the human and mouse genomes.

cDNA cloning

An alternative cloning strategy is to begin the process by selecting only those DNA sequences that are transcribed to give an mRNA molecule. These sequences are presumed to correspond to protein-encoding genes. This is done by extracting the mRNA from cells as outlined earlier (Section 16.3). A complementary DNA or cDNA is then made of each mRNA using reverse transcriptase, an enzyme prepared from retroviruses, which synthesizes a DNA molecule complementary to an RNA template. The single-stranded cDNA is converted into a double stranded molecule using DNA polymerase. Each double stranded DNA molecule is then inserted into a vector and cloned in the normal manner. Each clone prepared in this way is called a cDNA clone and the entire collection of clones derived from one mRNA preparation constitutes a **cDNA library**.

The major advantage of using cDNA molecules is that uninterrupted coding sequences of a gene are cloned. Eukaryotic genes usually consist of short coding sequences of DNA called exons that are separated by normally much longer non-coding sequences called introns. During the production of a specific mRNA, those portions of it that were formed by transcribing the introns are removed and the coding sequences (from the transcribed exons) are spliced together to give a continuous sequence. However, the mRNA transcripts of many genes are subject to alternative splicing, so one gene can encode a number of alternatively spliced and therefore different mRNAs. Thus, a cDNA library often contains many of the alternatively spliced mRNAs produced from a given cell line or tissue.

BOX 16.6 Vectors

Plasmids are small, only 3–20% the size of the bacterial chromosome, meaning they can carry inserts only 1–20 kbp in size. Their relatively small size also makes them comparatively easy to isolate from the bulk of the DNA present in the cell (Sections 16.3 and 16.9). Also, they have high multiple copy numbers, up to 500 plasmids per cell in some cases, and so the inserted DNA of the recombinant plasmid may be greatly amplified. The presence of selectable markers, such as the genes for antibiotic resistance or the *lac Z* gene (see below), means that bacterial cells transformed by the recombinant DNA can be easily detected because those cells that do not take up the vector will be unable to grow in a medium containing the antibiotic. Thus, a number of cloning vectors have been derived from plasmids.

The pUC vectors were introduced in the early 1980s and are based on fragments of DNA derived from naturally occurring *Escherichia coli* plasmids. Figure 16.28 shows the structure of the pUC18 plasmid. The pUC19 vector is identical except its genes are organized in the opposite direction. The pUC vectors have an origin of replication allowing them to be replicated in high copy numbers. They contain the gene for β-lactamase, which confers resistance to the antibiotic, ampicillin (Amp^R) and allows transformed cells, which have acquired the plasmid, to be identified. Additionally, pUC18/19 also contains a portion of the *E. coli lac* operon, which allows cells containing the plasmid to synthesize the amino terminal portion of β-galactosidase. pUC vectors may be used with appropriate host cells that have been modified to express the gene for only the carboxyl terminal portion of the enzyme. In the presence of a suitable inducer, for example, isopropylthio-β-galactoside (IPTG), those cells containing a pUC plasmid will produce the two fragments of enzyme and possess β-galactosidase activity. Such cells are able to hydrolyse the artificial substrate 5-bromo-4-chloro-3-indolyl-β-galactoside (X-gal) to form a blue-coloured product. Thus, colonies of the cells will appear blue when grown on nutrient agar containing IPTG and X-gal. However, pUC18 also contains a multiple cloning site or polylinker region near the *lac* gene, which contains the recognition sites for a number of different REs (Figure 16.28). The formation of a recombinant plasmid means the inserted fragment of DNA in this region prevents the formation of an active β-galactosidase. Hence, transformed cells possessing the recombinant DNA will be unable to hydrolyse X-gal and their colonies will appear white and so are easily identified.

The F plasmid or factor contains 'fertility genes', which enables a bacterium possessing it (F⁺ cells) to produce sex pili during bacterial conjugation and act as genetic donors to those cells lacking the factor (F⁻ cells). Thus, the plasmid and its genetic information is transferred between bacterial strains. Unlike smaller plasmids, only one or two copies of the F plasmid are present in each bacterial cell. Plasmid vectors based on the F plasmid of *Escherichia coli*, called bacterial artificial chromosomes (BACs), can be used to clone pieces of DNA as large as 300 to 10^3 kbp. The low numbers of BACs in bacterial cells are thought to contribute to the stability of the recombined BAC, despite the large sizes of the DNA inserts. This stability and their relative ease of handling, means BACs are useful vectors for building DNA libraries of, for example, the human genome.

The ability of bacteriophages or phages to replicate inside bacterial cells has allowed a number of them to be developed as cloning vectors. Examples of phages used include the Enterobacteria phage λ or lambda phage, the filamentous bacteriophage M13,

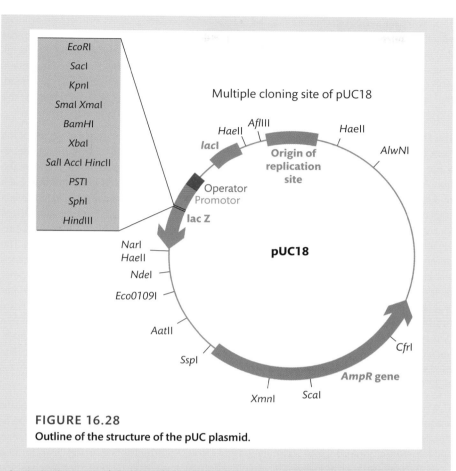

| EcoRI |
| Sacl |
| Kpnl |
| Smal Xmal |
| BamHI |
| Xbal |
| Sall Accl Hincll |
| PSTI |
| Sphl |
| HindIII |

Multiple cloning site of pUC18

HaeII   AflIII   HaeII

lacI

Origin of
replication
site

AlwNI

Operator
Promotor

lac Z

Narl
HaeII
NdeI
Eco0109I

AatII

SspI

Xmnl   Scal

pUC18

AmpR gene

CfrI

FIGURE 16.28
Outline of the structure of the pUC plasmid.

bacteriophage Mu (phage Mu) and phage P1. Phage P1 has been modified to form phage artificial chromosomes (PACs) that, like BACs, can be used to clone relatively large pieces of DNA. Cosmids are a type of hybrid plasmid that contain part of the DNA from the lambda phage to which suitable origin of replication from other phages or plasmids and a gene for selection, such as antibiotic resistance, have been added. Thus, they can be used to clone DNA inserts as large as 30–50 kbp of DNA and so can be used to build genomic libraries.

Polymerase chain reaction

Mullis devised the polymerase chain reaction (PCR) in 1983 and was awarded the Nobel Prize in Chemistry in 1993 for this work. The PCR is an elegantly simple *in vitro* method to increase or amplify the number of relatively short strands of specific DNA fragments in a sample. The fragments are normally 0.5 to 5 kbp in length, although longer ones up to 40 kbp can also be amplified. The DNA fragments may be complete genes, although they more usually comprise only a portion of one. Thus, the technique complements the recombinant DNA cloning strategy.

The PCR is used to replicate a DNA sample (template DNA) using a DNA polymerase. This enzyme copies the template to produce new complementary DNA strands. DNA polymerases cannot begin to synthesize a strand *de novo*, but only extend an existing piece of DNA.

Thus, two primer DNA strands are needed to initiate the copying process. Primers are artificial oligodeoxynucleotides less than 50 nucleotides long that are complementary to sequences that flank the region of the template DNA of interest. Hence, the primer determines the beginning of the region to be amplified. Primers are usually made to order by commercial suppliers who must be supplied with the required sequence. Some DNA polymerases can proofread, that is correct any mistakes in the newly formed strand, which ensures that the fidelity of the sequence is preserved. Other polymerases do not, however, have this property.

The PCR consists of a series of cycles, each of which doubles the number of molecules of the DNA template. Prior to the first cycle, the DNA is often heated to 90–96 °C for 5–10 minutes in a 'hot start' designed to break the hydrogen bonds connecting the two strands and ensure the template and primer DNA molecules are fully separated. This separation is called **melting**. Each of the subsequent cycles consists of three identical steps, as shown in Figure 16.29. In the first step, double-stranded DNA is heated for 0.5–3 minutes at 94–96°C, which is normally sufficient for melting. In the second step, the temperature is reduced to 50–65°C for 0.5–6 minutes during which the primers bind or anneal to their complementary sequence of the templates. The primer must be present in amounts that are in excess of the target DNA, otherwise its strands will simply rejoin. The design of the length of the primer requires careful consideration. Primer melting temperature increases with the length of the primer. The optimum length for a primer is generally 20–40 nucleotides, which will have melting temperatures of 56–75 °C. If primers are too short they will anneal at random positions on the relatively long template and result in a non-specific amplification. However, if the primer is excessively long, the corresponding melting temperature would be above 80°C and this could reduce the activity of the polymerase. The third step is the extension of the primer by the polymerase. This usually requires 0.75–2 minutes at 72°C. The extended portions of DNA are complementary to the template strand.

The high temperatures used in PCR mean that DNA polymerases from thermophilic organisms are preferred. The Taq polymerase from *Thermus aquaticus*, often abbreviated to Taq pol, or simply Taq, is widely used although it has the disadvantage of lacking proofreading capabilities and therefore can introduce errors (mutations) of 1 in 400–500 nucleotides in the newly formed DNA. Polymerases such as *Pwo* or *Pfu*, obtained from Archaea, have proofreading mechanisms that significantly reduce mutations and are used in 'long-range' PCR of up to about 30 kbp.

The first cycle results in two double stranded DNA molecules, which are usually overextensions of the target sequences. Each is composed of one of the original strands plus the newly formed complementary strand and associated primer. Thus, the amount of DNA present has been doubled. Cycles of PCR are usually repeated 20–30 times, with each cycle doubling the amount of DNA present (Figure 16.30). Thus, the yield of DNA increases exponentially with each cycle: after 30 cycles the original amount is amplified over 2^{30} or 10^9-fold. A PCR experiment normally terminates with a 10 minute incubation at 72°C to ensure that all of the new DNA molecules are fully extended by the polymerase.

Cloning by PCR is extremely sensitive to contamination and extremely stringent conditions are necessary to prevent unwanted template DNA being coamplified with the sample. The products of PCR are identified by determining their base sequences (Section 16.6) and/or their size using agarose or polyacrylamide gel electrophoresis against standards of known size (Section 16.5).

The great advantages of PCR are the increase in sensitivity it provides by amplifying the amount of DNA present in a sample and the ease with which its step can be automated in a **thermocycler**. These are instruments that can be programmed to heat and cool the reaction tubes to the appropriate temperatures, for the desired times and the required numbers of cycles. Samples of DNA isolated from even a single cell or that present in samples many years old can be amplified and then analysed.

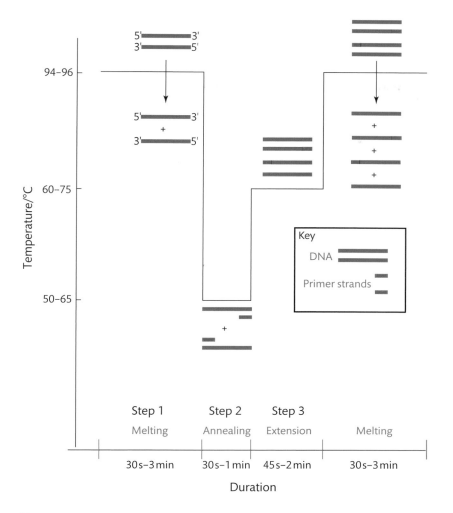

FIGURE 16.29
Three steps of the polymerase chain reaction. See text for details.

In addition to its application in many areas of molecular biological research, PCR has four major clinical uses:

1. The identification of infectious disease organisms for diagnostic purposes
2. The detection of variations and mutations in hereditary diseases
3. Detecting acquired mutations that lead to cancers
4. Tissue typing.

1 2 3 4 5 6 7 8

FIGURE 16.30
Agarose gel electrophoresis separation of a polymerase chain reaction amplification. Lane I shows the positions of fragments of DNA of known sizes. Lane 2 shows a 390 bp fragment of DNA, which forms the positive control. Lanes 3 to 7 show the results of amplifying this DNA by 10, 20, 25, 30, and 35 cycles of polymerase chain reaction. Lane 8 shows the results of the negative control, where the DNA was replaced by distilled water in the amplification.

Polymerase chain reaction is especially useful in diagnosing diseases caused by organisms that are difficult or impossible to culture. Its ability to amplify small amounts of DNA means that PCR-based tests can identify sources of infection more accurately, reliably, rapidly, and cheaply than previous methods. For example, PCR is the basis of sensitive and specific tests for *Helicobacter pylori*, the main causative agent of stomach ulcers. Three different sexually transmitted disease agents, Herpes, papilloma viruses, and Chlamydia, can be detected on a single swab using PCR tests. Indeed, even specific strains of papilloma viruses that predispose individuals to cervical cancer can be identified. Diagnostic tests are also available for the viruses involved in AIDS, viral hepatitis, and viral meningitis. In 2002, 47% of meningococcal infections in England and Wales were diagnosed using PCR tests for meningococcal DNA in clinical samples. Bacterial infections in middle ear fluid from children suffering otitis media are detectable by PCR, indicating an active infection, even when standard culture methods for the bacterium fail. The Lyme disease bacterium, *Borrelia burgdorferi*, is often difficult to diagnose accurately using its general symptoms but PCR can amplify, and therefore identify, its DNA in samples of body fluids. Cloning by PCR has largely replaced Southern blotting for diagnosing genetic diseases.

Reverse transcriptase PCR

Reverse transcriptase-PCR (RT-PCR) is a technique that allows an *RNA* molecule to be amplified but as complementary *DNA* copies, whose preparation was described above. The first step is to hybridize an oligodeoxynucleotide primer to the target RNA molecule. This primer may be specifically designed to hybridize to the mRNA from a specific target gene. However, oligo(dT) can be used as a universal primer for RT-PCR of mRNA molecules because it binds to the endogenous poly(A) tail they all possess. An RNA-dependent DNA polymerase or reverse transcriptase, obtainable from retroviruses, forms a cDNA to the target RNA molecule by extending the primer. The cDNA can be amplified by PCR, generally using gene-specific primers to generate copies of it. The amplification of contaminating genomic DNA can be minimized by treating the RNA preparation with RNase-free DNase, which will degrade it.

Reverse transcriptase-PCR is a sensitive and versatile procedure. It is used to:

- Clone the 5′ and 3′ termini of mRNA molecules
- Generate large cDNA libraries, that is copies of the mRNA molecules present in cell or tissue extracts
- Identify mutations (polymorphisms) by analysing mRNA molecules transcribed from the defective gene.

Real-time polymerase chain reaction

Cross reference
The automated use of PCR is discussed in Chapter 17 'Laboratory automation'.

Real-time polymerase chain reaction is also called real time PCR, quantitative real-time PCR (qPCR), and kinetic PCR. Real-time PCR is a technique based on the PCR that not only amplifies the target DNA, but also simultaneously quantifies the amount present. The amount can be expressed as the absolute number of copies or as the relative amount of target DNA compared to a so-called normalizing gene. Real-time PCR follows the general principles of PCR outlined above. However, the amount of amplified DNA is determined as it accumulates in the reaction in 'real time' following each cycle by one of two common methods. One method uses a fluorescent dye that intercalates into double-stranded DNA. The other method uses modified deoxyoligonucleotide probes that fluoresce when hybridized with their complementary DNA. In both cases, the amount of fluorescence is proportional to the amount of DNA formed.

Real-time PCR is often used in combination with RT-PCR to determine how many molecules of a specific mRNA are present in a sample. This gives the relative expression of its gene at a particular time, or in a particular cell or tissue type.

16.10 **DNA microarrays**

DNA microarrays are glass microscope slides, nylon filters, or silicon chips to which large numbers of DNA fragments have been attached (Box 16.7). Each fragment is an oligodeoxynucleotide whose sequence is a probe for a specific gene. The densest DNA microarrays contain
probes for all of the approximately 21,000 human genes packaged into an area smaller than
a postage stamp! Thus, despite their small size, microarrays allow thousands of hybridization
reactions to be performed simultaneously. Hybridization reactions use mRNA molecules isolated from the cell and used directly or more stable cDNA copies (Section 16.9) of them. In
either case, their sequences are complementary to the probes of the microarray. Each sample
mRNA or cDNA is radiolabelled with ^{32}P (or ^{33}P) or tagged with a fluorescent dye. The labelled
samples are allowed to hybridize to the probes on the microarray. Note how the probes differ
from those we described for other techniques, such as FISH and blotting, in not carrying the
label, which in this case is attached to the sample molecules. The microarray is thoroughly
washed to remove any unbound (that is non-hybridized) mRNA or cDNA molecules. The targets can be identified by autoradiography or by its fluorescent label. Usually, the fluorescent
cDNA from the experimental samples is labelled, for example with a dye that fluresceces red
when excited, and is mixed with a reference sample of cDNA fragments labelled with a dye that
shows green fluorescence. Thus, if the amount of a specific mRNA expressed from a particular
gene is increased relative to that of the reference sample, the resulting spot on the microarray
is red. Conversely, if the expression of the gene is decreased in comparison, the spot is green.
However, If there is no change compared to the reference sample, the spot is yellow (Figure
16.31). These initial hybridization intensities require sophisticated computerized analyses to
normalize, summarize, and analyse the differential gene expression patterns. However, using
such an internal reference, allows the expression of many genes to be accurately assessed.

Since their introduction in 1995, DNA microarrays have proved useful in analysing the expression of numerous genes by allowing the concentrations of *all* the mRNA molecules produced
by a cell to be simultaneously monitored. Hence, we can now identify and study the patterns
of gene expression that underlie both normal and pathological cell physiology and investigate
the differential expression of genes as cells grow, divide, and differentiate. For example, microarrays are used to identify and classify tumours based on the patterns of gene expression they
exhibit. This should help identify new targets in tumours for chemotherapy and the development of novel anticancer drugs. Microarrays can also be used to quickly identify disease-causing
microorganisms by hybridizing the DNA isolated from infected tissues to a microarray of
genomic DNA sequences from potential pathogens. At the time of writing, microarrays require
rigorous operating procedures making them difficult to use in every day clinical applications.

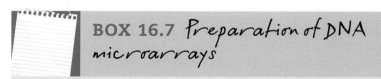

BOX 16.7 *Preparation of DNA microarrays*

DNA microarrays are prepared using longer fragments of DNA produced by PCR (Section
16.9) and then spotted onto the slides by a robot. Shorter artificially synthesized oligodeoxynucleotides are attached to the surface using techniques similar to those that are used
to etch circuits onto computer chips. In both cases, the sequence and position of every
probe on the chip is known. Thus, any nucleotide fragment that hybridizes to a probe can
be identified as the product of a specific gene by detecting its location on the microarray.

Cross reference

You can read about *antibody* microchips in Chapter 15 'Immunological techniques'.

FIGURE 16.31

Photograph showing a portion of a microarray following an experiment on gene expression. Genes whose expressions have been increased during the experiment (up-regulated) compared to the control are indicated by red spots. In contrast, those genes whose expression decreased (down-regulated) appear as green spots. Yellow spots show genes whose expression was unchanged during the experiment.

Furthermore, they are expensive to use; both in the cost of the microarrays themselves and in their need for highly trained personnel. It is generally expected that as their costs diminishes, microarray analyses will be powerful diagnostic tools in the clinical sciences. However, as the costs of DNA sequencing continues to fall, this more robust technique is likely to be the method of choice given that the two techniques have many similar biomedical applications.

SELF-CHECK 16.8

Huntington disease, myotonic dystrophy, and spinobulbar muscular dystrophy are all associated with increased numbers of trinucleotide repeats in their associated genes. Which method(s) could conveniently be used to confirm the status of potential carriers or patients?

 SUMMARY

- Molecular biology is generally concerned with the structures, functions, and interactions of the two major groups of macromolecules, the nucleic acids deoxyribonucleic acid (DNA), and ribonucleic acid (RNA), and proteins. Deoxyribonucleic acid and RNA carry biological information. Proteins perform many of the activities in cells and form some of their structural features

- Nucleic acid molecules or fragments of molecules are able to interact and recognize one another: adenine hydrogen bonds to its complementary bases of thymine in DNA and uracil in RNA, and guanine hydrogen bonds with its complement, cytosine. Protein molecules have discrete sites on their surfaces that are complementary to specific small ligands or to relatively small areas on the surfaces of other proteins and nucleic acid molecules. Thus, proteins also generally function by recognizing and specifically interacting with other biological molecules

- Nucleic acids can be purified from organisms using a variety of techniques. Restriction endonucleases can digest the isolated DNA into smaller-sized fragments suitable for analysis and for use in a number of techniques of clinical interest

- Molecular biology techniques used in biomedical science that rely on the complementary binding nucleic acids to provide a convenient way of recognizing and isolating specific base sequences within fragments of DNA or RNA molecules include the sequencing of isolated DNA, Southern and Northern blotting, fluorescence *in situ* hybridization (FISH), the cloning of DNA by recombination and PCR technologies, and DNA microarray analysis

■ One group of proteins used extensively in clinical tests are the immunoglobulins or antibodies. The binding of antibodies to their complementary antigens forms the basis of Western blotting and immunoassay techniques such as enzyme-linked immunosorbent assays (ELISAs), which are used in all branches of the biomedical sciences

■ The chapter discusses the underlying basic principles behind these techniques and provides general descriptions of their associated experimental protocols. In addition, we also highlight a number of essential precautions that must be observed if the reagents are to be handled and the procedures carried out safely

FURTHER READING

● Brown TA. *Gene cloning and DNA analysis: an introduction*, 5th edn. Blackwell, Oxford, 2006.

● Craig N, Cohen-Fix O, Green R, Greider C, Storz G, Wolberger C. *Molecular Biology: Principles of Genome Function*. Oxford University Press, 2010.

● Dreier J, Störmer M, Kleesiek K. *Real-time polymerase chain reaction in transfusion medicine: applications for detection of bacterial contamination in blood products*. *Transfusion Medicine Reviews*, 2007; **21**: 237–254.

● Elliott W, Elliott D. *Biochemistry & Molecular Biology*, 4th edn. Oxford University Press, Oxford, 2009.

● Kricka LJ. *Stains, labels and detection strategies for nucleic acid assays. Annals of Clinical Biochemistry*, 2002; **39**: 114–129.

● Lo YMD. (Ed.) *Clinical Applications of PCR*, 2nd edn. Humana Press, NJ, 2006.

● Maraqa L, Donnellan CF, Peter MB, Speirs V. *Clinicians' guide to microarrays. Surgical Oncology*, 2006; **15**: 5–10.

● Mikhailovich V, Gryadunov D, Kolchinsky A, Makarov AA, Zasedatelev A. *DNA microarrays in the clinic: infectious diseases*. *BioEssays*, 2008; **30**: 673–682.

● Nicholl DST. *An Introduction to Genetic Engineering*, 3rd edn. Cambridge University Press, Cambridge, 2008.

● Sambrook J. *Molecular Cloning: A Laboratory Manual* (3 volume set), Cold Spring Harbor Laboratory Press, Cold Harbor, 2001.

● Speers DJ. *Clinical applications of molecular biology for infectious diseases. Clinical Biochemist Reviews*, 2006; **27**: 39–51.

● Volpi EV, Bridger JM. *FISH glossary: an overview of the fluorescence in situ hybridization technique. Biotechniques*, 2008; **45**: 385–386.

● Williams SA, Slatko BE, McCarrey JR. *Laboratory Investigations in Molecular Biology*. Jones and Bartlett, Sudbury, 2007.

● Wiltgen M, Tilz GP. *DNA microarray analysis: principles and clinical impact*. *Hematology*, 2007; **12**: 271–287.

QUESTIONS

1. Which of the following statements is/are **UNTRUE**?

   (a)   RNA contains ribose residues

   (b)   RNA contains deoxyribose resides

   (c)   RNA contains the bases adenine, guanine, uracil, and cytosine

   (d)   RNA molecules do not possess any double stranded structure

   (e)   RNA contains the bases adenine, guanine, uracil, and thymine

2. Which of the following statements is/are **UNTRUE**?

   (a)   Proteins are largely concerned with the storage and transmission of genetic information

   (b)   Proteins can form structural elements of cells and tissues

   (c)   Proteins form when amino acid residues are joined by phosphodiester bonds

   (d)   Proteins can act as catalysts

   (e)   A protein composed of 336 amino acid residues would contain 337 peptide bonds

3. How does SDS help disrupt the plasma membrane?

4. How is contaminating RNA removed from a DNA preparation?

5. What is the major use of ethanol in the preparation of samples of DNA?

6. Pancreatic ribonuclease is an endonuclease that shows b-type specificity at nucleotide residues on the 5′ side of pyrimidine nucleotides. List the possible terminal nucleotides of its products.

7. Which of the following base sequences would not be potential recognition sites for restriction endonucleases? Explain your choice.

   (a)   5′... GAATTC... 3′
         3′... CTTAAG... 5′

   (b)   5′... CATATG... 3′
         3′... GTATAC... 5′

   (c)   5′... GACGGAT... 3′
         3′... CTGCCTA... 5′

   (d)   5′... CAATTG... 3′
         3′... GTTAAC... 5′

   (e)   5′... CATTAG... 3′
         3′... GTAATC... 5′

8. Why is polyacrylamide gel electrophoresis unsuitable for analysing plasmids?

9. If the proteins you wished to analyse by Western blotting are (a) acidic (anionic) and of relatively low M_r and (b) relatively hydrophobic and of M_r 65,000, respectively, what type of membrane would give optimal binding?

10. Outline a test based on Western blotting that would help in the diagnosis of infection with the AIDS virus. Assume that a sample of serum is available from the patient.

Answers to self-check and end of chapter questions are available in the Online Resource Centre accompanying this book.

 Go to www.oxfordtextbooks.co.uk/orc/glencross

17

Laboratory automation

Tim James

Learning objectives

After studying this chapter, you should be able to:

- Identify the major benefits of automation
- List the types of specimens that are processed in a high volume automated laboratory
- Describe the common automation units including those associated with the pre- and post-analytical processes
- Recognize the role of automation in the efficiency and productivity of all biomedical disciplines

Introduction

Laboratory automation has been defined by the Association for Laboratory Automation, the leading organization for laboratory automation, as 'a multi-disciplinary strategy to research, develop, optimize and capitalize on technologies in the laboratory that enable new and improved processes and products'. All clinical laboratories utilize automation albeit to a variable extent, dependent on the discipline and requirements of the service. The application of this automation has produced improvements in efficiency and effectiveness of testing and improved consistency of analysis.

The investigations and analyses undertaken by hospital laboratories can be categorized in a number of ways. Traditionally this has been based on the association of any given laboratory test with a defined biomedical science discipline, and conventionally these are haematology, clinical biochemistry, immunology, microbiology, histology, and cytology. Automation is evident in each of these disciplines and you can find discipline-specific descriptions of automation throughout the accompanying volumes in this book series. However, the availability of analytical systems with increased sophistication in terms of both robotics and underlying information technology has yielded a model of laboratory operation that crosses the traditional discipline boundaries.

Specimen reception

↓

Sample receipt and registration

↓

Sample preparation

↓

Sample presentation to analysers

↓

Sample analysis

↓

Sample archiving

↓

Sample storage

↓

Sample disposal

FIGURE 17.1

The basic steps undertaken for most samples in clinical laboratories utilize a common process path. Most of these steps can be automated.

Figure 17.1 outlines the basic pathway that a sample may undergo within most clinical laboratories. The sample will arrive in the laboratory and be registered, usually by booking the specimen into the laboratory computer system. The sample is then prepared for analysis, presented to analysers, and undergoes analysis. Following analysis the sample will be archived, stored and will eventually, after a defined period of retention, be disposed in an appropriate and safe manner.

Automation of the central analytical step usually using instrumentation of varying complexity is apparent in all hospital laboratories. However, the preceding steps and those following analysis can also be automated. Whilst the timescale and complexity of steps may vary between each of the individual disciplines, the underlying process is the same. Consequently, there has been a convergence of automation and technology across the conventional disciplines resulting in the development of what is termed the **core automated** or the **blood sciences** laboratory. This chapter will review automation in the context of general automation features and their associated benefits.

17.1 **Benefits of automation**

Cross reference

Audit is discussed in Chapter 19 'Quality assurance and management'.

The most apparent benefits of full automation compared with manual or semi-automated analysis are the improved speed and efficiency in processing samples to produce test results for clinical management. It has been estimated that growth in workload of clinical laboratories is between 5 and 10% per annum. The management of this expanding workload has only been possible through greater utilization of automation of increasing sophistication and capacity. However, the benefits are much broader than this, as detailed in Table 17.1. The ability of automation to process samples using a common procedure has resulted in simpler, more consistent sample handling and this has been associated with a reduction in processing errors. A consistent finding in laboratory audits is that the majority of errors, in some cases it has been found that greater than 80%, occur in the pre-analytical phase. This is the step most associated with manual sample handling. Equally, manual laboratory procedures may involve several steps, where the sample may need to be manipulated and analysed.

TABLE 17.1 General benefits of automation

Benefit	Impacts
Efficiency benefits	Improved sample throughput
	Reduced sample splitting
	Reduced sample retrieval requirements
	Reduced staff time required for manual sample handling
Improved health and safety	Reduced exposure of staff to clinical specimens and therefore clinical infection
	Reduced risk of repetitive strain injury
Error reduction	Reduced risk of samples being mis-aliquoted into the wrong secondary containers
	Reduced risk of samples being misplaced as the sample location can be monitored at all stages of the laboratory process
Improved patient care	Reduced volume of blood and number of tubes collected
	Reduced and more consistent turnaround times for test results resulting in improved patient management

Misidentification of samples and errors can occur at each of these steps even when there are significant quality measures in place. Evidence suggests that automated processes, whether they are pre-analytical, analytical, or post-analytical, are systematic and associated with lower error rates.

Automation has produced a number of improvements with respect to health and safety.

Manual handling of uncapped, open samples can expose staff to a range of hazardous organisms. Automating laboratory processes reduces the number of occasions an individual may need to handle specimens and thereby reduces the risk of exposure to these hazards. Equally, in a laboratory setting where several thousand specimens are manipulated, there is a risk of repetitive strain injury. Replacing the manual manipulation of samples by an equivalent automated process will remove this risk. For example, many automation systems can uncap and recap tubes, steps that if undertaken repeatedly by the same member of staff carry the risk of straining hand, wrist, and shoulders, and temporary or permanent musculoskeletal injury.

A consideration discussed widely in laboratories is the time taken for a test result to be available to the clinician after the specimen is taken from the patient. This is commonly referred to as the **turnaround time** (TAT). Most laboratories use some measure of TAT as a quality indicator of efficiency. The TAT can be measured in two ways, the *total* TAT and the *within laboratory* TAT. Both can be influenced by laboratory automation. The total TAT is the time taken from the collection of the sample to the time the report is available to the clinician to interpret the result. The within laboratory TAT is the time taken from the receipt of the specimen in the laboratory to the laboratory authorization of the result.

All TAT requirements vary with clinical setting and the particular investigation, for example, the TAT requirement for glucose analysis in an unconscious patient with diabetes will need to be rapid and is measured in minutes. In comparison, the TAT requirement for cholesterol testing undertaken to assess cardiovascular risk in a healthy, ambulatory patient can be longer and will be measured in hours. Reduced and more consistent TATs are apparent with increased

Cross reference
You can read more about health and safety in Chapter 4.

levels of automation and this may improve the timeliness of clinical decisions and patient treatment.

Cross reference

You can read about accuracy and precision in Chapters 5 'Statistics and handling data' and 19 'Quality assurance and management'.

SELF-CHECK 17.1

How might a reduced test turnaround time help patient care?

Sequential generations of automated analysers have produced more consistent and reproducible results, as evidenced by reduced imprecision. They are more robust, as measured by the proportion of time the analyser is in operation. The related changes and benefits have included:

- A reduction in the specimen volume required for individual assays
- A reduced usage of reagent
- Increased analyser capacity relative to physical size

The lower specimen volumes required by newer automated analysers compared with manual techniques and older analysers translates into the advantage that a greater number of tests may be undertaken from a single clinical sample. This is particularly important in children and neonates where blood collection may be difficult, but is also beneficial in other patient groups where blood collection is challenging, such as patients with learning difficulties or those with fragile veins, such as the very elderly.

The reduced reagent volume has a great number of associated advantages. Total reagent costs will be reduced if less is used and less chemical waste is generated thereby reducing the environmental impact. Reagent storage requirements, both on the analyser itself and within laboratory storage facilities are reduced if lower volumes are required. This further reduces the space requirements for both room temperature and refrigerator storage.

Modern automated instrumentation is often considered physically large, but the analytical capacity, as measured either as samples or tests per hour, relative to their size has increased significantly. Therefore, whilst most laboratories have been modified in design and layout to accommodate automation, the total laboratory space requirement has often remained the same. The components utilized within analysers have become increasingly sophisticated, and particularly with respect to liquid handling, the accuracy and precision of sample and reagent delivery is both highly precise and robust. The consequences of these improvements are that total assay imprecision has improved and methods are highly reproducible. Most instrument manufacturers have a programme of monitoring to identify patterns of analyser component failure. This informs component supplier requirements and subsequent instrument design. Increasingly remote diagnostics, where a manufacturer will monitor critical instrument parameters electronically using the internet, may even predict analyser deterioration before it actually impacts on test results. The best design features utilized within automation systems are maintained and copied by competitors, and therefore it is possible to see common features in a range of different suppliers' instrumentation. The overall analyser component failure rate has reduced and this has produced automated instrumentation that is operational for a greater proportion of time. This reduces demands on both laboratory operators and service engineers.

Automation allows flexible use of staff time as large automated systems require less human intervention for operation and a greater number and proportion of laboratory investigations can be undertaken by automated systems. This could be considered to be an opportunity to reduce overall operating costs, but the majority of laboratories will deploy any staff released by automation and redirect them to analytically more demanding laboratory areas or to improving laboratory quality. However, as automation hardware is expensive, its introduction into clinical laboratories requires careful consideration of the clinical necessity and staff saving against capital expenditure.

Key Points

Test turnaround times (TATs) are key indicators of laboratory performance. Turnaround time requirements vary with respect to investigation/test and clinical context. Automation can lead to significantly reduced and more consistent TATs.

17.2 Collection of suitable samples and delivery to the laboratory

The majority of the samples processed in a core automated laboratory will be blood specimens taken from a vein by phlebotomists. A smaller proportion, usually less than 10%, of the laboratory investigations are undertaken on other clinical specimens, for example urine. Clinicians often require a number of assays to be undertaken simultaneously and therefore, multiple assays may be required on a patient at the same time. In the UK the average test to request ratio of a typical hospital laboratory is between 6 and 8, however this will vary depending on the local arrangements and requirements as described in Box 17.1.

Key Points

A laboratory test may be considered an individual investigation or assay, for example, measuring blood glucose concentrations. A request is regarded as one or more tests asked for by the clinician at a single point in time, for example, a patient who is tired all the time may require a number of different tests to assess possible causes. A request is not the same as a profile, which is usually a number of clinically or metabolically related tests grouped together, for example, a liver function profile, which will include several tests associated with this organ's functions.

Cross reference

Blood sampling is described in Chapter 7 'Samples and sample collection'.

It would be convenient if all blood tests could be processed using a single specimen tube. However, many blood constituents require specific preservation conditions to stabilize the test component. Consequently, a number of different tube types are required depending on

BOX 17.1 How to calculate the test to request ratio

The average test to request ratio is simply the number of tests divided by the number of requests. A laboratory that undertakes 750,555 requests and 4,742,200 tests per annum therefore has a test to request ratio of 6.31.

TABLE 17.2 Commonly used collection tubes and associated colour coding. See also Table 7.2.

Tube type	Additive	Commonly used tube cap colours
Serum	Clot activator	Red, brown
Heparinized plasma	Lithium heparin	Green, orange
EDTA blood	Sodium EDTA*	Lavender, pink
Citrated plasma	Sodium citrate	Blue, green
Fluoride oxalate plasma	Sodium fluoride, potassium oxalate	Grey, yellow

*EDTA = ethylenediaminetetraacetic acid.

the combinations of tests required. Commonly available specimen tubes used in routine diagnostic laboratories on automated systems are presented in Table 17.2 together with commonly observed colour coding systems used to identify the different additives.

A serum collection tube contains no anticoagulant and the sample will form a clot: this sample can be centrifuged to yield **serum**. Serum requires time, usually about 30 minutes at room temperature, for the clot to form. To reduce the clotting time, many serum tubes are manufactured containing an additive to accelerate clotting so that the sample may be centrifuged more rapidly. Serum is suitable for a wide range of tests including the vast majority of biochemistry, immunology, and virology assays. The tube often contains a gel that has a density that is greater than that of serum, but lower than that of the clot. Therefore, when the sample tube is centrifuged, the contents are distributed into three components: a clot at the bottom of the tube; the gel in the middle, forming a barrier, and the serum on the top (see Figure 12.14). The majority of automated instrumentation analyses this serum directly from the original sample tube, a process called **'primary sampling'** as distinct from transferring the serum into a secondary container. Many automation systems can produce a secondary aliquot of the specimen, but it is more convenient in terms of efficiency not to do this. Most automation systems benefit significantly from primary sampling.

SELF-CHECK 17.2

Why is it necessary to use a number of different specimen tubes when undertaking laboratory investigations?

Sodium ethylenediaminetetraacetic acid (EDTA) is an anticoagulant by virtue of its ability to chelate Ca^{2+}, which is essential for the clotting process. Blood collected into a tube containing EDTA will not clot and a whole blood sample is produced that preserves the cellular constituents. This tube type is most frequently used for the analysis of standard haematology investigations such as the full blood count and erythrocyte sedimentation rate. It is also a suitable specimen tube for a range of other investigations including the analysis of glycated haemoglobin, a common test used for the monitoring of diabetes mellitus.

Sodium citrate, like sodium EDTA, binds Ca^{2+} and prevents coagulation. This tube may be centrifuged to yield citrated **plasma** for coagulation studies. A critical factor for this tube type is the ratio of blood to anticoagulant, which is essential for the accurate analysis of coagulation parameters. To ensure the correct volume of blood is added to each collection tube an indicator line is marked onto the side of the tube as a guide.

Lithium heparin is an effective anticoagulant as it inhibits the pro-thrombin to thrombin step of clotting. Centrifugation of this sample type produces heparinized plasma which is used for many investigations undertaken in clinical biochemistry. The advantage of this sample type over serum is the ability to centrifuge almost immediately after collection. An additional advantage of lithium heparin blood is that the plasma volume obtained is greater than the comparative yield from an equivalent volume of blood converted into serum. Whilst only a small effect, this can be beneficial for paediatric specimen and their naturally smaller sample sizes. It is common for lithium heparin tubes to contain a gel to enable primary sampling in a similar manner to that described for serum above.

The fluoride–oxalate blood collection tube contains two additives: sodium fluoride which binds Mg^{2+}, a co-factor of several enzymes associated with the glycolytic pathway thereby inhibiting glycolysis, and potassium oxalate which binds Ca^{2+} and prevents coagulation. This tube is usually associated with the analysis of glucose and lactate, assays that are adversely affected by continued glycolysis, which will occur in the other specimen tubes.

The development of laboratory automation systems has been helped considerably through standardization of specimen container shape and size and manufacturers produce the range of tubes described above in a standard format with respect to their physical dimensions and capping mechanism. The improved standardization of tube shape and size has improved sample presentation to analysers, the manipulations needed to remove and replace caps and their transfer by robotic arms. The two commonest diameters of specimen tube are 13 and 15 mm, and the commonly encountered tube heights are 75 and 110 mm. Many laboratories attempt to standardize the size of tube used by the automated system so, for example, all the laboratory users are supplied with tubes of one size, for example, 13 × 75 mm.

When the sample tubes have been collected from the patient they may be transported to the laboratory. The process of conveying the specimen from clinical to laboratory area within a hospital has been automated through the use of pneumatic air tubes. These tubes connect the clinical areas to the laboratory, and convey the specimen safely and promptly to enable faster total turnaround times.

Key Points

The application of automation to clinical laboratories was greatly enhanced with the introduction of separator gels that are incorporated into the blood collection tubes during manufacture. These gel-containing tubes allow processing of the plasma and serum samples through all steps of the analytical process, from collection through to disposal, in a single vessel, removing the requirement for specimen transfers.

17.3 Specimen reception considerations in the core automated laboratory

Most samples still arrive in the clinical laboratory with their paperwork in the form of a request card that contains information about the patient, the specimen, and the investigations required. This information is conventionally transcribed by manual data entry into the laboratory computer system. Improved efficiency and accuracy in this process is achieved if the data entry process is partly or completely electronic. Whatever method of sample registration is utilized, the interface between the specimen reception area and the automated instrumentation is a critical factor in overall laboratory efficiency. Automation alone does not produce the benefits detailed in Section 20.2; it is its utilization in the overall laboratory process that is critical.

Specimen reception is the usual location for an adhesive sample bar code to be applied to each sample tube thereby allowing its unique identification and routing through the laboratory automation. An alternative process exists in which the bar code is applied at the point of collection in the clinical or phlebotomy area. The accurate reading of the bar code label is required at many points in the automation process from the specimen reception workstations, as well as all of the interconnected devices and analysers. Consequently, the bar code quality in terms of clarity to bar code readers needs to be high enough to allow consistent reading and identification at each step. The format may be a simple number (numeric) or may allow inclusion of one or more letters into the specimen number (alpha-numeric).

The interface between specimen reception and automated systems can be described using three basic models presented in Figures 17.2 (a–c):

1. A traditional interface model
2. A semi-automated interface model
3. A totally automated interface model

Model 1, Figure 17.2a, represents the traditional laboratory in which the specimen reception process includes the step of sorting specimens into a range of racks appropriate and suitable for presentation to different analysers or laboratory areas. The disadvantage of this approach is that the specimen reception work station may be relatively cluttered and that multiple manual steps may be required to present samples to analysers. Model 2, Figure 17.2b, a model in which the specimen reception work station is simplified by using a single sample rack into which the bar coded specimens are placed. The single rack is then presented to a stand-alone automated sample management system that will then distribute the specimens into different analyser racks. The subsequent presentation to the analysers is manual. This model may be particularly beneficial if a laboratory utilizes analysers from more than one manufacturer that cannot be connected through a single tracking system. The system is also beneficial if laboratory space and/or layout is constrained and analysers are in different laboratory rooms and areas. Stand-alone automation can be utilized to improve the efficiency of multi-disciplinary laboratories but are most often found in clinical biochemistry laboratories. Figure 17.2c, represents a system in which both the sorting and conveyance of samples to analysers is achieved with a track system. Again, the work of the specimen reception is simplified by having only one rack in which to place all the specimens. In reality, specimen reception areas usually operate a mix of these models depending on the automation available (primarily whether the laboratory uses stand-alone sample management systems as opposed to tracks, as described in Section 17.4), the testing repertoire, and laboratory organization. Most automated laboratories review their specimen reception processes when new or different automation is introduced, with the primary aim of streamlining all the steps, reducing variation, and removing unnecessary steps.

Cross reference

Automation is also explored in Chapter 2 'Automation' in the companion text, *Clinical Biochemistry*.

SELF-CHECK 17.3

Specimen reception is the only place where bar code labels are applied to specimen tubes. True or false?

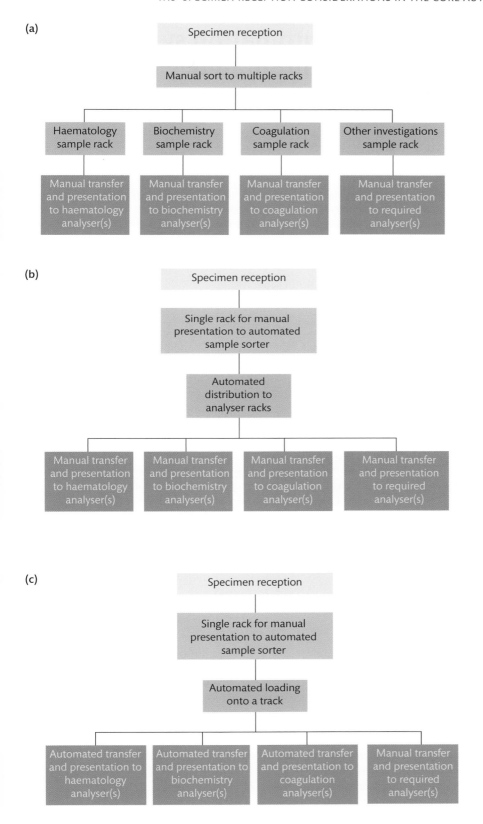

FIGURE 17.2

Specimen reception, automation interface models. (a) Samples are sorted and presented to individual analysers manually. This is complex with respect to the processes on specimen reception. (b) Samples are placed into a single rack that is presented to a standalone automated sample manager that distributes the samples to analyser racks. Racks are removed and presented to the analysers manually. (c) Samples are placed into a single rack that is loaded onto the input rack of an automated tracking system and automated sorting, distribution, and presentation to analysers.

17.4 **Tracked automation systems and the core automated laboratory**

A tracked automation system in a core automated laboratory can be considered as a group of modules or units each with a specific function. The number and type of units/modules within the automation system is dependent on the repertoire of assays and the number of samples/tests requiring analysis. There is also a general requirement for instrumentation back-up, particularly if continuous analysis is required. Since no two clinical laboratories are identical, a range of different combinations of instrument can be found connected to tracking systems to meet the requirements of the specific laboratory. You can see an outline design of an automated tracked system in Figure 17.3.

The modules/units of an automated tracking system can be categorized as:

- Modules that prepare samples for analysis
- Modules that analyse
- Modules that undertake post-analytical processes

Preparation modules include sample input systems, centrifugation units, and tube decappers. The introduction of samples into an automated system requires the specimens to be placed into a rack that can then be loaded onto a sample manager or input module connected to the track. The rack capacity varies depending on the design of the system and may be as low as 10 or as high as 100. Most systems also offer the facility to introduce one or a group of samples that can be given priority to ensure they have a faster turnaround time. The subsequent movement of the samples around the automation system may use pushing and pulling mechanisms to position small racks of samples or more commonly robotic arms (Figure 17.4) that can lift and convey individual specimens. The robotic arm usually moves the sample into an individual specimen holder that is positioned on the track and is referred to as a carrier or a puck. This carrier or puck moves around the track and will be directed to the modules and analysers as required for each particular sample. As the specimen progresses around the track the position of the tube may be monitored by re-reading the bar code at bar code reading stations at set intervals around the track. This enables a constant check to be made on the most

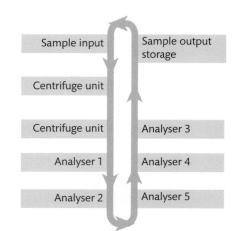

FIGURE 17.3

A schematic of an example tracking system and layout. The modules and units for any given system will vary depending on local laboratory requirements.

FIGURE 17.4
An example of a robotic arm placing a sample onto a tracking system.

appropriate route for the specimen tube to take. Alternatively, each of the pucks/carriers can be identified with a radio frequency tag that is monitored as it progresses around the track. When the sample is placed into each puck an association is established between them and this will be maintained until the sample is removed.

The majority of investigations on blood specimens require the plasma or serum component of blood to be obtained for which a centrifugation step is required. This can be achieved with a centrifuge unit integrated into the automation system. The centrifuge unit will weigh specimen tubes and load them into centrifuge buckets in an even manner to achieve a balanced load. The buckets are then placed into the centrifuge and spun for a defined period of time, usually set between 8 and 10 minutes, with application of a force of approximately 1000g.

Cross reference
You can read more about centrifugation in Chapter 12.

SELF-CHECK 17.4

How can the progress of a specimen tube on an automation system be followed?

The effect of the centrifugation is to bring the serum or plasma to the upper part of the collection tube where it can be sampled. Following centrifugation, the buckets are lifted from the centrifuge and the spun samples removed. Whilst some analysers have probes that can penetrate the rubber caps of the tubes to remove the plasma/serum/blood samples, most require the caps to be removed. Automated decappers provide this function and these can be found in three general formats:

1. As part of a stand alone pre-analytical system

2. As part of an integrated centrifuge-decapper module

3. As a distinct device fitted onto a track system

The general movement of a decapper is a hold, twist, and lift movement. Once removed, the caps are dropped into a waste receptacle. The waste receptacle may have varying levels of sophistication to alert the operator to the requirement to periodically empty the waste container. Once the cap has been removed the tube proceeds to the next stage of the automation process. In some instances, it may be necessary to produce aliquots of the primary sample as part of the standard automated process on the particular instrument, or because the specimen requires splitting for analysis across several analytical platforms. Sample aliquoting may also be required for referral to another laboratory.

SELF-CHECK 17.5

How do automated decappers reduce the risk of musculoskeletal injury to staff?

Each sample will be routed to one or more of the analysers connected to the track. In many systems the sample remains on the track and the analyser sample probe will extend out and over the track to a pre-set, defined position. The sampling probe will then be lowered until it detects the surface of the liquid sample, it will progress beneath the surface to a preset depth and liquid will be sampled and transferred into the analyser. This sample presentation is sometimes called 'point in space' sampling. Alternatively, a robotic device can remove the sample from the track and place it within a distinct and separate sampling area from which the analyser can sample.

The organization of the analytical modules reflects the balance of testing across each of the standard automated chemistry, immunoassay, haematology, coagulation, and urinalysis analysers. Table 17.3 illustrates some common testing areas undertaken on such systems.

TABLE 17.3 Typical automated analyser units and examples of the associated test repertoire

Traditional discipline	Analyser unit	Assays for:
Clinical biochemistry	General chemistry	Sodium, potassium, chloride, bicarbonate, urea, creatinine, calcium, phosphate, albumin, total bilirubin, alanine transaminase, alkaline phosphatase, C-reactive protein, glucose, total cholesterol, HDL cholesterol, triacylglycerol, uric acid, magnesium, total protein, creatine kinase, aspartate transaminase, lactate dehydrogenase, gamma-glutamyl transferase, amylase, lipase, iron, transferrin, conjugated bilirubin, lactate, paracetamol, salicylate, ethanol, zinc, ammonia, bile acids, angiotensin converting enzyme
	Immunoassay	Thyroid stimulating hormone, total and free thyroxine, total and free tri-iodothyronine, prostate specific antigen, alpha-feta protein, carcinoembryonic antigen, β–human chorionic gonadotrophin, Ca-125, Ca 19-9, cortisol, testosterone, oestradiol, progesterone, androstendione, follicle stimulating hormone, adrenocorticotrophin hormone, growth hormone, insulin-like growth factor-1, drugs of abuse assays
Haematology	Haematology analyser	Full blood count, red cell count, white cell count, haematocrit, haemoglobin
	ESR analyser	ESR
	Slide maker	Preparation of blood films
	Immunoassay	Vitamin B12, serum folate, red blood cell folate, erythropoetin
	Coagulation	Pro-thrombin time, activated pro-thrombin time, international normalized ratio, fibrinogen
Immunology	Immunoassay	Autoantibodies, allergens, complements C3 and C4, immunoglobulins
Virology	Immunoassay	Hepatitis A markers, hepatitis B markers, hepatitis C markers, HIV, CMV, rubella, syphillis

Each of these individual analysers will generally have a large test capacity ranging from a few hundred tests per hour through to several thousand. In Figure 17.3 the analysers may be the same or any combination of analyser type. Further detailed descriptions of the attributes of the individual analyser types can be found in the discipline-specific books in this series.

Post-analysis samples may be presented to a storage system. This may hold a variable number of specimens, usually in numbered racks within which the specimen is allocated to a specific, *x y* co-ordinate based position. The storage capacity may be as high as 15,000 samples and it is sometimes refrigerated to reduce specimen evaporation and deterioration. Samples may be re-accessed for further analysis if re-runs or additional tests are required and in these circumstances the samples may be retrieved manually or automatically reloaded onto the track, and re-routed to the designated analyser.

Practical considerations for large automation systems include the requirements for services. These include:

- Suitability of space, including the structural integrity of flooring as the systems may be very heavy, walking space around the instrumentation, and access for routine use and maintenance
- Water, which may be both a normal tap water quality requirement and/or de-ionized from a deionization system
- Power supply – large tracking systems will require a specific, dedicated power supply. It is standard practice to protect the power supply to each unit with an uninterruptible power supply (UPS) in case of power failures
- Drainage of waste – the waste drainage is best achieved with floor drainage rather than pumps
- Information interchange – most automated systems have a complex array of computers controlling different aspects of the system operation

Key Points

The majority of modules connected to a track will be conventional automated analysers that utilize a particular technology suitable for the analyses being undertaken. The other modules undertake sample processing functions, such as centrifugation, archiving, and storage.

17.5 Automation in wider laboratory settings

The biomedical science disciplines most often associated with automation are clinical biochemistry and laboratory haematology. However, all of the biomedical science disciplines, even those that have traditionally been associated with the manual skills and dexterity of biomedical scientists, such as histology, cytology, and microbiology, now have significant automation in use within clinical laboratories. The benefits of automation in any of the laboratory areas mirror those detailed in Section 17.1. Many laboratory processes require a series of independent manual steps and at each one there is potential for sample misidentification. Automation of the more traditionally manual areas of biomedical science is often associated with a reduction in the number of these manual steps at which this misidentification may occur.

This can therefore be seen as a significant improvement in patient safety, as it reduces laboratory errors. The introduction of automation in any of the laboratory areas also requires a re-assessment of laboratory operation. This often involves moving towards a laboratory work flow that is continuous, rather than as a batch mode and may also involve laboratory opening for longer periods of the day to maximize the benefits of automation.

In histology, several discrete stages of the preparation of histology slides have been automated. Automated tissue processors for preparation of tissues for sectioning, automated staining instruments, and automated cover slippers are all used routinely. The benefits of these individual automated devices are consistency and efficiency. Linking each of these individual discrete automated elements together into integrated single systems using robotics is developing. This type of automation is also an integral part of automated liquid based cytology.

While virology has utilized immunoassay technology and analysers (as described earlier) since the 1980s, the automation of bacteriology processes have only been evident since around 2005. The application of clinical samples to growth media within plates, the allocation of the plates to appropriate incubation conditions, and subsequent retrieval for visual assessment of colony growth has conventionally been undertaken manually. Modular instrumentation (Figure 17.5) can be used to automate these processes. Staff work at designated work stations where they are presented with the pre-labelled plate sets for innoculation and sample application. Once prepared the plates can then be conveyed automatically to incubators. After defined incubation periods the plates are automatically re-presented for review at which time point they will also be photographed. This enables comparison of plate growth at different time points and provides greater flexibility to when and where plates are read and interpreted.

Cross reference
You can read more about liquid based cytology in the companion volume *Cytopathology* in this book series.

FIGURE 17.5
The Kiestra system: an example of advanced automation in the bacteriology laboratory.

Automated systems in blood transfusion laboratories can be used for the majority of routine testing including blood grouping and sub-grouping, antibody screening as well as cross-matching. These systems can be highly reliable, sensitive, and specific.

In molecular diagnostics, the polymerase chain reaction (PCR) is a powerful analytical technique with a wide range of clinical applications in infectious diseases, cancer and genetic disorders. The instrumentation used for PCR has a range of designs related to the thermocycler, the assay features, the detection system and an increasing sophistication with respect to automation. Real-time PCR, in which the amplified genetic material is quantified after each cycle, is becoming a standard application. The automated instrument platforms have a growing repertoire of applications.

In several clinical laboratory areas there has been a high degree of professional expertise in pattern recognition, for example in reviewing histology and blood slides or identification of colony formation on a microbiological plate. The use of digital images in such disciplines can be considered an aspect of automation and allows remote review of images. It is also possible to develop pattern recognition software and this is an emerging field.

Cross reference
The PCR is described in Chapter 16 'Molecular biology techniques'.

SELF-CHECK 17.6

Which of the following laboratory areas can benefit from automation?

1) Coagulation
2) Immunology
3) Microbiology
4) Virology
5) Haematology

 # SUMMARY

- All laboratory disciplines utilize laboratory automation to improve efficiency of processing.

- The extent of laboratory automation use will depend on the size and complexity of the clinical laboratory service.

- Key aims of automation are process efficiency and consolidation.

- Typical clinical laboratory workload growth of between 5 and 10% per annum has been managed by increasing levels of automation, rather than by increasing staff numbers.

- The design of modern automation systems aims to minimize human intervention in the total analytical process.

- The test turnaround times for large automated track systems are generally considered to be faster and more consistent than separate analytical units.

- Good design features within automation systems are maintained and copied by competitors and, therefore, it is possible to see common features in a range of different suppliers' instrumentation.

■ Pre- and post-analytical processes such as centrifugation, decapping, and recapping of specimen tubes, aliquoting, archiving, and storage of samples can be integrated into a single automated process by using tracking systems.

■ Automation is prevalent well beyond the highly automated biomedical science areas, and many microbiology, histology, cytology, and blood transfusion laboratories have a range of automated instrumentation.

 FURTHER READING

● **Hawker CD. Laboratory automation: total and subtotal.** *Clinical and Laboratory Medicine* 2007; **27:** 749–770.

● **Hubl W, Zogbaum M, Boyd JC.** *et al.* **Modular analytics: A new approach to automation in the clinical laboratory.** *Journal of Automation Methods and Management in Chemistry*, 2005; **2005:** 8–25.

● **Melanson SEF, Lindeman NI, Jarolim P. Selecting automation for the clinical chemistry laboratory.** *Archives of Pathology and Laboratory Medicine.* 2007; **131:** 1063–1069.

● **Wheeler M. Overview of robotics in the laboratory.** *Annals of Clinical Biochemistry.* 2007; **44:** 209–218.

● **Sarkkozi L, Simson E, Ramanathan L. The effects of total laboratory automation.** *Clinical Chemistry,* 2000; **46:** 751–756.

The *Journal of Automated Methods and Management in Chemistry* is a publication that produces articles of relevance and can be found on the link: http://www.hindawi.com/journals/jammc/ (accessed 20 February 2009).

Useful Website

■ www.labautomation.org/about/index.cfm

 QUESTIONS

1.   Why do blood specimens require centrifugation on an automated system?

2.   Sodium fluoride is a suitable preservative for glucose because:

    (a)   It inhibits coagulation

    (b)   It promotes glycolysis

    (c)   It inhibits the formation of thrombin from pro-thrombin

    (d)   It inhibits glycolysis

    (e)   It promotes coagulation

3.   What is the average test to request ratio for a laboratory that undertakes 606,531 requests and 5,365,852 tests per annum.

4. How do automation units reduce the risk of infection to laboratory staff?

5. **TRUE** or **FALSE**? If a re-analysis of a test is required samples may be retrieved from the automation system's sample storage area.

6. **TRUE** or **FALSE**? The test turnaround time may be measured in two ways, the total turnaround time and the within laboratory turnaround time.

7. The gels added to blood collection tubes to enable primary sampling have:

    (a) A density higher than serum

    (b) A density higher than plasma

    (c) A density higher than red blood cells

    (d) A density higher than a blood clot

    (e) A density between serum and plasma

Answers to self-check and end of chapter questions are available in the Online Resource Centre accompanying this book.

 Go to www.oxfordtextbooks.co.uk/orc/glencross

18

Point of care testing

Jan Still and Lynda Petley

Learning objectives

After studying this chapter, you should be able to:

- Describe the basis and regulations that underpin POCT
- Discuss the requirements for POCT
- Discuss the differences between internal quality control, external quality assessment, and quality assurance in POCT
- Outline the equipment procurement process
- Describe the audit information required to minimize risk in the provision of a POCT
- Discuss the provision of a patient focused POCT service

Introduction

Point of care testing (POCT) is the provision of a diagnostic pathology testing service outside the traditional clinical laboratory setting and physically closer to the patient. This type of testing may be carried out by healthcare professionals in the ward, clinic, or GP surgery, by high street pharmacists, or by the patients themselves. This chapter describes point of care testing and its impact on the provision of pathology services, and outlines those aspects that must be considered when providing an effective POCT service.

Point of care testing is not a new concept, but earlier technologies paved the way for new devices in response to the demand for a more patient centred health service. Thus, urine dipstick testing has been an accepted practice in healthcare for many years and blood glucose meters were first introduced in the 1970s. Technological advances over the years have widened both the test repertoire and software capabilities available to POCT devices, with blood gas analyser manufacturers leading the marketplace in terms of quality standards for POCT. Indeed, POCT blood gas analysis is a well-established practice, having been developed in response to clinical requirements. Modern POCT now consists of a variety of devices from simple hand-held meters and test kits, some available for purchase and use by patients, to more sophisticated portable and bench top analysers, which provide a variety of tests in all disciplines of pathology.

This chapter introduces the basic concept of rapid testing for the benefit of the patient at the point of care. It illustrates how POCT is governed and regulated, and describes the requirements for a well-managed, reliable POCT service, and how such a service can be established and maintained. The training and competence of POCT practitioners is discussed, and the dangers associated with a lack of or poor quality training are highlighted.

18.1 **Standards and guidelines**

Every area of pathology laboratory testing is subject to regulation, inspection, and **accreditation**, including the POCT service. The aim of regulation is to ensure a safe and reliable service, offering quality assured, consistent, and accurate results from ***in vitro* diagnostic devices** (IVD) that are fit for purpose and used by suitably trained, **competent** individuals. The organization, management, and documentation of the POCT services are defined to ensure best practice and the competency of users. Point of care testing requires separate or additional requirements to the traditional types of analytical investigations because most or all of the analytical processes take place away from the direct overview of the clinical laboratory. Both standards and guidelines are necessary to ensure that POCT is performed to an appropriate level of quality, thereby ensuring the safety of the patient. Standards define what must be complied with, whereas guidelines are recommendations for *best* practice.

The European-wide directive, *Guidance notes on In-vitro diagnostic medical devices directive 98/79/EC* (applicable to member states of the EU and now also part of UK law) regulates all POCT IVDs, including analysers, meters, single use kits, associated reagents, and consumables. This directive was amended in February 2006 and deals specifically with the safety, quality, and performance of IVDs on the European market. All manufacturers must meet the essential requirements listed in Annex 1 of the directive to ensure that IVDs do not compromise the health and safety of patients or users, and perform at the levels claimed by their manufacturers. *In vitro* diagnostic devices that meet the requirements of the directive are denoted by the display of the **CE mark** (Figure 18.1). It is a requirement for all POCT IVD devices used in community or clinic settings to have this mark.

Pathology laboratories are subject to inspection and subsequent accreditation to standards produced by the International Organization for Standardization (ISO); its regulatory document for clinical laboratories is ISO 15189:2003 *Medical laboratories – Particular requirements for quality and competence*. A POCT specific adjunct to this document is ISO 22870:2006 entitled *Point of Care Testing Requirements for Quality and Competence*, which is a comprehensive guide to all aspects of POCT and how it should be organized as a quality assured, well managed, and accurately documented service.

Key Points

ISO 22870:2006 is a standard applicable to clinical laboratories and is an adjunct to ISO 15189:2003. It is not possible to assess against ISO 22870 alone and it is not suitable for use in community or clinic situations either.

FIGURE 18.1

The CE mark consists simply of the initials in the form shown. The marking on the device must be clearly visible, legible, and indelible throughout the expected life of the device. For complex products, the initials are followed by the identification number of the body involved in the control of its production. All the letters and numbers must have substantially the same vertical dimensions of not less than 5 mm. They may be in any colour provided it is legible and unlikely to be confused with any other markings.

Accreditation bodies incorporate the ISO requirements into their own standards for medical laboratories. All UK laboratories are inspected to these standards every 4 years by the Clinical Pathology Accreditation (CPA) UK Ltd now merged with United Kingdom Accreditation Service (UKAS). If its standards are met, the laboratory will be issued with a Certificate of Accreditation. This is a mark of quality, safety, and performance for that laboratory.

Any failure to achieve the necessary standards through deficiencies or non-compliances found during a laboratory inspection will result in the laboratory being given a period of time in which to rectify them. Continuing unresolved failures may result in the loss of or non-award of a Certificate of Accreditation. This, in turn, may have an adverse effect on the laboratory's ability to recruit and retain professional staff or to provide its analytical services to users or other health care providers. A number of other professional organizations have produced their own guidelines, including: the Institute for Biomedical Science (IBMS) *Point of Care Testing Guidance on the Involvement of the Clinical Laboratory* (2004) and the Medicines and Healthcare products Regulatory Agency (MHRA) *Management and Use of In-vitro Diagnostic Point of Care Testing Devices DB 2002 (03)*.

Cross reference

More detail about accreditation and international standards can be found in Chapter 19 'Quality assurance and management'.

These guidelines, like the ISO standards, are intended for laboratory use but the MHRA has also produced more informal leaflets to guide POCT users, such as MHRA *Point of Care Testing Top Ten Tips*. These cover tests such as those to estimate the concentrations of glucose and cholesterol in blood samples and for urinalysis, and are useful for practice nurses and other non-laboratory based users.

18.2 **Advantages and limitations of point of care testing**

From the advent of medicine, successful clinical outcomes have relied upon accurate diagnosis followed by appropriate intervention. Consequently, laboratory services have become more centralized in recent years to help increase efficiency by economies of scale. This has resulted in clinical laboratory testing often being remote from the clinician and patients, which may lengthen the turn around times for some important or urgent tests.

Advantages of POCT

Point of care testing can redress many of the problems associated with traditional testing in the clinical laboratory. For example, the clinician can obtain pathology results during the consultation. Having access to the information required to make a diagnosis, or change the management of the patient during the consultation offers many benefits, which we will discuss in the rest of this section.

Cost benefits

The cost per test in POCT is often higher than when the same analysis is conducted in a laboratory. However, the impact of a rapid result may offset this extra unit cost. The following list describes some of the financial benefits that can be achieved by POCT:

- Timely interventions improve patient outcomes, potentially reducing the length of stay as an inpatient
- It can reduce the inappropriate administration of antibiotics and other prescription drugs, and the use of blood products

- Point of care testing can determine a patient's immune state and help decide whether expensive drugs require purchasing. The efficient and timely management of chemotherapy drugs eliminates wastage

- Individuals who are less experienced can competently perform POCT, enhancing their role and releasing other professional staff for more appropriate duties

- The enhanced disease management benefits the patient and reduces costly complications

- Over the counter testing kits, for example, fertility monitoring kits and pregnancy testing kits cut costs associated with their provision by the health service

- Diagnostic testing carried out on finger prick capillary blood samples provides a cost saving as trained phlebotomists, syringes, sample tubes, and pathology request forms are not needed

- The use of POCT devices in, for example, the prison service may reduce the need for costly secure transport of prisoners to the local hospital for treatment, ensuring that only those with a genuine need are referred

- Travel and parking costs for patients are reduced. With POCT both the test and the visit to the clinician to review results can be completed in a single visit

Estate management

The use of POCT has implications for the management of hospital and human resources, and affects the strategic management of many hospital departments. Point of care testing can affect the whole patient experience where pathology results may be a rate-limiting factor. This contributes to the efficient use of bed space and clinician time, and can help shorten waiting lists.

In outpatient departments, the use of a 'One Stop Shop' approach reduces repeat visits, minimizes the flow of people visiting the hospital site (see Case Study 18.1). Fewer visits may also reduce cross-infections as patient contact is reduced.

Point of care devices sited within the hospital reduce the requirements for transportation of samples to the laboratory. They also reduce pressure on the clinical laboratory through workload reductions and can provide some temporary 'back up' service in the event of laboratory analyser failure. The introduction of local units that use POCT in satellite laboratories can reduce Accident and Emergency (A&E) Department admissions. This enables patient triage and observation to be performed in a community hospital or 'polyclinic' setting.

CASE STUDY 18.1 Appointments

An audit of average turnaround times at a district general hospital showed that the time taken from a patient booking in at the Outpatient Haematology Clinic to his or her **full blood count** result being available from the POCT analyser was eight minutes. This included taking the blood sample from the patient and delivering the result to the consultant. Had this test been carried out in the laboratory, the turnaround time would include transport of the specimen to the laboratory, booking it in, processing the sample, testing the sample, and authorizing the result.

Prior to the implementation of POCT, the patient was required to make two visits to the Outpatient dept – one for the blood test, and a second follow-up appointment with the consultant for the results.

(a) What impact could this have for the patient?

(b) What are the possible benefits to hospital staff?

Management of acute conditions

The management of rapid onset or acute conditions often involves the use of 'bench top' POCT analysers in a 'rapid response laboratory' setting. In these situations, the test results give a base line from which the clinician can monitor, refer, or discharge the patient. Diagnostic testing is also performed by the healthcare professional on the sample that they obtained, reducing the frequency of samples becoming lost or the result being attributed to the wrong patient. This assumes correct, relevant information on the sample is correctly entered into the analyser.

Smaller sample volumes are required for POCT compared with conventional laboratory samples. This is better for the patient as it minimizes blood loss related to phlebotomy. Furthermore, the analysis of fresh samples reduces inaccuracies caused by sample deterioration.

Management of chronic conditions

The management of longer-term or chronic conditions does not require the same level of complexity of devices as required in the acute setting. In chronic conditions, both diagnosis and prognosis are already established, and so only appropriate tests required to manage the patient's condition are necessary. However, the accurate monitoring of chronic diseases enables better treatment and reduces the burden of complications caused by poor patient management.

Cross reference

The management of the chronic disease, diabetes mellitus is described in Chapter 13 of the accompanying volume, *Clinical Biochemistry*.

The patients may be able to monitor themselves at home using a POCT device, increasing their understanding of their own disease and giving them more responsibility in their own disease management.

Chronic conditions can also be monitored in a GP surgery, health centre, or even local pharmacy, releasing hospital resources for the diagnosis and treatment of acute conditions or patients.

Patient satisfaction

Patient satisfaction may be often overlooked in the design of health service provision but must always be a major consideration. The provision of an efficient POCT service, delivered at the right time in an appropriate setting improves the patient experience. Test results provided during the consultation reduce worry and stress of patients who are likely to be concerned about their condition by allowing them to ask about the implications of their results and to make better informed lifestyle choices. Patient experiences are illustrated in Case Study 18.2.

Meeting government directives

The NHS throughout the UK is expected to be cost-effective and may be required to achieve government-defined health promotion or prevention targets. For example, in England, patients presenting at A&E departments must be treated within 4 hours of arrival. Point of care testing is of obvious help in this situation by offering a rapid pathology diagnostic service within the department. Other targets include the 18-week pathway, where the patient can expect to wait no longer than 18 weeks from referral to the start of their treatment. While the laboratory may provide an efficient service, any requirement for laboratory testing will result in repeat appointments with the clinician guiding the treatment plan. The use of POCT may eliminate the need for repeat appointments and so speed the patient progress.

Recent trends require the primary care sector to offer outpatient services to relieve the burden on secondary care (hospital) departments. This has necessitated a different approach in the provision of pathology services and POCT may prove a more effective way of providing test results.

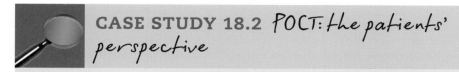

CASE STUDY 18.2 *POCT: the patients' perspective*

Cross reference
Arterial blood sampling is described in Chapter 7 'Samples and sample collection'.

The use of portable blood gas analysers allows respiratory practitioner nurses to measure blood gases in a patient's home. The following comments are from two patients who experienced this form of POCT.

Patient Mrs P. I am really grateful that I don't have to have the needle in my wrist for arterial blood. The nurses take a small sample of blood from my ear, which is not at all painful, and are able to tell me the result in 2 minutes. If I am unlucky enough to need to be admitted to hospital I dread the arterial stab – it is very traumatic for me.

Patient Mr R. It would be very difficult for me to get to the hospital to the oxygen clinic. I used to attend, but found that the worry about being on time made my breathlessness worse. The nurses always told me not to worry, but I did. They now visit me at home with the portable analyser and can let me know the result at once without the need to attend the hospital.

SELF-CHECK 18.1

Indicate which of the following statements are true or are false.

(a) POCT is useful in managing chronic and acute conditions.

(b) Patients do not like point of care testing.

(c) Patients must not perform their own self-checks.

Limitations of point of care testing

Analysis using POCT is often seen as a quick and easy alternative to standard clinical laboratory testing. However, a report from the British In-Vitro Diagnostics Agency (BIVDA) in 1997 found that hospital *managers* were twice as confident in the usefulness of POCT as *clinicians*. This may be because managers concentrate on factors such as rapid turn around times and lower costs, whilst clinicians have a much better understanding of what can go wrong in practice. A POCT solution may be sought when the problem is really about issues such as transport to the laboratory, or speed in obtaining test results. If these difficulties are addressed, then the laboratory could, in fact, be the most cost efficient, as well as reliable option. There are many potential and actual limitations in POCT, of which some are readily apparent, but others may only become evident in practice. Where POCT devices are in use, safeguards and protocols must be rigorously adhered to in order to maintain the reliability of results and ensure patient safety. The clinical laboratory must always be used for the confirmation of unexpected or spurious results.

If a particular POCT test becomes commercially available, there must be a clear clinical need that its use, rather than continuing with the laboratory-based analysis, will benefit the patient. There is little advantage in producing a test result in minutes if a doctor will not see it in hours or even days. Easy availability may result in 'over-testing' and duplication of effort by continuing requests to the laboratory for the same test, resulting in overall increased costs. Infrequent use of a POCT device will result in increased user error and wastage.

Before introducing any POCT device, consideration must be given as to who is going to do the testing, where and when. Nursing or clinical staff must have the necessary time to do the analysis and documentation, or patient care may be compromised. Staff will require training, competency checks, and annual update training. This all takes time and resources. Siting of the equipment must be considered from a health and safety perspective and for the convenience of the user. Maintenance and troubleshooting of equipment may divert laboratory staff from other duties, while poor maintenance may result in breakdowns, inaccurate results, and increase the risks of infections. Equipment may be damaged by misuse, especially if unauthorized users can gain access to it.

Unless staff complete written logbooks or data is automatically downloaded from the equipment, information may be lost. Thus, there may be a lack of a permanent record of POCT results. Users may not be correctly identified due to personal passwords being stolen or simply passed on to others. Errors in **transcription** may occur when test results are written into the patient's notes, and printouts may become damaged, lost, or degrade with time. There is often a lack of traceability: staff in a hurry or pressed for time may omit entering patient identifiers into the analyser so that reports cannot be traced to the correct patient. Reports and printouts that are identifiable may be left in view, leading to issues of patient confidentiality.

When a POCT result is obtained, there may be a lack of 'thinking time' for consideration of that result and for reflection on further action, which may lead to rushed and inappropriate decisions. This can have serious consequences for the patient, such as a misdiagnosis, which may result in **litigation** (see Section 18.11).

The main source of error with any item of POCT equipment is, quite simply, the user. Table 18.1 shows some of the most frequently encountered errors in POCT. Constant observation and the monitoring of all POCT devices is necessary to avoid these errors and ensure that POCT provides a safe and cost effective addition to the laboratory service. Further information on the use of standard operating procedures (SOPs) for POCT may be found in Section 18.9, but Case Study 18.3 describes what may happen when an SOP is not followed and the potential outcomes for a patient.

SELF-CHECK 18.2

State the major advantage(s) of using POCT.

18.3 **POCT committee and POCT policy**

The POCT committee and POCT policy provide the cornerstones for the POCT service. Establishing a multi-disciplinary POCT committee will help ensure that all POCT processes are compliant with the Trust's own POCT policy and with national guidelines, and ensures good **clinical governance**. The committee also ensures effective management in controlling POCT staff training and the use of resources, and so contributes to the production of reliable results that will benefit the patient. The committee must be accountable to the Trust Board, and maintain links with other relevant committees. Indeed, there may already be an established committee, such as a Medical Devices Committee, or Medical Equipment and Procurement Committee, whose aims and interests are broadly similar and relevant. It may prove effective to incorporate POCT into this existing structure, cutting down on

TABLE 18.1 Sources of error in point of care testing

Stage	Error
• Pre-analytical	• Wrong patient
	• Wrong type of sample container
	• Wrong type of sample (urine or blood?)
	• Wrong time for sampling (fasting? pre- or post-dose for drugs)
	• Poor sampling techniques resulting in poor quality samples affected by factors such as haemolysis or the introduction of air bubbles in blood gas samples
	• Delays in analysis and failure to mix samples
• Analytical	• Failure to follow instructions or standard operating procedure (SOP)
	• Lack of training
	• Lack of maintenance or servicing
	• Failure to carry out quality control (QC) or appreciate when it is incorrect
	• Failure of calibration
	• Failure to identify the patient on the analyser
	• Failure to understand error codes or warning messages
• Post-analytical	• Failure to check unexpected results with the laboratory
	• Failure to understand 'out of range' flags or implausible results
	• Failure to consult reference ranges or making inappropriate assumptions about them
	• Entering results into the wrong patient's notes
	• Transcription errors when entering results into patient's notes
	• Clinical decisions made on poor quality or incorrect results
	• Delay in acting on results
	• Loss of results

CASE STUDY 18.3 POCT and SOPs

A 27-year-old female presented to an A&E department with severe abdominal pain. She was not taking any form of contraception. However, she stated her periods were often irregular. A urine pregnancy test was negative. She was treated for mild food poisoning and discharged. The patient was readmitted as an emergency the following day with a ruptured **ectopic pregnancy**. She underwent surgery, received massive blood transfusions, and spent weeks in intensive care. She later sued for damages.

The nurse carrying out the pregnancy test had not received training and had not followed the SOP. A false negative result was obtained because too much urine was used and had flooded the test kit.

(a) Could this situation have been avoided?

(b) Is this case an example of a pre-analytical error?

the number of separate meetings and combining established committee members with new recruits. A typical organizational chart for POCT clinical governance resembles that shown in Figure 18.2.

Clinical laboratories have key functions in the promotion, management, and regulation of a POCT service. The initial impetus for setting up a POCT committee may come with the appointment of a POCT manager or co-ordinator, or quality manager. This person is usually a senior laboratory scientist, and will be the driving force behind the service. The committee should meet on a regular basis and must be as multi-disciplinary as possible. This allows for shared knowledge and experience, and a wide cross-section of opinions. The committee will need a chairperson who is committed to the concept of POCT and willing to be its figurehead. This person must have authority, respect, and drive. A senior clinician or pathologist may be ideally suited for this role. Other members must include the POCT co-ordinator, quality manager, and representatives from each of the pathology disciplines. Clinical representatives from surgical and medical directorates, as well as nursing and midwifery staff will involve those most likely to be active users of POCT devices in wards and departments. Given that the committee may make decisions regarding the purchase or replacement of POCT devices, a representative from Finance to give advice on economic considerations and funding of the service can be helpful. A committee member from the NHS Logistics Department to supply information when selecting and acquiring devices is also useful. Pharmacists who are involved in the purchase of relevant items, such as blood glucose strips, pregnancy tests, and also in their storage and issue to wards should also be invited members.

Clinical governance and **clinical risk** departments should also be represented on the POCT committee and an advisor on training is desirable. Devices may require repair and maintenance, and laboratory staff and biomedical and clinical engineering departments will have relevant input and experience. Valuable insights may come from patient representatives, and if possible, general practitioners, and community nurses. The committee should aim to reflect

FIGURE 18.2
Clinical governance and organizational chart for point of care testing.

the interests of all POCT service users. Secretarial support will be required for taking minutes and preparing agenda items.

Once the committee is appointed, it must agree and develop draft **terms of reference**. These will define the clinical governance, in terms of scope and responsibility of the committee; how it is constituted; accountability and direction; and how the committee relates to the healthcare trust structure.

Once formed, the committee must now develop and publish a trust-wide POCT policy with regard to the elements contained in the standards and guidelines, which we discussed in Section 18.1. The policy must be robust, concise and comprehensive, and provide clear guidance for users. It is recommended that the policy and the minutes be published electronically and be easily accessible. The policy should be subject to an annual review, with an updated version produced at least every three years. The policy must be ratified by the trust board and be accessible both in hard copy and electronically by all staff. The policy should define the organization, management, and funding for POCT, and include the extent to which patient self-testing is supported. The document should refer to European and national legislation, and to the accreditation standards of the laboratory and the trust.

The procedures and required documentation for selecting and procuring equipment, including the submission of a detailed business case, must be clear and relevant (Figure 18.3). These documents should show that consideration has been given to the processes involved in the ordering, supply

Week 1

Establish project plan, prepare and post advertisement on OJEU (open for 37 days)

↓

Produce tender document and specification, include means of evaluation

↓

Prepare and issue pre-qualification questionnaire (PQQ)

↓

Receive PQQ shortlist and issue invitation to tender documents (out for minimum of 40 days)

↓

Tenders received, round 1 short-listing processed

↓

Carry out evaluations before final short-listing

↓

Issue lease tender (allow 40 days)

↓

Receive leasing tenders

↓

Issue intent to award (10 day standstill period applies)

↓

Completion of award, including posting of award into OJEU

Week 36

FIGURE 18.3

Flow chart summarizing the steps involved in approving the commissioning of a new point of care testing device. OJEU, *Official Journal of the European Union*.

and storage of reagents and other consumables, and the cost of both internal quality control (IQC) and external quality assessment (EQA). The POCT committee should ensure that procedures are in place for the safe location and decontamination of equipment. Point of care testing devices must not be purchased, loaned, or installed without prior approval by the POCT committee.

Quality objectives are necessary to ensure correct performance and to maintain safety standards, and these must be regularly evaluated and reviewed. The POCT committee must have the authority to act in the event of poor performance or similar risk issues, including unauthorized use, password theft, or wilful misuse. Action may include withdrawal of the service, or the recommendation of disciplinary action for serious or repeated breaches.

Training objectives and methods of assessing competency must include the provision of standard operating procedures (SOPs) for each POCT IVD or test, including risk assessments. You can read more about some aspects of training in Section 18.8. The policy must define the various staff grades authorized to carry out the POCT procedure, and detail their responsibilities. The policy must define how records of all POCT results are kept and stored. This may include electronic transmission from POCT sites direct to an electronic patient record. Such connectivity must be available for all POCT IVDs. Connectivity is discussed in more detail in Section 18.7. Finally, mechanisms for peer review and shared experience must be defined, with an annual review report of the POCT service provided for the trust board.

SELF-CHECK 18.3

Which of the following records would be essential when preparing for a forthcoming accreditation inspection of the POCT service by CPA?

(a) Staff sickness records

(b) POCT Policy

(c) POCT Committee minutes

(d) Laboratory SOPs

(e) POCT SOPs

18.4 Procurement

The procurement of POCT devices is best managed by the clinical laboratory in co-ordination with the supplies department, once the source of funding has been agreed. Funding could be from a specified budget or from specific capital funds, charitable funds, or as part of a defined project. Procurement must be undertaken in accordance with the hospital trust policy and procedures for the procurement of equipment and services, and the process must also comply with the hospital trust standing financial instructions, public procurement law, and European Union legislation.

Compliance with all the procedures and legislation ensures that best value for money is achieved and procurement is in line with the trust's strategic and business plans.

The guidelines for purchasing equipment were discussed in Section 18.3. Before the decision to procure any POCT equipment is taken, the POCT manager must undertake the following:

- A cost benefit analysis
- A patient benefit analysis
- An equipment evaluation
- A connectivity evaluation
- Site evaluation.

Cost benefit analysis is a technique for deciding the economical outcome of a change of practice. It is derived from the sum of the costs of an action subtracted from a sum of the benefits. While the costs are usually simple to identify, the benefits of moving a test from the clinical laboratory to the point of care can be subjective and difficult to forecast, as the impact of such changes are often far reaching. Patient benefit analysis, as the term implies, is an assessment of the impact that a change has in practice on patient outcomes. The analysis must consider the effect on the whole patient experience, as small alterations in the patient pathway may have unforeseen negative effects. Speeding the patient through one area may cause a bottleneck in another, if this has not also been improved by similar efficiencies.

An evaluation of equipment for POCT use is described in Section 18.5. However, in addition to assessing the POCT device itself, the stability, and credibility of the supplier must also be examined to ensure the long-term viability of the support, maintenance, and availability of consumables throughout its life. Connectivity evaluation requires the involvement of the hospital information technology (IT) department from the outset in the purchase of any POCT device. This is necessary to ensure that the hospital network can support any IT solutions offered by the manufacturers. Health and safety aspects must be included in the site evaluation (see Section 18.6), hence, it can be useful to invite the laboratory health and safety officer to visit and offer advice prior to any purchase.

The results gathered from all the evaluations must answer the following questions before proceeding with the purchase of the POCT device:

- How much extra cost will be incurred during the lifetime of the new device?
- What will be the effects on the laboratory, the POCT service provision, and the staff involved?
- What will be its impact on the patient experience and outcome?
- Will the new technology offer a safe and reliable service?
- Is a complete audit trail in place that provides information when required?
- Is there an appropriate infrastructure and back-up from the provider?

The results from these analyses and evaluations can be used to formulate a business case to identify the requirement for the goods. A robust business case must include a service level agreement that outlines the support offered by the clinical laboratory. This will vary according to resources and service needs, but should include a series of core requirements so that the evaluation of the whole business case can be considered. The roles of the laboratory will include the maintenance, quality control (QC), and assessment of the device, and include the training and competency assessment of staff expected to use it. The financial information to be considered should include the funding stream for the device(s) including delivery, the costs of consumables over its lifetime, maintenance and service contracts (see Section 18.6), expenses associated with training staff, the disposal of clinical waste, and a depreciation and replacement strategy.

Case Study 18.4 shows what may happen if POCT devices are procured without taking into account all of the factors that a POCT committee or co-ordinator may consider.

All contracts from the public sector, which are valued above a defined threshold for the whole lifetime costs, need to be advertised in the *Official Journal of the European Union* (OJEU) to comply with EU legislation. Information on the process can be accessed from the website http://www.ojec.com.

18.5 **Evaluating a POCT device**

Evaluations are required to ensure that any equipment purchased for POCT is fit for purpose and suitable for safe and effective use at the location for which it is intended. All medical IVD

CASE STUDY 18.4 Grateful relatives

A female patient's relatives wished to thank the ward for the care she has received during a long hospital stay. The ward Sister asked them if they could raise the money to purchase a full blood count analyser, as delays in receiving results from the clinical laboratory had sometimes resulted in postponed treatments. The relatives managed to raise the full cost of the instrument. No one discussed this purchase with the POCT co-ordinator, clinical laboratory, or the consultant haematologists.

What are the potential disadvantages of purchasing a POCT instrument by this procedure?

(a) The above scenario could result in which of the following:

    i)   An unsuitable instrument could be purchased

    ii) Running costs could exceed budget

    iii) Untrained staff could use the instrument

    iv) Quality control and EQA could be overlooked

(b) How could the problem of slow turnaround time have been approached more fruitfully?

---

kits and reagents in use within member states of the European Union now carry a CE mark (Section 18.1) that indicates their performance has been tested and verified in accordance with the claims of the manufacturer. This does not imply that a particular device will meet the needs and requirements of day-to-day clinical use. It must be borne in mind that what works beautifully in a manufacturer's demonstration in a well-ordered laboratory may well be unsuitable for a busy critical care area. It is essential to carry out a thorough evaluation of the device prior to its purchase, to ensure that the users will have confidence in their ability to use it in a safe and competent manner.

An initial evaluation of an IVD may have been undertaken by a national device evaluation centre, such as the Centre for Evidence-based Purchasing (CEP), which is now part of the NHS Purchasing and Supply Agency (PASA). Reports are free to NHS sites, and details can be found at: http://www.pasa.nhs.uk/cep. If there has not been such a national evaluation of a device one is interested in, then information on its performance may be found in a literature search of peer-reviewed articles.

Evaluating the operation of POCT equipment

In a **competitive tender** situation, each proposed device should be placed in the POCT site for a set period. This will enable users to familiarize themselves with the device, comment upon its design, performance, and how well it fits their requirements. Planning is essential to ensure that such trials take place with a minimum of delay, so that if more than one device is under consideration, users can make valid comparisons between them. Staff must be trained by the manufacturer to an agreed set of training objectives and allowed free access to the equipment. Results obtained during trials should not, however, be used for clinical decision-making.

If the equipment under trial requires an additional blood sample to be taken from the patient, then approval from the hospital ethics committee is required with informed patient consent. This procedure is required because the additional duplicate sample will not be used for the direct benefit of the patient, and he or she has the right not to have their blood used in such a way.

Users should have access to evaluation forms that they can complete after using each piece of equipment. They should be asked to score each separate device for overall ease of use. This may include what sample types are required and how they are handled, and how passwords and patient identifiers are entered. The speed of use and time from obtaining a sample to acquiring results can be assessed, as can the quality of reports and ease of data recall. Instructions help menus and manuals must be clear and easily understandable. Quality control and maintenance procedures should be evaluated by staff undertaking them, and troubleshooting interventions should be concise and relevant.

Evaluation forms should be identical for each item and collated at the end of the trial. Laboratory staff should also have access to the equipment, or a separate trial should be arranged within the clinical laboratory. The preferences and comments of all concerned should be carefully considered, to help avoid reaching unsound conclusions. It would be unwise to purchase any item that none of the users liked, on the basis of other criteria, such as economy. Evaluations can avoid expensive mistakes.

Diagnostic kits

Reports indicating the relative performances and costs may also be available for specific diagnostic kits; comparative reports are particularly useful in product selection. Having provisionally chosen a kit, the views, and preferences of their users must be sought with regard to its ease of use and suitability for purpose. Potential sources of user error must be identified. Considerations to take into account include the conditions for kit storage, the sizes of the packs, possible wastage, and inevitably costs. It may be financially beneficial to consolidate the purchasing of diagnostic kits to a single supplier if possible. Both IQC and EQA are necessary to ensure batch-to-batch conformity. Precision must be assessed by performing at least 20 replicate tests and calculating the coefficient of variation (CV) for the results, which should be less than 5%.

Cross reference
Statistical analysis is covered in more detail in Chapter 5 'Statistics and handling data'.

Acceptance testing

Once delivered, the device must be checked by the appropriate department to ensure it is electrically safe. Its acquisition must be entered into the asset register of the trust and it requires a clear label. The supplier should install the device at its working site (Section 18.6) and thoroughly check it to ensure its performance is as specified. Manufacturer defined quality control checks must be carried out and replicate tests performed on laboratory samples as described above for diagnostic kits. An appropriate statistical analysis of the data is essential to ensure that the results concur with those of the clinical laboratory instrument routinely used in testing such samples. The acceptance forms must only be signed when everything is satisfactory.

Evaluation is not a one-off procedure. Continuous evaluation through IQC, EQA, and audit is necessary to ensure that the product performs satisfactorily throughout its life.

18.6 **Use of POCT devices**

The servicing, acquisition, and storage of spare parts for any POCT device should have been considered during its procurement process in consultation with the POCT committee. Without proper forethought and planning, additional problems may become apparent when the device is implemented. Insufficient space, lack of adequate facilities such as light, ventilation, and power, or an inability to store and provide spares can hamper the POCT service, cause inefficiencies, and lead to waste.

Siting of equipment

Point of care testing equipment should be located in safe, designated sites, close to the clinical area, and where the risks to the health and safety of patients and staff are minimal. The recommendations of the manufacturer should act as a guide to the siting of all POCT appliances.

The site must conform to all Health and Safety, and Fire Safety regulations. There must be sufficient space for the equipment and consumables, and room to carry out maintenance procedures. Work surfaces should be at a convenient height and capable of supporting the weight of the instrument. If it produces a significant output of heat, then air conditioning and ventilation may be required. The equipment should be arranged in an effective manner with consumables, documents including user manuals, spares, centrifuge, refrigerators or freezers readily to hand. Adequate lighting, power supplies, and network ports must be provided. The mains supply should be on a protected circuit and the equipment provided with an integral uninterruptible power supply (UPS). Facilities for the disposal of wastes, including sharps, clinical, and non-clinical waste are necessary. Hand washing and possibly eye washing provisions should be readily accessible.

These points must be considered prior to purchase, and included in any plans for new or modified premises to prevent problems only being discovered later, when they may prove costly and difficult to remedy.

Stock control

Appropriate stock control ensures that the POCT service does not fail through a lack of essential supplies, which are used economically preventing wastage. Commercial solutions are available that use barcodes or chip recognition systems to track supplies and to re-order them automatically. Supplies should be delivered to a secure central point for collection.

Immediately on arrival, new stock should be checked to ensure that it is as ordered, has been delivered in the right quantity, and has a long shelf-life. Supplies must be stored in accordance with the manufacturer's instructions to prevent wastage. Thus, for example, there must be adequate freezer or refrigerator space for the volume of stock ordered; regular temperature checks must be carried out and logged to ensure supplies are kept in optimum conditions. Batch and lot numbers of stock should be recorded, together with details of when these are brought into use. Stock rotation is necessary to ensure that reagents and consumables are used before their expiry dates. Out-of-date stock, particularly reagents and diagnostic kits, may deteriorate and produce invalid results. Never write on unopened packs because manufacturers will not accept the return of incorrect stock if the packages or boxes are marked. If they must be marked, it is better to use removable stickers. Regular inventories are necessary to ensure that the recorded levels of stock are accurate. It is all too easy to return an empty box to the store and confuse the system!

Maintenance and servicing

All maintenance and servicing of POCT devices must be carried out in accordance with their manufacturer's instructions to ensure they remain in good working order. Lack of or poor maintenance may result in incorrect results, safety risks, equipment failure, and invalidation of the manufacturer's warranty. All equipment should be entered into an asset register that should detail all maintenance procedures and how frequently they must be carried out. The manufacturer, supplier, laboratory staff, or clinical engineering staff may carry out maintenance, and responsibility for it should be noted in the register. Maintenance logbooks should be provided. These must be updated at each service or this information recorded electronically in the analyser itself. The logbook provides information about what has been done and when, and thus helps pinpoint any recurrent problems. Any POCT device that does not perform to its specification must not be used until the problem has been rectified. Serious faults may require the return of the equipment to the manufacturer for repair or replacement. It may be necessary to withdraw the POCT service while such repairs are in progress, with the clinical laboratory providing the necessary back-up service.

Users should know who to contact and how to report breakdowns. They should be aware of alternative arrangements for dealing with samples whilst analysers are 'down'. Manufacturers should specify response times and frequency of service calls.

All newly purchased equipment has a warranty period, which is usually its first year of use. The manufacturer must carry out servicing during the period of the warranty. Care must therefore be taken not to invalidate the warranty by misuse, adaptation, or unauthorized interference with the equipment. Service agreements should include provision of a backup instrument if required. If equipment is required urgently, it may be possible to arrange an immediate exchange policy.

For small items, such as blood glucose meters, it is sensible to have a supply of spares in stock for their rotation or for immediate use to replace defective instruments. This arrangement can be stipulated in their procurement specification. However, it is not sensible to have spare parts for large and expensive analysers, so any repairs must be swift and efficient. Returning equipment to the manufacturer for repair may be expensive once the costs of the inspection fee, spare parts, labour, and delivery are included. It is often economical to take out a service contract for a fixed fee, which covers the costs of several repairs.

Items that are returned for repair require a decontamination certificate stating whether the equipment has been exposed to body fluids, infected material, or other biological or radioactive hazards. It must also describe the decontamination process used to render it safe. In-house repairs also require a decontamination certificate.

'Repair and return' requests must state the equipment type, its serial number, the date, the site where it is used, the reason it requires repair, and a name and address to which it must

Cross reference
Aspects of health and safety are covered in Chapter 4.

be returned when working. Unless the equipment is under warranty, the manufacturer may require an order number to guarantee payment. Items must be securely packaged and dispatched as soon as possible. Valuable items should be sent via recorded delivery.

SELF-CHECK 18.5

If you find that reagents are not being stored appropriately on a regular basis, what could be done to solve this recurring problem?

18.7 Connectivity

Connectivity facilitates data capture without the requirement for manual transcription. Hence, data from POCT systems may be saved in the laboratory information system (LIS), in the Hospital Information System (HIS) as part of the patient's electronic record, or to 'stand alone' servers with a robust back up system. However, prior to 2000, POCT connectivity had been unregulated and, consequently, of variable quality. In January of that year, the Connectivity Industry Consortium (CIC) was formed and devised a document describing the standards required for POCT connectivity. This document formed the basis for the subsequent development of POCT connectivity requirements, including plug-and-play facilities, which allow the bi-directional control of multiple devices using a single data capture platform. Bi-directionality is the transfer of information from the analyser to the laboratory and back again. This allows the POCT co-ordinator to send the same information to many devices, for example to update the set-up, or to a single device to order a specific procedure, or lock a specific analyser that has developed a problem to prevent its use until the fault has been rectified. The management of POCT devices is improved by access to a bi-directional link to the devices, which allows the POCT co-ordinator to view the analyser status and patient results in 'real time', and enables remote **recalibration** and QC. The ability to view remotely the status of an analyser, including its 'on board' consumables and maintenance schedules, reduces the time spent visiting POCT sites. Communication is central to the safe and effective use of pathology results, particularly in the case of a remotely generated result. The ability of a biomedical scientist to view and verify a result in real time offers the opportunity to respond quickly to abnormal results, give advice to the clinician, and reduce risk to the patient in the event of an error in analysis. Linking into the whole patient record allows for a continuous and chronologically accurate record of any given test analysis. As we move towards a true electronic record of patient data, this collation of test results will be necessary for continued patient care by those who access the records. Hospital specific networks, the NHS central system, the Internet or other commercial packages all provide connectivity depending on the capabilities of POCT devices. Manufacturers that currently offer connectivity for POCT devices design specific software that may allow access to devices from other manufacturers, but with limited functionality. This is likely to change, as connectivity becomes more a requirement than an option. Whichever solution is available, the hospital IT department will be involved in the maintenance of the connection. It must also ensure that any connectivity package is compatible with any HIS or LIS upgrades to guard against it becoming obsolete.

The cost of connectivity for POCT devices is often significant and, indeed, may be more than the initial cost of the device. There may also be ongoing expenses for the maintenance or licence of such links and this should always be identified in the 'revenue costs' part of the business case (Section 18.3). However, the benefits of connectivity for any POCT device outweigh the costs when the total service provided is considered.

Connectivity also provides an audit (Section 18.10) trail allowing the recall of patient results, identification of the user, and verification of performance, lot numbers and expiry dates of consumables, as well as continuing analyser status and intervention history. All of these data are requirements of CPA and the NHS Litigation Authority (NHSLA). Audit data is required to minimize risks to the patient. In addition, such data protects the hospital trust as patient mis-management can lead to expensive litigation cases (Section 18.11). Although failures in IT and information corruption may occur, the biggest challenge from connectivity to any POCT co-ordinator is ensuring that correct patient details are entered by the user. Accurate identifiers are essential to integrate POCT results with laboratory records for a given patient since the correct patient information is necessary to recall results and demonstrate a quality assured patient management system in a legal environment. However, incorrect or incomplete information may enter the system because the patient details are simply unknown. This may happen in a major incident, on the admission of an unconscious patient, or where senility or confusion results in incorrect information being given by the patient. In can also happen when healthcare staff obtain a sample and simply forget to take the patient's details with them to the analyser. Unauthorized users may also enter made-up patient identifiers in mandatory fields to obtain results, bypassing the safeguards.

18.8 Training for POCT

Training and competency certification is the key to the success of any POCT service. The intro-duction of any point of care test needs to be accompanied by a SOP. This SOP must be plainly written and explain any scientific terminology and acronyms. The POCT co-ordinator, in con-junction with the ward sister, practice development nurse, or hospital training department must offer a comprehensive training package that covers:

- Pre-analytical techniques
- Analytical techniques
- Post-analytical actions
- Preventative actions

Pre-analytical techniques include assessing reagent requirements, such as their storage, stabil-ity, and handling, the need to label samples and how to obtain samples of the necessary quality. Substances that interfere with the analysis and invalidate the results should also be described. How to recognize device failure, and health and safety issues should also be included in the training package. Information about how the POCT device functions is an essential require-ment of any training on analytical techniques. Quality issues should include QC and external quality assessment requirements (Section 18.9).

Cross reference
You can read more about quality and quality assurance in Chapter 19 'Quality assurance and management'.

The operational ranges, device and analyte **reference ranges**, together with common user errors must be highlighted. It may be pertinent to include examples of serious errors or the harm caused by poor practices. Post-analytical instruction includes training in recording results, valida-tion, and verification of results, and recognition of error codes. This is essential since corrective actions are necessary if analytical results fall outside the range of the analyser or outside the refer-ence limits set by the laboratory manager or clinician. Descriptions of the risk reduction aspects of good record keeping are useful. The importance of compliance, an aspect of POC testing often disregarded, is apparent when an audit trail is requested during, for example, litigation.

Preventative action in the event of an analytical error is the users responsibility, including removing the device from use and alerting the laboratory. It is essential to emphasize to train-ees that the results they generate are *their* responsibility. Results from any POCT device that are inconsistent with the clinical signs and symptoms of the patient require checking by the clinical laboratory before management or treatment of the patient is initiated.

Feedback information on a training session is a constructive method of auditing quality and effectiveness. Unless trainees are observed performing POCT tests in a clinical setting, it is not possible to regard them as competent. Competency can also be lost over time if the test or technique is only performed irregularly. Competency is always the responsibility of the user. Although training update intervals may be fixed, the requirement for such training is particular to the individual, and depends on how often he or she performs the procedure and their ability to recall instructions. The use of eye-catching posters or short form guides placed near the POCT analyser can be effective *aide-mémoire*. The over-use of this resource, however, can result in the 'wallpaper syndrome' where users become blind to information provided in this way. When designing a training programme it is useful to be aware of the learning styles of the audience and include interactive practical demonstrations wherever possible.

Monitoring the error logs of analysers can assist in pinpointing users who may require further training. The POCT co-ordinator, when aware of an individual with questionable competency (as illustrated in Case Study 18.5) is responsible for alerting the ward sister of the situation. Incompetent users must be prevented from using the equipment until they are retrained and are fully competent. However well trained the POCT team is in the use of their equipment, a significant part of their duties is to be out and about in wards and departments, to help maintain contact with staff, answer questions and occasionally spot something untoward.

18.9 **Quality and POCT**

Quality assurance (QA) and its associated processes, quality control (QC), internal quality control (IQC), and external quality assessment (EQA) are all essential aspects of laboratory quality management. They are perhaps even more important when considering POCT, as testing activities may take place outside of a laboratory and be conducted by other healthcare professionals or even the patient themselves. All these quality measures are designed to help ensure the accuracy of testing and to be certain that appropriate clinical action is taken on the basis of test results.

Cross reference

You can read more about quality assurance and quality management in Chapter 19 'Quality assurance and management'.

Effective quality assurance poses a challenge to the POCT co-ordinator, and is a principal aspect of the role. The implementation of a system underpinned by adequate documentation will cover:

- Standardization of methods
- Monitoring and maintenance of equipment

CASE STUDY 18.5 *Incompetency*

The POCT co-ordinator received a phone call from a trained nurse who complained that a pregnancy testing kit was not working correctly. On investigation, it was found that the nurse had filled the sample well with urine, rather than the specified amount as detailed in the SOP.

What actions should be taken to assist someone whose competence is in doubt, as is the case with this nurse?

- Equipment downtime
- Equipment repair
- Audit of consumables
- Training and competency
- Quality control
- External quality assessment

The ISO 22870:2006 guidelines mentioned in Section 18.1 outlines a comprehensive approach to QA.

All POCT devices must be subjected to QC as determined by the manufacturer. This must be performed at regular device-specific intervals in order to monitor the basic functioning of the device. There may be several levels of QC material available for devices, to assess performance at low, normal, and high confidence levels. Quality control is designed to produce a result on an instrument that can be compared with the recommended acceptable range. This involves testing a solution or sample produced by the manufacturer of the instrument, stored according to their instructions. The recommended ranges set by the manufacturers are often wide to allow for batch-to-batch variations of consumables. An individual result falling within this range may offer no indication of instrument bias or precision.

Analysers such as blood gas analysers, can be programmed to perform this type of QC automatically from solutions stored in the analyser and can 'lock out' the device should it fail to achieve the expected performance. This prevents any inappropriate clinical management of the patient caused by inaccurate results and highlights the need for corrective maintenance of the analyser. For devices where automatic QC is not available, the operator must perform QC testing manually and take action in the event of its failure. These procedures must be recorded as evidence of a robust response designed to prevent use of the device prior to corrective maintenance.

Electronic quality control

Electronic QC involves challenging the POCT device (analyser) with an electronic stimulus designed to produce a known measurable response from the detector. This method of QC has the advantage of being integral in the performance of a measurement and is not user dependent. Where available, the POCT device should be 'quality controlled' using this procedure prior to use. The disadvantage of this method is that it may not mimic the analysis of a real sample and so may not replicate the mechanical aspects of sample flow through the testing system. Thus, the resulting QC data are qualitative, rather than quantitative and give the user no guidance as to the performance of the device in clinical use.

External quality assessment

External quality assessment (EQA) involves the analyses of identical samples, usually supplied by an EQA scheme provider, to various sites that use the same POCT device. Collating the results means the performances of individual analysers may be assessed. Information regarding the accuracy and precision of a device within the distribution of results is a basis for any necessary corrective actions. An EQA report is a retrospective 'snapshot' of device performance, which can be used to proactively manage the device and highlight problems, or to indicate adverse trends. Registration of POCT devices on a suitable EQA scheme is mandatory for CPA accreditation.

Cross references
You can find out more about the different types of data in Chapter 5 'Statistics and handling data'.

Accreditation is more fully explained in Chapter 19 'Quality assurance and management'.

18.10 **Audit**

Regular audits are essential to assess the performance and clinical usefulness of all POCT devices. CPA requires horizontal and vertical audit information (see, for example, Figure 19.4).

A vertical audit is a 'single process, all aspects' audit, such as the selection of a single test and examining a range of parameters as listed in Table 18.2. In contrast, a horizontal audit selects one element of the vertical audit, and expands the information relevant to it. Hence, it is a one process, all aspects procedure. An example of a horizontal audit surrounding a maintenance log may contain those points given in Table 18.2.

Accreditation will require the POCT co-ordinator to collate analyser workload figures and cost per test, registers of training and equipment, logbooks listing any equipment replaced or periods when it was not functional (downtime logbook), and stock control information.

SELF-CHECK 18.6

Information about a POC test, such as how to analyse for blood glucose, could be provided by which of the following?

a) Asking a user
b) Watch a user perform the test
c) Examine the record book of the device
d) Ask a patient
e) Check the patient records against the logbook

TABLE 18.2 Points to note when performing vertical and horizontal audits

Vertical audit	Horizontal audit
Patient ID	Device-specific training records
Date and time of test performance	Competency records
	Job description
Identity of user performing the test	Individual's contract
Analyser serial number	Evidence of continuing professional development (CPD) if appropriate to the grade
Observe a test being performed	
Lot number of consumables	
Evidence of correct storage of consumables	
Training record of user performing the test	
Competency record of user performing the test	
Analyser error log	
Analyser maintenance log	
Quality control records	
External quality control records	

18.11 **Problems, incidents, and litigation**

A well-managed POCT service should ensure that competently trained users obtain high quality results that benefit the patient. However, errors and incidents will inevitably occur from time to time. Such events must be investigated, remedied, documented thoroughly, and steps taken to avoid any repetition. This enables users to learn from such mistakes or malfunctions and to notify others of potential and actual risks.

Error logbooks

Fully completed error logbooks are used to record incidents, their seriousness, any remedial actions taken, and when such actions have been completed. Each item of POCT equipment or diagnostic kit must have its own logbook. Records must define the type of error, whether pre-analytical, analytical, or post-analytical, and if attempts at corrective action were effective. Case Study 18.6 shows an example of an error and the corrective action taken to prevent it happening again. For serious errors, an incident form is required. A record must be made of notification to the POCT committee, the Risk Management committee, or an external body, such as the MHRA.

The error logbook of a POCT analyser can be stored with its maintenance logbook but it may be better to keep them centrally with the POCT co-ordinator. This enables recurrent errors to be recognized and possible patterns of incidents or faults detected. These may identify a particular individual causing repeated errors, who requires retraining, or indicate that the siting or practical use of the equipment is causing unforeseen problems.

Reporting of incidents

All NHS Trusts will have their own incident reporting forms, and ward staff may complete these following incidents involving patients. Serious incidents that involve harm may require

CASE STUDY 18.6 When is total haemoglobin, really total haemoglobin?

External quality assessment results from an analyser used to measure total haemoglobin were found to be increasingly inaccurate when compared with results from the clinical laboratory. An incident was reported where cross-matched blood had been requested on a patient whose tHb was determined as 79 g dm^{-3} (below critical limits) using a POCT analyser on the ward, but was found to be 141 g dm^{-3} (the reference range for men is 129–165 and for women 123–147 g dm^{-3}) by the laboratory. An investigation showed that the **cuvettes** used with the POCT instrument were placed on a shelf in direct sunlight. The adversely high temperatures had caused them to deteriorate. Moving the instrument and cuvettes away from the window to a site with a constant room temperature cured the problem.

(a) Could the patients have been harmed by this incident?

(b) How could this problem have been detected earlier?

Cross reference

More information about health and safety is found in Chapter 4.

a Reporting of Injuries, Diseases, and Dangerous Occurrences Regulations (RIDDOR) report, a notification to the MHRA if the device was suspected to or known to be faulty, and to the National Patient Safety Agency (NPSA). All such incidents must be recorded in the trust risk register.

If the incident involved POCT as described in Case Study 18.7, then a copy of the report must be forwarded to the POCT co-ordinator, and investigated to establish what went wrong and how it might have been prevented. If equipment malfunctions and results were found to have a clinical impact, then other users of similar POCT devices require immediate notification. The manufacturer, MHRA, and NPSA must also be notified. If further investigation showed a risk of harm or danger, then the MHRA may issue a Safety, Alert or Hazard notice to warn all possible users of the problem. Each trust has a Safety Alert Broadcast System (SABS) which cascades the safety notice to all possible users.

SELF-CHECK 18.7

(a) In the incident described in Case Study 18.7, whose responsibility was it to ensure this nurse was competent? (b) Would this incident have been subject to a RIDDOR report?

Litigation

The primary concern of healthcare is always the safety of the patient. A robust and effective system of error and incident reporting is necessary to minimize risk and maximize patient protection. The risk of litigation is particularly relevant with regard to POCT, where unauthorized users may not be fully trained or competent, may fail to follow instructions, or may adapt procedures without authorization. Trusts are insured against any adverse consequences arising from litigation. Any that are able to demonstrate good practices in training, documentation, and competence can obtain a reduction in their insurance premiums. This scheme is the NHSLA for Risk Management Standards. It has three levels, each requiring the trust to submit to a rigorous assessment. If the Trust meets these standards, it receives a considerable reduction in its payments, with satisfying each level securing a greater reduction. The POCT service can help the trust obtain these reductions by ensuring its policies and guidelines are adhered to.

CASE STUDY 18.7 *Whose fault?*

An agency nurse was asked to monitor a diabetic patient who has been placed on an insulin infusion prior to surgery. This routine procedure is necessary to stabilize the concentration of glucose in the patient's blood. The concentration of glucose in capillary blood is measured using a glucose meter at hourly intervals. Although the nurse was unfamiliar with the meter provided, she had used a number of different types of glucose meters previously. For each test, she pressed the button to start the meter and recorded the result. All results were found to be stable and within the reference range. Unfortunately, the patient was later found in a **hypoglycaemic coma** and died of complications.

Suggest a plausible explanation as to why the patient died.

Human errors

User error is the major cause of incidents, which often arise from lack of training, misinterpretation of instructions, or a failure to follow the relevant SOP. Making ill-considered assumptions on the capabilities of POCT devices, a poor understanding of the chemistry underlying the test, or simply not thinking sufficiently about the procedures associated with it can all lead to serious accidents. For example, blood gas analysers give a measurement of the pH of blood samples. Some users extrapolate this to mean that the analyser is in fact a pH meter and is able to determine the pH in almost any fluid. Even if the user manages to obtain a result on a non-blood fluid, they will not have any idea of reference ranges, effects of dilutions, the limitations of the analysis, or the presence of interfering substances. False assumptions about the underlying chemistry of an analytical test may lead to incorrect conclusions.

The unauthorized purchases of POCT equipment can by-pass all the carefully constructed safeguards, increase the risk of harming the patient and incur unforeseen costs. You can read about some of the problems such actions may cause in Case Study 18.4. Constant watchfulness and awareness is necessary to pick up incorrect and appropriate practices, as illustrated in Case Study 18.8. Quality assurance is unlikely to identify random errors, but regular, thorough training and competency testing may eliminate most of them.

CASE STUDY 18.8 Blood in vomit?

Urinalysis strips are universal, simple, and easy to use. A consultant haematologist was called to a patient who was alleged to be vomiting blood. He was shown the contents of the vomit bowl and was unable to see any blood. The nurse was asked why she thought the contents contained blood. She has used a urine dipstick and it showed positive for blood. The patient was really suffering from an obstructed gastrointestinal tract (GIT) and vomiting **bile**, which had stained the strip green. The positive colour for blood in urinalysis is also green.

(a) Can urine strips be used to test other biological fluids?

(b) Could this error have led to an incorrect diagnosis?

SELF-CHECK 18.8

A POCT co-ordinator noticed three used urine drugs of abuse screening kits in a ward sluice room. Such kits detect drug metabolites, rather than the drugs themselves. It appeared that there were suspicions among the ward staff that a mother was adulterating her children's milk, orange juice, and water with her prescribed **methadone**. The screening kits had been used to test the drinks.

Suggest the most plausible of the following as to why methadone would not have been detected.

(a) There was insufficient methadone in the drinks.

(b) The drinks would interfere with the test.

(c) The nature of the kits.

SUMMARY

■ Point of care testing is the use of an analytical device for clinical testing outside the laboratory and close to the patient

■ Point of care testing is governed and regulated by European legislation and national guidelines. An accreditation body conducts inspections according to these standards and its own guidelines every four years. An accreditation certificate is awarded to POCT services that comply fully with these standards

■ To achieve a well-managed and reliable POCT service, NHS Trusts must have their own multi-disciplinary POCT committee, responsible for devising its own local POCT policy. This defines the service, and how it is to be managed and organized, and how all matters related to its services are to be documented

■ Training and competence are key elements in ensuring the safety of patients by performing quality controls and external quality assessments of every IVD. Regular audit and monitoring is necessary to ensure compliance and good performance

■ Point of care testing may be seen as a rapid process providing results almost immediately and so speeding the diagnosis and treatment, it does have potential pitfalls in terms of pre-analytical, analytical, and post analytical errors

■ Procuring a new IVD must be based on clinical need. Care is necessary in examining cost benefits to the whole patient episode and not merely cost per test. Evaluations of new devices, their acceptance testing, maintenance, and audit are part of the role of the POCT co-ordinator to ensure the safety and continuing satisfactory performance of the device

■ Connectivity enables various devices to be monitored from a remote location, and corrective actions to be initiated through a computer link. Connectivity benefits the POCT service by allowing results to be downloaded to the electronic patient and laboratory computer records

■ Errors and incidents must be recorded and reported to appropriate outside bodies where necessary. Human error may result in patient harm and litigation

FURTHER READING

● **Becker KM, Whyte JJ.** *Clinical Evaluation of Medical Devices: Principles and Case Studies*, **2nd edn. Humana, New Jersey, 2005.**

All that is necessary to know to evaluate a POCT device. The case studies are particularly useful in avoiding pitfalls and unexpected outcomes.

● **Bubner TK, Laurence CO, Gialamas A, Yelland LN, Ryan P, Willson KJ, Tideman P, Worley P, Beilby JJ. Effectiveness of point-of-care testing for therapeutic control of chronic conditions; results from the POCT in General Practice Trial.** *Medical Journal of Australia*, **2009; 190: 624–626.**

This short paper summarizes the first full randomized controlled trial of the clinical effectiveness of POCT in Australian general practice. It is an encouraging endorsement of some POCT tests with an interesting approach.

- Nichols JH, Christenson RH, Clarke W, Gronowski A, Hammett-Stabler CA, Jacobs E, Kazmierczak S, Lewandrowski K, Price C, Sacks DB, Sautter RL, Shipp G, Sokoll L, Watson ID, Winter W, Zucker ML. Executive Summary. **The National Academy of Clinical Biochemistry Laboratory Medicine Practice Guidelines: Evidence based practice for Point of Care Testing.** *Clinica Chimica Acta*, 2007; **379**: 14–28.

  Excellent discussion of what makes an effective application of POCT, some more US based but a good overview.

- Price CP, Hicks JM. (eds) *Point-of-Care Testing*, 2nd edition. AACC Press, Washington, 2004.

  This detailed and comprehensive work includes chapters on instrumentation, methodology, and POCT in primary care. It requires some prior knowledge but is an excellent source on all aspects for those wishing to enquire further.

- Price CP, St John A. (eds) *Point-of-Care Testing for Managers and Policymakers: From Rapid Testing to Better Outcomes*. AACC Press, Washington, 2006.

  A work intended for those in management and government and not an easy read but a useful overview from a manager's perspective.

- Westgard JO. *Basic QC Practices*, 2nd edn. AACC Press, Washington, 2002.

  Written by the foremost authority on quality control and the originator of 'Westgard Rules', this work takes the reader from simple basics to more complex statistics.

Useful Websites

- http://www.wales.nhs.uk/documents/PFSD-August-2008-Ebook.pdf

- http://www.acbi.ie/downloads/guidelines-for-point-of-care-testing.pdf

- The Institute for Biomedical Science (2000) Point of Care Testing Guidance on the Involvement of the Clinical Laboratory. Document Ref PR/07.04. Available at: **http://www.ibms.org**

- The Medicines and Healthcare Products Regulatory Agency (2002) Management and Use of In-vitro Diagnostic Point of Care Testing Devices DB 2002 (03). Available at **http://www.mhra.gov.uk**

- Guidelines on Point of Care Testing (2004) Royal College of Pathologists. Available at **http://www.rcpath.org**

- ISO 15189: 2003 Medical Laboratories – Particular requirements for quality and competence

- ISO 22870: 2006 Point of Care Testing (POCT) – Requirements of quality and competence.

 QUESTIONS

1. Which of the following records is/are necessary to prepare for an accreditation inspection of a POCT service by CPA?

   (a) Staff sickness records

   (b) POCT policy

   (c) POCT Committee minutes

(d)   Laboratory SOPs

(e)   POCT SOPs

2.   You have just taken delivery of a new POCT analyser. Which of the following may be involved in installing the device?

(a)   Porters

(b)   Clinical engineering

(c)   Infection control

(d)   Risk management

(e)   Cleaners

3.   You find several boxes of reagents have been delivered. Although marked *Store at 2 to 4 °C*, they have been left in a warm room for at least a day. Which of the following is the most appropriate action?

(a)   Discard all the reagents immediately

(b)   Perform a quality control check on a sample of the reagent, having refrigerated the remainder

(c)   Inform the supplier and request them to replace the reagents

4.   (a)   How would you establish which POCT devices are in use in your hospital?

(b)   In which areas would you be most likely to find such devices?

5.   How would you deal with any unauthorized or unexpected devices found? What actions could you consider?

6.   You find that a particular user has been responsible for several instances of damage to devices, has a poor record of training, and refuses to take part in the EQA scheme. What actions do you take?

7.   A Consultant surgeon has asked for a POCT device to be placed in a theatre. The same test can be done in the laboratory in less than 30 minutes. How would you establish whether there was a case of clinical need and what other factors should you consider before procurement?

Answers to self-check and end of chapter questions are available in the Online Resource Centre accompanying this book.

 Go to www.oxfordtextbooks.co.uk/orc/glencross

Quality assurance and management

Elaine Moore

Learning objectives

After studying this chapter, you should be able to:

- Describe the features of a quality management system
- Explain the general principles underlying quality assurance and quality control
- Discuss the differences between clinical sensitivity and specificity
- Outline the different types of documentation found in the medical laboratory
- Outline the standards and regulatory requirements that affect the work of the laboratory
- Define and use tools and techniques required for the continual improvement of quality assurance and management

Introduction

Quality and **quality assurance** (QA) are the processes and practices undertaken by a laboratory that are necessary to ensure a test on a clinical sample gives the correct result and that this finding is delivered to the appropriate clinician in a timely and efficient manner. Quality is essentially about consistency; that is getting procedures right first time and on every subsequent occasion. Quality assurance is all the planned and systematic actions necessary to give adequate confidence that a product or service satisfies the necessary level of quality.

There are a number of QA activities (Figure 19.1) that a laboratory can undertake to assure its service users that the level of quality it offers is of an adequate standard. These include implementing a **quality management system** (QMS), training staff, showing compliance with current regulations and standards, as well as activities such as internal and external quality controls (QCs), and training and competency assessments and audits. The clinical laboratory also needs to ensure that associated clinicians, doctors, and patients are involved in the laboratory service.

FIGURE 19.1
Diagram showing many of the facets of quality assurance.

List three of the facets of quality assurance (QA)?

This chapter will provide you with an overview of how these types of activities lead to QA and quality management relevant to the clinical laboratory.

19.1 Quality control

In addition to QA, **quality control** (QC) is necessary to ensure accurate and consistent clinical results. Quality control is the operational techniques, activities, and requirements necessary to reach the desired level of quality. In QC, the emphasis is on testing and stopping the release of incorrect results from clinical tests. Examples of QC include the positive control and the known negative control which are tested alongside a patient sample to ensure the accuracy of the test results. Another example of QC is the checking of histological slides to ensure that the quality of the preparation and the staining of the tissue elements are of an acceptable standard for diagnostic purposes.

Quality assurance does not necessarily eliminate the need for QC, as this is only one aspect of QA and some products are so critical that testing and checking as undertaken in QC is still necessary, just in case other QA activities fail. Quality testing can be classed as either internal or external.

Internal quality control

Internal QC is essentially performed in the clinical laboratory that is it is 'in house'. Internal QC is concerned with sampling and testing to ensure that the necessary and relevant tests on a clinical sample are, indeed carried out, and that the test results are not released prior to their quality being judged as satisfactory. Thus, the QC ensures that the results supplied by the laboratory are within acceptable limits. The clinical laboratory also performs the tests on control samples alongside clinical samples from patients. Given the composition and concentrations of the controls are known, they act as built-in checks on the process of analysis. Some assays require both high and low concentrations of the analyte to be determined to ensure, for example, that a chemical analyser or a procedure is operating correctly across the range of concentrations that may be encountered in the samples from patients. If the results for the controls vary from their known values, laboratory staff will immediately know that there is a problem that must be rectified before further testing of clinical samples can proceed.

Internal QC is used in manual techniques, such as special staining procedures of tissue sections and immunocytochemistry analyses in histopathology and cytopathology laboratories. Known positive and negative controls are performed simultaneously with clinical samples to ensure

Cross references

You can read more about immunocytochemistry in Chapter 15 'Immunological techniques' and the accompanying volumes, *Histopathology* and *Cytopathology*.

Chapter 3 'Communications in laboratory medicine'.

that the method is both working appropriately and being performed correctly by the staff. Quality control can also be used to determine the specificity and sensitivity (see below) of an individual test. Laboratory assessors from bodies such as Clinical Pathology Accreditation (CPA), which we introduced in Chapter 3, will usually examine QC records as part of the inspection or accreditation process.

External quality assessment

External quality assessment schemes (EQA) assess a laboratory's performance against a defined set of criteria. Laboratory accreditation assessors will look for evidence that a laboratory participates in relevant EQA schemes and that the laboratory's performance is acceptable. All laboratories that perform tests on clinical samples are expected to participate in EQA schemes.

In EQA schemes, an outside agency, commonly known as a scheme organizer, checks the accuracy of a laboratory's test results. This is achieved by the scheme organizer sending known test samples to the participating laboratory for analysis. The laboratory then must test the EQA samples in exactly the same way it would test specimens from patients. The results of these EQA test samples are then sent to the scheme organizer by the laboratory. The data from all the participating laboratories are then collated and presented as a report. These reports show the performance of a laboratory in two ways, by comparison with peer laboratories and by performance against a minimum standard. These reports will also endeavour to highlight areas of good or poor practice and indicate any trends identified when using certain reagents, kits, machines, or methods.

Each laboratory is given a unique code to ensure anonymity and scheme organizers will offer help, advice, or assistance to those laboratories identified as performing poorly by either measure.

Laboratories can participate in a variety of EQA schemes (Table 19.1). In the United Kingdom, many of these schemes are organized by the UK National External Quality Assessment Service (NEQAS).

In addition to QC issues relating to a clinical test or investigation, it is necessary to be aware of their limitations, which are described by the clinical sensitivity and clinical specificity of the test in question.

TABLE 19.1 Examples of UK national external quality assessment schemes

Scheme organizer	Scheme	Comment
UK NEQAS	Blood coagulation	Assesses samples of plasma for blood clotting disorders.
UK NEQAS	Foeto-maternal haemorrhage	Examination of slides to identify the presence of foetal cells in the mother's blood.
UK NEQAS	Biochemistry and sweat testing	Assesses Cl^- and Na^+ in samples.
UK NEQAS	AAFB microscopy	Examination by microscopy for the presence of acid and alcohol fast bacilli.
UK NEQAS	Immunocytochemistry in histopathology	Assesses immunocytochemistry in formalin-fixed, paraffin-processed sections of human tissue and alcohol-fixed cytosporin samples.
UK NEQAS	Histopathology technical external quality assessment (HTEQA)	Assesses the quality of staining methods.

Clinical sensitivity and specificity

To be effective, a medical test is expected to detect abnormalities with a high degree of confidence. How likely is it that an individual who has a positive test has the disease? What are the chances that an individual has a certain disorder even though the test for it was negative? Sensitivity is a measure of the ability of a test to identify those individuals who have a disorder after testing positive. Specificity is a measure of the ability of a test to identify individuals who are healthy after testing negative. Unfortunately, clinical tests are neither 100% specific nor 100% sensitive. Indeed, the two factors show an inverse relationship to each other.

Clinical sensitivity

Clinical sensitivity is the ability of a test to correctly identify individuals who have a given disease or condition. For example, a certain test may have been shown to be 90% sensitive. If 100 people are known to have a certain disease, the test that identifies that disease will correctly do so for 90 of those 100 cases. The other 10 people tested will not show the expected positive result for the test. For that 10%, the finding of a 'normal' result is a misleading *false negative* result. The sensitivity of a test is particularly important when seeking to *exclude* the presence of a disease. For example, a patient can be reassured they do not suffer from acquired immunodeficiency syndrome (AIDS) by demonstrating that antibodies to human immunodeficiency virus (HIV) are not present in blood samples. Screening for HIV antibodies often involves the use of an enzyme-linked immunoassay (ELISA) test, which has greater than 99% sensitivity. However, a sample from a person may give a false negative result if the test is performed too soon (less than 6 weeks) after the initial infection. Such a result would give the person involved a false sense of being disease-free when in fact they are not. The more sensitive a test, the fewer false negative results will be produced.

Cross reference

Chapter 15 'Immunological techniques'.

Clinical specificity

Clinical specificity is the ability of a test to correctly exclude individuals who do not have a given disease or condition. For example, a certain test may be 90% specific. Thus, if 100 healthy individuals are tested, only 90 of those 100 healthy people will be found to be disease-free. The other 10 people (who do not have the disease) will appear to be positive for that test. For that 10%, their 'abnormal' findings are a misleading *false positive* result. Patients who have been told they are suffering a particular illness when they are not may be subjected to unnecessary treatment, which may be potentially painful, and additional expense, and unwarranted anxiety. The more specific a test, the fewer false positive results it produces.

Cross reference

You can read more about clinical sensitivity and specificity in Chapter 5 'Statistics and handling data'.

19.2 Quality management

Quality management is any system where the management has direct control of an organization with regard to quality. Quality is not a *series* of events and therefore is *not* something with defined start and end points, but is a never ending cycle of events that requires the support and co-operation of all the people in the organization. Thus, QM is an approach, attitude, or culture within an organization. All organizations must ensure that the quality of services they provide are maintained and continually improved.

Every stage of a process, system, or project is a possible source of poor quality. The potential to provide poor quality is greatest in complicated systems such as a hospital. However, *learning organizations* have systems, mechanisms, and processes that enhance the services they provide.

They are able to adapt to their environments and apply the results of their learning to achieve better results. Hence, they continually evolve and enhance their capabilities. In 2000, the Department of Health published the document *An Organization with a Memory*, which advocated the NHS needs to learn from its mistakes and also from the mistakes of others. The recommendations that an expert group outlined in this document are designed to ensure that lessons from the past are used to reduce the **risk** to patients in the future. Thus, the NHS has recognized it needs to learn from **errors** and mistakes, and to continually refine and improve its systems and processes; in short to become a *learning* organization. Naturally, all adverse events cannot be eliminated from complex modern healthcare, but by continually learning from past errors, hospitals will be better able to achieve their stated objectives in a sustained manner, not only for the benefit of the hospital, but also for patients and the community they serve.

The International Standards Organization (ISO) develops and publishes standards for international organizations to adopt. One such document, ISO 9001:2008 outlines requirements of a QMS. A QMS is a model that may be employed to give guidance to the selection of appropriate QC activities. It is possible to apply a QMS to most areas, stages, and processes involved in the provision of a product or service. The aim of a QMS is to prevent non-conformity or a non-conformance to a set standard. Non-conformity is any deviation from that which is required. An effective QMS will try to satisfy customers or consumers by concentrating on the service or process. One way of achieving this is by customer surveys and monitoring their complaints. A QMS also documents processes designed to achieve consistency, providing a goal for system control and maintaining the required level of quality at each stage in the process.

There are eight requirements of a quality management system, as defined in ISO 9001:2008. These are:

1. Customer focus
2. Leadership
3. Involvement of people
4. Process approach
5. Systems approach
6. Continual improvement
7. Actual approach to decision making
8. Mutually beneficial supplier relationships

This list forms the basis for any QMS in the clinical laboratory; they all contribute to improving the quality of care for patients.

Customer focused QM meets the needs of the clinicians who treat the patients. Clinical laboratories achieve this through user surveys, monitoring complaints and compliments, and by analysing incidents.

Leadership is provided by management commitment from consultants and senior laboratory managers. If senior management are not involved in the quality process, then it will not work. Leadership is undertaken through the provision of quality policy and health and safety statements, and also through the actions and attitudes of senior managers. Within a pathology laboratory, the senior team may consist of a clinical director, a pathology service manager, and a lead or head biomedical scientist or clinical scientist for each department.

High quality service provision only arises from the commitment of everyone in the organization. It needs everyone to be involved and aware of the systems in place. Indeed, it is often the case that the people performing specific tasks have a better understanding of problems than the manager. The organization needs a well motivated work force who feel valued and who take pride in their work: excellence in the service provided is the responsibility of everybody concerned.

A process approach means individual areas must not be looked at in isolation; rather, the whole process must be examined. Often changing something in one area will have a knock-on effect in another area. For example, changing the patient request form to suit one department may mean vital information is not provided for another department. Once the process has been analysed it is necessary to consider the whole system (that is, a systems approach). For example, a large project set out in the 2004 NHS (England) improvement plan was the 18-week pathway, which had the aim of ensuring that by 2008 all patients are seen and treated by a hospital within 18 weeks of referral by their general practitioner. This approach requires the collaboration of a number of different services such as pathology, radiology, and pharmacy, and was to a great extent successful.

Continual efforts must be made to improve the clinical laboratory service and to improve the accuracy and timeliness of results as an aid to the diagnostic process. Many tools can be used when continually improving the laboratory service; these include *audit* (Section 19.5), *Lean thinking*, *six sigma*, *cause and effect*, and *process redesign*. However, no single improvement tool provides all the answers to improve any one situation and it is essential to select the tool that best fits the one in hand.

Factual approach to decision making means making decisions based on facts and the analysis of data. This allows the clinical laboratory to match the needs of the service users with business planning. For example, to select equipment that not only deals with the current number of tests, but also takes into account predicted future increases in workload. The use of audit, incident reporting, and statistical analysis to predict future trends all help in making the right decision.

Mutually beneficial supplier relationships improve services in a number of ways. Thus, if the laboratory and the suppliers of laboratory equipment, and its maintenance and service contracts, reagents and consumables work together this will improve the quality of the laboratory provision.

Quality management must be led by senior management who usually employ personnel to implement and maintain the QMS and the Clinical Governance framework. These personnel are usually called a quality manager or clinical governance manager. Thus, alongside the eight principles listed above, it is necessary to have a clear clinical governance framework in place in a hospital or clinical laboratory. Clinical governance is a form of QM that has been implemented in some organizations to improve the quality of care for patients. It is the system by which NHS organizations, such as hospital trusts, are accountable to continuously improve the quality of the services they provide and to safeguard high standards of care. The organization can achieve an environment in which clinical excellence flourishes. Clinical governance requires the creation of a culture, as well as systems and methods of working that ensure opportunities for quality improvement are identified in all the organization's services and that over time there is a major increase in the quality of care provided throughout the NHS. It therefore provides a better experience for carers and staff.

The cost of adverse events, errors, and incidents is increasing in the NHS, and clinical governance provides an opportunity to focus upon these problems. Specific types of adverse events are repeated in the NHS, which demonstrates that although the NHS is committed to becoming a learning organization, it has yet to fully achieve that goal.

Improving the quality of medical laboratories

Adverse healthcare events, such as mistakes made in hospitals and in the clinical laboratory can lead to serious harm to a person or even cost lives. How many times have you seen stories in newspapers or reported on the television similar to those listed in Figure 19.2? When dealing with hundreds of samples a day it is easy to forget that there is a person who is waiting to receive treatment as a result of a laboratory investigation. A pathology laboratory that has efficient processes in place is one that strives to deliver the correct test result to the right patient in a timely and efficient manner. This ensures that an appropriate diagnosis is made and effective treatment can be given.

Patients pay price for NHS errors

PATIENTS PAY the ultimate price for NHS errors, says watchdog

Thousands of patients are feared to be dying needlessly every year because of poor communication between hospital staff, faulty equipment and a lack of skills.

Times online
July 26, 2007

PATIENT TOLD "SORRY WE MADE A MISTAKE"

Diagnosed with a life threatening cancer a patient was told she might have just months to live.

After five operations to excise a cyst under the gum, remove the lower jaw and teeth, and reconstruct the face with bone taken from the lower leg, the patient heard some shocking news: The slides from the biopsy of her cyst had been contaminated by cells from another patient, and she had never had cancer in the first place.

Patient wristband errors could have been fatal

The National Patient Safety Agency (NPSA), an agency set up to monitor and collect data on errors and incidents, has issued new guidelines on patient wristbands after more than 2,900 errors were attributed to cases of mistaken identity. The NPSA has received reports of patients being placed in the wrong wards and given the wrong medication and blood due to errors on wristbands.

Some of these mistakes could have been lethal.

FIGURE 19.2
Newspaper articles relating to the breakdown of healthcare and the harm caused to patients.

Once effective processes have been established, the focus is to continually improve the operation of the laboratory. This could mean establishing more efficient ways of working, reducing costs, or improving training opportunities. To improve we must learn; learn from mistakes, from analyses of data, and from others. The process of continual improvement is a never ending cycle.

19.3 Documentation in the clinical laboratory

Laboratory standards, such as ISO 15189:2007 *Medical laboratories – Particular Requirements for Quality and Competence* and ISO 22870:2006 *Point of Care Testing Requirements for Quality and Competence* outline a set of criteria that need to be fulfilled by clinical laboratories (Section 19.4). These standards also include the requirement for a QMS, as outlined in ISO 9001 – *Quality Management Systems – Requirements*. When implementing such standards in a laboratory, written, that is documented, evidence is needed to show that the laboratory is meeting the requirements specified in the standards. The minimum requirements of a QMS for a clinical laboratory are a quality policy (QP), quality manual, and standard operating procedures (SOPs) for the processes in place (see Figure 19.3). There are also requirements to have procedures to control the documents and to control changes made to any processes.

Quality policies (QPs) are short, one page, documents that outline the scope of the service the laboratory intends to provide. A QP contains details of the laboratory's commitment to its users and staff, and outlines the organization's **quality objectives**. It is usually signed by one of the organization's senior managers such as its clinical director. The QP is communicated and available to all staff in the laboratory, usually by displaying it on a notice board and including it in the laboratory quality manual.

A quality manual is a document that outlines how the objectives stated in the QP are to be achieved. Documented sections within the quality manual are linked or referenced to the standard(s) with which the organization is trying to comply. A quality manual contains organizational-specific information, including its organizational structure, processes, and procedures, and the resources needed to implement quality and meet the objectives described in the QP.

Standard operating procedures (SOPs) are written instructions that describe in a clear and concise manner how procedures or methods are to be performed. They include such details as to how individual tasks are to be carried out, that is what is to be done, by whom and when.

FIGURE 19.3
Triangle showing the possible hierarchy of documentation in a quality management system.

They accurately describe the process and methods that are in use in the laboratory and contain essential information such as health and safety instructions and the correct use of equipment and controls. It is necessary to have up-to-date SOPs, that correctly describe what is done to ensure that no step is missed and that the quality of the results is not compromised.

Cross reference

Chapter 4 'Health and safety', discusses risks, risk assessment, and COSHH.

SELF-CHECK 19.2

What is a standard operating procedure and why is it needed?

Document control is an essential part of any effective QMS. Documents associated with the QMS will change with organizational changes, changes in practices, technology, and improvements in processes. Control of these changes is necessary to ensure that the procedures correctly describe the processes, are readily available, and are up to date. For example, if the procedure to issue blood changes, these changes need to be documented so that all staff are aware of the new procedure to follow. If out-of-date procedures are followed, mistakes, such as using incorrect times to incubate microbiology samples may occur, and lead to an incorrect diagnosis.

Change control in the clinical laboratory aims to ensure all changes are assessed and approved by management before they are implemented. For example updating a computer program without considering the effect across the network may render the system unusable in some areas. Thus, changes to procedures or to equipment must be introduced in a controlled and co-ordinated manner to prevent time, effort, and resources being wasted.

19.4 **Standards and regulatory requirements in the medical laboratory**

Clinical laboratories are subjected to rigorous inspections by a number of different assessment and accrediting bodies. Accreditation is the process by which an organization gains recognition that its activities and products have attained a certain level of compliance against a set of defined standards. In the United Kingdom, the United Kingdom Accreditation Service (UKAS) is the sole national accreditation body recognized by government to assess against national and international standards as published by the British Standards Institute (BSI) and the International Organization for Standardization (ISO). The British Standards Institute develops, publishes, and markets standards in the UK. **Standards** are codes of best practice that improve safety and efficiency. The ISO is the world's largest developer and publisher of international standards such as ISO 9001, Quality management systems – requirements. The Clinical Pathology Accreditation Ltd (CPA UK Ltd), which is part of UKAS, works to accredit medical laboratories to a set of standards that incorporate ISO 15189, *Medical laboratories – Particular requirements for quality and competence* and ISO 9001 mentioned above. The Human Tissue Authority (HTA) and the Medicines and Healthcare products Regulatory Agency (MHRA) are two other assessment bodies that undertake inspections of clinical laboratories.

Cross reference

We introduced UKAS in Chapter 18 'Point of care testing' and the CPA in both Chapter 3 'Communicatios in laboratory medicine' and Chapter 18.

Clinical Pathology Accreditation standards

Organizations that declare their adherence to a defined standard of practice and have their ability to attain these standards independently confirmed, that is they are 'accredited' can reassure their users of the quality of service they provide. Clinical Pathology Accreditation provides a means by which this can be achieved both by clinical laboratories and also by providers of external quality assessment (EQA) schemes. The CPA has the authority to grant accreditation status to clinical laboratories that meet its standards. The standards adhered to by CPA are based on ISO 15189 *Medical laboratories – Particular Requirements for Quality and Competence,* which is an international standard that incorporates some elements of ISO 9001. Theses standards cover eight major areas:

1. Organization and quality management system (A standards)
2. Personnel (B standards)
3. Premises and environment (C standards)
4. Equipment, information systems, and materials (D standards)
5. Pre-examination process (E standards)
6. Examination process (F standards)
7. Post-examination process (G standards)
8. Evaluation and quality assurance (H standards)

Organization and QMS encompasses areas, such as the needs and requirements of users, the QP document, the quality objectives and plans of the laboratory, the quality manual, document control procedures, and control of process and quality records.

Personnel relates to areas concerned with personnel management, staff induction, records and meetings, staff training, and education. In contrast, premises and environment, as the name implies, encompass the work areas, staff and patient facilities, and also health and safety in the laboratory. Equipment, information systems and materials covers the procurement and management of equipment, materials, data, and information.

The pre-examination process encompasses information for users, specimen collection, and handling, requirements for specimen request forms and transport of specimens, while the examination process sets standards relating to the selection and validation of the examination and the quality of the examination that is performed on the specimens. Post-examination process covers areas, such as the final report and how results are reported.

Evaluation and QA investigates how the laboratory assesses user satisfaction, the laboratory audit process, and quality improvement processes in place in the laboratory.

CPA assessment process

The assessment *process* involves the laboratory enrolling with the CPA, which then inspects the laboratory every four years. The inspection process is undertaken by trained assessors who perform an external audit (Section 19.5) of the laboratory over 1 or 2 days. The inspection team consists of a regional assessor and at least two peer assessors. The regional assessors are employed full time by UKAS and co-ordinate the assessment visits. Peer assessors are drawn from practising consultant pathologists, clinical scientists, and biomedical scientists. They accompany the full time assessors and act as experts in specific subjects such as immunology or biochemistry.

Blood safety and quality regulations

The Blood Safety and Quality Regulations (BSQR) changed two European Blood Directives (Directives 2002/98/EC and 2004/33/EC) into UK law. The UK has a duty to translate the European Directives into national legislation. The BSQR require hospital blood banks and blood establishments to maintain a quality system based on the principles of good laboratory practice (GLP), which is a set of principles that provides a framework within which laboratory work is performed. The Medicines and Healthcare Products Regulatory Agency (MHRA) assesses the compliance of hospital blood banks, and blood establishments to the BSQR.

The laboratory is required to submit an online compliance report to the MHRA, which then assesses the online report against the BSQR. The MHRA may decide to visit the laboratory and undertake an onsite assessment. This visit is usually conducted over one day by a single assessor. The assessor checks that there is accurate and complete **traceability** of blood and blood products. He or she will also examine the records kept by the hospital laboratory and wards, and check records and documents of satellite sites that receive and store blood and blood components.

One of the BSQR requirements is that the laboratory notifies the MHRA and serious **hazards** of transfusion (SHOT) of any *serious adverse blood reactions and events* (SABRE). This is undertaken through an online reporting system that can be accessed through the MHRA website. The reporting of SABREs allows the MHRA and SHOT teams to identify any trends in blood transfusion incidents.

Human tissue act

The Human Tissue Act (England) came into force in 2004. It arose from concerns raised by events such as those at Bristol Royal Infirmary (BRI, 1984–1995) and the Alder Hey Children's Hospital Liverpool (prior to 1999) where it was established that organs and tissues from children had been removed, stored, and used without proper consent from next of kin. Subsequent government reports showed that the storage and use of organs and tissues from both adults and children had been widespread in the past. It became clear that current law in this area was not as clear and as consistent as it might have been for professionals or families. The purpose of the Human Tissue Act is therefore to provide a legal framework for issues relating to the taking, storage, and use of human tissue and organs and whole body donation. The Act made consent the foundation for the lawful storage and use of human bodies, body parts, organs, and tissue, and the removal of tissue from the deceased, and regulates the use of tissues and cells that are to be used for transplant and stem cell research from living donors.

The Act outlined the need for a regulatory authority to be appointed: the Human Tissue Authority (HTA) is the regulatory body appointed in the UK. The HTA ensure that the activities it regulates are undertaken in a proper manner, and provides advice and guidance related to the Human Tissue Act and also to the EU Tissue and Cells Directives, which is also part of the Quality and Safety Regulations (see Chapter 4). These laws ensure that human tissue is used safely and ethically, and with proper consent.

The HTA ensures these laws are followed by setting standards that are clear and reasonable, helping people to understand the legal requirements, and providing codes of practice, advice, and support in nine different areas. These are anatomy, human application, stem cells and cord blood, public display, research, post-mortem, coroners, transplants, and DNA.

The HTA issues codes of practice that give guidance to each of the different areas. The codes differ for each, but all broadly cover the need for consent, implementation of governance, and quality systems including control of documentation and records. For example, the post-mortem standards cover consent for a hospital post-mortem, governance and quality systems, premises, facilities and equipment, and disposal of clinical material.

The laboratory or mortuary is required to complete an online compliance form which is sent to the HTA. The HTA then assess the online report against the appropriate set of standards. If the report is deemed satisfactory, the HTA will issue the laboratory or mortuary with an HTA Licence. It is unlawful to undertake certain activities, such as post-mortems, storage of the

Cross reference
You can read more about the Human Tissue Act and related issues in Chapter 3 'Communications in laboratory medicine'.

Cross reference
Chapter 4 'Health and safety'.

deceased, and storage of body parts without this licence. As part of its role to ensure compliance with the Human Tissue Act, the HTA will inspect the laboratories, mortuaries, and other establishments that fall under its remit. An inspection team consisting of two or more people will usually spend 1 or 2 days in the laboratory conducting an audit against the HTA standards. The HTA require that a named individual, known as the *designated individual*, takes responsibility for ensuring that suitable practices are used when undertaking licensed activities such as post-mortems. Following an inspection, a report is written and any areas that need improving are identified together with a timescale for their rectification.

The Human Tissue (Scotland) Act was introduced in Scotland in 2006. The HTA Codes of Practice also cover the Human Tissue (Scotland) Act in most, but not all areas. Where the HTA refers to a coroner in England and Wales, the equivalent person in Scotland is the *Procurator Fiscal*.

Point of care testing, requirements for quality and competence

Cross reference

Chapter 18 'Point of care testing'.

Point of care testing (POCT) can be defined as testing at or near the site of patient care, which increases the likelihood that the patient will receive the result of a clinical test in a timely manner. Points of care testing is often used in A&E departments, neonatal care units, GP surgeries, or even in the home of a patient (see Chapter 18).

The ISO standard, ISO 22870:2006 *Point of Care Testing (POCT), Requirements for Quality and Competence* outlines the requirements for POCT. The clinical laboratory must ensure that POCT is co-ordinated effectively and efficiently in the hospital environment. Thus, POCT policies and procedures need to be in place that outline how the equipment must be used. Staff must be trained on use of the equipment and records of this training maintained. Medicolegal considerations and the requirements of the Data Protection Act must be borne in mind when establishing these procedures. The laboratory must be involved in any purchases of new equipment for use outside of the laboratory, as well as being involved in training staff to use it correctly. Most clinical laboratories have a dedicated person to look after POCT equipment, and liaise with ward and departmental staff about its correct use and cleaning. The laboratory also has a duty to ensure that the equipment is serviced in accordance with manufacturer's requirements and recommended local and/or national QA schemes (NEQAS).

Point of care testing is not assessed or accredited in isolation, but as part of the overall laboratory accreditation process.

Key Points

Currently, POCT that takes place in the community, such as at GP surgeries or patient homes, is neither assessed nor accredited.

Additional standards affecting clinical laboratories

Clinical laboratories are likely to be part of a larger organization, such as a hospital. In such cases they may be subject to hospital regulatory requirements such as:

- Standards for Better Health
- Clinical negligence scheme for trusts

Standards for Better Health came into being from April 2005 as a new performance framework for the NHS.

This framework outlined the level of quality all organizations providing NHS care are expected to meet or aspire to in England. These standards provide a common set of requirements that are applied across all healthcare organizations to ensure that health services are both safe and of an acceptable quality, and provide a framework for continuous improvement in the overall quality of care patients receive. Thus, the standards describe the level of quality that health-care organizations, including NHS Foundation Trusts, and private and voluntary providers of NHS care, are expected to meet in terms of safety, clinical and cost effectiveness, governance, patient focus, accessible, and responsive care, care environment and amenities, and public health. Standards for Better Health form a key part of the performance assessment by the care quality commission.

Clinical negligence scheme for trusts (CNST) standards are assessed by the NHS Litigation Authority (NHS LA). CNST is an insurance scheme covering litigation against a hospital. It requires the development and implementation of clinical incident reporting systems. NHS Trusts must have basic systems in place across some of the organization to attain even the most basic CNST standards (level 1). There must be a widespread comprehensive system in place to reach the highest level (3). Attainment of the highest level of the standard indicates the most sophisticated processes for incident reporting and risk detection and prevention and thereby results in the smallest premiums payable by the hospital.

SELF-CHECK 19.3

What are the main bodies that inspect clinical laboratories?

19.5 Tools and techniques for continual improvement

We will now outline some of the tools used to check or assure that the processes and methods used in a clinical laboratory comply with the required standards. These tools include audit, incident reporting, and root cause analysis (RCA).

Audit

ISO9001 defines an audit as 'A systematic and independent examination to determine whether quality activities and related results comply with planned arrangements and whether these arrangements are implemented effectively and are suitable to achieve objectives'. Hence, audit is an integral part of any QMS and is a requirement for accreditation by the CPA (Section 19.4) and other clinical laboratory and hospital-wide standards. All audits are a means for continually improving the service provided by the laboratory.

The concept of audit is to gather information by means of observation, interview, and sampling to identify areas where improvements can be made. This is essential if a process is to continually improve. An audit will also identify those processes that are working well, and so give an opportunity to learn from good practice and to allow that knowledge to be transferred to other processes. Thus, audit enables organizations to evaluate their processes, determine deficiencies, and generate cost effective and efficient solutions to those problems.

Key Points

Audit is used to measure an organization's ability 'to do what it says it is going to do'.

All audits are performed to check practices against procedures and to thoroughly document any differences. However, audits are only sampling exercises and so they cannot confirm that all aspects of a process are being complied with at all times.

Audits fall into four main types:

1. Vertical audit
2. Horizontal audit
3. Witness or examination audit
4. Self-assessment audit

You can see an illustration of some audit methods in Figure 19.4.

A vertical audit examines a single item, following it from start to finish. For example, this could be the journey a single clinical sample follows once it is received by the laboratory to the result of the tests performed on it being issued to the relevant clinicians. In contrast, a horizontal audit examines one element in a process that is performed on more than one item.

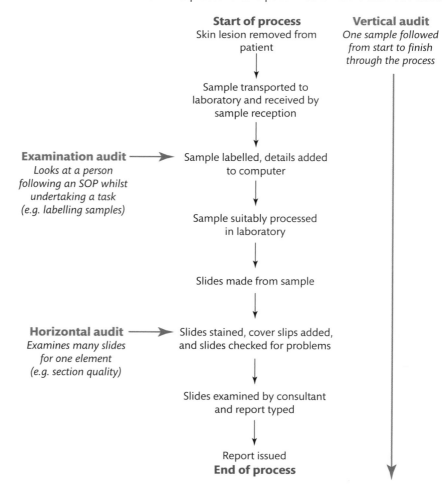

FIGURE 19.4
Diagram showing the different methods used when undertaking an audit. See also Table 18.2.

This could, for example, be an examination of 20 request forms to ensure that details of all 20 patients have been entered correctly on the computer system. A witness audit examines a person undertaking a task. This may be observing a person labelling samples to check that he or she have read and are following the relevant SOP. Self-assessment audit is a careful evaluation that is usually performed by the organizations own management, which results in an opinion or judgement as to its effectiveness and efficiency.

Audits are classified as internal, external and cooperative types. Internal audits are conducted by clinical laboratory staff themselves. For example, a biomedical scientist in haematology could conduct an internal audit of the microbiology laboratory. An internal vertical audit would be one conducted by laboratory staff on a clinical sample subjected to the laboratory process. The purpose of an internal audit is to confirm that everything is in order and as expected and to identify any **non-compliance** with the standards that are applicable to the laboratory. It will identify any deficiencies in the QMS and make recommendations of any actions that need to be taken to improve the laboratory. Naturally, the type and content of the internal audit will vary according to the size and activities of the laboratory concerned.

An external audit is conducted by an outside person or body, such as the CPA, who have an interest in accrediting the laboratory. An external horizontal audit is conducted by an external person(s) on a specific element such as staff training records. A co-operative audit is conducted between the laboratory and another party for mutual benefit, such as clinical audits, user satisfaction surveys, or benchmarking activities.

Before any audit can be undertaken, it is necessary to prepare an audit plan. The plan is called an audit schedule and contains information about what audits are to be undertaken, when they will be undertaken and who will perform the audit. The audit schedule is discussed and agreed by senior management at the beginning of the year. You can see an example of an audit schedule in Table 19.2.

SELF-CHECK 19.4

What methods of audit do external inspectors undertake when visiting the laboratory?

Performing an audit

The four stages in an audit are:

1. Prepare the audit checklist
2. Undertake the audit
3. Write a summary of the audit
4. Undertake **corrective action** to improve the laboratory process

TABLE 19.2 **Audit schedule**

Audit area	January	February	March
Audit of laboratory equipment	Horizontal audit on equipment such as analysers, centrifuges, microscopes used in the laboratory.		
Staff records		Horizontal audit on CPD records of staff.	
Transport of samples			Vertical audit on receipt of samples from general practioners to the issuing of results of clinical tests.

An audit checklist is a list of the questions to be asked relating to the set standard(s). Hence, if the set standard was: 'There shall be up to date information for users, a suitable checklist question is, 'Is there up to date information available for users?' The audit checklist must also contain the names of the auditors (the person undertaking the audit), the proposed date of the audit, a unique reference number and the area(s) being audited. The department will appoint an auditee; a person who will assist the auditor(s), accompanying them around the department to verify factual information. An audit is the process of going into a department and asking the questions in the checklist. Essentially, these questions will be of the type: WHAT ...? WHERE ...? WHY ...? WHO ...? HOW ...? SHOW ME ...? Evidence to support the answers to these questions must be examined; the auditor(s) must *see* the evidence. If it is not seen, it does not exist! Audit is all about seeing the evidence of compliance with the standard being audited.

After an audit has been completed, a summary of *all* the findings are documented. The summary outlines areas of compliance, that is where the laboratory meets the standard, partial compliance, where the laboratory has some evidence of meeting the standard, and points of non-compliance, where evidence of meeting the standard was not provided. This is an essential stage in an audit and should not be missed out. It gives an opportunity to highlight areas of good practice, what the laboratory is doing well, in addition to identifying areas for improvement. Areas that need to be improved are issued with a non-compliance form, which outlines the area of non-compliance against the standard, what corrective actions need to be taken, by whom and when. When all identified non-compliances against the standards are corrected, the audit can be said to be closed. However, the audit process is a never ending cycle that leads to continual improvement in the laboratory provision and in the service it provides to laboratory users. You can see an illustration of the cycle in Figure 19.5.

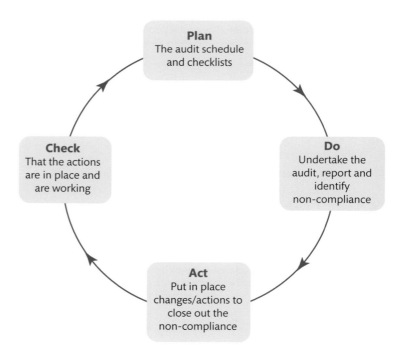

FIGURE 19.5

The audit cycle is a never ending process of improvement.

Incident reporting and root cause analysis

Incident reporting is a major tool for examining what happened when an error or problem occurred. Adequate and appropriate incident reporting systems are necessary to provide a core of sound, representative information on which to base any analysis of errors or problems and to make recommendations to prevent their repetition. Experience in other sectors, such as the airline industry (Figure 19.6), has demonstrated the value of systematic approaches to recording and reporting adverse events.

It is possible to identify a number of barriers that can prevent active learning, but the NHS can draw valuable lessons from particular approaches used in other sectors. The first approach is one of 'safety cultures', where open reporting and balanced analyses are encouraged in principle and by example. This can have a positive and quantifiable impact on the performance of organizations. In contrast, 'blame cultures' may encourage people to cover up errors for fear of retribution. This acts against identifying the true causes of failure, because they focus heavily on individual actions and largely ignore the role of underlying systems. Thus, organizational culture is central to every stage of the learning process: from ensuring that incidents are identified and reported through to embedding the necessary changes deeply into practice. The second approach is 'local risk reporting systems', which provide the bedrock for onward reporting to regional or national systems. These are now more widely used in the NHS. Incident reporting systems are now in place across most NHS organizations.

The National Patient Safety Agency (NPSA) helps to improve the quality of safe patient care by informing, supporting, and influencing healthcare organizations and individuals working in health. It analyses national reports on patient safety incidents, identifies risks, and recommends actions. The NPSA collates incident data and issues guidance to the NHS on how to improve processes and systems, thereby reducing risks and harm to patients and staff alike.

Adverse incidents involving medical devices, including laboratory equipment, are reported to the MHRA. Information is logged on a central database, which contains details of over 48,000 incidents. Incidents are assigned to a specific level of investigation depending on the risks involved. The outcomes of these investigations are subject to formal reviews. Patterns or clusters of incidents can thus be identified, subjected to further risk assessment procedures, and investigated where necessary.

When an incident reveals equipment-related safety problems, the MHRA will produce a *hazard* or *safety* notice for distribution to appropriate users so that action can be taken to withdraw or modify the device.

NHS told to learn from its mistakes

YOU WOULD NOT just decide to forget about recovering the black box after an air crash. So why should it be thought so strange to want to learn from every accident in healthcare?' *Sir Liam Donaldson speaking on patient safety to Health Service non-executive directors in Bristol, 20 November 2001*

FIGURE 19.6
Newspaper article that emphasizes the need to learn from incidents involving patient safety.

Detecting and accurately recording errors is a fundamental step in learning from experience. It is common sense to recognize that knowledge of what is wrong is required before steps can be taken to put it right.

Reporting of accidents and incidents must be seen as professional behaviour rather than as disloyalty to other staff or the NHS. All organizations, including the NHS must operate a genuine two-way communication system, not just a 'top down' system from management. Communication systems need to be in place to allow people to see what has changed as a result of an incident or the reporting of a near miss. An organization needs to recognize staff concerns. The emphasis of management should be on personal and service benefits, rather than on threats. Within the NHS, more emphasis needs to be placed on creating an informed culture; to raise the awareness of the costs of not taking risk seriously. This includes the human costs, to staff and patients, as well as financial ones.

Not all unsafe systems produce bad outcomes all the time. However, more attention must be focused on **healthcare near misses**, where an accident is just avoided, in addition to the incidents themselves. This approach can remove the emotion associated with an incident and allow learning to take place more effectively. It is also easier to keep **near miss** data anonymous, itself a factor that encourages reporting. It is essential that concerns, near misses, and incidents can be reported without fear.

The potential for disasters may exist but for any number of reasons those may not occur, or occur very rarely. Most accidents have the potential to produce serious injury but do not do so in practice; either because of some intervention or compensation or simply through good fortune. Confining analysis to and learning from only those events that resulted in serious harm risks skewing learning towards a small proportion of accidents and missing important lessons that could prevent future adverse events. These points were also discussed in Chapter 4.

Industry has estimated that for every major injury, 29 minor ones and 300 accidents with no injury occur. Thus, to some extent the effectiveness of a reporting system can be judged by the relative proportion of minor incidents to more serious accidents: the greater the proportion of minor incidents reported, the 'better' the reporting system is working. This is illustrated in Figure 19.7.

Root cause analysis

Root cause analysis (RCA) is a powerful tool that identifies, records and visualizes the possible causes of a problem, error, or incident. It is undertaken as an investigation into the error to

Cross reference
You can read more about professional behaviour and duties in Chapter 2 'Fitness to practise'.

Cross reference
Chapter 4 'Health and safety'.

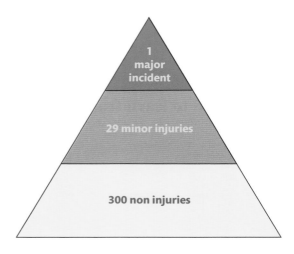

FIGURE 19.7
Triangle of incident relationships showing the proportion of near misses to harm to or death of a patient. See also Figure 4.6.

determine its basis that is its *root cause*. When an incident has occurred it is imperative that an investigation identifies its root cause to ensure that improvements can be made and to prevent the same incident reoccurring. The basis of RCA is to break down the problem into smaller, more manageable tasks, which allows the problem to be identified and described. This approach will assist in finding a solution or corrective action that will prevent the same problem reoccurring.

Many methods can be used when undertaking RCA, including a cause and effect diagram. Cause and effect diagrams provide a logical means to analyse a problem to get at its root cause. Look at Figure 19.8. This show you an example of how a cause and effect diagram can be used to determine the cause of why a laboratory analyser has failed and is giving inaccurate results for tests on patient samples. Although the cause of the failure is initially unknown, various possibilities need to be examined in detail to determine why the equipment failed. These possible causes can be mapped on the cause and effect diagram as shown and each one analysed until the root of the problem is discovered. An action plan can then be put in place to ensure the same problem does not occur again.

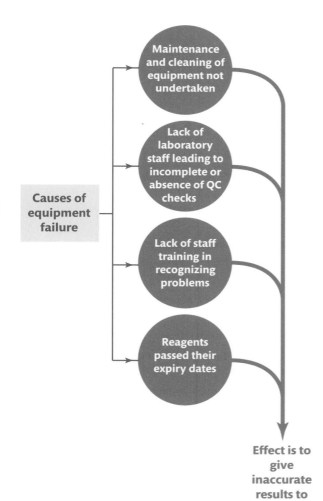

FIGURE 19.8

Cause and effect diagram illustrating how an equipment failure should be investigated.

SUMMARY

- The necessity to provide services of appropriate quality in clinical laboratories, as well as the need to satisfy the users of laboratory services has led to increasing numbers of regulations governing their operations and assessment

- Standards against which clinical laboratories are assessed include ISO 15189 *Medical Laboratories, Particular requirements for quality and competence,* which includes the standards contained in ISO 9001 – *Quality Management Systems – Requirements*

- Clinical laboratories are subjected to inspection and assessment by a number of different bodies, namely the Clinical Pathology Accreditation, The Human Tissue Authority and the Medicines and Healthcare Products Regulatory Authority

- Quality assurance, quality control, and the implementation of quality management systems has allowed clinical laboratories to continually evaluate, monitor, and improve their processes and systems, and deliver a high quality laboratory service to their users

- A variety of techniques, such as audit, can be used by the laboratory to internally assess their compliance to set standards

- Following good laboratory practices ensures the integrity of the sample audit trail in the laboratory and allows the cycle of continual improvement to be embedded into the laboratory culture

FURTHER READING

- **Burnett D. *A Practical Guide to Accreditation in Laboratory Medicine*. ACB Venture Publications, London, 2002.**

  This book clearly explains the accreditation process and the accompanying documentation needed by clinical laboratories for the inspection process.

- **Department of Health *An organization with a memory*. Department of Health, London, 2000.**

  This is a report on learning in the NHS from adverse events.

- **MHRA *Rules and Guidance for Pharmaceutical Manufacturers and Distributors*. Pharmaceutical Press, London, 2007.**

  Section II: Guidance on good manufacturing practice is worth dipping into.

- **Vorley G, Tickle F. *Quality Management, Tools and Techniques*, 5th edn. Quality Management & Training (publications) Limited, London, 2002.**

  Sections 3 and 4 give accounts of quality management systems, ISO9000, and audit.

Useful Websites

- www.cpa.org
- www.hta.gov.uk
- www.iso.org
- www.ukneqas.org.uk
- www.ibms.org

- www.mhra.gov.uk/index.htm
- www.npsa.nhs.uk
- www.nhsla.com
- www.dh.gov.uk
- www.ukas.com
- www.thecqi.org
- www.bsi-global.com

 QUESTIONS

1. Which, if any, of the following has a responsibility to control quality in a medical laboratory?

   (a) The quality manager

   (b) Management

   (c) The staff

   (d) Laboratory manager

   (e) All of the above

2. Indicate which of the following statement(s) is(are) **TRUE**?

   (a) Quality Assurance, Quality Control, and the implementation of Quality Management Systems has allowed medical laboratories to continually evaluate, monitor, and improve their processes and systems thereby delivering a high quality laboratory service to their users.

   (b) Quality Assurance, Quality Control, and the implementation of Quality Management Systems has allowed the medical laboratory to eliminate all errors thereby delivering a high quality laboratory service to their users.

   (c) Quality Assurance, Quality Control, and the implementation of Quality Management Systems has allowed medical laboratories to employ more staff to undertake audits in order to monitor their processes and systems and improve patient care.

   (d) Quality control means clinical laboratories do not have to check their test results before releasing them to the clinicians.

   (e) External Quality Assessment involves outside agencies checking the clinical test results before they are sent to the requester.

3. Outline the process of quality control as performed in the laboratory. Indicate how it differs from external quality assessment?

4. What is meant by *document control*? State why document control is necessary?

5. What techniques can be employed to improve the laboratory service?

6. Outline the structure of a *learning organization*?

7. Why is it necessary to report concerns about incidents that occur in the laboratory?

8. Give an example of how a cause and effect diagram (Figure 19.8) can be used to investigate an error?

9. Why is it necessary to have an audit trail outlining the path a clinical sample takes through a laboratory?

Answers to self-check and end of chapter questions are available in the Online Resource Centre accompanying this book.

 Go to www.oxfordtextbooks.co.uk/orc/glencross

20

Personal development

Georgina Lavender

Learning objectives

After studying this chapter, you should be able to:

- Describe what is meant by personal development and continuing professional development
- Discuss opportunities and events that in everyday life contribute to personal development
- Discuss the types of continuing professional development needed to satisfy the requirements for registration with the Health Professions Council
- Describe the opportunities provided by the Institute of Biomedical Science for professional development
- Identify opportunities within the workplace to extend personal development opportunities
- List the mandatory training that must be provided by an employer
- Select and utilize appropriate study skills for different situations
- Describe the structure of a generic performance appraisal scheme
- Explain how the performance appraisal and development structure within the NHS is of benefit to a biomedical scientist

Introduction

Personal development is any means by which an individual is able to expand on his or her existing knowledge, skills, and talents in all aspects of their lives. Newly qualified biomedical scientists leaving behind the academic environment (Figure 20.1) and entering professional life must not imagine that this is the end of their learning. Personal development is ongoing and can be applied to anything that he or she chooses to do.

Anyone with a degree even remotely connected with science is aware that science is constantly evolving, and that today's innovations and discoveries are often old news and outdated tomorrow. Anybody listed on a professional register is required to continually reaffirm and review their knowledge base within their chosen profession; biomedical scientists are no exception. Indeed, biomedical scientists are expected to be able to demonstrate a designated level of **continuing professional development** (CPD) to remain on the **Health Professions Council** (HPC) register and every 2 years at the renewal of their registration must declare they

Cross reference
The HPC CPD requirements were discussed in Chapter 2 'Fitness to practise'.

FIGURE 20.1
Undergraduate biomedical science students learning basic laboratory skills in a university laboratory.

have been undertaking CPD to a predefined standard and crucially must be able to provide evidence to support this claim.

There is a saying that *it is never too soon or too late for learning* and this applies to both *professional* study and development and to that done for pleasure. **Lifelong learning** is often a concept that many people adopt and embrace without ever realizing that they are doing so. Lifelong learning is associated with many unexpected places, a newspaper article, a television documentary, even helping children with homework or trying out a new recipe in the kitchen! This learning can be carried across to biomedical science also, and a keen biomedical scientist will easily be able to meet professional development targets, almost without realizing that they are doing so, as there are so many opportunities within the working environment and elsewhere for the lifelong learning that will constitute CPD.

A framework for higher education qualifications exists across Europe, the European Qualifications Framework (EQF), which is being adopted and has been referenced to the different systems within England, Wales, and Northern Ireland by the Qualifications and Curriculum Development Agency (QCDA). Scotland is undertaking a similar referencing exercise to its qualifications too. These systems describe the educational level of an award or describes the expected achievement of knowledge and skills of the student at the end of a course of study (see also Box 20.1).

Of the eight EQF levels, the principal ones of interest to biomedical scientists are:

- EQF Level 5 (Intermediate (I) level or levels 8/9 in Scotland: Foundation degree, HND)
- EQF Level 6 (Honours (H) level or level 10 in Scotland: Bachelors degree)

BOX 20.1 Qualification levels and biomedical science training

A biomedical science undergraduate would be described as working towards an EQF Level 6 qualification, whereas a **trainee** biomedical scientist with an honours degree, but not yet HPC registered would be described as already being in possession of an EQF Level 6 qualification and being able to work towards an EQF level 7 qualification. These are indicators of what you are already able to achieve and what you would be able to work towards. These factors are important when choosing your study options.

- EQF Level 7 (Masters (M) level or level 11 in Scotland: Masters degree)
- EQF Level 8 (Doctoral (D) level or level 12 in Scotland: Doctoral degree)

This chapter will examine the personal and professional development of biomedical scientists (Figure 20.2) and, hopefully, help you to find and utilize a range of information resources that are there for the taking. All healthcare professionals have a responsibility to participate, alone or with work colleagues, to improve their knowledge and skills as a practising biomedical scientist.

20.1 Continuing professional development

Continuing professional development may be defined as the ongoing learning that is carried out throughout your professional career. It is needed in order to remain up to date in all aspects of a chosen profession. It is also necessary for promotion and is often a mandatory requirement to remain on a professional register. The two bodies with responsibilities for regulating biomedical science, the Institute of Biomedical Science (IBMS) and the Health Professions Council (HPC), have their own definitions for CPD.

Key Points

The IBMS defines CPD as a process of lifelong learning, which enables you to expand and fulfil your personal and professional potential, as well as meet the present and future needs of patients, and deliver health outcomes and priorities. It assures that you meet the requisite knowledge and skills levels that relate to your evolving scope of professional practice.

The definition of CPD by the HPC is the way health professionals continue to learn and develop throughout their careers so they keep their skills and knowledge up to date and are able to work safely, legally, and effectively.

FIGURE 20.2

A group of professional biomedical scientists working in a clinical laboratory.

What exactly do these definitions mean to a newly qualified and registered biomedical scientist? They mean that, as a professional person, on a professional register, a biomedical scientist has a legal responsibility to meet the standards of the HPC and also to remain abreast of current trends and developments within his or her working environment.

Requirements of the Health Professions Council

It is a requirement of the HPC that all **registrants** including biomedical scientists must undertake CPD to remain on their register. Whenever he or she is asked to renew their registration, they must declare that they are complying with the HPC standards relating to CPD. The HPC is able to ask for evidence from registrants selected at random, for factual support that they are meeting the set standards. The standards of CPD state that a registrant on the HPC register must:

- Maintain a continuous, up-to-date and accurate record of CPD activities
- Demonstrate that CPD activities are a mixture of learning activities relevant to current or future practice
- Seek to ensure that CPD has contributed to the quality of their practice and service delivery
- Seek to ensure that CPD benefits the service user
- Present a written profile containing evidence of CPD upon request

The HPC has extensively listed activities that constitute CPD (Table 20.1), together with examples of suitable supporting documents. Of the total registrants, 2.5% will be asked to submit their personal CPD records at the biennial point of renewal of registration.

SELF-CHECK 20.1

Why is continuing professional development necessary for biomedical scientists?

20.2 Opportunities for personal development in the workplace

An induction period is normally provided with any new post or even a change in workplace. Information regarding the new employer is often given within the first few weeks of employment to all new employees to introduce them to specific organizational philosophies and policies. Induction should always be treated as an opportunity for personal development. Indeed, a formal induction programme needs to be given at the start of working at a new organization.

Induction courses are also an introduction to other employees. Presentations are ideal opportunities to find out more about your place of work, and ask seeking questions of a new employer. Frequently, presentations are given by senior members of staff and are often an excellent way to find out who does what within a large organization and to find out what they look like. It can save embarrassment at a later date if you can identify your senior managers!

Certain aspects of induction training have common themes and have to comply with legal employment and health and safety issues:

- Introduction to health and safety
- Safe manual handling
- Fire policy and evacuation of the premises

- Child protection
- Cardiopulmonary resuscitation

Introduction to health and safety

Any introduction to health and safety will include an outline of the Health and Safety at Work Act, which was described in Chapter 4. This act places legal obligations on both employers and employees. Employers will outline the local policies and procedures that are in place to comply with this legislation and highlight the responsibilities of the employee to uphold these local rules. All employees should also be given the information needed to research their own health and safety concerns, such as access to health and safety publications (Figure 4.2).

TABLE 20.1 **Examples of opportunities for professional development in the workplace**

Work-based learning
Learning by doing
Case studies
Reflective practice
Audit of patients
Coaching from others
Discussions with colleagues
Peer review
Gaining and learning from experience
Involvement in the wider work of your employer (for example, being a representative on a committee)
Work shadowing
Secondments
Job rotation
Journal club
In-service training
Supervising staff or students
Visiting other departments and reporting back
Expanding your role
Significant analysis of events
Filling in self-assessment questionnaires
Project work
Evidence from learning activities undertaken as part of your progression on the NHS Knowledge and Skills framework (KSF)

Manual handling

Local policies will define what manual handling training is necessary for employees as part of their legal obligations. All manual handling must be carried out safely to protect the employee against serious injury. Manual handling is often subdivided into lifting people and lifting inanimate objects such as heavy boxes, awkward items of furniture, rubbish containers, or computer equipment. However, as laboratory staff are rarely required to lift patients, the manual handling in regard to direct patient care rarely applies.

Fire policy and evacuation of premises

It is essential that all employees are aware of the local procedure to be followed if they detect a fire and need to raise the alarm or if they have to evacuate the premises in response to a fire alarm. Among the first tasks of a new employee is to become familiar with fire exit routes, fire alarm points, and the fire fighting equipment available in the workplace.

Child protection

Local policies to protect both children and adults working with children against the physical, mental, and emotional abuse of children are in place. Naturally, all local policies, like health and safety policies, must comply with the wider legal framework.

Cardiopulmonary resuscitation

Training in cardiopulmonary resuscitation techniques is often a requirement for anyone employed in healthcare professions. All employees within healthcare must be aware of their individual responsibilities should they find a patient requiring urgent medical assistance. Within the NHS, different levels of training apply to different groups of staff.

Human resource issues

Issues related to human resources are also opportunities for employers to introduce policies that may relate to:

- Sickness and absence procedures
- Payroll queries
- Occupational health and vaccination procedures
- Holidays
- Dealing with the press
- Disciplinary and grievance procedures

Sickness and absence procedures

Sickness and absence procedures give clear instructions about the behaviour expected if employees are too ill to attend work or if there is an unforeseen emergency that prevents them from attending work. It is essential that employees follow such policies and are made aware of the consequences to the department if they fail to comply. Staff sickness is monitored by human resource and occupational health departments, and also may affect the pay of the individual concerned.

Payroll queries

Pay is always an emotive subject and new employees must understand the information given on their payslip. They must also be aware of the procedures in place to clarify pay issues, for example, what should be done if there is a discrepancy between pay given and that expected. Most employers will explain taxation and national insurance procedures and their own system for giving extra payments, for example, when overtime has been worked.

Occupational health

Occupational health is the department that is devoted to the health and wellbeing of employees. Naturally, they are involved in health and safety aspects of the workplace, set policies for preventative medicine (health screening and vaccinations), and take a non-biased approach when employees may become unfit for their duties or return to work following prolonged illness.

Holidays

Holiday entitlement is usually written into a contract of employment, but employers and departments are likely to have their own individual policies regarding booking holidays in advance and to record any leave taken. Annual leave is the predominant reason for taking holiday from work, but there are other reasons that an employee may be able to take time off work without encroaching on their holiday entitlement. It is therefore necessary to be aware of one's full holiday entitlement. Other discretionary leave may include jury service, carer leave, and bereavement leave.

Dealing with the press

Most employers have a policy in place that employees should follow when dealing with the press. Frequently, employees are not able to talk to the press without appropriate permission and to do so is a disciplinary matter. This also applies to all organizational human resource issues.

Disciplinary and grievance procedures

All employers have their own disciplinary and grievance procedures, which must be consistent with current employment law. Employees must be made aware of basic employment law but also of their employers' individual policies, as there are certain channels to follow if they are unfortunately embroiled in a difficult situation. Cases that go to court or to an employment tribunal are frequently won or lost on procedural issues, rather than on what has actually happened; it is therefore desirable to resolve disciplinary and grievance issues at the earliest possible stage.

20.3 Training and development

Most employers have a training and development department that provides access for all staff to a variety of generic training courses, appropriate to their roles within the organization. It is always worthwhile to be aware what training may be on offer from the employer outside the immediate laboratory environment. Many courses are available on request, and it may be stating the obvious, but often you will only get what you ask for. When agreeing a personal development programme, using training facilities that are readily available within the organization can reduce cost and is more likely to be agreed.

Following the initial broad induction given to a new employee that was described in the previous section, a specific induction programme is usually provided by the individual's department. This covers all of the statutory requirements, health and safety, visitors to the department, confidentiality, dealing with telephone calls, as well as specific human resource issues that relate to the department. These points would include who to contact when telephoning the department to say you will be late, who to liaise with when arranging holidays. All of these more informal pieces of information are specific to a place of work and are opportunities for personal development, as not only are employers never exactly the same, it is unlikely that two departments within a single organization will be identical.

Leading on from what might seem an initial overload of information, it is now time to get down to applying yourself to your professional work!

To satisfy the requirements of Clinical Pathology Accreditation UK Ltd (CPA) standards associated with training, it is necessary for every department to keep training records that provide evidence that an individual has been trained and has had their competence assessed. The proof of training and evidence of its success will form a large part of any biomedical scientist's personal development portfolio. Reading the background of many of the more interesting issues and finding information about many of the daily laboratory procedures and workload also gives many opportunities for **reflective learning** (Section 20.7).

It is clear that personal development is an ongoing process that arises naturally during the course of a professional person simply doing his or her job at work. This is an age of evidence collection: it is essential to document all learning and development as they arise and to keep appropriate records to satisfy statutory legislation, the needs of the employer, and those of the individual for either personal interest or professional registration (Section 20.1).

Cross reference
Chapter 3 'Communications in laboratory medicine'.

Cross reference
You can read more about CPA in Chapter 19 'Quality assurance and management'.

Additional employer-based CPD activities

Many employers offer a variety of other opportunities for professional development and any available should be sought out. Many clinical laboratories provide access to presentations, lectures of interest by visiting speakers, opportunities to visit other departments in the organization to find out more about its wider activities in a more informal setting than that presented at initial induction. Again, just because something is not immediately offered, it does not mean that it is not possible or that it does not already exist. Medical staff are usually happy to explain interesting clinical cases and biomedical scientists are often able to accompany them on visits to the wards. However, it is often up to the biomedical scientist to take the initiative and ask!

Tutorial programmes

If an ongoing tutorial programme is not available in the workplace, the obvious question is, *why not?* Is it simply because no one has bothered to organize one? It may be that several individuals would be prepared to speak to their peers on their favourite subject. It need not be too difficult to find a venue and put together a rotation of several lectures covering the most recent advances in the pathology disciplines. Newly qualified staff could discuss their degree projects, those currently studying postgraduate qualifications could pass on knowledge of a particularly exciting area of research. Often a project such as tutorial organization only requires a leader, who could be any member of staff with the right amount of enthusiasm, participation is likely to follow.

Journal clubs

Journal clubs involve an analysis of an appropriate article that has been published in a relevant journal. A particular journal article is selected in advance and all participants read this article and enter into a (usually) lively debate about it, or a single participant reads an article and then presents a review of it to the other members of the club for discussion. Members of a journal club usually participate equally. Again, you do not have to be a senior member of staff to instigate such activities. The skills required often have nothing to do with laboratory management and much more to do with gentle persuasion and being able to stimulate enthusiasm in colleagues.

20.4 **Professional bodies and personal development**

All biomedical scientists in employment will have many opportunities to develop professional contacts with others in the field. These contacts present opportunities for a variety of ways for personal development, a number of which you can see listed in Table 20.2.

Naturally, the majority of biomedical scientists belong to their own professional body, the IBMS (Figure 1.2), which offers a variety of services to its members (Figure 20.3). The IBMS

TABLE 20.2 A number of the personal development activities associated with professional contacts

Professional activities
Involvement in a professional body
Membership of a specialist interest group
Lecturing or teaching
Mentoring
Being an examiner
Being a tutor
Branch meetings
Organizing journal clubs or other specialist groups
Maintaining or developing specialist skills (for example, musical skills)
Being an expert witness
Membership of other professional bodies or groups
Giving presentations at conferences
Organizing an accredited course
Supervizing research
Acting as a national assessor
Promotion

FIGURE 20.3
Membership files of the Institute of Biomedical Science.

(see also Box 20.2) recognizes the crucial role of CPD and has many resources available to its members in all membership classes.

The areas where the IBMS most directly contributes to CPD are education, journal-based learning (JBL), structured reading schemes, and by organizing regional and national meetings.

Cross reference
You can read more about CPD activities in Chapter 2 'Fitness to practise'.

Education and the Institute of Biomedical Science

The IBMS is active in a range of educational activities (but see also Box 20.3) including accrediting a number of postgraduate professional qualifications. These were introduced in Chapter 2.

These qualifications are widely recognized by both employers and by higher educational institutions. Success in studying for any of them indicates that a biomedical scientist is willing to further their personal development. If you want to progress to the higher levels of the profession, such further study is increasingly necessary to demonstrate knowledge and skills at high levels of competence. Thus, participation in postgraduate learning is advisable, although

BOX 20.2 *Biomedical scientists and professional bodies*

The IBMS is not the only professional organization to which a biomedical scientist can belong. There are common interest organizations that are not necessarily professional bodies. Amongst these are the British Blood Transfusion Society and the British Society for Immunology. All these specialist societies and groups organize series of lectures and meetings for like-minded individuals, and attendance at any such event is an opportunity for CPD.

BOX 20.3 Non-IBMS providers of education in the biomedical sciences

Any opportunity to learn new skills or learning about other disciplines constitute CPD. Of course, even taking a postgraduate qualification in the same area of practice, such as a biomedical scientist choosing to study for a Masters Degree in their own subject, constitutes PD.

Many providers of education for biomedical scientists provide courses other than the established graduate and postgraduate degree courses. There are also specialist courses available that are appropriate for the non-science based areas of biomedical science.

Many universities provide a series of short courses for special interest groups or to allow biomedical scientists to venture out of their particular disciplines. Thus, for example, haematologists may take short courses in clinical chemistry. There are also opportunities to attend courses in management topics. These may be short courses in specific areas such as audit, finance, or tutoring, or an extended postgraduate management course lasting several years.

Many colleges and universities offer not only taught courses but also opportunities for distance and on-line learning in areas such as quality management, point of care testing and training and development of others.

not yet mandatory. However, completion of the registration portfolio *is* mandatory for initial registration with the HPC. Once registered, there are many opportunities for professional education.

The IBMS as a professional body has developed a comprehensive set of professional qualifications. The *Specialist Diploma* is awarded for the successful completion and assessment of the specialist portfolio, which is based on detailed knowledge and the presentation of evidence collected to meet a set of standards in one of the following subjects:

- Cellular pathology
- Cytopathology
- Clinical biochemistry
- Clinical immunology
- Haematology and hospital-based transfusion practice
- Histocompatibility and immunogenetics
- Medical microbiology
- Transfusion science
- Virology

Building on from the Specialist Diploma is the *Higher Specialist Diploma*. This is aimed at practising biomedical scientists who have skills and theoretical knowledge at EQF level 7. They are normally actively performing in a complex role, and are continuously developing a clinical, scientific, or management responsibility for a laboratory service. The Higher Specialist Diploma is open to suitable candidates who already have an accredited MSc or an approved

M-level (EQF level 7) qualification and who are able to pass a series of written examination papers following a period of self-directed study. The Higher Specialist Diploma is available in a series of clinical laboratory specialities and, in addition, Management and Leadership.

The IBMS also offers a series of *Certificates of Expert Practice*, which enable biomedical scientists with highly specific sets of skills and knowledge within their own specialist area to demonstrate their command of a sub-speciality.

Moving forward from these certificates are further specific IBMS qualifications in subspecialities at differing educational levels. These include *Diplomas of Expert Practice* and *Advanced Specialist Diplomas*. These diplomas are designed to provide evidence of the highest level of professional knowledge and skills.

Differing levels of knowledge and skills may be required within a particular job description or person specification for a specific post. However, the only mandatory qualification for a biomedical scientist is the completion of an HPC approved degree or the registration portfolio alongside an IBMS accredited degree to meet the threshold standards of proficiency for registration with the HPC. All other qualifications and educational activities in the wider sense (Table 20.3), while desirable in terms of proof of personal development and as aids for promotion are not mandatory and the decision to complete further studies after gaining a BSc degree is the personal choice of the individual.

SELF-CHECK 20.2

What does the IBMS suggest a new biomedical science graduate do next by way of professional qualifications if they are already in full-time employment?

Journal-based learning

In JBL, the participant is expected to read a current paper published in a recognized scientific journal in a named area of biomedical science. The reference to the paper is supplied by the CPD provider to the participant. However, unless the paper is in an open-access journal or in a journal belonging

TABLE 20.3 **Educational activities constituting personal development**

Formal and educational qualifications
Additional attendance at courses
Further education
Research
Attending conferences
Writing articles or papers
Going to seminars
Distance learning
Attending courses accredited by a professional body
Planning or running a course

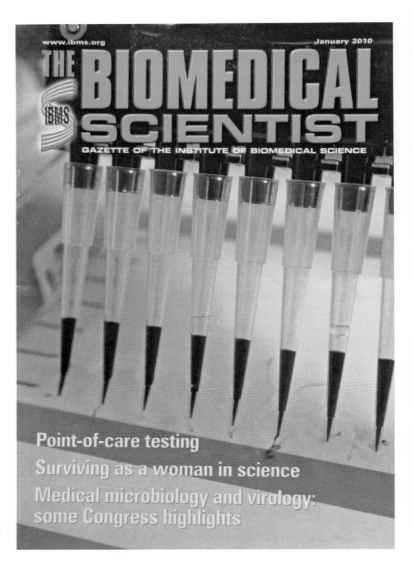

FIGURE 20.4
Front cover of a past issue of *The Biomedical Scientist*. Copyright © Institute of Biomedical Science.

to the CPD provider as in Figures 2.4 and 20.4, the participant must obtain their own copy of the relevant paper (see Section 20.9). Any such papers provided by a CPD provider would be an expensive process and may be a breach of the copyright. The chosen paper may be either scientific or relate to a topical management issue. Typically, a CPD provider such as the IBMS then gives a list of 20 questions related to the article, which are usually printed in *The Biomedical Scientist*, *British Journal of Biomedical Science*, or on the IBMS website. Each question can be answered as either true or as false. The answers to all the questions can be found in the information given in the paper. The participant can submit his or her answers to the IBMS by post or via its website (Figure 2.5). The answers are marked, with 17 correct answers constituting the minimum pass mark.

Structured reading

Structured reading, another activity instigated by many CPD schemes including the IBMS, is an essay writing programme organized on a defined timeline (see also Section 20.8). A series of essay titles are published quarterly in *The Biomedical Scientist*. The essay titles cross all pathology disciplines, including management topics. The questions are searching and require up-to-date research of current literature, some of which is supplied with the title.

The participant is invited to attempt any of the titles, so may decide to take on the extra challenge of detailed learning in a discipline that may not necessarily be their specialist subject and will be expected to write a postgraduate level, properly referenced essay. The essay is marked by an IBMS-appointed peer assessor who will grade it as a pass, borderline, or fail. A pass grading is proof that the topic has been studied in detail and the subject matter understood.

Key Points

Structured reading is also good preparation for professional qualifications such as a Higher Specialist Diploma.

Professional meetings

The IBMS organizes and hosts both regional and national meetings (Figure 20.5). These meetings may be small events for a specialist interest group in a local area or, at the other end of the spectrum, may be a full multidiscipline conference at a central venue lasting several days. The former types of events are often arranged by a willing volunteer for a small group of people who may have a common interest. Larger meetings are usually sponsored by commercial companies, and attendance gives opportunities to see the latest products and technologies and to network with fellow biomedical scientists from other hospitals, as well as listening to a formal lecture, usually by experts in their own field, or to develop ones expertise by talking to senior, more experienced participants.

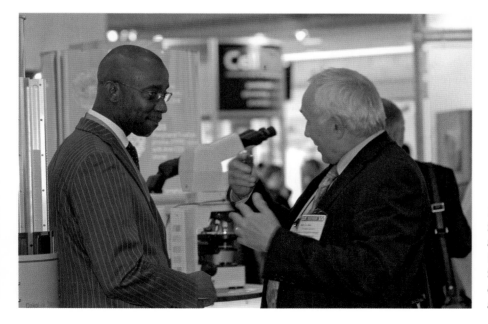

FIGURE 20.5
Scientific meetings such as those organized by the Institute of Biomedical Science, provide excellent opportunities to discuss all aspects of the subject.

20.5 Commercial concerns and personal development

Many of the commercial concerns that supply equipment, consumables, and services provide both short and long courses. It is possible to learn presentational and telephone skills, time management, and many other skills that may not be specific to biomedical science or even employees in broader healthcare organizations of the public or private sectors. Many companies will offer generic courses on business premises, or many even come to a workplace and give courses to several members of staff in the same department. This is a cost-effective method for the employer to train several members of staff simultaneously, so again, if there is nothing already in place where you work, it might be worth making a suggestion to your managers.

Companies that supply scientific instruments and laboratory computer systems also offer training. Often the courses come free of charge with the installation of newly purchased equipment, but additional courses can usually be purchased as separate items. When a new instrument is installed, the person who attends the training will usually become its 'key operator', and be the link between the laboratory and the company that supplied the instrument. This arrangement usually works well until the key operator leaves to work elsewhere. The remaining staff are then left to muddle through; their success in this depends on how much knowledge was given to them before the key operator left. This situation presents an opportunity for another member of staff to receive full training from the company and again, this is an excellent opportunity to gain in-depth knowledge into the inner workings of laboratory equipment and thus for personal development.

Many of the major instrument and laboratory information technology (IT) suppliers also organize meetings and user groups to update the knowledge of laboratory workers. Not only is this an opportunity to network with colleagues, it brings together users of common equipment from different sites and so is often an excellent opportunity to assess the different ways in which one's own instrument can be utilized.

Trade unions

Trade unions have a commitment to personal development and lifelong learning. Many offer courses for their own members that focus on basic skills for lay persons and the development of union representatives. Funding is available through the trade unions to develop basic skills such as numeracy and literacy, and where English is not the first spoken language, opportunities to improve spoken English. All employees should be able to attain a basic standard of education to help their progress in the workplace; trade unions are able to act as vehicles to help achieve this.

Many professionals are not only members of a professional body, but also of an appropriate trade union. The professional activities undertaken by the trade unions often support and complement those undertaken by professional bodies, and are rarely in complete contradiction to one another. Trade unions offer many excellent courses in, for example, employment and health and safety law, presentation, and negotiating skills.

The trade union philosophy is that their members in the workplace should have access to the same level of information that is available to the employer. All management skills, regardless of how they were acquired, are opportunities for personal development. The government frequently asks for the support of trade unions to disseminate information, knowledge, and skills to their members. Indeed, many of the lower level educational courses run by trade unions for their membership are able to take place only because of initial government sponsorship.

20.6 Opportunities for self-guided personal development

Opportunities for personal development do not just occur in the workplace or a formal education situation. Watching television, reading a newspaper or a magazine, or surfing the internet will present professional information, even if you are not immediately aware of it. Learning in this way is equally important and a brief paragraph of writing after reading an article in a Sunday newspaper or watching a television documentary is perfectly acceptable as a record for additional or unstructured CPD.

20.7 Reflective learning

Reflective learning is a mechanism of critically examining an event, situation, or experience after it has taken place, thinking about what has happened, the context in which it happened, and what can be learnt from the whole experience. Personal development constitutes somewhat more than listening to lectures and attending training courses. The most important aspect of personal development happens after the learning event when there is an opportunity to really use what has just been learnt. Reflective learning is an excellent mechanism to really think about the new skills and knowledge acquired and identify where they can be applied in everyday procedures.

Several templates are available to help with reflective learning but, in essence, an individual should start by asking themselves a series of questions, as shown in Table 20.4, and then try to use their new knowledge to answer them.

TABLE 20.4   Question relevant to reflective learning

Possible questions after an incident in the laboratory	Possible questions after a period of training	Possible questions after a conference
What actually happened?	What was the purpose of the training?	What was the purpose of the conference?
Were all policies and standard operating procedures (SOPs) followed?	What did I learn?	What did I learn?
If not, what was the deviation?	Was the learning useful?	Was the learning useful to me?
Was the deviation avoidable?	Can I change my current practices to incorporate what I have learned?	Will the learning be useful in my own place of work?
Was the outcome better, worse, or the same as the expected outcome?	Will my future practices benefit from these changes?	Do I need to look at any of the issues in more detail to identify where they apply in my own place of work?
What lessons have been learned from the event?		Do I need to inform anyone else of my new knowledge in order to be able to make changes?
Should the SOP be changed?		Will the changes be of benefit to the service?

Clearly, each period of personal development may be equally important in terms of acquiring new knowledge and skills that may enhance the activities of the department or allow changes to the service it provides to be carried out more smoothly. However, the most beneficial development to health services occurs when the personal development of employees brings about changes that provide real benefits to the treatment and experiences of patients. Reflective practice is a means of exploring how new personal knowledge and skills can be used for the enhancement of the patient pathway. Thus, reflective practice can run alongside and enhance many of the other ways of CPD already mentioned; so attending a conference is not only an opportunity to learn, for example, cutting edge science, but also to reflect afterwards as to how the newly acquired knowledge can be used to enhance the workplace for the benefit of all.

20.8 **Study skills**

Study skills can be put into practice informally and formally. Skills for general study and personal development are used almost all of the time and in everyday learning. For formal assessments of study, when an individual is asked to provide evidence of study such as submitting coursework or sitting a written examination, more structured skills are needed. Four basic groups of study skills are outlined in Table 20.5; each of course overlaps considerably.

General study skills are necessary for even the most basic academic study and are taught as part of the school curriculum. For those wanting to return to study in later life, the lack of general study skills is an enormous disadvantage over their younger peers. Younger biomedical scientists and laboratory employees often have an advantage over some of their more mature colleagues.

The basic study skills listed in Table 20.5 should be acquired easily by anyone working towards or in possession of an H-level (EQF level 6) qualification (see Introduction). Indeed, if you are working towards an honours degree in biomedical science, or are already a graduate with an approved degree for registering with the HPC, or a graduate needing to obtain additional units for HPC registration, you will most likely be familiar with them. The first consideration is the basic skills that should be familiar to anyone who has recently been part of any education system, regardless of the level of academic achievement. These are the skills needed for any learning, academic or vocational qualifications, and even for leisure pursuits and hobbies. We will now expand on a number of them.

TABLE 20.5 Basic study skills

Basic study skills for everyday life	General study skills for learning	Study skills for producing coursework	Revision and exams
Reading	Literary skills	Producing coursework	Assessing objectives
Writing	Collecting data	Gathering information	Revision
Verbal communication	Essay writing	Research	Exam technique
Using a library	Case studies		
Information technology skills	Report writing		
	Citing references		
Working as a team member	Using the web		
Numeracy	Mathematical skills		

Literary skills

Literary skills include the abilities to read, use language in listening and speaking, and to write. Good communication skills are essential when working in a laboratory. Appropriate levels of English language skills, written, and verbal, are needed for the safe communication of results, instructions, and information. Basic literacy is essential for biomedical scientists at every level. It is unlikely that anyone could maintain a laboratory position without literary skills equivalent to that taught at secondary school. However, a higher level of English language and grammar is needed to study for an H-level qualification (see Introduction), so if you are working towards or are in possession of a biomedical science degree, this will give others an indication of your ability to communicate in English. H-level qualifications require more than the ability to communicate in the required standard of English, you must also be able to undertake academic work, such as taking notes, writing essays and projects, present data, and produce reports at an appropriate level.

Staff in a potentially dangerous working environment, such as a laboratory, must be able to read health and safety notices, SOPs, and be able to differentiate between identification of samples, chemicals, record sheets, and manuals. Writing skills are necessary to enter data at numerous points and write safety notices, and policy documents must be read and understood.

Reading

Academic reading is a different activity to reading for pleasure: it requires an effective use of time to obtain the maximum benefit from a book, article, or webpage in the shortest possible time. The five steps to effective reading are:

1. Survey
2. Question
3. Read
4. Recall
5. Review

Academic reading is a means of enhancing ones understanding of or gaining basic knowledge about a new topic, or finding alternative views about a particular topic and to help with further studies. The reading, however, needs to be selective. Thus, skim reading the first few paragraphs of a book, studying its index or the summaries of chapters will all give clues as to the relevance of the contents. Something written in complicated English with indecipherable jargon is not likely to be worth the time spent trying to unravel it. It should be easy to grasp the key facts and to differentiate between these and the author's own ideas.

It is also pertinent to look at the date of publication and the identity of the author; scientific writing can quickly become outdated and authors can become discredited by their peers.

Academic publications may be expensive and are therefore frequently borrowed, rather than purchased unless it is a key book or journal. Workplaces, friends, and libraries (Figure 20.6) are inexpensive sources of academic books, particularly if you are not going to use them frequently. This does mean that they have to be returned so there must be a maximum reading benefit in a fixed amount of time. While it is useful to make notes of key points while reading, it is counter-productive to copy out vast chunks of text with limited understanding of the topic.

Note taking

Notes may have to be taken from books, during demonstrations in the laboratory, and of course, in lectures and tutorials (Figure 20.7). This is an excellent way of staying focused on

FIGURE 20.6
Undergraduate students
reading and working in a
university library.

(a)

(b)

FIGURE 20.7
Photographs of undergraduate
students studying in (a) a
formal lecture and (b) the
more informal setting of a
tutorial.

what you are supposed to be learning, except when you are expected to write reams of dictated text! Notes should be concise and be an aid to self study. There is no need to write verbatim from a lecturer or speaker.

Many ways of taking notes have been suggested; it is best to try different techniques to find which one works for you. Notes may be linear, in sentences, or expressed only as key words. They may take on more of a picture form, with a central theme in the middle of a page and ideas and topics radiating outwards. You may wish to develop your own form of shorthand. In essence, the only person who has to comprehend and use your notes is you, so it should not matter how illegible they are to others as long as they aid your understanding of a topic, help you to remember the points, and act as a revision prompt.

Essay writing

Essays are not simply collections of written facts; they need structure and to be a response to a specific title or theme, and demonstrate a depth and breadth of knowledge on a specific subject. They should be written in English prose with correct spelling and grammar.

Essays should be planned from the outset and not be a collection of summary paragraphs. They require a clear introduction in the opening paragraph, followed by the main portion or body of the essay, and to end with a summary or concluding last paragraph. The body of the essay should be a collection of related paragraphs, all linked to the title of the essay. Wherever possible, use examples to illustrate the points of the essay. Care should be taken to avoid merely listing facts. A 'good' essay will demonstrate knowledge of a subject from more than one viewpoint and contain a critical appraisal of a variety of concepts or arguments.

SELF-CHECK 20.3

What features would be shown by an essay that gained high marks?

Case studies and writing reports

Case studies and report writing are one step on from essay writing. Case studies are commonplace in biomedical science and the success of a well presented case study is demonstrated by being able to present the key facts of the topic in an interesting and understandable manner, while removing much of the peripheral and often irrelevant material. A case study will include some mention of the theory behind the case and there should be a blend of theoretical knowledge with factual data.

Report writing is also commonplace in biomedical science and many organizations ask for them to be written in a particular format or house style. Most reports require an executive summary or overview of the report presented as short, concise, bulleted points. The body of the report should include an introduction (including terms of reference if appropriate), the findings, the conclusions, and any recommendations leading on from the conclusion. References and appendices should be at the end, and are best kept separate so as not to detract from the main text. Many reports in science rely heavily on the presentation of data, and this should be presented clearly with the appropriate statistical analysis and conclusions for any non-mathematicians who may be required to read the report.

Numeracy

Basic numeracy is an essential skill for anyone working within an area of science, and biomedical science in no exception. The mathematics that is required for biomedical science centres

Cross reference
Chapter 5 'Statistics and handling data'.

on manipulation of numbers in basic calculations, and a reasonable knowledge of the collection and interpretation of statistical data. The fact that you are already in possession of, or working towards, a biomedical science degree gives others a good indication of your current mathematical abilities.

Using libraries

Libraries are essential to anyone studying for their CPD. It is worth getting to know the library and its librarians, particularly if you use it regularly, either for study or for pleasure. Understanding the layout of a library and the services available to you will make using the library more effective and less stressful. Contrary to the myths and stereotypes surrounding librarians, the reality is that most are knowledgeable and willing to help when approached for advice. In other words, it will pay dividends to make friends with yours!

Libraries are usually calm, quiet places and often offer a better study environment than home or the immediate workplace. They are not only a source from which books can be borrowed, but also have reference material, journals, and periodicals that can be used on the premises. If a particular book or journal is not readily available, they can usually be obtained through the inter-library services, so that if one library is unable to provide a copy of something in particular another can usually be contacted to supply the required book or article. Libraries are able to supply copies of articles from scientific journals and if you want to take part in the CPD activity of JBL (Section 20.4) you will have to source the journals needed to participate.

SELF-CHECK 20.4

What facilities are likely to be provided by local authority libraries?

Information technology

The ability to use IT is now as important as basic numeracy and literacy. Children are taught IT skills as part of everyday school life and it features high on the national curriculum from the age of 4 years upwards. Some students of biomedical science may find themselves out of their depth compared to others, as IT skills have only been taught as a compulsory part of education for a relatively short time. Information technology is an essential aid to biomedical science. Databases are part of everyday life and the use of spreadsheets and word processing for study is virtually universal. Anyone who has not received training in the use of IT to store, retrieve, and present information is extremely disadvantaged. The electronic transfer of information and the use of electronic methods of communication are central to the operations of the NHS. Information technology skills are frequently offered by employers as part of staff development and there are many providers of adult education in the community that offer a variety of courses for the beginner to the advanced student.

Team work

In the past, traditional teaching focused on the individual, but with the introduction of key skills in compulsory education, working as an active member of a team has become a skill that employers seek in their employees. Thus, further and higher educational institutions promote working with others as a key skill. The advantages of teamwork are that while everyone has clear and common objectives, an appropriate mix of people will ensure that everyone will be able to make unique contributions to the efforts of the team as a whole. Teamwork is not confined to the everyday functioning of the workplace but may extend into study and common interest groups, multidisciplinary case conferences, and leisure activities.

Academic study skills

Academic study skills require assessment of course work and examination papers to monitor progress, and all students must spend some time finding out how to get the best results based on the requirements of the awarding body. Remember academic institutions are there to pass students, rather than fail them, as a high pass rate reflects well on the teaching of the college or university. Lecturers and tutors are in the business of helping their students to achieve the highest possible standards. All help and advice given by tutors should be gratefully received and stored away for future good use.

Under- and postgraduate courses usually rely on two types of assessment: coursework and a written examination. Coursework must have defined learning criteria in the form of aims and assessed outcomes, and students should be familiar with the principles by which the coursework will be assessed. The content of coursework sometimes allows the student a degree of freedom in choosing the topic and when this is the case, it is obviously better to base it on an area of personal interest wherever possible. Written examinations assess knowledge and skills at the end of a block of learning. Examinations should demonstrate what the student knows; they are not designed to highlight what is not known. They do, however, require careful revision and preparation. Taught courses follow a specified syllabus or curriculum; knowledge of its contents will give an indication of what material will be examined and the form the examination will take.

Assessment by portfolio is becoming increasingly popular, as this involves collecting evidence to show that you have met specific learning objectives. Most biomedical scientists who wish to register with the HPC should have had experience in completing the registration portfolio themselves. More experienced members of the profession have probably been involved in helping others to collect evidence and construct a portfolio in a clear and logical way. If you have not yet completed your initial portfolio for professional registration, you will become familiar with this method of assessment during your time as an undergraduate. Formative assessment is the regular and on-going assessment during a course of study. Summative assessment is the final or end-point assessment stage.

20.9 Evidence supporting personal development

It is essential to keep records of all activities that can be offered as proof of personal development when asked. Table 20.6 lists many activities that provide evidence that a biomedical scientist has been able to learn new skills or improve him or herself in some way. It is advisable to keep copies of all documents that you have prepared or have been involved in the preparation of, and to retain attendance certificates, copies of presentation handouts, notes made during seminars, or summary documents following reading or research.

TABLE 20.6 Types of evidence showing the ability to learn new skills

Materials you may have produced	Materials obtained from others	Materials showing reflection and evaluation of your learning and work
Information leaflets	Testimonies, testimonials	Adapted documents arising from appraisals, supervision reviews
Case studies	Letters from service users, carers, students, or colleagues	Documents produced following local or national schemes related to CPD
Critical reviews	Course certificates	Evaluations of courses and conferences attended
Adaptations of student notes		Personal development plans (PDPs)
Policies or position statements		Approved claims for credit for previous learning or experience
Discussion, procedural, and course programme documents		
Documents about national or local processes		
Recent job applications		
Reports on project work, clinical audits, reviews		
Business plans		
Guidelines for dealing with patients		
Course assignments		
Action plans		
Presentations, journals articles, and research papers		
Questionnaires		
Funding applications		
Induction materials for new members of staff		
Learning contracts		
Contributions to the work of a professional body or of a special interest group		

It is clear from Table 20.6 that not all types of professional development are accessible to all grades of staff, but individuals should be aware and take advantage of, every available opportunity to professionally develop themselves.

20.10 **Performance appraisal and personal development plans**

Many employers, both in the public sector and in the commercial world, integrate the personal development of employees into their organizational culture. They ensure that employees have a framework that they can work within to formally develop themselves in ways that are of benefit to both the employer and the individual.

There is an organizational structure for targets to be set and performance against the targets is monitored and reviewed. Usually a structured mechanism is established where an employee is able to meet with his or her manager (or a representative of the employer) on a regular basis to assess their past performance. The first of these meetings normally occurs soon after the employee begins working in their new post and then usually at regular six-monthly or annual intervals, in accordance with employment policy at the particular place of work. These targets must be non-discriminatory and achievable, so that you (as an employee) are able to pass comment and agree to them.

Performance appraisal is not an opportunity for employers to set unreasonable targets or unfair additional tasks onto an individual that are not appropriate to an existing job description for a post. Nor should it be used as part of a disciplinary process. While it is an opportunity to identify poor performance, it should be done so with the intention of offering suitable training and support to enable the individual concerned to improve and work towards meeting targets.

The NHS has introduced a programme or national strategy, called the Knowledge and Skills Framework (KSF), to standardize knowledge and skills across all the staff groups that it currently employs. This scheme does not apply to biomedical scientists working within the private health sector or commercial world, but many of these employers already have similar schemes in place. The principles of KSF are that staff in different departments and even in different organizations, but who essentially have the same role, are measured against a series of generic standards appropriate to their grade of employment, with the aim of ensuring consistency throughout the NHS in terms of expectations and rewards.

All NHS posts have a series of KSF standards that employees have the opportunity to work towards completing. Performance appraisals are usually carried out by an immediate line manager. This should be a two-way process in which both the appraiser and appraisee participate equally. Performance appraisal is a structured and formal period where the NHS employee has the opportunity to discuss how he or she is performing in line with the set KSF standards, and other local targets specific to the position in the laboratory. It is also the time to identify where extra help may be needed or to highlight an area that the employee may wish to develop further. This is then formally written in to a **personal development plan** (PDP). Having made reasonable requests in the PDP, the employer, both the NHS Trust and the line managers, then have a duty to help the person concerned achieve the areas of personal development highlighted. Performance appraisals are also conducted to ensure that requests from employees for opportunities for personal development are being met.

The system of performance appraisal and PDP is therefore a two-way process: it should benefit both the employer and employee. It should be an opportunity for the employer to encourage and support staff to develop themselves as individuals, while giving added value to the service that the member of staff is able to provide; in short it is part of the cycle of continuing professional development of the individual (Figure 20.8).

FIGURE 20.8
Overview that emphasizes the never ending process of continuing professional development.

SUMMARY

- Personal development is an ongoing process

- The Institute of Biomedical Science has a variety of schemes and qualifications tailored to the needs of biomedical scientists and, while not mandatory, it is advisable to take part in a variety of continuing professional development activities

- The Health Professions Council has the mandatory requirement that all its registrants must be continually involved in professional development activities and be able to provide evidence to support this when asked

- There are always opportunities for personal development wherever you choose to seek them: at work, home, within organizations and academic institutions

- The skills needed to develop both professionally and personally are varied and extensive, and can be adapted to suit any style of learning

FURTHER READING

- **Dryden G, Vos J. *The Learning Revolution*. The Learning Web, Torrance, CA, 2001.**

  An inspirational guide for anyone involved in learning, which explores various aspects to effective learning in an innovative and thought provoking manner. Suitable for teachers and students alike.

- **Institute of Biomedical Science. *Good Professional Practice for Biomedical Scientists*, 2nd edn. Institute of Biomedical Science, London, 2005.**

- **Wood J. The roles, duties and responsibilities of technologists in the clinical laboratory. *Clinica Chimica Acta*, 2002; 319: 127–132.**

- **Pitt SJ, Cunningham JM. *An introduction to biomedical science in professional and clinical practice*. Wiley-Blackwell, Chichester, 2009.**

  Chapters 1 and 8 'Introduction to a career as a biomedical scientist' and 'Development of knowledge and competency for biomedical scientists', respectively, are worth browsing through.

- **Wilson G. Continuing professional development: current understanding and practice. *Biomedical Scientist*, 2010; 54: 168–169.**

  Useful short review.

Useful Websites

- www.sussex.ac.uk/academicoffice/documents/qaafram
- www.ibms.org
- www.bized.co.uk/reference/studyskills
- www.dh.gov.uk/en/Publicationsandstatistics/Publications/PublicationsPolicyAnd Guidance
- www.hpc-uk.org. This is the official website of the Health Professions Council. Publications detailing the latest professional documents relating to CPD requirements are available to download.

  The following documents are particularly useful:

  - *Continuing Professional Development and your Registration.*
  - *Continuing Professional Development and your Registration: Appendix 1*
  - *Continuing Professional Development and your Registration: Appendix 2*
  - *Demonstrating Competence through CPD*
  - *Standards of Conduct, Performance and Ethics*

QUESTIONS

1. The Institute of Biomedical Science does NOT offer:

    (a) journal-based learning

    (b) specialist M-level academic qualifications directly related to aspects of biomedical science

(c)   access to a full-time doctoral programme

(d)   assessment of essays based on structured reading

(e)   information regarding suitable courses leading to accredited biomedical science degrees

2.   Performance appraisal is an opportunity to discuss one's progress with a line manager. Which of the following is NOT suitable for discussion at this type of meeting?

(a)   Aspects of your everyday work that you enjoy and wish to develop

(b)   Aspects of your everyday work that you find particularly difficult

(c)   Funding for a postgraduate course at a local university

(d)   Introduction of a new method that you have read about in a journal

(e)   The behaviour of another member of staff that particularly irritates you

3.   Which of the following is NOT true of the Knowledge and Skills Framework?

(a)   It applies to professional and non professional groups of people

(b)   Is linked to targets set in a performance development plan

(c)   Applies to all employees in the NHS

(d)   Has been promoted by the Department of Education of the UK

(e)   It has been published by the Department of Health

4.   Indicate whether the following statements are **TRUE** or **FALSE**

(a)   The highest academic qualification that a biomedical scientist can achieve is at M-level

(b)   Libraries that are under the control of local authorities only provide a lending service to the local community and do not routinely offer internet access

(c)   The Health Professions Council has a clear policy regarding activities for continuing professional development that are acceptable if you wish your name to remain on the Biomedical Scientist register

(d)   Only activities that are approved and regulated by an employer are considered to be professional development

(e)   Biomedical scientists are not required to meet minimum standards for numeracy and literacy

5.   List the formal qualifications that are offered by the Institute of Biomedical Science that are appropriate to a single, specialist discipline.

Answers to self-check and end of chapter questions are available in the Online Resource Centre accompanying this book.

Go to www.oxfordtextbooks.co.uk/orc/glencross

Glossary

Absorption spectrum graph illustrating the fractions of incident electromagnetic radiation absorbed by a material over a range of different wavelengths.

Acceleration voltage the potential difference in volts necessary to produce electrons from the tungsten filament.

Accreditation an inspection and examination procedure to ensure participants comply with published practice and documentation standards.

Achromatic a lens corrected for two colours only.

Acquired immunity or adaptive response one that produces antibodies specific to antigens on pathogens and develops an immunological memory.

Activity a measure of an ion's ability to affect chemical equilibria and reaction rates.

Activity the rate of decay of a radioactive nuclide.

Adhesion molecules mediate the binding of cells to each other or to other biological surfaces.

Adjuvant substance that enhances the immune response to an antigen.

Adsorption accumulation of molecules from a solution on to the surface of an adjoining solid without penetrating its interior.

Adverse healthcare event incident or omission arising during clinical care that causes physical or psychological injury to a patient.

Adverse incident any event that may affect the reliability of a result issued by the laboratory or any delay in the diagnosis and treatment of a patient following laboratory investigations.

Affinity chromatography form of column chromatography that relies on the specific binding of the analyte of interest to an immobilized ligand.

Affinity measure of the strength of antibody–antigen binding.

Agglutination clumping together of cells or particles by antibody binding to antigens on their surfaces.

Allergy inappropriate immune response to normally harmless environmental antigens.

Amperometry a form of voltammetry that involves the reduction or oxidation of an electroactive molecular species at a constant applied potential.

Ampholytes synthetic polyamino-polycarboxylic or polysulphonic acids of low M_r that differ from one another in the values of their isoelectric points (pI).

Analyser polarizing filter placed in the light path fixed within the body of the microscope. It is used together with a sub-stage polarizer.

Analyte specific substance in a sample under investigation to be identified and/or quantified.

Anions negatively charged ions that are attracted to the **anode**.

Anode plate disc in the electron gun of an electron microscope that is maintained at earth potential and causes the cloud of electrons produced by the filament of the gun to be accelerated into the microscope column.

Anode the negatively charged electrode.

Anthropometry study of human body measurement related to capabilities.

Antibody immunoglobulin with the ability to bind to a specific antigen.

Antigen molecule that can elicit the production of and bind to an antibody.

Antigenic determinant or epitope site on an antigen that is recognized by an antibody.

Antihuman globulin polyclonal antibodies produced against an epitope on a human immunoglobulin molecule.

Antinuclear antibodies antibodies to nuclear proteins.

Antiserum one that contains antibodies and may be used as an analytical reagent.

Apochromatic a lens corrected for three colours and which is the best quality type of lens available.

Approved Codes of Practice explain legal regulations and the detailed requirements necessary to comply with them.

Astigmatism the inability of a lens to bring light or electrons passing through different parts of the lens to a common focus.

Atomic number the number of protons in the nucleus.

Autoimmune disease condition in which the immune system of an individual reacts against self antigens.

Autoradiography sensitive method of producing a photographic image of a specimen containing radioactively labelled material(s).

Avidin-biotin system Avidin is found in egg white and has a high affinity for biotin which is a co-enzyme. Together they can form complexes with antibodies, enzymes, radioisotopes and fluorochromes.

Avidity ability of an antibody to bind multiple antigens.

B lymphocyte type of white blood cell that can form a plasma cell.

Balancing of a rotor consists of distributing the centrifuge tubes in the rotor, i.e. the weight, evenly around the centre of rotation.

Barrier filter a filter used in the fluorescence microscope, which is only capable of transmitting light in the visible spectrum and which prevents the transmission of ultraviolet light.

Becquerel the SI unit that has superseded the Curie, and is the amount of a radioactive nuclide that gives 1 disintegration per second.

Bile green coloured product of the liver stored in the gall bladder.

Binding site variable part of an antibody molecule that can bind to an epitope on an antigen.

Biomedical science is the application of the natural sciences, to medicine, in the study of the causes, consequences, diagnosis, and treatment of human diseases.

Biomedical scientist an individual who practises biomedical science and is entitled to use the legally protected job title. Only those individuals who have met the threshold requirements of the Health Professions Council may use this term to describe themselves within the UK.

Biosensor an analytical device that incorporates a biological component, such as an enzyme, and an electrochemical transducer.

Birefringence the ability of a substance to split a ray of light into two components, the ordinary and extraordinary ray.

Blood sciences laboratory a commonly used term for an automated laboratory in which the predominant analyses are blood investigations and which encompasses the traditional disciplines of biochemistry, haematology, coagulation, immunology, and virology. Sometimes called a core automated laboratory.

Blotting transfer of the molecule of interest from an electrophoretic agarose or polyacrylamide gel to a piece of nitrocellulose, paper, or a nylon membrane.

Calibration a process using calibrants of known concentration or value to set the performance parameters of a device.

Cannula flexible plastic tube carried on a needle that is inserted into the body to obtain a sample of fluid or introduce medication.

Capacity is the weighing range of a balance, for example up to 200 g, 1000 g, or more, respectively.

Cathode the positively charged electrode.

Cations positively charged ions that are attracted to the **cathode**.

cDNA library collection of clones (transformed cells) each containing a recombinant DNA prepared using mRNA molecules extracted from a single source (organism).

CE mark a mark placed on an IVD to show that it is compliant with European Directives, and is fit for the purpose stated by the manufacturer.

Centrifugal field an acceleration directed outwards from the centre of rotation.

Centrifugation tubes containers for samples used in the rotor during centrifugation.

Centrifugation term applied to the mechanical process of separating mixtures by applying centrifugal forces.

Centrifuges mechanical devices used to separate or isolate substances suspended in a fluid by spinning them about a central axis to produce a suitable centrifugal force.

Certificate of Competence an award made to individuals who have completed an Institute of Biomedical Science accredited honours degree and their registration portfolio. This award allows an individual to apply to the Health Professions Council to be admitted to the biomedical scientist register.

Chemokine protein that stimulate the migration and activation of cells.

Chromatic aberration the inability of a lens to bring light or electrons of different wavelengths to a common focal point.

Chromatography collective term for a family of analytical techniques used to separate the components of mixtures of molecules for their identification and possible estimation of their concentrations in the original mixture based on differences in their partitioning between two immiscible phases.

Chromophore atom, molecule, or part of a molecule that becomes excited by absorption of electromagnetic radiation.

Chronic ill health a long-term illness or condition that may not be curable, but can be managed using laboratory tests and their results.

Clinical governance a method by which a procedure can be controlled, documented, and audited.

Clinical risk a measure of anything that harms, or potentially harms, a patient.

Clone population of cells derived from a single stem cell.

Coherent light rays of the same amplitude and wavelength, which are in phase with one another.

Combination electrode a pH electrode in which the external reference electrode is built into the glass electrode in a concentric double barrel arrangement.

Commensals a term used to describe all the natural bacteria that live on and in a healthy person. The main areas for commensal organisms are the skin, mouth, upper respiratory, gastrointestinal, and urogenital systems.

Competent ability of an individual to perform a task repeatedly to an agreed level of quality.

Competent having the ability to undertake a procedure correctly and according to the SOP.

Competitive tender where companies bid for a specified business contract.

Complement system group of proteins that act directly or with antibodies to kill pathogens.

Continuing professional development any learning associated with your chosen profession, rather than your personal development. The two are not exclusive.

Continuous polyacrylamide gel electrophoresis uses the same type of buffer to dissolve the sample as that present in the gel and in the compartments of the electrophoresis equipment. Compare **discontinuous polyacrylamide gel electrophoresis**.

Core automated laboratory an alternative name for the blood sciences laboratory.

Corrective action that taken to eliminate the cause of a detected nonconformity or other undesirable situation.

Corrosive substance that will destroy or irreversibly damage another that it comes into contact with.

Control of substances hazardous to health (COSHH) set of laws that require employer control materials that are hazardous

Coulomb kg^{-1} the SI unit of exposure to radiation.

Coulomb is a unit of electrical charge equal to the charge transferred by a current of 1 ampere s^{-1}.

Council for Professions Supplementary to Medicine predecessor of the Health Professions Council.

Critical illumination achieved when a substage condenser is adjusted so that the light it emits comes to a **focal point** at the level of the specimen.

Curettings scrapings of tissues.

Curie a pre SI unit, which is the amount of a radioactive nuclide that gives 3.7×10^{10} disintegrations per second.

Cuvettes specialized transparent container in which the absorbance of a solution is measured.

Cytocentrifuge or slide centrifuge a low-speed centrifuge with a modified rotor used to separate cells and deposit them as a monolayer onto the slides for cytological examination.

Cytokine protein that directs the behaviour of other cells.

Dalton a unit equal to one twelfth of that for ^{12}C, which is sometimes used to express the **mass** of an atom or molecule.

Decalcifiying agents solutions of, for example ethylenediaminetetraacetic acid (EDTA), formic, or hydrochloric acid that reduce the mineral content of samples of tissues.

Delta check a facility that allows a current result to be compared with previous results from an individual.

Destaining the removal of excess stain following staining to give a gel of clear background.

Differential centrifugation technique that separates particles during sedimentation mainly on differences in their sizes.

Discontinuous polyacrylamide gel electrophoresis uses a buffer that differs in composition and pH to dissolve the sample to that present in the gel and in the compartments of the electrophoresis equipment. Compare **continuous polyacrylamide gel electrophoresis**.

Dissertation involves a literature review and a written presentation of the findings of that review. It is a component of an honours degree programme.

DNA cloning the preparation of many identical copies of a DNA molecule.

DNA dependent DNA polymerase enzyme that replicates DNA. Usually abbreviated to **DNA polymerase** or **DNA pol**.

DNA pol abbreviation of **DNA dependent DNA polymerase**.

DNA polymerase abbreviation of **DNA dependent DNA polymerase**.

Double junction reference electrode an external reference electrode that has an additional chamber between the reference electrode and the external solution.

Dress code a clear set of written instructions that specifies what can and cannot be worn in a place of work.

Dusts fine particles of solid substances or pellets that may be harmful or flammable.

Ectopic pregnancy implantation of the foetus outside the uterus, often within the Fallopian tubes, and a life-threatening condition.

Effective dose of radiation the product of the absorbed dose and the relative biological effectiveness.

Electric field the potential difference or voltage applied between two electrodes. In electrophoresis it is normally quoted in units of V cm^{-1}.

Electroactive a molecular species that can undergo oxidation and reduction.

Electrode potential a spontaneous potential difference that is established across an electrode/electrolyte interface due to oxidation or reduction events occurring within the half-cell.

Electrolytic cell an electrochemical cell in which reactions are not spontaneous, only occurring if an external voltage is applied across the two electrodes.

Electromagnetic radiation spectrum the full range of electromagnetic radiation.

Electroosmosis occurs in support media that possess negatively charged groups on their surface and so attract hydrated **cations** present in the buffer. When an **electric field** (*E*) is applied the movement of cations and associated water molecules is towards the **cathode**. Electro-osmosis may be sufficiently strong that weakly charged **anions** can be carried towards the cathode.

Electrophoretic mobility velocity of an ion per unit of **electric field**.

Eosinophils granulocytes that are associated with combating parasitic infections and causing allergic reactions.

Epitope or antigenic determinant site on an antigen that is recognized by an antibody.

Ergonomics study of designing a job, equipment, and workplace to fit the worker.

Error failure to complete a planned action as intended or the use of an inappropriate plan of action to achieve a given aim.

Exciter filter one used in a fluorescence microscope which transmits ultaviolet light of the required wavelength and prevents transmission of unwanted visible light.

External reference electrode a separate electrode that generates a stable electrochemical potential in potentimetric analysis. Requires electrical continuity with the working electrode.

First aid provision of initial care for an illness or injury.

First voided urine the initial urination of any given day.

Flammable substance that will easily ignite causing fire.

Fluorescence virtually instantaneous emission of radiant energy by a chromophore following the absorption of radiation of a higher energy.

Fluorochrome fluorescent dye used to label or stain a biological substance.

Fluors chemicals that give off fluorescent energy when they interact with radiation.

Focal length the distance between the centre of the lens and its **focal point**.

Focal point the point at which a lens brings all of the light passing through it to a common focus.

Frictional coefficient property of a particle sedimenting through a solvent and equal to the frictional resistance opposing its movement divided by its sedimentation velocity.

Fronting production of an asymmetrical peak during column chromatography by a slow rise at the beginning of the peak and a sharp fall after the peak.

Full Blood Count number of haematological measurements of blood indicating the numbers and sizes of various types of blood cell.

Fume mist or vapour containing very small metallic particles.

Galvanic cell an electrochemical cell in which reactions occur spontaneously at the electrodes when they are connected externally by a conductor.

Gel electrode a pH or ion selective electrode in which the electrolyte of the reference electrode is retained within a gel layer, enhancing portability, but meaning that the electrolyte cannot be refilled as in conventional electrodes.

Gel filtration is a form of column chromatography that separates molecules according to differences in their sizes and to some extent shapes.

Gradient elution use of a buffer of gradually changing pH or one whose concentration of salt progressively increases to wash materials out of a chromatography column. Compare **isocratic elution.**

Granulocyte type of leukocyte that has granules in its cytoplasm containing enzymes that can digest microorganisms.

Gray the SI unit equal to the absorption of 1.0 J of radiant energy per kilogram of tissue.

Ground state the lowest energy and therefore the most stable state of a subatomic particle, atom, or molecule.

Group III pathogens one of the four categories of pathogenic microorganisms recognized by the Advisory Committee for Dangerous Pathogens.

Haematoma a localized swelling consisting of blood that has leaked from veins and capillaries into the tissues.

Haemolysis breakdown of erythrocytes leading to release of free haemoglobin into the plasma.

Haemolysis breakdown of erythrocytes affecting analysis.

Half-cell an electrochemical system comprising an electrode surrounded by an electrolyte.

Half-life the time required for half of the atoms in a radioactive sample to decay.

Harmful any chemical that causes or is likely to cause harm.

Hazard substance, activity or process that may cause harm.

Health and safety guidance (HSG) documents written materials concerned with legal and best practices relating to health and safety produced by the HSE.

Health Professions Council (HPC) one of the statutory regulatory bodies operating within the UK health services. Its principal function is to protect the public by regulating a number of healthcare professionals.

Health state of well being in body and mind.

Healthcare near miss situation in which an event or omission, or a sequence of events or omissions arising during clinical care fails to develop further and so injury to a patient does not occur.

Hep2 cells human epithelial cells derived from tumours and which are cultured for use in research and diagnostic tests.

High performance liquid chromatography any of a variety of column chromatography methods that use the relatively high pressures produced by pumps to force the mobile phase through the column to increase the resolution and gain faster separation times.

HPC register list of all those individuals who have met the threshold requirements of the HPC and have been admitted to the register.

Hybrid DNA molecules comprised of portions from two or more different sources. Also called **recombinant DNA**.

Hybridoma hybrid cell formed by fusing two different types of cells, for example a specific antibody-producing B lymphocyte with a myeloma cell to produce monoclonal antibodies.

Hydrophobic effect the tendency of molecules in an aqueous environment to fold so that their hydrophobic portions are buried in the interior of the molecule.

Hydrophobic interaction chromatography form of column chromatography that relies on the specific binding of the analyte of interest to an immobilized hydrophobic ligand.

Hyperkalaemia refers to a K^+ concentration that is higher than the quoted laboratory reference range.

Hypoglycaemic coma a state of unconsciousness caused when blood glucose falls to a dangerously low level.

Immiscible phases ones that do not mix.

Immobilised enzyme an enzyme that is either bound to a surface or entrapped within a structure.

Immune complex consists of an antibody-antigen.

Immune surveillance capacity of the immune system to continually monitor for and detect foreign antigens.

Immune tolerance failure of the immune system to respond to an antigen.

Immunoassay any analytical method that uses an antibody to detect and/or quantity an analyte in a sample.

Immunodeficiency situation in which the immune response is reduced or absent.

Immunofluorescence methods employing the fluorescence microscope to identify the presence of specific antigens within tissue samples using fluorescently labelled antibodies.

Immunoglobulins group of large protein molecules that includes all types of antibodies and the membrane-bound receptors on B lymphocytes.

Immunophenotyping technique used to categorize cells by the proteins expressed on the surfaces of their plasma membranes.

Improvement notice legal order to improve practice issued by the HSE.

In vitro literally "in glass": occurring in the laboratory. Compare *in vivo*.

In vivo within the body. Compare *in vitro*.

Incident individual occurrence or event.

Innate immunity that which is present at birth.

Institute of Biomedical Science a professional body whose principal aims are to promote biomedical science and its practitioners.

Internal reference electrode an electrode found within the glass bulb of a pH electrode.

Ion pair consists of a positively charged ion and a free electron, which are formed during ionization events.

Ion selective field effect transistor a field effect transistor with the gate modified to respond to specific ions.

Ion-exchange chromatography use of a charged resin to separate ions on the basis of differences in their charge densities.

Ionic strength adjustor a solution of known high ionic concentration. Proprietary adjustors may also contain pH adjustors and decomplexing agents or agents to remove species that interfere with the measurement of the ion of interest.

Ions atoms or molecules that possess an electrical charge.

Irritant substance that causes irritation.

Isocratic elution use of a buffer of constant composition and pH to progressively wash materials out of a chromatography column. Compare **Gradient elution**.

Isopotential point the pH at which a pH electrode, or the ion concentration at which an ion selective electrode, generates zero potential, and where temperature has no effect on potential.

Isopycnic density gradient centrifugation technique to separate particles based mainly on differences in densities during their sedimentation through a column of liquid whose density increases from the top of the centrifuge tube towards its bottom.

Isotopes the forms of an atomic species of an element that have the characteristic atomic number of the element, but a varying numbers of neutrons and therefore different mass numbers.

Journal club an arrangement between a group of like-minded individuals who agree that one member should be read a particular journal paper and then present the contents to the other members of the group.

Köhler illumination achieved where the substage condenser is used to focus the maximum amount of light onto the specimen.

Laboratory information management systems computer databases that hold complete patient records of incoming requests and outgoing results and allow for the electronic exchange of laboratory data between computers.

Lens piece of glass ground into a spherical or ovoid shape that produces a magnified image. Two or more lenses may be used in combination to produce higher magnifications.

Lifelong learning term used to encompass any new skill or knowledge that you may acquire at any time during your life.

Linear energy transfer is the rate of ionization along the path a radiant particle takes through body tissues.

Lipaemia high levels of fats such as cholesterol and triacylglycerols, especially found after eating a meal with high fat content resulting in cloudy or cream coloured plasma.

Litigation legal action against an individual or organization.

Lymphocyte leukocyte that mediate the activities of antibodies and cellular elements of the immune response.

Macromolecules those with M_r greater than 5,000.

Magnification the process of enlarging something only in appearance, not in physical size.

Mass number is numerically equal to the sum of the numbers of neutrons and protons.

Mass is a measure of the amount of matter present in an entity.

Mast cells cells found in connective tissue that play a major role in allergic reactions.

Mediator a soluble electroactive molecule that can accept electrons from an enzyme and then diffuse to an electrode surface and donate the electrons to that electrode.

Melting separation of the two strands of DNA by breaking the intrachain hydrogen bonds.

Metal oxide semiconductor field effect transistor the electrical transistor found in nearly all electrical devices.

Metal-chelation chromatography form of column chromatography that relies on the specific binding of the analyte of interest to an immobilized metal ion.

Methadone a heroin substitute used to wean addicts off the drug.

Michaelis–Menten equation one description of the kinetics of an enzyme-catalysed reaction.

Microscope an instrument that forms an enlarged image of an object that would often be too small to be seen with the naked eye.

Microscopy the use of a microscope to examine and analyse objects that would often be too small to be seen with the naked eye.

Mobile phase the fluid phase used in chromatography, can be a gas or liquid.

Modified glass electrode a type of ion selective electrode that relies on a glass bulb to generate a potential.

Molality is way of expressing concentration where the **mass** of the **solvent** is used rather than the total volume of **solution**. A 1.0 molal solution would contain 1.0 **mole** of **solute** per kilogram of **solvent**.

Molar mass is the **mass** of all the elementary entities in one **mole** of a substance.

Molarity a way of expressing concentration as the number of **moles** of **solute** in a total volume of **solution** of one dm^3 or litre.

Mole the amount of a substance that contains as many elementary entities, such as photons, atoms, or molecules, as there are atoms in 0.012 kg (12 g) of ^{12}C. This value is called Avogadro's number and is equal to 6.022×10^{23}.

Molecular biology study of biology in terms of the structure and functions of the atoms and molecules concerned.

Molecular sieving the separation of ions during electrophoresis by porous support media that restrict the flow of ions. Smaller ions experience a lower frictional resistance and so migrate through the gel faster and travel further than larger ones for a given time.

Monochromatic radiation electromagnetic radiation of a single wavelength.

Monochromator a device for separating a beam of electromagnetic radiation into its component parts on the basis of their wavelengths.

Monoclonal antibody single type of antibody that binds to a single epitope and is produced by a single clone of B lymphocytes.

Monocyte type of leukocyte that are the precursors of macrophages or mononuclear phagocytes, which phagocytose pathogenic organisms.

Natural killer cell lymphocytes that are part of the innate immune response and have a role in the destruction of some tumour cells.

Near miss unplanned event that did not result in injury, illness, or damage, but had the potential to do so.

Nephelometry analytical technique in which the intensity of light scattered by a suspension is measured at an angle to the incident beam.

Nernst equation an equation that describes the potential developed by a potentiometric electrode.

Neutrophils phagocytic granulocytes that engulf and destroy pathogenic organisms.

Nomograms arrangements of three scales, such that a line connecting a value on one scale to that on one of the others gives the required value on the third scale. Nomograms relating the radius of rotation, speed of the rotor, and RCF are useful in centrifugation.

Noncompliance non-fulfilment of a requirement or standard.

Non-verbal communication any form of communication that does not involve the spoken or written word.

Nucleoside chemical combination of a base and sugar.

Nucleotide chemical combination of a phosphate(s) and a nucleoside.

Nuclide an alternative name for the nucleus.

Numerical aperture quantitative measure of the light gathering capacity of a lens. The greater the numerical aperture, the more efficient the lens.

O rings gaskets formed of a flat ring of rubber or plastic used to make joints air and water tight.

Occult blood small amounts of blood in faeces that are indications of possible disease, but whilst invisible to the naked eye may be detected using a biochemical test.

Ohm's law a description of the relationship between the current passing between any two points of a conductor, the potential difference between the two points, and the resistance of the conductor.

Orbitals the specific energy levels to which subatomic particles such as electrons are restricted.

Oxidation a reaction in which there is a loss of electrons.

Palindrome word, phrase or sequence that reads the same in both directions.

Paraprotein identical immunoglobulin or monoclonal antibody molecules produced by plasma cells in the bone marrow in excessive amounts in the condition multiple myeloma.

Partial specific volume (of a particle) change in volume that occurs when the particle is added to a large volume, that is, excess of solvent.

Partition or distribution coefficient numerical descriptor of the way in which a substance distributes at equilibrium between two immiscible phases.

Pathobiology is the study of disease processes and their effects on humans. It is a principal subject component of an honours degree programme in biomedical science.

Pathogen a disease-causing organism.

Pathology the study of disease (pathological, relating to disease).

Pellet sedimented material found at the bottom of centrifuge tubes or bottles following centrifugation.

Performance appraisal a formal interview held between an employee and their line manager to assess the workplace performance of the individual, both retrospectively and looking to the future.

Personal development plan a strategy agreed at a performance appraisal that includes the future training and educational needs of the employee.

Personal development a strategy by which an individual is able to expand on their existing knowledge and skills, in their personal life and in their professional life.

Personal protective equipment a series of protective clothing or devices designed to minimise or completely remove the potential for an individual to be exposed to harm during work activities.

Phagocytic cells see monocyte and neutrophil.

Phase plate device within the substage condenser of a phase contrast microscope that causes light to be retarded by ¼ of its wavelength.

Phlebotomist a person trained to collect blood samples in a safe and appropriate manner.

Phosphor coating on the viewing screen of an electron microscope that emits visible light when excited by electrons to produce an image of the sample on the viewing screen.

Photons 'particles' or discrete packages of electromagnetic radiation.

Plasma cells dedicated antibody producing cells derived from B lymphocytes.

Plasma the bodily fluid obtained following centrifugation of a blood sample that has been collected and preserved using an anticoagulant such as lithium heparin.

Plasmid extrachromosomal circular and double stranded DNA molecule found in bacteria and some yeasts.

Point of care testing (POCT) is the provision of a diagnostic pathology testing service outside the traditional clinical laboratory and physically closer to the patient.

Polarizer filter placed in the light path of the microscope beneath the substage condenser which only allows the transmission of plane polarized light.

Polarography a form of voltammetry that employs a dropping mercury electrode as the working electrode.

Polyclonal antibodies mixture of different types of antibodies produced when a number of B lymphocyte clones respond to different epitopes on the same antigen or a mixture of antigens.

Polymer membrane electrode a type of ion selective electrode that incorporates an ion exchange material in an inert matrix such as PVC, polyethylene, or silicone rubber to generate a potential.

Polymorphisms occurrence of two or more distinct forms of the same kind of structure, be it organism or molecule.

Polynucleotide polymers of nucleotides therefore alternative name for nucleic acid.

Potential gradient (E) the electric field found by dividing the potential difference between two electrodes by the distance separating them.

Potentiometry electrochemical techniques that employ galvanic cells and measure the spontaneous potential of the electrode with minimal current flow.

Precision a measure of the closeness of a series of measurements of the same substance.

Primary fluorescence substances or tissue elements that naturally exhibit fluorescence under UV light (also called autofluorescence).

Primary immune response response of the immune system on first encountering an antigen, usually this results in the production of an IgM antibody.

Primary sampling the analysis of a sample using aliquots from the initial sampling tube.

Probe labelled single-stranded polynucleotide that is complementary to, and used to detect and characterize specific nucleotide sequences in RNA and DNA samples by forming DNA–DNA, RNA–RNA homoduplexes, or RNA–DNA heteroduplexes.

Professional body organization of like-minded individuals who have specialised knowledge which sets standards of practice and examinations on that knowledge.

Prohibition notice legal order to cease activity issued by the HSE.

Project a piece of individual work involving a literature review, research, and the written and oral presentation of the findings. It is a principal component of an honours degree programme.

Proteomics the study of the complete complement of proteins encoded by a genome.

Prozone phenomenon occurs when high concentrations of antigen or antibody produce a false positive result due to saturation of binding sites.

Pyogenic bacteria Are bacteria that in the process of infection form pus, which is composed of dead neutrophils.

Quality Assurance Agency for Higher Education body (QAA) in the UK that helps ensure HE qualifications are of a suitable standard.

Quality assurance the planned and systematic actions necessary to give adequate confidence that a product or service satisfies the necessary level of quality.

Quality control charts graphical representations that clearly show how laboratory analytes or instruments are performing.

Quality control procedures internal laboratory checks carried out to ensure that a scientific method or analytical instrument is working properly and producing results that are reliable and valid.

Quality control the techniques and activities used to reach the desired level of quality.

Quality management system establishment of relevant quality objectives and the policy necessary to achieve those objectives.

Quality management activities of the overall management that determine the quality policy objectives and responsibilities and implement them by means such as quality planning, control, assurance and improvement within the quality system.

Quality objective anything sought or aimed for related to quality.

Quality consistently obtaining the correct answer.

Quenching phenomenon where the emitted fluorescence is less than expected because of the presence of contaminating molecules in the solution.

Rad the pre-SI unit (from *radiation absorbed dose*) of absorbed dose of radiation that gives an energy absorption of 1×10^{-2} J kg^{-1}.

Radiation the emission or transfer of particles or electromagnetic waves.

Radioactivity the spontaneous emission of particles or electromagnetic waves from the unstable nuclei of some atomic species.

Radioisotopes the unstable nuclei of some atomic species that spontaneously emit radiation.

Radiopharmaceuticals radioactively labelled chemical agents that distribute in the body.

Radiotherapy the medical use of ionizing radiation given as part of the treatment of certain cancers.

Radius of rotation distance of a particle from the centre of rotation.

Rate-zonal density gradient centrifugation technique to separate particles based mainly on differences in their sedimentation coefficients during sedimentation through a column of liquid whose density increases from the top of the centrifuge tube towards its bottom.

Readability the number of decimal places to which a balance weighs accurately.

Reagent books a means to record the reagent changes and lot numbers that are associated with a clinical test or instrument.

Real image a representation of an object (source) in which the perceived location is actually a point of convergence of the rays of light that make up the image. If a screen is placed in the plane of a real image the image will generally become visible on the screen.

Reanneal reassociation of two complementary DNA strands to form the original double helical structure.

Recombinant DNA one whose molecules are comprised of strands from two different sources.

Reduction a reaction in which there is a gain of electrons.

Reference electrode within potentiometric devices, an electrode that generates a stable electrochemical potential and thereby enables measurement of the potential difference with the working electrode.

Reference range a range of values found within the 95th percentile of the population for a given substance. Sometimes incorrectly referred to as a "normal" range.

Reflective learning assessing an event, its strengths and weaknesses, what has been learned from it, and how this could be put to future use.

Refractive index a numerical measure of the ability of a substance to deviate light rays.

Registrant a biomedical scientist whose name appears on the HPC register.

Relative biological effectiveness (RBE) is defined as equal to a dose of X-rays that produces a given effect divided by the dose of a specific radiation that produces the same effect.

Relative centrifugal field one expressed as multiples of the earth's gravitational field (*g*), which is an acceleration of 981 cm s^{-1} per s.

Reports issued by a laboratory are the permanent record of the results and conclusions for a single or series of laboratory investigations.

Requests the investigations that a laboratory is asked to carry out on patient samples.

Resolution ability to separate (or resolve) completely one analyte or one point from another.

Resonance absorption of radio frequency (RF) radiation by nuclei with low-energy spins that changes their orientation to that of a higher-energy spin state.

Restriction endonuclease endonuclease that catalyses the hydrolysis of phosphodiester bonds at identical positions within a palindromic site of DNA.

R_f (relative to front value) distance travelled by any component relative to that moved by the solvent during chromatographic analyses.

Rhesus blood group blood group system defined by the presence of the D antigen on erythrocytes.

Risk likelihood, high or low, that somebody or something will be harmed by a hazard.

Roentgen non-SI unit of exposure to radiation equal to the quantity of ionizing radiation that produces one electrostatic unit of electricity in one cubic centimetre of dry air at 0°C and standard atmospheric pressure. This is equivalent to the amount of radionuclide that produces 1.61×10^{15} ion pairs kg^{-1}.

Roentgen equivalent man a non-SI unit of effective dose of radiation equal to the amount of radiation that gives a dose in humans equivalent to that of 1 rad of X-rays.

Rotors specialized containers that hold liquid samples in centrifuge tubes and which are rotated during centrifugation.

Safety prevention of physical injury.

Salt bridge a solution of ions, often saturated potassium chloride, which enables electrical continuity between two half cells.

Scintillant a liquid containing fluors that produce flashes of light when the fluors interact with β-particles.

Scope of practice range of tasks that a registrant has been trained to undertake and been assessed as competent to perform.

Secondary fluorescence the binding of fluorochromes to substances or tissue elements in order to make them visible under UV light.

Secondary immune response response of memory plasma cells to rapidly produce specific antibodies at a higher intensity against an antigen the immune system has previously encountered.

Sedimentation coefficient (of a particle) ratio of its velocity to the applied centrifugal field ($v/\omega^2 r$) when sedimenting during centrifugation.

Selectivity ability of a chromatography column to distinguish and separate two similar analytes.

Sensing electrode the electrode that generates a variable potential at its surface with the analyte, dependent on analyte concentration. Also called a working electrode.

Serum the fluid derived following centrifugation of a blood sample that has been allowed to clot.

Sievert an SI unit of effective dose of radiation equal to the amount of radiation that gives a dose in humans equivalent to 1 Gy of X-rays.

Sodium dodecyl sulphate polyacrylamide gel electrophoresis (SDS-PAGE) perhaps the most widely used form of polyacrylamide gel electrophoresis. It is particularly useful in assessing the purity of proteins, to determine their M_r and was used extensively in sequencing DNA molecules.

Solid-state electrode a type of ion selective electrode that incorporates one or more salt crystals to generate a potential.

Solute the substance dissolved in a **solvent** to form a **solution**.

Solution a liquid consisting of a **solute** dissolved in a **solvent**.

Solvent the liquid used to dissolve a **solute** when making a **solution**.

Specific activity the amount of radioactivity per unit mass.

Spectroscopy the study of the interactions between electromagnetic radiation and matter.

Staining the detection of separated components following electrophoresis by converting them to visible, coloured complexes. Compare **destaining**.

Standard curve method of plotting the measurements from a known series of concentrations to determine that of an unknown sample.

Standard operating procedure clear set of written instructions to ensure that all staff act in a consistent manner when carrying out all aspects of laboratory work.

Standards codes of best practice that improve safety and efficiency.

Stationary phase the non-mobile phase used in chromatography; can be a solid or an immobilized liquid.

Supernatant liquid found above the pellet that contains unsedimented material following centrifugation.

Svedberg units unit of sedimentation coefficient (s) equal to 1×10^{-13} seconds.

System set of interdependent elements that interact to achieve a common aim. These elements may be human, equipment and technologies.

T lymphocyte type of lymphocyte that develops in the thymus and has a variety of functions in the immune response.

Tailing production of an asymmetric peak during column chromatography characterized by a normal rise before the peak but a slow fall after the peak.

Tare the operation used to set the readout of a balance to 0 g (zero).

Teamwork a means by which a group of individuals all work together to provide a common outcome.

Terms of reference rules governing how a committee is appointed, how it acts, and how this is documented.

Thermocycler automated instrument for performing polymerase chain reaction (PCR).

Titre Is a measure of antibody concentration based on serial dilutions to a visible end point e.g. colour intensity, agglutination or fluorescence.

Toxic substance capable of causing injury or damage to an organism.

Traceability value or measurement of a standard in relation to stated references, which allows comparisons through an unbroken chain.

Tracking dye small coloured substances that pass through the pores in an electrophoresis gel and so act as a marker for the electrophoretic front. Bromophenol blue is often used.

Trainee someone who is receiving instruction, and is not yet deemed competent to work alone.

Transcription synthesis of a complementary RNA molecule using a single strand of DNA as template.

Transcription transfer of results from a print-out or direct from a device into the patient's notes.

Transducer within a biosensor, the electrochemical device that converts a biological or biochemical signal or response into a quantifiable electrical signal.

Transformation process of a cell taking up recombinant DNA.

Transition change between any two energy levels that occurs when an atom or molecule absorbs an amount of energy exactly equal to that between the two relevant states.

Transport emergence card written instructions carried by drivers when transporting dangerous goods that give details of the substance, its category name, UN number and class, what to do and who to contact in an emergency, and first aid information in case of spillage.

Troubleshooting sheets a means to record problems experienced with clinical tests or instruments and to record how particular problems are solved.

Turbidity degree of opacity or cloudiness of a fluid caused by suspended particles.

Turnaround time that taken for a test result to be available to a clinician after the sample has been obtained from the patient.

Ultracentrifuge one capable of producing an RCF in excess of 100 000 g.

Vacutainer tubes containers used for blood sampling via venepuncture in which a vacuum has been created prior to sealing. Thus, blood enters the tube from the vein by gentle suction.

Valency of an antibody the number of other molecules it can bind to simultaneously.

Vapour phase between the liquid and gaseous states of matter.

Vectors carrier of the DNA into a recipient cell for cloning. Vectors are usually plasmids or bacteriophage DNA.

Vernier scale a device used on a microscope to accurately identify the position of specific points within a specimen.

VIR triangle Ohm's Law relates voltage (V) to current (I) and resistance (R). The equation that describes Ohm's Law may be written in three distinct ways and this is best remembered using the VIR triangle, so that when any two of the values are known, the third unknown may be calculated using this triangle.

Virtual image is an image in which the outgoing rays from a point on the object always intersect at a point. A simple example is a flat mirror where the image of oneself is perceived at twice the distance from oneself to the mirror. That is, if one is half a metre in front of the mirror, one's image will appear to be at a distance of 1 metre away (or half a metre inside or behind the mirror).

Voltammetry electrochemical techniques that employ electrolytic cells and measure the current passing through an electrode when an external potential is applied.

Wavelength the distance (measured in the direction of propagation) between two points in the same phase of any two consecutive cycles of a wave.

Weight the force (measured for example in Newtons, grams, or pounds) reflecting the effect of gravity at a particular location upon a **mass**.

Welfare condition of well being, happiness, and comfort.

Wired enzyme electrode an electrode that incorporates an immobilized enzyme within a network of fixed electroactive centres that are able to shuttle electrons from the enzyme to the electrode surface.

Working electrode the electrode that generates a variable potential at its surface with the analyte, dependent on analyte concentration. Also called a sensing electrode.

Workplace inspection audit of a workplace in terms of its health and safety.

Workplaces exposure limit extent to which a person may be safely exposed to a hazardous substance without endangering health.

Index